Ernst & Sohn
A Wiley Brand

Georgios Gaganelis, Peter Mark, Patrick Forman

Optimization Aided Design

Reinforced Concrete

- numerous examples e.g. columns, beams, deep beams, corbels, cantilevers, frame corners, pylons, parabolic trough solar collectors, fiber reinforced concrete
- the book is suitable for graduates, young professionals and for teaching & research
- useful introduction to optimization methods for practicing engineers

2022 · 184 pages · 200 figures · 6 tables
Softcover
ISBN 978-3-433-03337-1   € 69*
eBundle (Softcover + ePDF)
ISBN 978-3-433-03338-8   € 99*

BESTELLEN
+49 (0)30 470 31–236
marketing@ernst-und-sohn.de
www.ernst-und-sohn.de/en/3337

* All book prices inclusive VAT.

The Partner for Sustainable Change

"We provide expertise in maintaining a healthy marine environment including shoreline management, coastal defence, marine ecology and flood protection. Our understanding of complex port and marine requirements enables us to deliver projects of varying sizes and complexities anywhere in the world."

Ports & Marine · Coastal Engineering · Flood Protection · Structural & Ground Engineering · Holistic Designs & Solutions

www.ramboll.com/transport

RAMBOLL Bright ideas. Sustainable change.

Alfred Steinle, Hubert Bachmann, Mathias Tillmann (Ed.)

Precast Concrete Structures

- completely revised by a new group of authors for this edition
- introduction to this subject and as a practical resource with examples for both structural engineers and architects

Building with precast concrete elements is one of the most innovative forms of construction. This book serves as an introduction to this topic, including examples, and thus supplies all the information necessary for conceptual and detailed design.

2. edition · 2019 · 356 pages · 335 figures · 40 tables

Softcover
ISBN 978-3-433-03225-1   **€ 79***

eBundle (Print + PDF)
ISBN 978-3-433-03273-2   **€ 99***

ORDER
+49 (0)30 470 31-236
marketing@ernst-und-sohn.de
www.ernst-und-sohn.de/en/3225

* All book prices inclusive VAT.

Recommendations of the Committee for Waterfront Structures Harbours and Waterways

Recommendations of the Committee for Waterfront Structures Harbours and Waterways

EAU 2020

*Issued by
the Committee for Waterfront Structures of the German Port
Technology Association and the German Geotechnical Society*

Translation of the 12th German Edition

10th Completely Revised Edition

Ernst & Sohn
A Wiley Brand

Publishers

The original German edition was published under the title *Empfehlungen des Arbeitsausschusses „Ufereinfassungen" Häfen und Wasserstraßen EAU 2020*

Hafentechnische Gesellschaft e. V. (HTG)
Neuer Wandrahm 4
20457 Hamburg
Germany

Deutsche Gesellschaft für Geotechnik e. V. (DGGT)
Hollestr. 1g
45127 Essen
Germany

Cover
@ bremenports

10th completely revised Edition

All books published by Ernst & Sohn are carefully produced. Nevertheless, authors, editors, and publisher do not warrant the information contained in these books, including this book, to be free of errors. Readers are advised to keep in mind that statements, data, illustrations, procedural details or other items may inadvertently be inaccurate.

Library of Congress Card No.:
applied for

British Library Cataloguing-in-Publication Data:
A catalogue record for this book is available from the British Library.

Bibliographic information published by the Deutsche Nationalbibliothek
The Deutsche Nationalbibliothek lists this publication in the Deutsche Nationalbibliografie; detailed bibliographic data are available on the Internet at http://dnb.d-nb.de.

© 2024 Ernst & Sohn GmbH, Rotherstraße 21, 10245 Berlin, Germany

All rights reserved (including those of translation into other languages). No part of this book may be reproduced in any form – by photoprinting, microfilm, or any other means – nor transmitted or translated into a machine language without written permission from the publishers. Registered names, trademarks, etc. used in this book, even when not specifically marked as such, are not to be considered unprotected by law.

Cover Design   Design Pur GmbH, Berlin
Translation   Philip Thrift, Hannover
Typesetting   le-tex publishing services GmbH, Leipzig
Printing and Binding   CPI books GmbH, Leck

Print ISBN   978-3-433-03392-0
ePDF ISBN   978-3-433-61133-3
ePub ISBN   978-3-433-61134-0
oBook ISBN   978-3-433-61133-3

Printed on acid-free paper.

Contents

Preface XV

1	**Safety and verification concept** 1	
1.1	Principles of the safety and verification concept for waterfront structures 1	
1.1.1	General 1	
1.1.2	Normative regulations for waterfront structures 1	
1.1.3	Geotechnical categories 3	
1.1.4	Design situations 3	
1.2	Verification for waterfront structures 4	
1.2.1	Principles for verification 4	
1.2.2	Design approaches 4	
1.2.3	Analysis of the serviceability limit state 5	
1.2.4	Analysis of the ultimate limit state 6	
	References 10	
2	**Ship dimensions** 11	
2.1	Sea-going ships 11	
2.1.1	Passenger ships and cruise liners 12	
2.1.2	Bulk carriers 12	
2.1.3	General cargo ships 12	
2.1.4	Container ships 13	
2.1.5	Ferries 13	
2.1.6	RoRo/ConRo vessels 13	
2.1.7	Oil tankers 14	
2.1.8	Gas tankers 15	
2.2	Inland waterway vessels 15	
2.3	Offshore installation vessels 19	
3	**Geotechnical principles** 21	
3.1	Geotechnical report 21	
3.2	Subsoil 21	
3.2.1	Mean characteristic values of soil parameters 21	
3.2.2	Layout and depths of boreholes and penetrometer tests 27	

3.2.3	Determining the shear strength c_u of saturated, undrained cohesive soils	28
3.2.4	Assessing the subsoil for the installation of piles and sheet piles and for selecting the installation method	31
3.2.5	Classifying the subsoil in homogeneous zones	34
3.3	Water pressure	35
3.3.1	General	35
3.3.2	Resultant water pressure in the direction of the water side	36
3.3.3	Resultant water pressure on quay walls in front of embankments with elevated platforms in tidal areas	38
3.3.4	Taking account of groundwater flow	39
3.4	Hydraulic heave failure	45
3.5	Earth pressure	49
3.5.1	General	49
3.5.2	Considering the cohesion in cohesive soils	49
3.5.3	Considering the apparent cohesion (capillary cohesion) in sand	49
3.5.4	Determining active earth pressure for a steep, paved embankment in a partially sloping waterfront structure	49
3.5.5	Determining the active earth pressure shielding on a wall below a relieving platform with average ground surcharges	50
3.5.6	Earth pressure distribution under limited loads	52
3.5.7	Determining active earth pressure in saturated, non-consolidated or partially consolidated, soft cohesive soils	53
3.5.8	Effect of water pressure difference beneath beds of watercourses	56
3.5.9	Considering active earth pressure and resultant water pressure, and construction guidance for waterfront structures with soil replacement and a contaminated or disturbed base of excavation	57
3.5.10	Effect of groundwater flow on resultant water pressure and active and passive earth pressures	60
3.5.11	Determining the amount of displacement required to mobilise passive earth pressure in non-cohesive soils	62
3.5.12	Measures for increasing the passive earth pressure in front of waterfront structures	63
3.5.13	Passive earth pressure in front of abrupt changes in ground level in soft cohesive soils with rapid load application on the land side	65
3.5.14	Waterfront structures in seismic regions	65
	References	69
4	**Loads on waterfront structures**	**73**
4.1	Vessel berthing velocities and pressures	73
4.1.1	Guide values	73
4.1.2	Loads on waterfront structures due to fender reaction forces	74
4.2	Vertical imposed loads	74
4.2.1	General	74
4.2.2	Basic situation 1	76
4.2.3	Basic situation 2	76
4.2.4	Basic situation 3	76

4.2.5	Loading assumptions for quay surfaces	76
4.3	Sea state and wave pressure	77
4.3.1	General	77
4.3.2	Description of the sea state	77
4.3.3	Determining the sea state parameters	78
4.3.4	Design concepts and specification of design parameters	82
4.3.5	Conversion of the sea state	83
4.3.6	Wave pressure on vertical quay walls in coastal areas	85
4.4	Effects of waves due to ship movements	90
4.4.1	General	90
4.4.2	Wave heights	91
4.5	Choosing a greater design depth (allowance for scouring)	94
4.6	Loads arising from surging and receding waves due to the inflow or outflow of water	94
4.6.1	General	94
4.6.2	Determining wave values	94
4.6.3	Loading assumptions	95
4.7	Wave pressure on piled structures	96
4.7.1	General	96
4.7.2	Method of calculation according to Morison et al. (1950)	98
4.7.3	Determining the wave loads on a single vertical pile	98
4.7.4	Coefficients C_D and C_M	100
4.7.5	Forces from breaking waves	100
4.7.6	Wave load on a group of piles	101
4.7.7	Raking piles	101
4.7.8	Safety factors	102
4.7.9	Vertical wave load ("wave slamming")	103
4.8	Moored ships and their influence on the design of mooring equipment and fenders	106
4.8.1	General	106
4.8.2	Critical wind speed	108
4.8.3	Wind loads on moored vessels	108
4.8.4	Loads on mooring equipment and fenders	110
4.9	Loads on bollards	110
4.9.1	Loads on bollards for sea-going vessels	110
4.9.2	Loads on bollards for inland waterway vessels	112
4.9.3	Direction of line pull load	113
4.9.4	Design for line pull loads	113
4.10	Quay loads from cranes and other transhipment equipment	113
4.10.1	Conventional general cargo cranes	113
4.10.2	Container cranes	113
4.10.3	Load specifications for port cranes	115
4.10.4	Notes	116
4.11	Impact and pressure of ice on waterfront structures, fenders and dolphins in coastal areas	116
4.11.1	General	116

4.11.2	Determining the compressive strength of ice	117
4.11.3	Ice loads on waterfront structures and other structures of greater extent	118
4.11.4	Ice loads on vertical piles	121
4.11.5	Horizontal ice load on a group of piles	121
4.11.6	Ice surcharges	122
4.11.7	Vertical loads with rising or falling water levels	122
4.12	Impact and pressure of ice on waterfront structures, piers and dolphins at inland facilities	123
4.12.1	General	123
4.12.2	Ice thickness	123
4.12.3	Compressive strength of the ice	124
4.12.4	Ice loads on waterfront structures and other structures of greater extent	124
4.12.5	Ice loads on narrow structures (piles, dolphins, bridge and weir piers and ice deflectors)	125
4.12.6	Ice loads on groups of structures	125
4.12.7	Vertical loads with rising or falling water levels	126
	References	126
5	**Earthworks and dredging**	**131**
5.1	Dredging in front of quay walls in seaports	131
5.2	Dredging and hydraulic fill tolerances	132
5.2.1	General	132
5.2.2	Dredging tolerances	133
5.3	Hydraulic filling of port areas for planned waterfront structures	135
5.3.1	General	135
5.3.2	Hydraulic filling of port above the water table	136
5.3.3	Hydraulic filling of port areas below the water table	137
5.4	Backfilling of waterfront structures	139
5.4.1	General	139
5.4.2	Backfilling in the dry	140
5.4.3	Backfilling underwater	140
5.4.4	Additional remarks	141
5.5	In situ density of hydraulically filled non-cohesive soils	141
5.5.1	General	141
5.5.2	Empirical values for in situ density	141
5.5.3	In situ density required for port areas	142
5.5.4	Checking the in situ density	142
5.6	In situ density of dumped non-cohesive soils	142
5.6.1	General	142
5.6.2	Influences on the achievable in situ density	143
5.7	Dredging underwater slopes	144
5.7.1	General	144
5.7.2	Dredging underwater slopes in loose sand	144
5.7.3	Dredging equipment	144
5.7.4	Execution of dredging work	145
5.8	Subsidence of non-cohesive soils	146

5.9	Soil replacement along a line of piles for a waterfront structure	147
5.9.1	General	147
5.9.2	Dredging	148
5.9.3	Cleaning the base of the excavation before filling it with sand	149
5.9.4	Placing the sand fill	150
5.9.5	Checking the sand fill	150
5.10	Dynamic compaction of the soil	151
5.11	Vertical drains to accelerate the consolidation of soft cohesive soils	151
5.11.1	General	151
5.11.2	Applications	152
5.11.3	Design	152
5.11.4	Design of plastic drains	153
5.11.5	Installation	154
5.12	Consolidation of soft cohesive soils by preloading	154
5.12.1	General	154
5.12.2	Applications	154
5.12.3	Bearing capacity of in situ soil	155
5.12.4	Fill material	156
5.12.5	Determining the depth of preload fill	156
5.12.6	Minimum extent of preload fill	158
5.12.7	Soil improvement through vacuum consolidation with vertical drains	158
5.12.8	Execution of soil improvement through vacuum consolidation with vertical drains	159
5.12.9	Checking the consolidation	159
5.12.10	Secondary settlement	159
5.13	Improving the bearing capacity of soft cohesive soils with vertical elements	160
5.13.1	General	160
5.13.2	Methods	160
5.13.3	Construction of pile-type load-bearing elements	162
	References	163
6	**Protection and stabilisation structures**	**165**
6.1	Bank and bottom protection	165
6.1.1	Embankment stabilisation on inland waterways	165
6.1.2	Slopes in seaports and tidal inland ports	170
6.1.3	Use of geotextile filters in bank and bottom protection	174
6.1.4	Scour and protection against scour in front of waterfront structures	176
6.1.5	Scour protection at piers and dolphins	185
6.1.6	Installation of mineral impervious linings under water and their connection to waterfront structures	185
6.2	Flood defence walls in seaports	187
6.2.1	General	187
6.2.2	Critical water levels	187
6.2.3	Excess water pressure and unit weight of soil	188
6.2.4	Minimum embedment depths for flood defence walls	189
6.2.5	Special loads on flood defence walls	189

6.2.6	Guidance on designing flood defence walls in slopes	190
6.2.7	Structural measures	190
6.2.8	Buried utilities in the region of flood defence walls	191
6.3	Rouble mound moles and breakwaters	191
6.3.1	General	191
6.3.2	Stability analyses, settlement and subsidence and guidance on construction	192
6.3.3	Specifying the geometry of the structure	192
6.3.4	Designing the armour layer	194
6.3.5	Construction of breakwaters	198
6.3.6	Construction and use of equipment	198
6.3.7	Settlement and subsidence	201
6.3.8	Invoicing for installed quantities	201
	References	201

7 Configuration of cross-sections and equipment for waterfront structures 205

7.1	Configuration of cross-sections	205
7.1.1	Standard cross-sectional dimensions for waterfront structures in seaports	205
7.1.2	Top edges of waterfront structures in seaports	207
7.1.3	Standard cross-sections for waterfront structures in inland ports	208
7.1.4	Upgrading partially sloped waterfronts in inland ports with large water level fluctuations	212
7.1.5	Design of waterfront areas in inland ports according to operational aspects	214
7.1.6	Nominal depth and design depth of the harbour bottom	215
7.1.7	Strengthening waterfront structures for deepening harbour bottoms in seaports	217
7.1.8	Embankments below waterfront wall superstructures behind closed sheet pile walls	221
7.1.9	Re-design of waterfront structures in inland ports	221
7.1.10	Waterfront structures in regions with mining subsidence	224
7.2	Equipment	227
7.2.1	Provision of quick-release hooks at berths for large vessels	227
7.2.2	Layout and design of and loads on access ladders	227
7.2.3	Layout and design of stairs in seaports	230
7.2.4	Armoured steel sheet pile walls	231
7.2.5	Equipment for waterfront structures in seaports with supply and disposal systems	235
7.2.6	Layout of bollards	241
7.2.7	Foundations to craneways on waterfront structures	243
7.2.8	Fixing crane rails to concrete	245
7.2.9	Connection of expansion joints seal in reinforced concrete bottoms to load-bearing steel sheet pile walls	251
7.2.10	Connection of steel sheet piles to a concrete structure	252
7.2.11	Steel capping beams for sheet pile waterfront structures	254
7.2.12	Reinforced concrete capping beams for waterfront structures with steel sheet piles	257

7.2.13	Steel nosings to protect reinforced concrete walls and capping beams on waterfront structures *261*	
7.2.14	Floating berths in seaports *263*	
7.3	Drainage *265*	
7.3.1	Design of weepholes for sheet pile structures *265*	
7.3.2	Design of drainage systems for waterfront structures in tidal areas *266*	
7.4	Fenders *268*	
7.4.1	Fenders for large vessels *268*	
7.4.2	Fenders in inland ports *283*	
7.5	Offshore energy support bases *284*	
7.5.1	General *284*	
7.5.2	Basis for design *284*	
7.5.3	Nautical requirements *285*	
7.5.4	Calculating the leg penetration of WTIVs *288*	
7.5.5	Maintaining and monitoring the jacking surfaces *291*	
7.5.6	Logistical requirements *292*	
7.6	RoRo berths *298*	
7.6.1	General *298*	
7.6.2	Loading assumptions for RoRo terminals *299*	
7.6.3	Kinematics *301*	
7.6.4	Classification of ship-to-shore facilities *303*	
7.6.5	Facilities and equipment on the land side *308*	
7.7	Jetties *312*	
7.7.1	Introduction *312*	
7.7.2	Design of jetties *313*	
7.7.3	Design of berthing and mooring facilities (ship-to-shore) *315*	
7.7.4	Structural elements of berths *317*	
7.7.5	Interaction between load-bearing structure and installations on deck *320*	
	References *322*	
8	**Sheet pile walls** *325*	
8.1	Materials and construction *325*	
8.1.1	Materials for sheet pile walls *325*	
8.1.2	Steel sheet pile walls – properties and forms *326*	
8.1.3	Watertightness of steel sheet pile walls *340*	
8.1.4	Welding steel sheet pile walls *342*	
8.1.5	Installation of steel sheet pile walls *346*	
8.1.6	Driving assistance *366*	
8.1.7	Monitoring pile driving operations *371*	
8.1.8	Repairing interlock declutching on driven steel sheet piling *373*	
8.1.9	Noise control – low-noise driving *377*	
8.1.10	Corrosion of steel sheet piling, and countermeasures *381*	
8.1.11	Risk of sand grinding on sheet piling *387*	
8.2	Design of sheet pile walls *387*	
8.2.1	General *387*	
8.2.2	Free-standing/cantilever sheet pile walls *391*	

8.2.3	Design of sheet pile walls with fixity in the ground and a single row of anchors *392*	
8.2.4	Design of sheet pile walls with a double row of anchors *396*	
8.2.5	Applying the angle of earth pressure and the analysis in the vertical direction *397*	
8.2.6	Taking account of unfavourable groundwater flows in the passive earth pressure zone *407*	
8.2.7	Verifying the load-bearing capacity of a quay wall *407*	
8.2.8	Selection of embedment depth for sheet piles *410*	
8.2.9	Determining the embedment depth for sheet pile walls with full or partial fixity in the soil *410*	
8.2.10	Steel sheet pile walls with staggered embedment depths *413*	
8.2.11	Horizontal actions on steel sheet pile walls in the longitudinal direction of the quay *415*	
8.2.12	Design of anchor walls fixed in the ground *418*	
8.2.13	Staggered arrangement of anchor walls *419*	
8.2.14	Waterfront sheet pile walls in unconsolidated, soft cohesive soils, especially in connection with non-sway structures *419*	
8.2.15	Design of single-anchor sheet pile walls in seismic zones *420*	
8.2.16	Sheet pile waterfronts on inland waterways *421*	
8.2.17	Calculation and design of cofferdams *422*	
	References *432*	
9	**Anchorages** *435*	
9.1	Piles and anchors *435*	
9.1.1	General *435*	
9.1.2	Displacement piles *435*	
9.1.3	Load-bearing capacity of displacement piles *437*	
9.1.4	Micropiles *438*	
9.1.5	Special piles *439*	
9.1.6	Anchors *439*	
9.2	Walings and pile and anchor connections *446*	
9.2.1	Design of steel walings for sheet piling *446*	
9.2.2	Verification of steel walings *447*	
9.2.3	Reinforced concrete walings to sheet pile walls with driven steel piles *448*	
9.2.4	Auxiliary anchors at the top of steel sheet piling structures *450*	
9.2.5	Sheet piling anchors in unconsolidated, soft cohesive soils *451*	
9.2.6	Design of protruding quay wall corners with round steel tie rods *454*	
9.2.7	Configuration and design of protruding quay wall corners with raking anchor piles *456*	
9.2.8	Prestressing of high-strength steel anchors for waterfront structures *458*	
9.2.9	Hinged connections between driven steel piles and steel sheet piling structures *460*	
9.3	Verification of stability for anchoring at the lower failure plane *469*	
9.3.1	Stability at the lower failure plane for anchorages with anchor walls *469*	

9.3.2	Stability at the lower failure plane in unconsolidated, saturated cohesive soils *471*	
9.3.3	Stability at the lower failure plane with varying soil strata *471*	
9.3.4	Verification of stability at the lower failure for a quay wall fixed in the soil *472*	
9.3.5	Stability at the lower failure plane for an anchor wall fixed in the soil *472*	
9.3.6	Stability at the lower failure plane for anchors with anchor plates *472*	
9.3.7	Verification of safety against failure of anchoring soil *472*	
9.3.8	Stability at the lower failure plane for quay walls anchored with piles or grouted anchors at one level *473*	
9.3.9	Stability at the lower failure plane for quay walls with anchors at more than one level *474*	
9.3.10	Safety against slope failure *475*	
	References *476*	
10	**Quay walls and superstructures in concrete** *481*	
10.1	General *481*	
10.2	Construction materials *482*	
10.2.1	Concrete *482*	
10.2.2	Steel reinforcement *484*	
10.3	Design and construction *484*	
10.3.1	Construction joints *484*	
10.3.2	Expansion joints *485*	
10.3.3	Jointless construction *485*	
10.3.4	Crack width limitation *485*	
10.4	Forms of construction *486*	
10.4.1	Concrete walls *486*	
10.4.2	Retaining walls *488*	
10.4.3	Block-type construction *488*	
10.4.4	Box caissons *491*	
10.4.5	Open caissons *492*	
	References *496*	
11	**Pile bents and trestles** *499*	
11.1	General *499*	
11.2	Configuration and design of a pile bent *499*	
11.2.1	General *499*	
11.2.2	Earth pressure loads *500*	
11.2.3	Load due to excess water pressure *501*	
11.2.4	Load path for piles *502*	
11.3	Design of pile trestles *503*	
11.3.1	Free-standing pile trestles *503*	
11.3.2	Special structures designed as pile trestles *505*	
11.3.3	Structural system and calculations *505*	
11.3.4	Construction guidance *506*	
11.4	Design of pile bents and trestles in earthquake zones *507*	
11.4.1	General *507*	

11.4.2	Active and passive earth pressures, excess water pressure, variable loads	507
11.4.3	Resisting the horizontal inertial forces of the superstructure	507
	References 508	

12	**Dolphins** *509*	
12.1	Design and construction	509
12.1.1	Dolphins – purposes and types	509
12.1.2	Layout of dolphins	509
12.1.3	Equipment for dolphins	510
12.1.4	Advice for selecting materials	511
12.2	Detailed design	512
12.2.1	Stiffness of the system	512
12.2.2	Structural behaviour	512
12.2.3	Actions	513
12.2.4	Safety concept	515
12.2.5	Soil—structure interaction and the resulting design variables	515
12.2.6	The required energy absorption capacity of breasting dolphins	520
12.2.7	Other calculations	521
	References 522	

13	**Operation, maintenance and repair of waterfront structures** *525*	
13.1	Operation of waterfront structures	525
13.1.1	General	525
13.1.2	Building information modelling (BIM)	525
13.2	Inspecting waterfront structures	526
13.2.1	Documentation	527
13.2.2	Structural inspections	528
13.2.3	Inspection intervals	529
13.2.4	Structural monitoring supported by measurements	530
13.3	Assessing the load-bearing capacity of an existing waterfront structure	531
13.4	Repairing concrete waterfront structures	533
13.5	Upgrading and deconstructing existing waterfront structures	533
13.5.1	Upgrading measures	533
13.5.2	Deconstruction in conjunction with replacement measures	535
	References 535	

Appendix A Notation 537

A.1	Symbols for variables	538
A.1.1	Latin lower-case letters	538
A.1.2	Latin upper-case letters	540
A.1.3	Greek letters	542
A.2	Subscripts and indices	543
A.3	Abbreviations	544
A.4	Water levels and wave heights	545

List of Advertisers 547

Preface

Eight years have passed since the publication of the 11th edition (9th English edition) of the *Recommendations of the Committee for Waterfront Structures* (known as the "EAU" in professional circles). Over those years, new developments have been described in the committee's annual, in some cases six-monthly, technical reports. This 10th English edition (the translation of the 12th German edition) represents a completely updated version of the recommendations of the "Waterfront Structures Committee", a body organised jointly by the German Port Technology Association (HTG) and the German Geotechnical Society (DGGT). I am certain that, once again, this edition will become a standard work of reference for every engineer concerned with ports, harbours and inland waterways.

The numbered recommendations grew over many years to become the main means of orientation in the EAU. In this new EAU 2020 edition, the content has been given a facelift and the recommendations restructured to provide the reader with a better, clearer arrangement of the chapters. The numbered recommendations are, therefore, no longer included in this new edition of the EAU. The new technical developments described in the committee's annual reports for the years 2013–2019 have been incorporated. Those developments concern topics such as vertical load-bearing capacity, line pull, offshore energy support bases and vessel sizes. There are also recommendations covering jetties and RoRo berths.

The "Waterfront Structures Committee" adheres to the principles for constituting committees laid down by the German Institute for Standardisation (DIN), i.e. appropriate representation of all groups with an interest and the provision of the necessary expertise. Therefore, the committee is made up of members from all relevant disciplines, who are drawn from universities, the building departments of large seaports, inland ports and national inland waterways, the construction industry, the steel industry and consulting engineers.

The following current and former members of the committee were involved in the preparation of EAU 2020:

- Univ.-Prof. Dr.-Ing. Jürgen Grabe, Hamburg (Chair since 2009)
- Ir. Tom van Autgaerden, Antwerp
- Dr.-Ing. Karsten Beckhaus, Schrobenhausen
- Ir. Erik J. Broos, Rotterdam
- Dipl.-Ing. Frank Feindt, Hamburg
- Dipl.-Ing. Francois Gaasch, Esch-sur-Alzette
- Ir. Leon A. M. Groenewegen, The Hague

- Dr.-Ing. Michael Heibaum, Karlsruhe
- Prof. Dr.-Ing. Stefan Heimann, Berlin
- Dipl.-Ing., M.Eng.Sc. Sebastian Höhmann, Hamburg
- Prof. Ir. Aad van der Horst, Delft
- Dipl.-Ing. Robert Howe, Bremerhaven
- Dipl.-Ing. Hans-Uwe Kalle, Hagen
- Dr.-Ing. Jan Kayser, Karlsruhe
- Dr.-Ing. Karl Morgen, Hamburg
- Dipl.-Ing. Hendrik Neumann, Hamburg
- Dipl.-Ing. Matthias Palapys, Duisburg
- Dipl.-Ing. Gabriele Peschken, Bonn
- Dipl.-Ing. Torsten Retzlaff, Rostock
- Dr.-Ing. Peter Ruland, Hamburg
- Dr.-Ing. Hartmut Tworuschka, Hamburg

In a similar way to the work of the DIN when preparing a standard, new recommendations are presented for public discussion in the form of provisional recommendations in the annual technical reports. After considering any objections, the recommendations are published in their final form in the subsequent annual technical report. The status of the *Recommendations of the Committee for Waterfront Structures – Harbours and Waterways* is, therefore, equivalent to that of a standard. However, from the point of view of its relevance to practice and also the dissemination of experience, the information provided goes beyond that of a standard; this publication can be seen more as a "code of practice".

The incorporation of the European standardisation concept has now been completed, and so this latest edition of the EAU complies with the notification requirements of the European Commission. It is registered with the European Commission under notification number 2019/655/D.

The fundamental revisions in EAU 2020 called for in-depth discussions with colleagues outside the committee, even the establishment of temporary working groups to deal with specific topics. The committee acknowledges the assistance of all colleagues who in this way made a significant contribution to developing the content of EAU 2020.

In addition, considerable input from experts plus recommendations from other committees and international engineering science bodies have found their way into these recommendations.

These additions and the results of the revision work mean that EAU 2020 corresponds to modern international standards. Specialists working in this field now have at their disposal an updated edition adapted to the European standards, which will continue to provide valuable help for issues in design, tendering, award of contract, engineering tasks, economic and environmentally compatible construction, site supervision, contractual procedures, operation, maintenance and repairs. It will, therefore, be possible to design and build waterfront structures that are in line with the state of the art and are based on consistent specifications.

The committee would like to thank all those who contributed to and made suggestions for this edition. It is hoped that EAU 2020 will attract the same resonance as previous editions.

I would also like to thank Ms Anne Hagemann, M.Sc., who has worked with the committee for quite some time.

I am also grateful to the publishers Ernst & Sohn for the good cooperation, the careful preparation of the many illustrations, tables and equations and, once again, the excellent quality of the printing and layout of EAU 2020.

Hamburg, October 2020 *Univ.-Prof. Dr.-Ing. Jürgen Grabe*

1
Safety and verification concept

1.1 Principles of the safety and verification concept for waterfront structures

1.1.1 General

A structure can fail as a result of exceeding the ultimate limit state of bearing capacity ("ultimate limit state – ULS", failure of the soil or the structure, loss of static equilibrium) or the limit state of serviceability ("serviceability limit state – SLS", excessive deformations).

1.1.2 Normative regulations for waterfront structures

The "Eurocodes" (EC) – harmonised directives specifying fundamental safety requirements for buildings and structures – were drawn up as part of the realisation of the European Single Market. Those Eurocodes are as follows:

DIN EN 1990:	Basis of structural design ("EC 0")
DIN EN 1991, EC 1:	Actions on structures
DIN EN 1992, EC 2:	Design of concrete structures
DIN EN 1993, EC 3:	Design of steel structures
DIN EN 1994, EC 4:	Design of composite steel and concrete structures
DIN EN 1995, EC 5:	Design of timber structures
DIN EN 1996, EC 6:	Design of masonry structures
DIN EN 1997, EC 7:	Geotechnical design
DIN EN 1998, EC 8:	Design of structures for earthquake resistance
DIN EN 1999, EC 9:	Design of aluminium structures

The Eurocodes "Basis of structural design" (DIN EN 1990) and "Actions on structures" (DIN EN 1991) with their various parts and annexes form the basis of European construction standards, the starting point for building designs throughout Europe. The other eight Eurocodes, along with their respective parts, relate to these two basic standards.

Verification of safety must always be carried out according to European standards. However, in some instances such verification is not possible with these standards alone; a number of parameters, e.g. numerical values for partial safety factors, have to be specified on a national level. Furthermore, the Eurocodes do not cover the entire range of German standards, meaning that a comprehensive set of national standards has been retained in Germany. However, this set of German standards along with its requirements may not contradict the regulations contained in the European standards, which in turn necessitated the revision of national standards.

For proof of stability according to the EAU, the standards DIN EN 1990, DIN EN 1991, DIN EN 1992, DIN EN 1993, DIN EN 1994, DIN EN 1995, DIN EN 1996, DIN EN 1997, DIN EN 1998, DIN EN 1999, and especially DIN EN 1997 (Geotechnical design), are of particular importance. DIN EN 1997-1 defines a number of terms and describes and stipulates limit state verification procedures. The various earth pressure design models for stability calculations are also included in the annexes for information purposes. A particular feature here is that three methods of verification using the partial safety factor concept are available for use throughout Europe.

The publication of DIN 1054:2010 ensured that any duplication of DIN EN 1997-1 was avoided, but specific German experience has been retained. This standard was combined with DIN EN 1997-1:2010 and the National Annex (DIN EN 1997-1/NA:2010) to create the EC 7-1 manual (2015).

The many years of experience with the specific boundary conditions of waterfront structures (e.g. greater tolerances for deformations compared with other engineering works) have led to the EAU containing a number of specific stipulations for the design of such structures that can deviate from those given in DIN EN 1997-1 and DIN 1054.

Those specific stipulations include, for example, the following:

- In some cases, lower partial safety factors for actions, action effects and resistances for the limit state of failure (Section 1.2.4, Tables 1.1 and 1.3).
- The determination of a characteristic resultant hydrostatic pressure by offsetting favourable and unfavourable hydrostatic pressures against each other where this is realistic (see Section 3.3.1).
- Simplified assumptions for hydrostatic pressure (see Section 3.3.2).
- Redistribution of active earth pressure independently of the method of construction for sheet pile walls (see Section 8.2.3.2).
- Increasing the theoretical anchor force by 15% for the robust construction of sheet pile wall components (see Section 9.2).

DIN EN 1997-2 covers the planning, execution and evaluation of soil investigations. As for part 1, this standard has been published together with DIN 4020:2010 and the National Application Document in the EC 7-2 manual (2011).

The execution of special civil engineering works is covered by European standards. In Germany, more specific information for such work is laid out in DIN SPEC publications.

Calculations for large-scale soil stabilisation measures (e.g. jet grouting, grout injection) in Germany are covered by DIN 4093.

Where standards are cited in the recommendations, the current version applies, unless stated otherwise. Standards quoted in the text are listed at the end of each chapter.

1.1.3 Geotechnical categories

The minimum requirements in terms of scope and quality of geotechnical investigations, calculations and monitoring measures are described by three geotechnical categories in accordance with EC 7: low (category 1), normal (category 2) and high (category 3) geotechnical difficulty. These are reproduced in DIN 1054, A 2.1.2. Waterfront structures should be allocated to category 2, or category 3 in the case of difficult subsoil conditions. A geotechnical expert should always be consulted.

1.1.4 Design situations

Load cases for verifying stability and allocating partial safety factors are defined in DIN 1054, Section 6.3.3. These result from the combinations of actions in conjunction with the safety categories for resistances. The following classifications apply to waterfront structures:

1.1.4.1 Design situation DS-P (persistent)
This design situation covers loads due to active earth pressures (separately for the initial and final states in the case of unconsolidated, cohesive soils) and excess water pressure in the case of the frequent occurrence of unfavourable inner and outer water levels (see Section 3.3.2), active earth pressure influences due to normal imposed loads and normal crane and pile loads, instantaneous surcharges due to self-weight and normal imposed loads.

1.1.4.2 Design situation DS-T (transient)
Transient situations, i.e. those related to a certain period of time, are allocated to design situation DS-T. They include, for example, situations during construction or repairs. For hydraulic engineering works, besides permanent actions and variable actions that occur regularly during the service life of the structure, which are all allocated to DS-P, transient actions include limited scour due to currents or ship propellers, excess water pressure in the case of rare occurrences of unfavourable inner and outer water levels (see Section 3.3.2) or wave loads according to Section 4.3.

1.1.4.3 Design situation DS-A (accidental)
This is as for design situation DS-T, but with extraordinary design situations such as unscheduled surcharges over a larger area, unusually extensive flattening of an underwater slope in front of the base of a sheet pile wall, unusual scour due to currents or ship propellers, excess water pressure following extreme water levels (see Section 3.3.2 or 6.2), excess water pressure following exceptional flooding of the waterfront structure, combinations of earth and hydrostatic pressures with wave loads resulting from waves that occur only rarely (see Section 4.3), combinations of earth and hydrostatic pressures with flotsam impact according to Section 6.2.5, all load combinations in conjunction with ice states or ice pressures.

1.1.4.4 Extreme case
When extremely improbable combinations of actions occur concurrently, then DIN 1054, Section A 2.4.7.6.1, A(4), A 2.4.7.6.3 and A(5), permit partial safety factors for actions and

resistances to be taken as $\gamma_F = \gamma_R = 1.0$. The combination factors are set to $\psi = 1.0$ according to Section 1.2.4.

Examples of this are the simultaneous occurrence of extreme water levels and extreme wave loads due to plunging breakers according to Section 4.3.6, extreme water levels and the simultaneous, total failure of a drainage system (see Section 6.2), combinations of three short-term events acting simultaneously, e.g. high water (highest astronomical tide, see Section 6.2), waves that occur only rarely (see Section 4.3) and flotsam impact (see Section 6.2).

1.2 Verification for waterfront structures

1.2.1 Principles for verification

A stability analysis of a waterfront structure must include the following in particular:

- Details of the use of the facility.
- Drawings of the structure with all essential, planned structural dimensions.
- Brief description of the structure including, in particular, all details that are not readily identifiable from the drawings.
- Design value of bottom depth.
- Characteristic values of all actions.
- Soil strata and associated characteristic values of soil parameters.
- Critical water levels related to the German NHN height reference system (previously mean sea level) or a local gauge datum, together with corresponding groundwater levels (no high water, no flooding).
- Combinations of actions, i.e. load cases.
- Partial safety factors necessary/used.
- Intended building materials and their strengths or resistances.
- All data regarding construction timetables and construction operations, with critical temporary states.
- Description of and justification for the intended verification procedures.
- Information about literature used and other calculation aids.

1.2.2 Design approaches

1.2.2.1 Analytical method

Geotechnical analyses according to the relevant standards generally make use of analytical models based on failure mechanisms. The critical failure planes in the subsoil are either specified or determined by examining variations. The level of safety to be verified should take into account the uncertainties of the earth pressure analysis, the soil investigation and the type of construction. Implicitly, maximum deformations, i.e. serviceability requirements, often have to be considered as well.

1.2.2.2 Numerical simulations

In the meantime, numerical methods of calculation, e.g. the finite element method (FEM), have become established for calculations for the limit state of serviceability (deformations).

An example of a comprehensive numerical simulation for the deformations of a quay structure caused by backfilling can be found in Mardfeld (2005). For earth structure, the analysis of the ultimate limit state can be carried out using the $\varphi' - c'$ reduction. Compared with conventional approaches such as the slip circle method, FEM has the advantage that the shear joint can be in any position, which means that more relevant results can be obtained than is the case when assuming planar or curved failure body geometries. The Z^* method is a good choice when assessing the load-bearing capacity in soil-structure interaction problems because the stresses in the components are determined for the serviceability limit state and then transferred to a conventional analysis. An analysis of the limit state based solely on FEM is currently the subject of debate. When it comes to modelling the ultimate limit state and integrating the safety factors, there are still no stipulations that apply to all cases. Numerical simulations call for modelling based on correct states of stresses and deformations, an adequately large section of the subsoil, the drainage conditions of the soil and, above all, material models for the undisturbed soil types that model the stress–strain behaviour phenomena relevant for the structure. For more information on this topic, please refer to *Numerik in der Geotechnik* (EANG 2014).

1.2.2.3 Observational method

The monitoring method according to DIN EN 1997-1 should be employed for complex structures in which the structural behaviour cannot be modelled with sufficient reliability or accuracy during the design. This involves taking measurements on the structure or in the subsoil and comparing these with predicted or warning values. Countermeasures or safety measures that are to be implemented if warning values are exceeded are inherent to the monitoring method. Deformations and forces obtained from numerical simulations form the basis for assessing in situ measurements.

1.2.2.4 Experiments

Experiments, trials and tests can be used to determine the structural behaviour of individual geotechnical elements and even complex geotechnical load-bearing structures. Tests can be carried out on full-size elements (e.g. trial loadings on piles or pull-out tests on anchors) or on scale models. The latter require compliance with the model laws that ensue from the similitude concept for engineering models if the observations made using the model are to be transferred to full-size components. The various laws of the different physical variables limit the transferability. This is particularly true for geotechnical applications, where the stress state in the subsoil has a crucial influence on the stress–strain behaviour but is very difficult to model. Tests on models can be carried out in a geotechnical centrifuge so that the soil is subjected to a realistic stress state, thus providing a correct representation of the pressure-dependent stress–strain behaviour of the subsoil. Further details of geotechnical centrifuge modelling can be obtained from Technical Committee TC 104 "Physical Modelling in Geotechnics" of the International Society for Soil Mechanics and Geotechnical Engineering (ISSMGE).

1.2.3 Analysis of the serviceability limit state

Deformation analyses must be carried out for all structures whose function can be impaired or rendered ineffective through deformations. The deformations are calculated with the characteristic values of actions and soil reactions and must be less than the deformations permissible for correct functioning of the component or whole structure. Where applicable, the calculations should include the upper and lower bounds of the characteristic values.

In particular, deformation analyses must consider the course of actions over time in order to allow for critical deformation states during various operating and construction stages.

1.2.4 Analysis of the ultimate limit state

Numerical proof of adequate stability is carried out for limit states STR and GEO-2 with the help of design values (index d) for actions or action effects and resistances, and for limit state GEO-3 with the help of design values for actions or action effects and soil properties.

Verification of safety is assessed according to the following fundamental equation:

$$E_d \leq R_d$$

where

E_d Design value of sum of actions or action effects
R_d Design value of resistances derived from sum of resistances of soil or structural elements

When analysing the limit state of loss of equilibrium (EQU) or failure due to hydraulic heave (HYD) or buoyancy (UPL), it is necessary to compare the design values for favourable and unfavourable or stabilising and destabilising actions and to verify compliance with the respective limit state condition. Resistances do not play a role in these analyses.

Six cases apply for analyses of the ultimate limit state of bearing capacity:

Loss of equilibrium of structure or ground	EQU
Loss of equilibrium of structure or ground due to uplift by water pressure (buoyancy)	UPL
Hydraulic heave, internal erosion or piping in the ground due to hydraulic gradients	HYD
Internal failure or very large deformation of the structure or its components	STR
Failure or very large deformation of the ground	GEO-2
Loss of overall stability	GEO-3

DIN EN 1997-1 permits three options for verifying safety, designated "design approaches 1 to 3". For approach 1, two groups of partial safety factors are taken into account and are used in two separate analyses. For approaches 2 and 3, a single analysis with one group of partial safety factors suffices.

In approaches 1 and 2, the partial safety factors are applied, in principle, to either actions or action effects and to resistances. However, DIN 1054 stipulates that the characteristic, or representative, effects $E_{Gk,i}$ or $E_{Qrep,i}$ (e.g. shear forces, reactions, bending moments, stresses in the relevant sections of the structure and at interfaces between structure and subsoil) are determined first and then the partial safety factors are applied. This is also referred to as design approach 2*.

In approach 3, the partial safety factors are applied to the soil parameters and to actions or action effects not related to the subsoil. Actions or action effects induced by the subsoil are derived from the factored soil parameters.

According to DIN 1054, design approach 2 (2*) should be used for the geotechnical analysis of limit states STR and GEO-2, and design approach 3 for analysing limit state GEO-3. The partial safety factors specified in DIN 1054 are reproduced in Tables 1.1–1.3.

Remarks:

- For the limit state of failure due to loss of overall stability GEO-3, the partial safety factors for shear strength are to be taken from Table 1.2, and pull-out resistances are multiplied by partial safety factors according to STR and GEO-2.
- The partial safety factor for the material resistance of steel tension members made from reinforced and prestressed steel for limit states GEO-2 and GEO-3 is given in DIN EN 1992-1-1 as $\gamma_M = 1.15$.
- The partial safety factor for the material resistance of flexible reinforcing elements for limit states GEO-2 and GEO-3 is given in EBGEO (2010).

Provided that greater displacements and deformations of the structure do not impair the stability and serviceability of the structure, as can be the case for waterfront structures, ports, harbours and waterways, the partial safety factor γ_G can be reduced for earth and water pressures in justified cases (DIN 1054, A 2.4.7.6.1, A(3)). This is exploited in the EAU by using the partial safety factors in the form of $\gamma_{G,red}$ (Table 1.1) and $\gamma_{R,e,red}$ (Table 1.3). Furthermore, a partial safety factor $\gamma_G = \gamma_Q = 1.00$ is used for action effects due to permanent and unfavourable variable actions in design situation DS-A.

When calculating a design value for actions F_d according to EN 1990, this value must either be stipulated directly or derived from representative values:

$$F_d \leq \gamma_F \cdot F_{rep}$$

where

$$F_{rep} = \psi \cdot F_k$$

γ_F Partial safety factor
ψ Combination factor

For permanent actions and the leading action of variable actions,

$$F_{rep} = F_k$$

applies.

A combination factor $\psi = 1.00$ is usually used for waterfront structures. To verify safety against buoyancy (UPL) and safety against hydraulic heave (HYD) the design values F_d are always calculated without considering combination factors.

Table 1.1 Partial safety factors for actions and action effects (to DIN 1054:2010, Table A2.1, with additions) for the ultimate and serviceability limit states.

Action or action effect	Symbol	Design situation DS-P	DS-T	DS-A
HYD and UPL: limit state of failure due to hydraulic failure and buoyancy				
Destabilising permanent actions[a]	$\gamma_{G,dst}$	1.05	1.05	1.00
Stabilising permanent actions	$\gamma_{G,stb}$	0.95	0.95	0.95
Destabilising variable actions	$\gamma_{Q,dst}$	1.50	1.30	1.00
Stabilising variable actions	$\gamma_{Q,stb}$	0	0	0
Flow force in favourable subsoil	γ_H	1.45	1.45	1.25
Flow force in unfavourable subsoil	γ_H	1.90	1.90	1.45
EQU: limit state of loss of equilibrium				
Unfavourable permanent actions	$\gamma_{G,dst}$	1.10	1.05	1.00
Favourable permanent actions	$\gamma_{G,stb}$	0.90	0.90	0.95
Unfavourable variable actions	γ_Q	1.50	1.25	1.00
STR and GEO-2: limit state of failure of structures, components and subsoil				
Action effects from permanent actions generally[a]	γ_G	1.35	1.20	1.00
Action effects from permanent actions for calculating anchorage[b]	γ_G	1.35	1.20	1.10
Action effects from favourable permanent actions[c]	$\gamma_{G,inf}$	1.00	1.00	1.00
Action effects from permanent actions due to earth pressure at rest	$\gamma_{G,EO}$	1.20	1.10	1.00
Water pressure in certain boundary conditions[d]	$\gamma_{G,red}$	1.20	1.10	1.00
Water pressure in certain boundary conditions for calculating anchorage[b]	$\gamma_{G,red}$	1.20	1.10	1.10
Action effects from unfavourable variable actions[e]	γ_Q	1.50	1.30	1.00
Action effects from unfavourable variable actions[f] for calculating anchorage[b]	γ_Q	1.50	1.30	1.10
Action effects from favourable variable actions	γ_Q	0	0	0
GEO-3: limit state of failure due to loss of overall stability				
Permanent actions	γ_G	1.00	1.00	1.00
Unfavourable variable actions	γ_Q	1.30	1.20	1.00

SLS: limit state of serviceability
$\gamma_G = 1.00$ for permanent actions or action effects
$\gamma_Q = 1.00$ for variable actions or action effects

a) The permanent actions are understood to include permanent and variable water pressure. Differing from DIN 1054:2010-12, $\gamma_G = 1.00$ applies in DS-A except when verifying anchorage.
b) The design of anchorages (grouted anchors, micropiles, tension piles) also includes verifying stability at the lower failure plane when dealing with retaining structures (Section 9.3).
c) If during the determination of the design values of the tensile action effect a characteristic compressive action effect from favourable permanent actions is assumed to act simultaneously, then this should be considered with the partial safety factor $\gamma_{G,inf}$ (DIN 1054, 7.6.3.1, A(2)).
d) For waterfront structures in which larger displacements can be accommodated without damage, the partial safety factors $\gamma_{G,red}$ for water pressure may be used if the conditions according to Section 8.2.1.3 are complied with (DIN 1054, A 2.4.7.6.1, A(3)).
e) Differing from DIN 1054:2010-12, $\gamma_Q = 1.00$ applies in DS-A except when verifying anchorage.
f) The permanent actions are understood to include permanent and variable water pressures.

Table 1.2 Partial safety factors for geotechnical parameters (DIN 1054:2010, Table A 2.2).

Soil parameter	Symbol	Design situation		
		DS-P	DS-T	DS-A
HYD and UPL: limit state of failure due to hydraulic failure and buoyancy				
Friction coefficient $\tan \varphi'$ of drained soil and friction coefficient $\tan \varphi_u$ of undrained soil	$\gamma_{\varphi'}, \gamma_{\varphi_u}$	1.00	1.00	1.00
Cohesion c' of drained soil and shear strength c_u of undrained soil	$\gamma_{c'}, \gamma_{c_u}$	1.00	1.00	1.00
GEO-2: limit state of failure of structures, components and subsoil				
Friction coefficient $\tan \varphi'$ of drained soil and friction coefficient $\tan \varphi_u$ of undrained soil	$\gamma_{\varphi'}, \gamma_{\varphi_u}$	1.00	1.00	1.00
Cohesion c' of drained soil and shear strength c_u of undrained soil	$\gamma_{c'}, \gamma_{c_u}$	1.00	1.00	1.00
GEO-3: limit state of failure due to loss of overall stability				
Friction coefficient $\tan \varphi'$ of drained soil and friction coefficient $\tan \varphi_u$ of undrained soil	$\gamma_{\varphi'}, \gamma_{\varphi_u}$	1.25	1.15	1.10
Cohesion c' of drained soil and shear strength c_u of undrained soil	$\gamma_{c'}, \gamma_{c_u}$	1.25	1.15	1.10

Table 1.3 Partial safety factors for resistances (according to DIN 1054:2010-12, Table A 2.3, with additions).

Resistance	Symbol	Design situation		
		DS-P	DS-T	DS-A
STR and GEO-2: limit state of failure of structures, components and subsoil				
Soil resistances				
Passive earth pressure and ground failure resistance	$\gamma_{R,e}, \gamma_{R,v}$	1.40	1.30	1.20
Passive earth pressure when determining bending moment[a]	$\gamma_{R,e,red}$	1.20	1.15	1.10
Sliding resistance	$\gamma_{R,h}$	1.10	1.10	1.10
Pile resistances from static and dynamic pile loading tests				
Base resistance	γ_b	1.10	1.10	1.10
Skin resistance (compression)	γ_s	1.10	1.10	1.10
Total resistance (compression)	γ_t	1.10	1.10	1.10
Skin resistance (tension)	$\gamma_{s,t}$	1.15	1.15	1.15
Pile resistances based on empirical values				
Compression piles	$\gamma_b, \gamma_s, \gamma_t$	1.40	1.40	1.40
Tension piles (in exceptional cases only)	$\gamma_{s,t}$	1.50	1.50	1.50
Pull-out resistances				
Ground or rock anchors	γ_a	1.40	1.30	1.20
Grout body of grouted anchors	γ_a	1.10	1.10	1.10
Flexible reinforcing elements	γ_a	1.40	1.30	1.20

a) Reduction for calculating the bending moment only. For waterfront structures in which larger displacements can be accommodated without damage, the partial safety factors $\gamma_{R,e,red}$ for passive earth pressure may be used if the conditions according to Section 8.2.1.2 are complied with (DIN 1054, A 2.4.7.6.1, A(3)).

References

Andrews, J.D. and Moss, T.R. (1993). *Reliability and Risk Assessment*. Burnt Mill: Longman Scientific & Technical.

EANG (2014). *Empfehlungen des AK Numerik in der Geotechnik*, (Hrsg. DGGT). Berlin: Ernst & Sohn.

EBGEO (2010). *Empfehlungen für den Entwurf und die Berechnung von Erdkörpern mit Bewehrungen aus Geokunststoffen (EBGEO)*, (ed. Deutsche Gesellschaft für Geotechnik e. V.), 2nd ed. Berlin: Ernst & Sohn.

Heibaum, M. and Herten, M. (2007). Finite-Element-Methode für geotechnische Nachweise nach neuer Normung? *Bautechnik* 84 (9): 627–635.

Mardfeldt, B. (2005). Zum Tragverhalten von Kaikonstruktionen im Gebrauchszustand. Dissertation. Veröffentlichungen des Arbeitsbereichs Geotechnik und Baubetrieb der TU Hamburg-Harburg, No. 11.

Richwien, W. and Lesny, K. (2003). Risikobewertung als Schlüssel des Sicherheitskonzepts – Ein probabilistisches Nachweiskonzept für die Gründung von Offshore-Windenergieanlagen. *Erneuerbare Energien* 13 (2): 30–35.

Schwab, R. and Kayser, J. (2002). Continuous model validation for large navigable lock. International Symposium on Identification and Determination of Soil and Rock Parameters, PARAM 2002, Paris.

Schweiger, H.F. (2017). Numerik in der geotechnischen Nachweisführung. In: *Mitteilungsblatt der Bundesanstalt für Wasserbau* Nr. 101. Karlsruhe: self-published.

Schuëller, G.I. (1981). *Einführung in die Sicherheit und Zuverlässigkeit von Tragwerken*. Berlin: Ernst & Sohn.

Standards and Regulations

DIN 1054: Subsoil – Verification of the safety of earthworks and foundations – Supplementary rules to DIN EN 1997-1.

DIN 4020: Geotechnical investigations for civil engineering purposes – Supplementary rules to DIN EN 1997-2.

DIN 4093 (2015): Design of strengthened soil – Set up by means of jet grouting, deep mixing or grouting.

DIN EN 1990 Eurocode: Basis of structural design.

DIN EN 1991 Eurocode 1: Actions on structures.

DIN EN 1992 Eurocode 2: Design of concrete structures.

DIN EN 1993 Eurocode 3: Design of steel structures.

DIN EN 1994 Eurocode 4: Design of composite steel and concrete structures

DIN EN 1995 Eurocode 5: Design of timber structures.

DIN EN 1996 Eurocode 6: Design of masonry structures.

DIN EN 1997 Eurocode 7: Geotechnical design.

DIN EN 1998 Eurocode 8: Design of structures for earthquake resistance.

DIN EN 1999 Eurocode 9: Design of aluminium structures.

Handbuch EC7-1 (2015): Handbuch Eurocode 7, Geotechnische Bemessung Band 1. Allgemeine Regeln.

Handbuch EC7-2 (2011): Handbuch Eurocode 7, Geotechnische Bemessung Band 2. Erkundung und Untersuchung.

2 Ship dimensions

2.1 Sea-going ships

The average ship dimensions given by way of example in Tables 2.1–2.8 can be used when designing and detailing waterfront structures, fenders and dolphins. Allowance must be made for the fact that these are average values that can vary by up to 10% either way. The figures for tankers, bulk carriers and container ships were taken from the Port of Rotterdam's "Table of design ships" and the Ships Register database. It is recommended that designers carry out their own research (Internet, Ships Register, etc.) if more detailed information is required.

Definitions of the conventional data on ship sizes:

- Ship size is based on the gross tonnage (GT), a non-dimensional quantity derived from the ship's total volume. The use of the unit of measurement that was the standard previously, the gross registered ton (a registered ton corresponding to 100 cubic feet, i. e. 2.83 m^3) has not been permitted since 1994 in accordance with an international agreement.

Table 2.1 Passenger ships and cruise liners.

Tonnage measurement	Cargo capacity [DWT]	Displacement G [t]	Overall length [m]	Length between perpendiculars [m]	Beam [m]	Maximum draught [m]
225 000	—	100 000	362	330	60.5	9.3
149 000	—	76 000	345	301	41.0	10.3
128 000	—	N/A	305	270	37.2	8.2
110 000	—	N/A	290	248	35.5	8.3
90 000	—	N/A	294	263	32.2	8.3
70 000	—	37 600	260	220	33.1	7.6
50 000	—	27 900	231	197	30.5	7.6
30 000	—	17 700	194	166	26.8	7.6
20 000	—	12 300	169	146	24.2	7.6

Recommendations of the Committee for Waterfront Structures Harbours and Waterways – EAU 2020, 10th Edition. Issued by the Committee for Waterfront Structures of the German Port Technology Association and the German Geotechnical Society.
© 2024 Ernst & Sohn GmbH. Published 2024 by Ernst & Sohn GmbH.

Table 2.2 Bulk carriers.

Tonnage measurement	Cargo capacity [DWT]	Displacement G [t]	Overall length [m]	Length between perpendiculars [m]	Beam [m]	Maximum draught [m]	Generation
—	400 000	468 000	362	350	65.0	23.0	Chinamax ULBC[a]
—	325 000	380 000	340	N/A	62.0	21.0	VLBC[b]
—	175 000	205 000	290	N/A	45.0	17.0	Capesize
—	75 000	96 000	229	N/A	32.3	13.5	Panamax
—	55 000	64 000	200	N/A	32.3	11.5	Handymax
—	30 000	35 100	170	N/A	27.0	9.4	Handysize
—	8 000	N/A	107	N/A	18.2	6.8	Coaster

a) Ultra Large Bulk Carrier.
b) Very Large Bulk Carrier.

- The deadweight tonnage (DWT) is given in tonnes and indicates the maximum cargo capacity of a ship fully equipped and ready for operation. There is no mathematical relationship between deadweight tonnage and vessel size.
- Displacement indicates the actual weight of a ship in tonnes including the maximum cargo capacity.
- There is no mathematical relationship between displacement and cargo capacity and/or vessel size.
- Container vessels are often assessed according to their loading capacity, which is specified in TEUs (twenty-foot equivalent units). TEU is an internationally standardised unit for counting ISO containers with dimensions of 6.058 m long × 2.438 m wide × 2.591 m high.

2.1.1 Passenger ships and cruise liners

When it comes to passenger ships and cruise liners, the trend in recent years has definitely been towards ever larger vessels. Therefore, new terminals should be designed around a correspondingly large typical vessel. It should be noted here that in some cases, the maximum beam (width) lies above the waterline. For example, the ships of the Oasis class are 47 m wide at the waterline, but 60 m wide above the waterline.

2.1.2 Bulk carriers

The trend towards ever larger bulk carriers appears to have reached a climax with the Chinamax Class (388 000–400 000 dwt). These 35 ships have been built since 2010 and are the largest of their kind in the world.

2.1.3 General cargo ships

There is no trend towards larger general cargo ships. In addition, new general cargo ships are becoming a rarity. Instead, container, semi-container and special ships for transporting heavy cargos are replacing or have already replaced this traditional type of vessel. Where necessary, the dimensions given in the other sections can be applied accordingly. Design data should be determined carefully.

2.1.4 Container ships

The beam (width) of a container ship depends on the maximum number of rows of containers that can stand side by side on deck.

The dynamic development in container ship sizes shows no sign of abating. The largest tankers and bulk carriers to date, with up to 70 m beam and 24 m draught, provide some kind of guideline. Design data must, therefore, be determined carefully. Vessels with a capacity exceeding 14 000 TEU are known as ultra large container ships (ULCS) or megabox ships.

Table 2.3 Container ships.

Cargo capacity	Displacement G	Overall length	Length between perpendiculars	Beam	Rows	Maximum draught	Number of containers	Generation
[DWT]	[t]	[m]	[m]	[m]		[m]	[TEU]	
250 000	335 000	430	412	66.5	26	18.0	N/A	ULCS (megamax 26)[a]
228 000	290 000	400	383	61.5	24	16.0	23 500	ULCS (megamax 24)
195 000	259 000	400	383	59.0	23	16.0	21 400	ULCS
157 000	209 000	366	N/A	48.4	20	15.0	14 000	New-Panamax
118 000	157 000	334	N/A	45.6		13.5	10 000	Post-Panamax
66 000	88 000	294	N/A	32.3		13.5	5 100	Panamax
39 000	51 000	222	N/A	30.0		12.0	2 800	Feeder
15 000	20 000	150	N/A	23.0		7.6	1 000	Coaster

a) This is a projected future ship; the other container vessel classes exist.

2.1.5 Ferries

The dimensions of ferries are hugely dependent on the areas in which they are used and on their purposes. Ferries are often built to serve just one, two or three routes and, therefore, frequently have dimensions specific to the ports they use.

2.1.6 RoRo/ConRo vessels

RoRo vessels transport wheeled cargoes, e.g. cars, trucks, and breakbulk cargoes (i.e. all kinds of non-containerised heavy goods) over short, medium and long sea routes.

A special type of RoRo vessel is, for example, the ConRo ship, a combination of container and RoRo ship.

2 Ship dimensions

Table 2.4 RoRo vessels.

Cargo capacity [DWT]	Displacement G [t]	Overall length [m]	Length between perpendiculars [m]	Beam [m]	Maximum draught [m]
56 700	34 000	296	287	37.6	11.5
44 000	N/A	265	250	32.3	12.3
30 000	45 600	229	211	30.3	11.3
20 000	31 300	198	182	27.4	9.7
15 000	24 000	178	163	25.6	8.7
10 000	16 500	153	141	23.1	7.5
7 000	11 900	135	123	21.2	6.6
5 000	8 710	119	109	19.5	5.8
3 000	5 430	99	90	17.2	4.8
2 000	3 730	85	78	15.6	4.1
1 000	1 970	66	60	13.2	3.2

2.1.7 Oil tankers

Table 2.5 Oil tankers.

Cargo capacity [DWT]	Displacement G [t]	Overall length [m]	Length between perpendiculars [m]	Beam [m]	Maximum draught [m]	Generation
442 000	509 000	380	N/A	68	24.5	ULCC[a]
319 000	373 000	333	N/A	60.0	21.0	VLCC[b]
165 000	193 000	274	N/A	50.0	15.6	Suezmax
105 000	123 000	244	N/A	42.0	13.4	Aframax
55 000	64 000	229	N/A	32.3	12.3	Panamax
45 000	53 000	183	N/A	32.3	11.3	Handymax
25 000	29 250	170	N/A	25.5	8.9	Handysize
8 000	9 400	116	N/A	18.0	7.1	Coaster

a) Ultra Large Crude Carrier.
b) Very Large Crude Carrier.

2.1.8 Gas tankers

Table 2.6 Gas tankers.

Cargo capacity [DWT]	Capacity [m³]	Displacement G [t]	Overall length [m]	Length between perpendiculars [m]	Beam [m]	Maximum draught [m]
128 900	266 000	180 000	345	332	53.8	12.0
100 000	155 000	125 000	305	294	50.0	12.5
70 000	110 000	100 000	280	269	45.0	11.5
50 000	77 000	75 000	255	245	38.0	10.5
20 000	30 500	34 000	195	185	30.0	8.5
10 000	15 000	19 000	148	135	26.0	7.0
5 000	5 000	8 000	110	100	18.0	6.8
2 000	2 000	3 500	90	75	13.0	5.5

2.2 Inland waterway vessels

See Tab. 2.7

Table 2.7 Classification for European inland waterways.

Type of inland waterway	Class of navigable waterway	Motor vessels and barges Type of vessel: general characteristics						Pushed convoys Type of convoy: general characteristics						Minimum headroom under bridges	Graphical symbol on map
		Designation	Max length L [m]	Max width B [m]	Draught d [m][9]	Tonnage T [t]		Formation	Length L [m]	Width B [m]	Draught d [m][9]	Tonnage T [t]		[m][b]	
1	2	3	4	5	6	7		8	9	10	11	12		13	14
With regional significance	West of River Elbe I	Barge	38.5	5.05	1.8–2.2	250–400								4.0	‖
	II	Campine barge	50–55	6.6	2.5	400–650								4.0–5.0	‖
	East of River Elbe I	Large Finow	41	4.7	1.4	180								3.0		
	II	BM 500	57	7.5–9.0	1.6	500–630								3.0	‖
	III	c)	67–70	8.2–9.0	1.6–2.0	470–700			118–132[d]	8.2–9.0[d]	1.6–2.0	1000–1200		4.0	‖‖

2.2 Inland waterway vessels | 17

Table 2.7 (Continued.) Classification for European inland waterways.

With international significance	IV	Johann Welker	80–85	9.50	2.50	1000–1500		85	9.50[e)]	2.50–2.80	1250–1450	5.25 or 7.00[d)]	
	Va	Large Rhine vessel	95–110	11.40	2.50–2.80	1500–3000		96–110[a)]	11.40	2.50–4.50	1600–3000	5.25 or 7.00 or 9.10[d)]	
	Vb	Enlarged large Rhine vessel	135	11.4	4.0			172–185[a)]	11.40	2.50–4.50	3200–6000		
	VIa	Max. Rhine vessel	135	17	4.0			95–110[a)]	22.80	2.50–4.50	3200–6000	7.00 or 9.10[d)]	
	VIb	c)						185–195[a)]	22.80	2.50–4.50	6400–12 000	7.00 or 9.10[d)]	
	VIc							270–280[a)] 195–200[a)]	22.80 33.00–34.20[a)]	2.50–4.50 2.50–4.50	9600–18 000 9600–18 000	9.10[d)]	
	VII[h)]							285	33.00–34.20[a)]	2.50–4.50	14 500–27 000	9.10[d)]	

Table 2.7 (Continued.) Classification for European inland waterways.

a) The first number takes into account the actual situation, whereas the second corresponds to future developments, although in some cases also existing situations.
b) Takes into account a safety margin of approximately 30 cm between the uppermost point of the ship's structure or cargo on deck.
c) Takes into account the dimensions of self-propelled vessels anticipated in RoRo and container traffic; the dimensions given are approximate values.
d) Designed for transporting containers: 5.25 m for vessels carrying two tiers of containers, 7.00 m for vessels carrying three tiers of containers, 9.10 m for vessels carrying four tiers of containers; 50% of the containers may be empty, otherwise ballasting is required.
e) On account of the greatest permissible lengths of vessels and convoys, a number of existing waterways can be assigned to class IV even though the maximum beam is 11.40 m and the maximum draught is 4.00 m.
f) Vessels on the River Oder and on waterways between the River Oder and the River Elbe.
g) The draught for a certain waterway should be defined in accordance with local circumstances.
h) Convoys consisting of a larger number of barges/lighters may also be used on certain class VII waterways. In such cases, the horizontal dimensions may exceed the figures given in the table.

2.3 Offshore installation vessels

Table 2.8 Selection of installation vessels for offshore works.

	Length [m]	Beam [m]	Draught [m]	Cargo capacity [DWT]	Displacement [t]
Jack-up installation vessels					
Voltaire	180	60	7.5	21 500	N/A
Innovation	147.5	42.0	7.33	11 166	22 313
Seabreeze class					
Victoria Mathias, Friedrich Ernestine	100.0	40.0	4.5	6315	18 000
Pacific Orca (6 jack-up legs)	161.0	49.0	5.5	8400	N/A
Offshore crane vessel					
Thialf (semi-submersible, 14 200 t max. lifting capacity)	201.6	88.4	max. 31.6	129 221	198 750
Sleipnir (semi-submersible, 20 000 t max. lifting capacity)	220.0	102.0	max. 32.0	155 702	273 700
Stanislav Yudin	183.3	40.0	5.5–8.9	5600	49 200
Oleg (now *Seaways*) *Strashnov*	183.0	47.0	8.5–13.5	48 000	77 200
Feeder vessels (Panamax class)					
Typical vessel	224.0	32.3	12.0	70 000	81 900
Barges	100.0	30.0	6.0	11 000	16 000
Other vessels					
Pioneering Spirit (pipe-layer)	370.0	123.75	27.0	499 125	max. 1 000 000
Solitaire	300.0	40.6	13.5	N/A	127 435
MS Regina Baltica (cruise ferry)	145.2	25.5	5.5	2830	11 900
Mooring boat	12.0	3.5	1.5	N/A	N/A

3 Geotechnical principles

3.1 Geotechnical report

The geotechnical report according to DIN EN 1997-1 (3.4) or DIN EN 1997-2 (6) summarises the results of the soil investigations, specifies characteristic design values and provides foundation recommendations.

The nature and scope of the soil investigations and their results must be recorded in the geotechnical report. The assessment of the subsoil also includes investigations to establish the presence of any contamination or whether there are any chemical substances in the ground that could damage concrete and/or steel.

The specified characteristic design values of the soil properties may have to be selected with reference to the intended calculation methods and may differ accordingly.

The findings gleaned from the geotechnical report are summarised as foundation recommendations for the specific structure. For waterfront structures, this also includes information on the installation of piles and sheet piles, as well as any obstacles to driving.

Together with the verification of stability and serviceability, the aforementioned contents form the basis of the draft geotechnical report according to DIN EN 1997-2 (2.8).

The geotechnical report also forms the basis for classifying the subsoil in homogeneous zones according to the German Construction Contract Procedures (VOB Part C).

3.2 Subsoil

3.2.1 Mean characteristic values of soil parameters

For preliminary designs, the characteristic values (index k) given in Tables 3.1–3.3 may be used as empirical values for a larger body of soil. Without verification, the values in the table may only be assumed for low penetration resistance or soft consistency.

Detailed and final designs should always be based on the soil parameter values determined by site-specific field investigations an laboratory tests. Wherever possible, the effective shear parameters φ' and c' of cohesive soils should be ascertained in triaxial tests on undisturbed soil samples.

3 Geotechnical principles

Table 3.1 Reference characteristic values for soil parameters, non-cohesive soils. In this table, the soil groups GU, GT, SU, ST and G$\bar{\text{U}}$, G$\bar{\text{T}}$, S$\bar{\text{U}}$, S$\bar{\text{T}}$ should be allocated to non-cohesive soils in accordance with DIN 1054, A 3.1.2.

No.	Soil type	Soil group to DIN 18 196[a]	Penetration resistance q_c [MN/m²]	Unit weight γ_k [kN/m³]	Unit weight γ'_k [kN/m³]	Compressibility[b] Initial loading[c] $E_s = v_e \sigma_{at} (\sigma/\sigma_{at})^{w_e}$ v_e	w_e	Shear parameters of drained soil φ'_k [°]	Hydraulic conductivity k_k [m/s]
1	Gravel, poorly graded	GE $C_U < 6$	< 7.5 7.5–15 > 15	16.0 17.0 18.0	8.5 9.5 10.5	400 to 900	0.6 to 0.4	30.0–32.5 32.5–37.5 35.0–40, 0	2·10⁻¹ to 1·10⁻²
2	Gravel, well- or gap-graded	GW, GI $6 \leq C_U \leq 15$	< 7.5 7.5–15 > 15	16.5 18.0 19.5	9.0 10.0 11.5	400 to 1,100	0.7 to 0.5	30.0–32.5 32.5–37.5 35.0–40.0	1·10⁻² to 1·10⁻⁶
3	Gravel, well- or gap-graded	GW, GI $C_U > 15$	< 7.5 7.5–15 > 15	17.0 19.0 21.0	9.0 10.5 12.0	400 to 1200	0.7 to 0.5	30.0–32.5 32.5–37.5 35.0–40.0	1·10⁻² to 1·10⁻⁶
4	Gravel with silt/clay mixture, fraction $d < 0.06$ mm is $< 15\%$	GU, GT	< 7.5 7.5–15 > 15	17.0 19.0 21.0	9.0 10.5 12.0	400 to 1200	0.7 to 0.5	27.5–32.5 32.5–37.5 35.0–40.0	1·10⁻⁵ to 1·10⁻⁶
5	Gravel with silt/clay mixture, fraction $d < 0.06$ mm is $> 15\%$	G$\bar{\text{U}}$, G$\bar{\text{T}}$	< 7.5 7.5–15 > 15	16.5 18.0 19.5	8.5 9.5 10.0	150 to 400	0.9 to 0.7	27.5–32.5 32.5–37.5 35.0–40.0	1·10⁻⁷ to 1·10⁻¹¹
6	Sand, poorly graded, coarse sand	SE $C_U < 6$	< 7.5 7.5–15 > 15	16.0 17.0 18.0	8.5 9.5 10.5	200 to 700	0.75 to 0.55	30.0–32.5 32.5–37.5 35.0–40.0	5·10⁻³ to 1·10⁻⁴
7	Sand, poorly graded, fine sand	SE $C_U < 6$	< 7.5 7.5–15 > 15	16.0 17.0 18.0	8.5 9.5 10.5	150 to 500	0.75 to 0.60	30.0–32.5 32.5–37.5 35.0–40.0	1·10⁻⁴ to 2·10⁻⁵

Table 3.1 (Continued). Reference characteristic values for soil parameters, non-cohesive soils. In this table, the soil groups GU, GT, SU, ST and $\overline{GU}$, $\overline{GT}$, $\overline{SU}$, $\overline{ST}$ should be allocated to non-cohesive soils in accordance with DIN 1054, A 3.1.2.

8	Sand, well- or gap-graded	SW, SI $6 \leq C_U \leq 15$	< 7.5 7.5–15 > 15	16.5 18.0 19.5	9.0 10.0 11.5	200 to 600	0.70 to 0.55	30.0–32.5 32.5–37.5 35.0–40.0	$5 \cdot 10^{-4}$ to $2 \cdot 10^{-5}$
9	Sand, well- or gap-graded	SW, SI $C_U > 15$	< 7.5 7.5–15 > 15	17.0 19.0 21.0	9.0 10.5 12.0	200 to 600	0.70 to 0.55	30.0–32.5 32.5–37.5 35.0–40.0	$1 \cdot 10^{-4}$ to $1 \cdot 10^{-5}$
10	Sand with silt/clay mixture, fraction $d < 0.06$ mm is < 15%	SU, ST	< 7.5 7.5–15 > 15	16.0 17.0 18.0	8.5 9.5 10.5	150 to 500	0.80 to 0.65	27.5–32.5 32.5–37.5 35.0–40.0	$2 \cdot 10^{-5}$ to $5 \cdot 10^{-7}$
11	Sand with silt/clay mixture, fraction $d < 0.06$ mm is > 15%	$\overline{SU}$, $\overline{ST}$	< 7.5 7.5–15 > 15	16.5 18.0 19.5	9.0 10.0 11.5	50 to 250	0.9 to 0.75	27.5–32.5 32.5–37.5 35.0–40.0	$2 \cdot 10^{-6}$ to $1 \cdot 10^{-9}$

a) Higher or lower values outside the ranges specified for parameters can occur in certain circumstances.

Code letters for primary and secondary components:

F organic silt/peat clay – alternatively e.g. gyttja
G gravel
H peat (humus)
O organic inclusions
S sand
T clay
U silt.

Code letters for characteristic physical soil properties:

W well-graded grain size distribution
E poorly-graded grain size distribution
I intermittently-graded grain size distribution

Plastic properties:

L lightly plastic
M medium plastic
A highly plastic

Degree of decomposition in peats:

N not decomposed or scarcely decomposed peat
Z decomposed peat.

Symbols

v_e stiffness factor, empirical parameter
w_e empirical parameter
σ load in kN/m^2
σ_{at} atmospheric pressure ($= 100$ kN/m^2)

b) The increase in v_e is always coupled with a decrease in w_e!

c) v_e values for reloading up to 10 times higher, w_e values tend towards 1.

24 | 3 Geotechnical principles

Table 3.2 Reference characteristic values for soil parameters, cohesive soils. Here, the soil groups GU, GT, SU, ST and $\overline{GU}$, $\overline{GT}$, $\overline{SU}$, $\overline{ST}$ should be allocated to cohesive soils in accordance with DIN 1054, A 3.1.3. The parameters for compressibility should preferably be determined by laboratory tests.

No.	Soil type	Soil group to DIN 18196[a]	Consistency in initial state, i.e. to DIN EN 14688-1	Unit weight γ_k [kN/m³]	γ'_k [kN/m³]	Compressibility[b] Initial loading[c] $E_s =$ $v_e \sigma_{at}(\sigma/\sigma_{at})^{w_e}$ v_e	w_e	Shear parameters of drained soil φ'_k [°]	c'_k [kN/m²]	Hydraulic conductivity k_k [m/s]
1	Gravel with silt/clay mixture, fraction $d < 0.06$ mm is $< 15\%$	GU, GT	Soft Firm Stiff	19.0 20.0 21.0	10.5 11.5 12	300 to 1 000	0.7 to 0.5	27.5 to 35.0	0 0 0	$1 \cdot 10^{-5}$ to $1 \cdot 10^{-7}$
2	Gravel with silt/clay mixture, fraction $d < 0.06$ mm is $> 15\%$	$\overline{GU}$, $\overline{GT}$	Soft Firm Stiff	19.0 20.5 22.0	10.0 11.0 12.0	50 to 200	0.8 to 0.6	25.0 to 30.0	0 5 10	$1 \cdot 10^{-6}$ to $1 \cdot 10^{-10}$
3	Sand with silt/clay mixture, fraction $d < 0.06$ mm is $< 15\%$	SU, ST	Soft Firm Stiff	19.0 20.0 21.0	10.5 11.5 12	120 to 400	0.7 to 0.6	27.5 to 35.0	0 0 0	$2 \cdot 10^{-5}$ to $5 \cdot 10^{-7}$
4	Sand with silt/clay mixture, fraction $d < 0.06$ mm is $> 15\%$	$\overline{SU}$, $\overline{ST}$	Soft Firm Stiff	19.0 20.5 21.5	9.0 10.0 11.0	40 to 120	0.8 to 0.6	25.0 to 30.0	0 5 10	$2 \cdot 10^{-6}$ to $1 \cdot 10^{-9}$
5	Inorganic cohesive soils with low plasticity ($w_L < 35\%$)	UL	Soft Firm Stiff	18.0 19.0 20.0	9.0 10.0 11.0	40 to 110	0.80 to 0.60	27.5 to 32.5	0 2–5 5–10	$1 \cdot 10^{-5}$ to $1 \cdot 10^{-7}$
6	Inorganic cohesive soils with medium plasticity ($35\% < w_L < 50\%$)	UM	Soft Firm Stiff	17.5 18.5 19.5	8.5 9.5 10.5	30 to 70	0.90 to 0.70	25.0 to 30.0	0 5–10 10–15	$2 \cdot 10^{-6}$ to $1 \cdot 10^{-9}$

Table 3.2 (Continued.) Reference characteristic values for soil parameters, cohesive soils. Here, the soil groups GU, GT, SU, ST and GŪ, GT̄, SŪ, ST̄ should be allocated to cohesive soils in accordance with DIN 1054, A 3.1.3. The parameters for compressibility should preferably be determined by laboratory tests.

7	Inorganic cohesive soils with low plasticity ($w_L < 35\%$)	TL	Soft Firm Stiff	20.0 21.0 22.0	10.0 11.0 12.0	20 to 50	1.0 to 0.90	25.0 to 30.0	0 5–10 10–15	$1 \cdot 10^{-7}$ to $2 \cdot 10^{-9}$
8	Inorganic cohesive soils with medium plasticity ($35\% < w_L < 50\%$)	TM	Soft Firm Stiff	19.0 20.0 21.0	9.0 10.0 11.0	10 to 30	1.0 to 0.95	22.5 to 27.5	5–10 10–15 15–20	$5 \cdot 10^{-8}$ to $1 \cdot 10^{-10}$
9	Inorganic cohesive soils with high plasticity ($50\% < w_L$)	TA	Soft Firm Stiff	18.0 19.0 20.0	8.0 9.0 10.0	6 to 20	1.0 1.0 1.0	20.0 to 25.0	5–15 10–20 15–25	$1 \cdot 10^{-9}$ to $1 \cdot 10^{-11}$

a) Higher or lower values outside the ranges specified for parameters can occur in certain circumstances.

Code letters for primary and secondary components:

F organic silt/peat clay – alternatively e.g. gyttja
G gravel
H peat (humus)
O organic inclusions
S sand
T clay
U silt.

Code letters for characteristic physical soil properties:

W well-graded grain size distribution
E poorly-graded size distribution
I intermittently-graded grain size distribution

Plastic properties:

L lightly plastic
M medium plastic
A highly plastic

Degree of decomposition in peats:

N not decomposed or scarcely decomposed peat
Z decomposed peat.

b) The increase in v_e is always coupled with a decrease in w_e!

Symbols:

v_e stiffness factor, empirical parameter
w_e empirical parameter
σ load in kN/m²
σ_{at} atmospheric pressure ($= 100$ kN/m²)

c) v_e values for reloading up to 10 times higher, w_e values tend towards 1.

3 Geotechnical principles

Table 3.3 Reference characteristic values for soil parameters, organic soils. The parameters for compressibility should preferably be determined by laboratory tests.

No.	Soil type	Soil group to DIN 18196[a]	Consistency in initial state, i.e. to DIN EN 14688-1	Unit weight		Compressibility[b] Initial loading[c] $E_S = v_e \sigma_{at}(\sigma/\sigma_{at})^{w_e}$		Shear parameters of drained soil		Hydraulic conductivity
				γ_k [kN/m³]	γ'_k [kN/m³]	v_e	w_e	φ'_k [°]	c'_k [kN/m²]	k_k [m/s]
1	Organic silt, organic clay	OU and OT	Very soft Soft Firm	14.0 15.5 17.0	4.0 5.5 7.0	[e] 5 20	[e] 1.00 0.85	17.5 to 22.5	0 2–5 5–10	$1 \cdot 10^{-9}$ to $1 \cdot 10^{-11}$
2	Peat[e]	HN, HZ	Very soft Soft Firm Stiff	10.5 11.0 12.0 13.0	0.5 1.0 2.0 3.0	[e]	[e]	[e]	[e]	$1 \cdot 10^{-5}$ to $1 \cdot 10^{-8}$
3	Mud[f] Digested sludge	F	Very soft Soft	12.5 16.0	2.5 6.0	4 15	1.0 0.9	[f]	0	$1 \cdot 10^{-7}$ $1 \cdot 10^{-9}$

a) Higher or lower values outside the ranges specified for parameters can occur in certain circumstances.

Code letters for primary and secondary components:
F organic silt/peat clay – alternatively e.g. gyttja
G gravel
H peat (humus)
O organic inclusions
S sand
T clay
U silt.

Code letters for characteristic physical soil properties:
Particle size distribution:
W well-graded grain size distribution
E poorly-graded size distribution
I intermittently-graded grain size distribution

Plastic properties:
L lightly plastic
M medium plastic
A highly plastic

Degree of decomposition in peats:
N not decomposed or scarcely decomposed peat
Z decomposed peat.

b) The increase in v_e is always coupled with a decrease in w_e!

Symbols:
v_e stiffness factor, empirical parameter
w_e empirical parameter
σ load in kN/m²
σ_{at} atmospheric pressure (= 100 kN/m²)

c) v_e values for reloading up to 10 times higher, w_e values tend towards 1.

d) For the compressibility of organic soils with a mushy consistency no reference values can be given.

e) The compressibility and shear parameter values for peat exhibit such a wide scatter that empirical values cannot be given.

f) The effective angle of internal friction of fully consolidated mud can be very high, but the value corresponding to the true degree of consolidation, which can only be determined reliably in laboratory tests, always governs.

The characteristic values of the shear parameters φ'_k and c'_k for cohesive soils apply to calculations for final stability (consolidated state, final strength).

Empirical values for the shear parameters of the undrained, initially loaded soil $c_{u,k}$ are specified in DIN 1055-2 (2010–2011).

3.2.2 Layout and depths of boreholes and penetrometer tests

3.2.2.1 General

The nature and extent of soil investigations, their layout and the depth of any such investigations must be determined by a geotechnical expert according to the provisions of DIN EN 1997-2 and DIN 4020.

The aim of boreholes is to investigate the stratification and obtain soil samples for soil mechanics tests in the laboratory. For investigation and monitoring of groundwater conditions, boreholes can be upgraded to groundwater monitoring wells.

Penetrometer tests allow the strength properties of the in situ soil types to be determined. With the help of empirical correlations, the soil types can be identified, and the values of soil properties derived.

The number and layout of boreholes and penetrometer tests must always be such that all the characteristics of the subsoil relevant to the planning are established and a sufficient number of suitable soil samples is obtained for the laboratory tests. When determining the number and type of boreholes and penetrometer tests, the results of earlier surveys in the form of geological maps and, where applicable, the findings of earlier boreholes and penetrometer tests should also be taken into account.

Geophysical surface measurements in conjunction with the boreholes and penetrometer tests can supply two-dimensional data on the geological profile, groundwater level and indications regarding any large obstacles in the subsoil.

Where major construction projects are involved, it can be useful to begin with principal boreholes and penetrometer tests to gain an overall picture and then to supplement these with intermediate boreholes and further penetrometer tests during the planning phase (see Figure 3.1).

The soil investigations can be supplemented by loading tests and trial embankments to enable a reliable assessment of the load-bearing behaviour of foundation elements and soil compaction options.

To assess the soil–structure interaction a number of model tests may be required.

The execution of and results from loading tests, trial embankments and model tests must be recorded in the geotechnical report.

3.2.2.2 Principal boreholes

Principal boreholes should preferably lie on the later axis of the structure (waterfront). For cantilever walls they should be drilled to a depth equal to approximately twice the difference in ground levels or as far as a known geological stratum. As a guide, the recommended borehole spacing is approx. 50 m; recommendations regarding their location and depth are specified in DIN EN 1997-2 (2.4.1.3) and DIN 4020. In specific cases, the positions and spacings of the boreholes must be adapted in line with the geological and constructional boundary conditions. Given that soil samples for soil mechanics tests in the laboratory must be at least grade 2 according to DIN EN ISO 22475-1, the principal boreholes must be designed as boreholes suitable for obtaining samples in liners.

3 Geotechnical principles

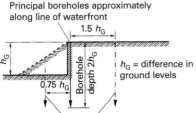

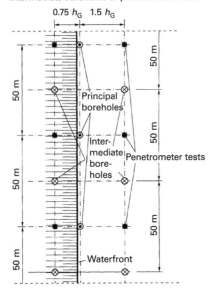

Figure 3.1 Example of layout of boreholes and penetrometer tests for waterfront structures.

3.2.2.3 Intermediate boreholes

Depending on the findings of the principal boreholes or the earlier penetrometer tests, intermediate boreholes are also sunk to the depth of the principal boreholes or to a depth at which a known, homogeneous soil stratum is encountered. The typical borehole spacing is again approximately 50 m.

3.2.2.4 Penetrometer tests

Penetrometer tests are generally executed according to the layout in Figure 3.1. As far as possible, they are sunk to the same depth as the principal boreholes. The relevant standards should be consulted regarding details of the equipment for and execution of penetrometer tests, along with their application.

In order to interpret the results of penetrometer tests, individual tests must be carried out directly adjacent to boreholes. In such cases the penetrometer tests must be performed prior to drilling the boreholes in order to prevent the results of the penetrometer test from being influenced due to any loosening of the soil during drilling.

3.2.3 Determining the shear strength c_u of saturated, undrained cohesive soils

If saturated cohesive soil is loaded without being able to consolidate (undrained conditions), its change in volume is negligible due to the low compressibility of the pore water at loads below its strength. The load generates excess pore water pressure only and no additional effective stresses in the soil skeleton. As a result, the angle of internal friction for saturated cohesive soils in undrained conditions is $\varphi_u = 0$. The strength is only described by the cohesion of the undrained soil c_u. In the case of partial saturation, part of the load can generate additional effective stresses in the soil skeleton; in such cases $\varphi' > \varphi_u > 0$.

3.2.3.1 Cohesion c_u of undrained soil

The cohesion c_u of undrained cohesive soils essentially depends on the following conditions:

- For normally consolidated soils, c_u is proportional to the effective vertical stress σ'_v, i.e. c_u increases linearly with depth:

$$\frac{c_u}{\sigma'_v} = \lambda_{cu}$$

- According to Jamiolowski et al. (1985), the cohesion constant is $\lambda_{cu} = 0.23 \pm 0.04$, although Gebreselassie (2003) states that values as low as $\lambda_{cu} = 0.18$ and even lower are possible. For north German marine clay, c_u is often very low, and any dependency on the vertical stress σ'_v is hard to measure with any certainty.
- For overconsolidated soil, c_u is likewise proportional to the effective vertical stress σ'_v, but is also determined from the stress history:

$$\frac{c_u}{\sigma'_v} = \lambda_{cu} OCR^\alpha$$

- The overconsolidation ratio (OCR) is the ratio of the stress σ'_{vc} at which the soil is consolidated and the current stress σ'_v.

$$OCR = \frac{\sigma'_{vc}}{\sigma'_v}$$

- Reference values for exponent α lie between 0.8 and 0.9.
- A number of different authors demonstrate that the cohesion c_u of an undrained soil depends on the stress path. Under triaxial compression (tc), c_u is greater than for triaxial extension (te) (Bjerrum 1973, Jamiolowski et al. 1985, Scherzinger 1991); $c_{u,tc}$ can be approximately 50% greater than $c_{u,te}$. Values for direct simple shear $c_{u,dss}$ lie in between for the same pore volume:

$$c_{u,tc} > c_{u,dss} > c_{u,te}$$

- Owing to the viscosity of cohesive soils, the cohesion c_u of an undrained soil depends on the rate of load application. This can be described using the shear rules of Leinenkugel (1976) or Randolph (2004), for instance. Leinenkugel's relationship is

$$\frac{c_u}{c_{u\alpha}} = \left[1 + I_{v\alpha} \ln\left(\frac{d\gamma/dt}{d\gamma_\alpha/dt}\right)\right]$$

- The viscosity index $I_{v\alpha}$ for the reference strain rate $d\gamma/dt$ can be determined, for example, by means of CU triaxial tests with an abruptly varying strain rate (step test) or by using one-dimensional creep tests. Reference values for $I_{v\alpha}$ can be found, for example, in Leinenkugel (1976) and Gudehus (1981).

3.2.3.2 Determining the cohesion c_u of an undrained soil

The cohesion c_u of an undrained soil can be determined in laboratory or field tests. There are two essentially different methods for such investigations: the recompression method and the stress history method.

3.2.3.2.1 Recompression method

In this method, c_u is determined by way of triaxial tests on soil samples that are reconsolidated prior to shearing with the stress that acts on the soil in situ (Bjerrum 1973). According to Seah and Lai (2003), however, this method overestimates the cohesion c_u of normally consolidated soil. Therefore, the recompression method is preferred for highly structured, brittle soils, e. g. sensitive clays, cemented soils and severely overconsolidated soils. The results of shear tests should always be checked by comparing them with the stress history.

3.2.3.2.2 The stress history and normalised soil engineering properties method (SHANSEP)

This method enables the cohesion c_u to be determined while taking into account the sample disorder, the anisotropy to a limited extent and the rate of load application. It is based on investigations carried out at MIT during the 1960s and was initially published by Ladd and Foott (1974). A revised version can be found in Ladd and DeGroot (2003). The SHANSEP method involves the following steps to specify the soil model:

- Carrying out a soil investigation, taking special samples (undisturbed soil samples) and compiling a soil profile based on the results of cone penetration tests and field vane shear tests.
- Determining the degree of overconsolidation in the laboratory based on compression tests and deriving the overconsolidation ratio (OCR).
- Determining the effective shear parameters φ'/c' and c_u in laboratory tests, normally by way of triaxial tests. Triaxial tests with anisotropic consolidation (CK_0) and subsequent undrained triaxial compression (UC with $\sigma_1 > \sigma_3$) and triaxial extension (UE with $\sigma_1 < \sigma_3$) are recommended. Stipulating the reconsolidation stress corresponding to the calculated OCR.
- Performing shear tests to determine the relationship between the OCR and the normalised shear strength c_u/σ'_v.
- Stipulating a c_u design profile for the cohesive strata that lies on the safe side.

3.2.3.3 Determining c_u in laboratory tests

The advantage of determining c_u in laboratory tests is that the test conditions can be reproduced in an ideal manner within larger test series. However, the disadvantage is that test specimens can never be obtained from boreholes without disturbing their structure and strength. In addition, test specimens are not continuous, meaning that the distribution of c_u over the stratum thickness is only ascertained at discrete points.

The test conditions are best checked in triaxial tests. CU triaxial tests supply both drained and undrained shear parameters because the pore water pressure is measured. When determining c_u by way of laboratory vane shear tests as well as unconfined compression tests, a distorted capillarity influence cannot be ruled out. For soft soils, c_u can also be determined by way of various pressure and fall cone tests.

3.2.3.4 Field tests

Determining c_u by way of cone penetration tests to DIN 4094-1 and vane shear tests to DIN 4094-4 supplies a profile of the cohesion c_u over the depth. Owing to the high shearing rate, the shear resistance τ_{fvt} in the vane shear test must be reduced by a factor μ, which depends on the plasticity index I_P:

$$c_{u,fvt} = \mu \tau_{fvt}$$

Details of the correction factor μ can be found in DIN EN 1997-2, Annex I.

Deriving c_u from the penetration resistance in cone penetration tests requires knowledge of the OCR of the soil. For example, the following applies for the CPTU test:

$$c_{u,cptu} = \frac{q_c - \sigma_v}{N_{kt}}$$

The factor N_{kt} depends on the cone geometry and OCR, and lies between 10 and 20.

The derivation of c_u from borehole-widening tests to DIN 4096 is less common.

Plate load tests to DIN 18 134 only supply a c_u value for soils near the surface.

3.2.3.5 Correlations

A number of authors have suggested correlations between c_u and the water content w, the consistency index I_C, the plasticity index I_P and the liquidity index I_L; Gebreselassie (2003) provides a detailed overview of this. It should be noted that these correlations apply, at best, to the soils examined and the corrsponding test conditions, and can, therefore, only be used as reference values.

3.2.4 Assessing the subsoil for the installation of piles and sheet piles and for selecting the installation method

3.2.4.1 General

In the first place, the material, form, size, length and angle of piles and sheet piles play a decisive role with respect to the installation of piles and sheet piles and the selection of the installation method. Important information on this can be found in Sections 8.1 and 9.1.2

In connection with these recommendations, it is especially important to note that when selecting the type of sheet pile (material, form), it is essential to take into account the stresses due to the installation procedure in the respective subsoil, in addition to the structural requirements and economic issues. As a result, the geotechnical report must also include an evaluation of the in situ subsoil with respect to the installation of piles and sheet piles (see also Section 3.1).

3.2.4.2 Assessment of soil types with respect to installation methods
3.2.4.2.1 General

The shear parameters have only a limited significance when it comes to describing the behaviour of the subsoil during the installation of piles and sheet piles. For example, rocky calcareous marl can exhibit relatively low shear parameters due to its fissuring but may, in fact, present difficult conditions in terms of pile driving.

3.2.4.2.2 Impact driving

Easy driving conditions are to be expected in soft or very soft soils such as moorland, peat, silt, marine clay, etc. Easy driving conditions are also to be generally expected in loose medium and coarse sands and gravels with no rock inclusions, unless there are embedded cemented strata.

Moderate driving conditions are to be expected in dense medium and coarse sands, fine gravel soils and firm clays and loams.

Difficult to very difficult driving conditions are to be expected in most instances of dense medium and coarse gravels, dense fine sandy and silty soils, embedded cemented strata, stiff to very stiff clays, cobbles and moraine strata, glacial till and weathered and soft to medium-hard rock.

Earth-moist or dry soils present a greater resistance to penetration during impact driving than those subject to buoyancy. This does not apply to saturated cohesive soils, and especially silts.

An increasingly high penetration resistance during driving can be expected with a number of blows $N_{10} > 30$ per 10 cm penetration in the heavy dynamic penetration test (DPH, DIN EN ISO 22476-2) or $N_{30} > 50$ per 30 cm penetration in borehole dynamic probing (BDP, DIN 4094-2). It can generally be assumed that driving is possible up to a number of blows $N_{10} = 80–100$ per 10 cm penetration (DPH). Driving with a higher number of blows can be possible in individual cases. For more information, see Rollberg (1976, 1977).

3.2.4.2.3 Vibratory driving

The skin friction and base resistance of the pile being installed are greatly reduced when using vibratory driving methods. As a result, the piles or sheet piles can quickly reach their required depth compared with impact driving. For more information, see Section 8.1.5.3.

Vibratory driving is particularly successful in sands and gravels with a rounded grain shape and in very soft or soft soil types with low plasticity. Vibrating is much less suitable for highly cohesive soils or sands and gravels with an angular grain shape. Dry fine sands and firm marl and clay soils are particularly critical as they absorb the energy of the vibrator without reducing the skin friction and base resistance.

If the subsoil is compacted during vibration, then its penetration resistance can increase to such an extent that the pile being installed can no longer reach the required depth. This risk arises, in particular, when piles and sheet piles are installed at close spacings and when using vibration in non-cohesive soils. Vibratory driving must be stopped in such cases (see Chapter 8). The use of auxiliary driving measures according to Section 8.1.6 may represent an option.

Above all, vibratory driving in non-cohesive soils may lead to localised settlement, the magnitude and extent of which depend on the power output of the vibrator, the section being driven, the duration of the vibratory driving and the soil. When working close to existing structures, checks must be made to establish whether any such settlement could cause damage. If required, the installation procedure must be adjusted accordingly.

3.2.4.2.4 Pressing

For pressing to be used, there should be no obstacles in the soil, or if there are any, they must be removed prior to driving.

Table 3.4 Pressing limits for steel sheet piles.

Soil parameter				Without driving aid	With driving aid
CPT peak pressure	q_b	MN/m²		< 20	< 35
CPT skin friction	q_s	MN/m²		< 0.1	< 0.3
DPH	N_{10}	—		< 25	< 40
Consistency index	I_c	—		< 1.0	> 1.0
Plasticity index	I_P	—		> 10	
Ratio[a]	I_f	—		< 1.0	> 1.0
Angle of friction	φ'	°		< 35	< 45

a) $I_f = (\max e - \min e)/\min e$ (higher compactability in the event of decreasing I_f).

Slender sections can generally be pressed hydraulically into cohesive soils without obstacles or into loose non-cohesive soils. Sheet piles can only be pressed into dense non-cohesive soils if the soil has been loosened beforehand. Empirical values according to Busse (2009) are given in Table 3.4.

3.2.4.2.5 Auxiliary driving measures

Water-jetting can ease driving in dense sands and gravels, as well as in firm and stiff clays in particular; indeed, it may be essential to enable driving in the first place.

Additional auxiliary driving measures include pre-drilling to loosen the soil or local soil replacement, etc. Rocky soils can be loosened by way of local blasting in such a way that the required depth can be reached using conventional impact driving and appropriate pile sections. For more information, see Section 8.1.6.2.

3.2.4.2.6 Drilling equipment, piles and installation methods

Driving equipment, piles and installation methods must be suited to the subsoil through which the piles are being driven; see Sections 8.1.5.3.4, 8.1.5.2.3, 8.1.5.3 and 5.13.

Slow-acting drop hammers, diesel hammers and hydraulic hammers are suitable for both cohesive and non-cohesive soils. Rapid-acting hammers and vibration equipment place less stress on the pile, but are generally only effective in non-cohesive soils with a rounded grain shape. Rapid acting hammers or heavy hammers with short drop heights should be preferred when driving in rocky soils, even when using pre-blasting to loosen the ground.

Interruptions during driving a pile, e. g. between initial and final driving, can make subsequent driving easier or harder depending on the soil type and water saturation, as well as the length of the interruption. Any changes to the penetration resistance should generally be identified and quantified by the way of tests in advance.

Assessing the subsoil for the installation of piles and sheet piles presumes appropriate experience and specialised knowledge of the installation methods. Experience of construction projects with similar subsoil conditions can, indeed, be very beneficial.

3.2.4.2.7 Testing installation methods and load-bearing behaviour in difficult conditions

If on construction projects with considerable embedment depths there are concerns that sheet piles cannot be driven to the depth required to satisfy the structural design without being damaged or other piles cannot reach the intended embedment depth to carry the loads, then test piles must be driven and pile loading tests carried out beforehand. At least two test piles should be driven for each installation method in order to obtain accurate information.

Testing the installation method may also be necessary in order to predict any settlement of the soil, as well as the spread and impact of vibrations as a result of the installation method.

3.2.5 Classifying the subsoil in homogeneous zones

In Germany, VOB Part C (German Construction Contract Procedures), September 2016 edition (VOB 2016), requires the subsoil to be classified into homogeneous zones for tendering and settlement of accounts in civil engineering and special civil engineering works. The following standards of VOB Part C are the main ones relevant here:

- DIN 18300 Earthworks
- DIN 18301 Drilling works
- DIN 18304 Piling
- DIN 18311 Dredging work
- DIN 18312 Underground construction work
- DIN 18313 Diaphragm walling
- DIN 18319 Trenchless pipelaying
- DIN 18320 Landscape works
- DIN 18321 Jet grouting work
- DIN 18324 Horizontal directional drilling works.

A homogeneous zone is a limited area comprising one or more soil or rock strata that – for the particular method of construction – exhibit comparable properties.

The description of the homogeneous zones is achieved by way of several properties and parameters that are specified separately for each method of construction in the respective VOB/C standard. An upper and lower bound (range) must be specified for every parameter. The VOB standards specify one or more standards or recommendations for every parameter, which apply for checking the parameters if required within the scope of the construction work.

For the contractor, the classification into homogeneous zones with details of properties and parameters forms the basis for planning the construction equipment and, based on that, their pricing. Furthermore, in the standard case, the construction work is invoiced according to homogeneous zones.

The strata contained within one homogeneous zone must have comparable properties with regard to the method of construction as tendered. The range for the parameters of a homogeneous zone results from the sum of the ranges of the strata in the zone. The permissible size of the range for the parameters within a homogeneous zone is not specified in any standard.

Criteria related to soil mechanics and construction operations and processes must be considered when specifying the homogeneous zones. Close cooperation between the designer and the geotechnical engineering expert is, therefore, essential.

The geotechnical report forms the basis for classifying the subsoil into homogeneous zones and specifying the ranges for the individual parameters. Those ranges should be defined based on in situ soil investigations and laboratory tests. However, empirical values should also be considered.

If the costs of testing to ascertain the in situ density and density of non-cohesive soils and the proportions of rocks and boulders are to be kept reasonable, then such tests are only possible near the surface. Generally, such parameters cannot be determined by way of tests within the scope of soil investigations. The ranges (upper and lower bounds) of these parameters must be estimated by correlating the penetrometer tests, employing empirical values or examining the genesis. Therefore, such work is always affected by great uncertainties.

In the case of methods of construction in which soil or rock is extracted and/or further processed, soil constituents with an environmental relevance must be taken into account when modelling the homogeneous zones. The further processing or disposal of contaminated soil is normally very expensive, and that can have serious consequences for the profitability of a construction project. It is frequently the case that contamination does not follow the bedding planes, and so it may prove necessary to allocate one stratum to several homogeneous zones corresponding to the contamination.

A construction project normally involves several trades employing different methods of construction. It might, therefore, be necessary to classify the subsoil into different homogeneous zones depending on those methods.

BAW (2017) contains further information for classifying the subsoil into homogeneous zones for waterways projects.

3.3 Water pressure

3.3.1 General

The loads determined in this section are allocated to the design situations as per Section 1.1.4. Regarding partial safety factors, Sections 1.1 and 1.2.4 must be observed.

Specifying the critical water pressure loads due to changing outer and groundwater levels, which for simplicity are applied as hydrostatic loads, requires an analysis of the geological and hydrological conditions of the area affected. Series of observations – over several years – of groundwater levels depending on outer water levels should be evaluated wherever possible.

The groundwater level behind a waterfront structure is very much affected by the soil stratification and the design of the waterfront structure. In tidal areas, the groundwater level for permeable soils tracks the tides in an attenuated manner to a greater or lesser extent. The relationship between groundwater and tidal water levels can be ascertained by way of measurements.

Groundwater levels are subject to temporal and spatial fluctuations, which need to be taken into account when determining the characteristic water levels.

Geotechnical analyses according to DIN 1054 require the partial safety factors for permanent loads to be applied to the characteristic loads resulting from the characteristic water levels, irrespective of their classification as permanent or variable loads.

Where realistic, geotechnical analyses according to DIN 1054 still require a resultant characteristic water pressure to be calculated from favourable and unfavourable water pressures without applying the partial safety factors separately. The corresponding partial safety factor for permanent loads is applied to this resultant characteristic water pressure depending on whether it is, on the whole, a favourable or unfavourable effect.

3.3.2 Resultant water pressure in the direction of the water side

The resultant water pressure $w_{ü}$ on a waterfront structure in the direction of the water side for a difference in height Δh between the critical outer water level and the corresponding groundwater level for the unit weight of water γ_{w} amounts to

$$w_{ü} = \Delta h \cdot \gamma_{w}$$

The resultant water pressure can be applied according to Figure 3.2 in permeable soils in non-tidal areas and Figure 3.3 in tidal areas and assigned to design situations DS-P, DS-T and DS-A. If the resultant water pressure is applied assuming weepholes, their permanent effect must be guaranteed. The approaches of Figures 3.2 and 3.3 are based on the assumption of a planar flow but without taking into account how waves might influence the resultant water pressure.

The water levels associated with Figures 3.2 and 3.3 must be applied as characteristic values.

Where there is a rapid and severe drop in the canal water level, the two load cases given below for the accidental design situation DS-A must be examined when designing sheet pile walls alongside canals (Figure 3.4):

a) Canal water level 2.00 m below groundwater level.
b) Canal water level is set at waterway bed with groundwater level 3.00 m higher.

Non-tidal area				
Situation	Figure	Design situations in accordance with Chapter 1.1.4		
		P	T	A
1 Minor water level fluctuations (≤ 0.5 m) with weepholes or permeable soil and structure	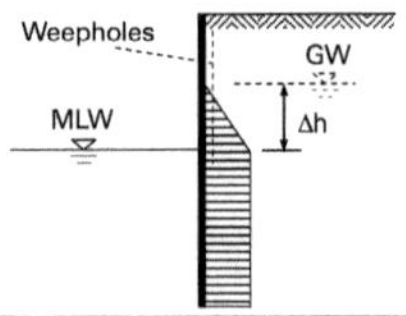	$\Delta h = 0.50$ m	$\Delta h = 0.50$ m	–
2a Major water level fluctuations (> 0.5 m) with weepholes or permeable soil and structure	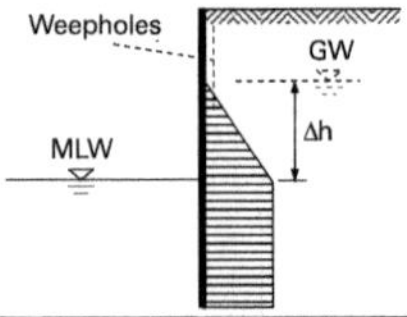	$\Delta h = 0.50$ m at common level	$\Delta h = 1.00$ m at unfavourable level	$\Delta h \geq 1.00$ m max. drop in outer water level over 24 h
2b Major water level fluctuations (> 0.5 m) without weepholes	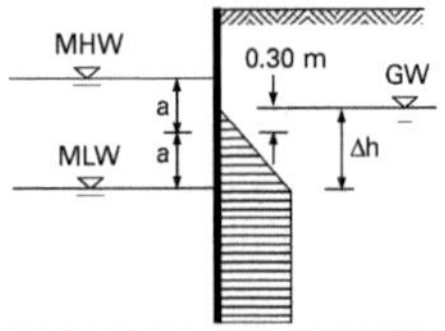	$\Delta h =$ $a + 0.30$ m	$\Delta h =$ $a + 0.30$ m	–

Figure 3.2 Approximate approaches for resultant water pressure on waterfront structures for permeable soils in non-tidal areas and standard situations (without significant wave effect).

3.3 Water pressure

Situation	Figure	Tidal area Design situations in accordance with R 18		
		P	T	A
3a Major water level fluctuations without drainage = normal case	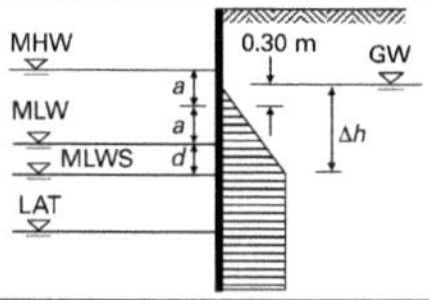 MHW, MLW, MLWS, LAT	$\Delta h = a + 0.30\ \text{m} + d$	–	–
3b Major water level fluctuations without drainage = limit case for extreme low water level	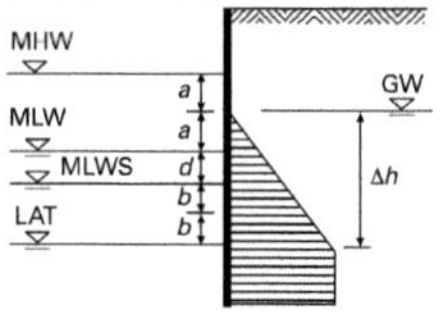 MHW, MLW, MLWS, LAT	–	–	$\Delta h = a + 2b + d$
3c Major water level fluctuations without drainage = limit case for falling high water	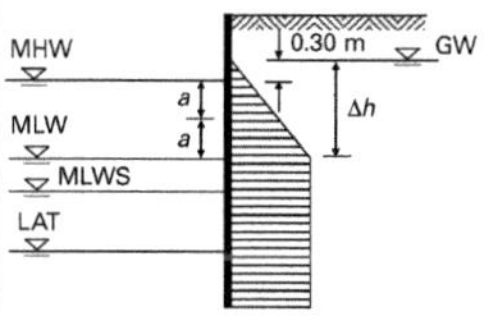 MHW, MLW, MLWS, LAT	–	–	$\Delta h = 0.30\ \text{m} + 2a$
3d Major water level fluctuations with drainage	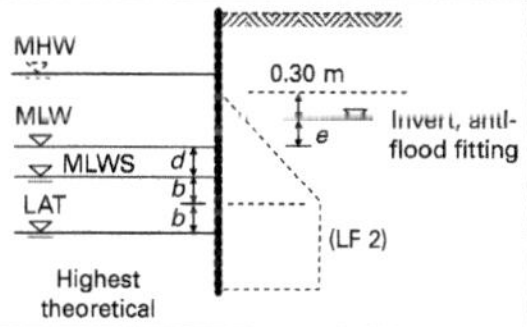 MHW, MLW, MLWS, LAT, Highest theoretical	$\Delta h = 1.00\ \text{m} + e$ for outer water level at MLWS	$\Delta h = 0.30\ \text{m} + b + d + e$	–

Figure 3.3 Approximate approaches for resultant water pressure on waterfront structures for permeable soils in tidal areas and standard situations (without significant wave effect).

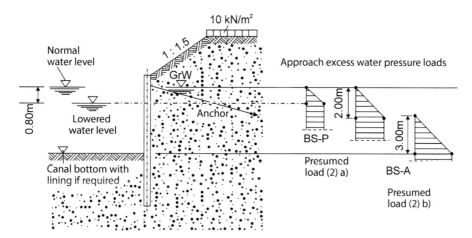

Figure 3.4 Approach of the water pressure for the sheet pile bank of an inland navigation channel.

The resultant water pressure to be used in design situation DS-P is the pressure that can be expected to occur frequently due to unfavourable canal and groundwater levels. This also includes a lowering of the canal water level in front of the sheet piling due to passing vessels (see Section 4.5).

Hydraulic backfilling to waterfront structures can lead to much higher water levels temporarily, which must be taken into account for the respective condition of the building works.

Flooding of the waterfront structure, stratified soils, highly permeable sheet pile interlocks or confined artesian pressure are cases that require special investigations to determine the water levels critical for the resultant water pressure (Section 3.3.4). If a natural groundwater flow is constricted or cut off by a waterfront structure to such an extent that there is a build-up of groundwater in front of the wall, this accumulation can be determined within the scope of a numerical model of the groundwater flow (Section 3.3.4.4).

The relieving effect of drainage systems according to Sections 7.3.2, 7.3.1 and 3.3.4 may only be considered if their effectiveness can be constantly monitored and the system can be restored at any time.

3.3.3 Resultant water pressure on quay walls in front of embankments with elevated platforms in tidal areas

3.3.3.1 General

On an overbuildt embankment with a permeable revetment in a tidal area, the retained soil drains via the base of the slope at low outer water levels. The position of the seepage line depends on the soil conditions, the size and frequency of the water level fluctuations and the inflow from the land.

The flow through the critical active earth pressure wedge results in an increase in the active earth pressure. The water pressure behind the quay wall can be higher than the hydrostatic water pressure, depending on the drainage options and the ensuing flows.

3.3.3.2 Approximate method for resultant water pressure

As a simplification, the increase in the active earth pressure and the water pressure acting on a quay wall that is caused by the current can be taken into account by way of an increased hydrostatic groundwater pressure. In relation to Sections 3.3.2 and 3.3.1, experience gained in the tidal regions of Germany's north coast shows that the approximate method according to Figure 3.5 can be used to determine the resultant water pressure on a quay wall where the subsoil exhibits fairly uniform permeability. Here, the increased hydrostatic groundwater pressure applied results from the groundwater level at the intersection of the seepage line and the failure plane of the critical earth pressure wedge. Figure 3.5 shows an example of this for two failure planes for the DS-T transient design situation, but this can be transferred to other design situations accordingly. When applying the resultant water pressure according to Figure 3.5, the permeability of the revetment should be such that it does not cause a build-up of groundwater. Increasing the active earth pressure on the sheet pile wall due to groundwater currents lies on the safe side when applying the resultant water pressure approach.

Odenwald and Schneider (2015) recommend a similar approach for the resultant water pressure acting on sheet pile walls alongside navigation channels in cuttings with a lateral

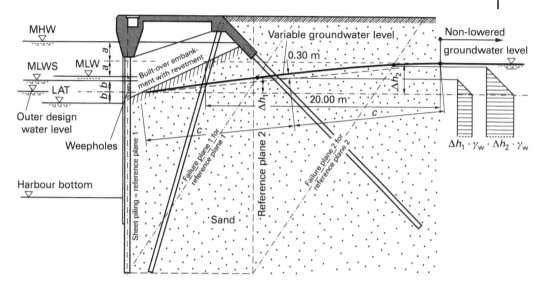

Figure 3.5 Application of the resultant water pressure for design situation DS-T for an embankment with elevated platform.

groundwater inflow. The water pressure and active earth pressure loads on sheet pile walls using different methods are compared in Odenwald and Ratz (2017).

3.3.4 Taking account of groundwater flow

3.3.4.1 General

Where quay walls and other hydraulic engineering works and their components are to be built in flowing groundwater, then the effects of such flowing groundwater on the active and passive earth pressures must be taken into account in the design and detailing.

Groundwater flows can be calculated using Darcy's law if the flow within the pore channels is laminar, which is usually the case in natural aquifers.

3.3.4.2 Principles of groundwater flow calculation

The laws governing laminar, steady groundwater flows are described by potential theory (see Verruijt 1982, for example). The solution to the differential equation for steady groundwater flows with given, suitable boundary conditions reveals the spatial distribution of the groundwater potential within the modelling area considered. Based on this, for steady flows in the vertical plane, it is possible to construct a flow net made up of flow and equipotential lines forming fields with a constant aspect ratio (Figure 3.7). The principles for calculating groundwater flows are presented in detail in Odenwald et al. (2018), for instance.

For steady flows, the flow lines can be interpreted physically as the paths of the water particles. The equipotential lines, which intersect the flow lines perpendicularly, represent lines joining equal hydraulic heads (equal groundwater potential) (Figure 3.7). Sections 3.3.4.7 and 3.3.4.8 explain how to calculate hydraulic heads for different situations.

3.3.4.3 Boundary conditions for calculating groundwater flow

It is necessary to define appropriate boundary conditions along the entire boundaries of the area being modelled in order to calculate steady groundwater flows in the vertical plane and, thus, determine the seepage below a quay wall. If there are no natural boundaries to the model (e.g. impermeable soil strata or parts of the structure), the boundaries must be defined such that the flow calculations for the groundwater potential distribution determined and the stress on the wall and the surrounding soil calculated from it yields results that are on the safe side. In particular, it is important to define the lateral boundaries to the model, for which, in most cases, there are no natural boundary conditions within the area that is meaningful for modelling the flows in the vertical plane under the wall. For calculations of steady flows in the vertical plane beneath a wall, it is usually sufficient to specify flow lines or equipotential lines as the boundaries of the model.

Examples of boundary flow lines are boundaries between permeable and impermeable soil strata, surfaces of impermeable structures and an unconfined groundwater level.

Examples of boundary equipotential lines are bottoms of watercourses and inlet slopes to permeable barrage structures.

Figure 3.6 shows examples of areas with their boundary flow and equipotential lines.

3.3.4.4 Application of numerical groundwater models to determine potential distribution

In numerical groundwater models, the entire potential field is divided into individual elements (discretised) and is represented by the hydraulic heads of a sufficient number of support points. These points are the corner points (finite element method, FEM) or the grid points (finite difference method, FDM) of individual small but finite areas. Boundaries and discontinuities (wells, springs, drains, etc.) in the flow field must be represented by nodes or element lines.

The use of numerical groundwater models presumes the correct choice of boundary conditions and geohydraulic parameters. Whereas groundwater models calculate the entire hydraulic potential field and derive the distribution of gradients, velocities, discharges, etc., from that, only individual variables can be calculated using simplified methods (according to Section 3.3.4.7, for example). However, such methods are often used successfully for preliminary designs.

3.3.4.5 Example of a flow net for a quay wall with vertical inflow and homogeneous soil

In this example, it is it assumed that there are different depths of water on both sides of the quay wall (Figure 3.7). With a vertical difference of 4.5 m between the water levels, the result is a primarily vertical flow towards the quay wall in the permeable, homogeneous soil. At a depth of −23.00 m, the upper boundary of an essentially impermeable soil stratum constitutes a boundary flow line marking the bottom edge of the flow net model. The lateral boundaries to the model are also in the form of boundary flow lines, although it is assumed that the positions of these boundaries do not have any significant influence on the groundwater potential in the region of the quay wall. The quay wall itself is assumed to be impermeable and, thus, constitutes another boundary flow line. On the other hand, the bottoms of the watercourses on either side of the quay wall at −8.00 and +5.00 m represent boundary equipotential lines with hydraulic heads of +2.50 and +7.00 m, respectively.

Figure 3.7 shows the resulting flow net determined from the equipotential and flow lines. The following rules apply here:

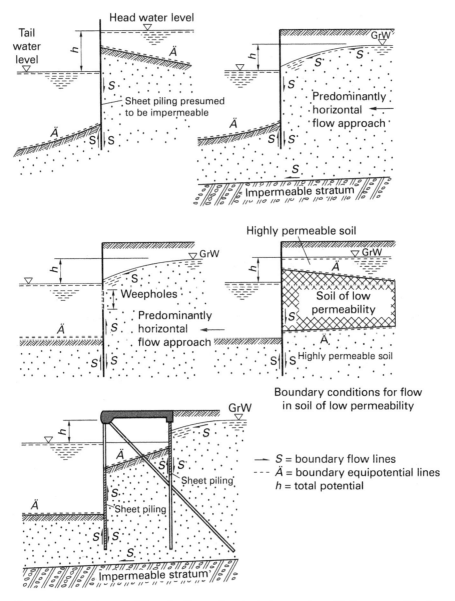

Figure 3.6 Boundary conditions for flow nets – typical examples with flow around the base of sheet piling.

- Flow lines are perpendicular to equipotential.
- The total potential difference Δh between the highest and lowest hydraulic potentials is divided into equal (equidistant) potential steps dh (15 steps in Figure 3.7, i.e. every step is equal to a potential difference of $4.50 \, \text{m}/15 = 0.3 \, \text{m}$).
- The number of equipotential steps and flow lines is chosen such that neighbouring equipotential and flow lines form squares bounded by curved lines in order to guarantee the geometrical similarity across the whole net (which is checked by drawing inscribed circles).

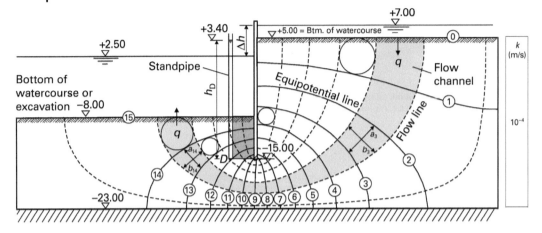

Figure 3.7 Example of a flow net for a quay wall in homogeneous soil with vertical inflow.

The concentration of the equipotential lines, and hence the higher hydraulic gradients, around the base of the wall can be seen clearly.

3.3.4.6 Approximate solution for the influence of flow on active and passive earth pressures

With a primarily vertical flow around a waterfront structure, the increase in the active earth pressure and the decrease in the passive earth pressure as a result of the flow can be calculated approximately by increasing the unit weight of the soil on the active earth pressure side and reducing it on the passive earth pressure side. The change in the unit weight of the soil due to the flow is the product of the vertical hydraulic gradient and the unit weight of water:

$$\Delta\gamma' = i \cdot \gamma_w$$

The increase in the unit weight on the active earth pressure side and the decrease on the passive earth pressure side can be determined approximately for exclusively vertical flow and homogeneous subsoil according to Brinch Hansen (1953) using the following equations:

- on the active earth pressure side:

$$\Delta\gamma' = \frac{\Delta h}{t_0 + \sqrt{t_o \cdot t_u}} \cdot \gamma_w$$

- on the passive earth pressure side:

$$\Delta\gamma' = -\frac{\Delta h}{t_u + \sqrt{t_o \cdot t_u}} \cdot \gamma_w$$

where in the equations and in Figure 3.8

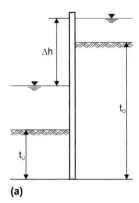
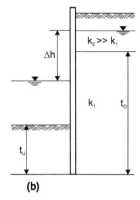

Figure 3.8 Sketches illustrating the approximate determination of the effective unit weight of water that is changed by the flow force in front of and behind a sheet pile wall for (a) unconfined water on the land side and (b) the water level below ground level on the land side with stratified subsoil; k_1, k_2 = permeability coefficients.

Δh difference in water levels on both sides of the wall (difference in potential)
t_o embedment depth of sheet pile wall on land side from bottom of watercourse to base of wall or depth of soil strata in which a decrease in potential occurs
t_u embedment depth of sheet pile wall on water side from bottom of watercourse to base of wall or depth of soil strata in which a decrease in potential occurs
γ' buoyant unit weight of soil
γ_w unit weight of water

The above equations apply when there is also soil underneath the base of the sheet piling which, in a flow situation, contributes to reducing the potential to the same extent as the soil in front of and behind the sheet piling.

In the case of a horizontal inflow, the residual potential at the base of the sheet piling increases considerably, so the above approximation should not be used in such cases.

In the real world, only rarely do we see the hydraulic boundary condition assumed in this example, i. e. a watercourse above ground level on the land side (Figure 3.8a) causing a vertical flow at the quay wall. However, a vertical flow at a quay wall can ensue in stratified subsoil where there is a readily permeable, groundwater-bearing stratum on top of a relatively impermeable stratum in which the quay wall is embedded (Figure 3.8b). In this case, the decrease in potential takes place almost exclusively in the relatively impermeable stratum, and the groundwater level in the readily permeable stratum above this on the land side is not normally lowered. Furthermore, an essentially vertical inflow can occur during a fast drop in the water level (e. g. at low tide).

3.3.4.7 Example of a flow net for a quay wall with horizontal inflow and homogeneous soil

This example considers the standard case of a horizontal flow towards a watercourse with a quay wall embedded in an aquifer consisting of homogeneous soil. The inflow is due to the fact that the watercourse is at a lower level than that of the groundwater on the land side (Figure 3.9). In this case, the boundary condition on the land side for the flow calculation is usually taken to be a boundary equipotential line that is not affected by the flow of groundwater towards the watercourse. (In this example, the hydraulic head of +7.00 m

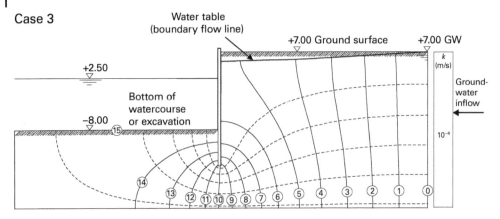

Figure 3.9 Example of a flow net for a quay wall in homogeneous soil with horizontal inflow.

corresponds to the ground level.) The distance of this boundary condition from the quay wall and the assumed hydraulic head along this boundary exert a considerable influence on the groundwater potential distribution, also near the wall, and, therefore, they must be specified carefully, lying on the safe side and taking into account the range of fluctuation in the level of the groundwater. An unconfined groundwater level, which also represents a boundary flow line in this case, ensues between the boundary condition on the land side and the quay wall. The other boundary conditions for the model correspond to those of the example for vertical flow towards the quay wall (Section 3.3.4.5 and Figure 3.7).

In the flow net shown in Figure 3.9, the potential difference $\Delta h = 4.50$ m between +7.00 and +2.50 m is divided into 15 steps $dh = 4.50 \, \text{m}/15 = 0.30$ m.

For example, the following critical hydraulic heads can be derived from the flow net:

- Hydraulic head at point F (base of wall):

$$h_F = 7.00 \, \text{m} - 10.4/15 \cdot 4.50 \, \text{m} = 3.88 \, \text{m} \quad (= 2.50 \, \text{m} + 4.6/15 \cdot 4.50 \, \text{m})$$

- Hydraulic head at point E (corner of failure body according to Terzaghi (Terzaghi and Peck, 1961) – see Section 3.4):

$$h_E = 7.00 \, \text{m} - 12.2/15 \cdot 4.50 \, \text{m} = 3.34 \, \text{m} \quad (= 2.50 \, \text{m} + 2.8/15 \cdot 4.50 \, \text{m})$$

3.3.4.8 Example of a flow net for a quay wall with horizontal inflow and stratified soil

This example considers the influence of a stratified soil with horizontal inflow (Figure 3.10). The boundary conditions from the previous example with homogeneous soil (Figure 3.7) will be retained. However, in addition, a 2 m thick, horizontal stratum between −10.00 and −8.00 m will be included, the permeability of which ($k = 10^{-5}$ m/s) is a factor of 10 lower than that of the soil above and below ($k = 10^{-4}$ m/s). As a consequence of this impermeable stratum, the groundwater level above this stratum is approximately horizontal. The quay wall is embedded 1 m in the stratum with lower permeability (the base of the wall at −9.00 m).

Figure 3.10 shows the flow net defined by the flow and equipotential lines. From this, we can see the concentration of equipotential lines and the ensuing higher hydraulic gradients in the stratum with lower permeability. The condition for this potential distribution is that

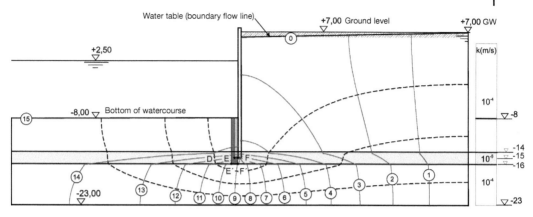

Figure 3.10 Example of a flow net for a quay wall in stratified soil with horizontal inflow.

the stratum with lower permeability has a horizon that extends sufficiently far beyond the wall. If this is not the case, the potential distribution at the base of the wall is determined by the flow around this stratum and not only by the flow within this stratum.

In the flow net shown in Figure 3.10, the potential difference $\Delta h = 4.50$ m between +7.00 and +2.50 m is again divided into 15 steps $dh = 4.50$ m/15 = 0.30 m.

For example, the following critical hydraulic heads can be derived from the flow net:

- Hydraulic head at point F (base of wall):

$$h_f = 7.00 \text{ m} - 9.5/15 \cdot 4.50 \text{ m} = 4.15 \text{ m} \quad (= 2.50 \text{ m} + 5.5/15 \cdot 4.50 \text{ m})$$

- Hydraulic head at point E (corner of first failure body examined – see Section 3.4):

$$h_E = 7.00 \text{ m} - 11.1/15 \cdot 4.50 \text{ m} = 3.67 \text{ m} \quad (= 2.50 \text{ m} + 3.9/15 \cdot 4.50 \text{ m})$$

- Hydraulic head at point F' below base of wall:

$$h'_F = 7.00 \text{ m} - 8.8/15 \cdot 4.50 \text{ m} = 4.36 \text{ m} \quad (= 2.50 \text{ m} + 6.2/15 \cdot 4.50 \text{ m})$$

- Hydraulic head at point E' (corner of second failure body examined – see Section 3.4):

$$h'_E = 7.00 \text{ m} - 9.5/15 \cdot 4.50 \text{m} = 4.15 \text{ m} \quad (= 2.50 \text{ m} + 5.5/15 \cdot 4.50 \text{ m})$$

3.4 Hydraulic heave failure

There is a risk of a hydraulic heave failure in front of a wall embedded in a flow net where the soil is loaded by the flow force of the groundwater acting upwards, which reduces the passive earth pressure. Failure occurs when the vertical component S_{dst} of this flow force within the relevant body of soil in front of the base of the wall exceeds the self-weight G'_{stb} of the body of soil affected by the buoyancy. The body of soil is then raised hydraulically (fluidisation), and the passive earth pressure is lost completely.

According to DIN EN 1997-1 in conjunction with DIN 1054, it is necessary to check that the failure body with an upward flow in front of the wall complies with the following condition for the HYD limit state:

$$S_{dst,k} \cdot \gamma_H \leq G'_{stb,k} \cdot \gamma_{G,stb}$$

3 Geotechnical principles

where

$S_{dst,k}$ characteristic value of flow force in body of soil with upward flow
γ_H partial safety factor for flow force for limit state HYD according to DIN 1054, Table A2.1
$G'_{stb,k}$ characteristic value of buoyant unit weight of body of soil subjected to upward flow
$\gamma_{G,stb}$ partial safety factor for stabilising permanent actions for limit state HYD according to DIN 1054, Table A2.1

The flow force $S_{dst,k}$ can be calculated with the help of a potential distribution. The flow force $F'_{s,k}$ is the product of the volume of the assumed hydraulic failure body, the unit weight of water γ_w and the mean vertical hydraulic gradient (flow gradient) in this body of soil.

Where there is an upward flow through homogeneous soil in front of the base of the wall, then according to DIN 1054, 11.5.(4), the flow force in a body of soil may be assumed to have a width generally equal to half the embedment depth of the wall. In this case, an approximate value for the characteristic vertical flow force $S_{dst,k}$ can be calculated as follows:

$$S_{dst,k} = \gamma_w \cdot \frac{\Delta h_F + \Delta h_E}{2} \cdot \frac{t}{2}$$

where

γ_w the unit weight of water
Δh_F the effective potential difference at base of wall (difference between hydraulic head at the base of the sheet piling and the level of the watercourse)
Δh_E the effective potential difference at the boundary of the failure body opposite the base of the wall
d the wall embedment depth

In the example of a wall in a flow net with homogeneous soil according to Section 3.3.4.7 and the hydraulic head determined from the flow net (Figure 3.9), the effective potential differences are

$$\Delta h_F = 3.88 \text{ m} - 2.50 \text{ m} = 1.38 \text{ m}$$
$$\Delta h_E = 3.34 \text{ m} - 2.50 \text{ m} = 0.84 \text{ m}$$

The characteristic flow force in the failure body, which is assumed to be rectangular, is

$$S_{dst,k} = 10 \frac{\text{kN}}{\text{m}^3} \cdot \frac{1.38 \text{ m} + 0.84 \text{ m}}{2} \cdot \frac{7 \text{ m}}{2} = 39 \text{ kN/m}$$

Assuming a unit weight of water when subjected to buoyancy $\gamma' = 11 \text{ kN/m}^3$, the characteristic weight of the failure body subjected to buoyancy is then

$$G_{stb,k} = \gamma' \cdot t \cdot \frac{t}{2} = 11 \frac{\text{kN}}{\text{m}^3} \cdot 7 \text{ m} \cdot 3.5 \text{ m} = 270 \text{ kN/m}$$

3.4 Hydraulic heave failure

DIN 1054, Table A2.1, gives the following partial safety factors for the HYD limit state in the persistent design situation DS-P:

$$\gamma_{G,stb} = 0.90$$
$$\gamma_H = 1.45 \quad \text{for favourable subsoil}$$
$$\gamma_H = 1.90 \quad \text{for unfavourable subsoil}$$

Assuming a subsoil that increases the risk of a hydraulic heave failure, the result for the HYD limit equilibrium condition is then

$$39\,\frac{kN}{m} \cdot 1.90 = 74\,\frac{kN}{m} \leq 243\,\frac{kN}{m^3} = 270\,\frac{kN}{m^3} \cdot 0.9$$

Owing to the small decrease in effective potential up to the bottom of the watercourse, compared with the embedment depth of the wall, and the ensuing small hydraulic gradient, this example has a clearly sufficient safety against hydraulic heave failure. The utilisation ratio of the design values is

$$\mu = \frac{74\,kN/m}{243\,kN/m} = 0.30$$

In the example of flow around a wall in stratified soil according to Section 3.3.4.8, it can be seen from the flow net (Figure 3.10) that the decrease in potential takes place essentially only within the stratum with lower permeability. The permeable stratum above this, in which the drop in potential is negligible, acts like a surcharge filter. In this case, the depth of this superimposed stratum may not be taken into account when determining the width of the assumed failure body according to DIN 1054, 10.3 A (1d).

When assessing the risk of hydraulic heave failure in stratified soil, it is necessary to find the critical lower boundary of the failure body with the highest risk. Where there is a stratum with lower permeability on top of a permeable stratum, the underside of the stratum with lower permeability is generally the critical lower boundary of the failure body.

However, a failure body beginning at the base of the wall is initially examined. The wall is embedded 1 m in the stratum with lower permeability, and so the width of the failure body is taken to be $t' = 0.5$ m. The effective potential differences at the two lower corner points F and E of this failure body are

$$\Delta h_F = 4.15\,m - 2.50\,m = 1.65\,m$$
$$\Delta h_E = 3.67\,m - 2.50\,m = 1.17\,m$$

The characteristic flow force in the assumed rectangular failure body is

$$S_{dst,k} = 10\,\frac{kN}{m^3} \cdot \frac{1.65\,m + 1.17\,m}{2} \cdot 0.5\,m = 7.1\,kN/m$$

The characteristic weight of the failure body subjected to buoyancy is

$$G_{stb,k} = \gamma' \cdot t \cdot t' = 11\,\frac{kN}{m^3} \cdot 7\,m \cdot 0.5\,m = 38.5\,kN/m$$

Again, assuming an unfavourable soil, for the limit equilibrium condition HYD results:

$$7.1\,\frac{kN}{m} \cdot 1.90 = 13.5\,\frac{kN}{m} \leq 34.7\,\frac{kN}{m^3} = 38.5\,\frac{kN}{m^3} \cdot 0.9$$

The utilisation ratio of the design values is, therefore,

$$\mu = \frac{11.6 \text{ kN/m}}{34.7 \text{ kN/m}} = 0.39$$

As the drop in potential in this example of flow around a wall in a stratified soil takes place mainly in the stratum with lower permeability, and the wall does not extend to the bottom of this stratum, it is also necessary to investigate a failure body that extends as far as the bottom of the stratum. The thickness of the stratum with lower permeability is 2 m, and so the width of the failure body is taken to be $t' = 1$ m. The effective potential differences at the two lower corner points F' and E' of this failure body are

$$\Delta h_{F'} = 4.36 \text{ m} - 2.50 \text{ m} = 1.86 \text{ m}$$
$$\Delta h_{E'} = 4.15 \text{ m} - 2.50 \text{ m} = 1.65 \text{ m}$$

The characteristic flow force in the assumed rectangular failure body is

$$S_{\text{dst,k}} = 10 \frac{\text{kN}}{\text{m}^3} \cdot \frac{1.86 \text{ m} + 1.65 \text{ m}}{2} \cdot 1.0 \text{ m} = 17.6 \text{ kN/m}$$

The characteristic weight of the failure body subjected to buoyancy is

$$G_{\text{stb,k}} = \gamma' \cdot t \cdot t' = 11 \frac{\text{kN}}{\text{m}^3} \cdot 7 \text{ m} \cdot 1.0 \text{ m} = 77 \text{ kN/m}$$

The HYD limit equilibrium condition is

$$17.6 \frac{\text{kN}}{\text{m}} \cdot 1.90 = 33.4 \frac{\text{kN}}{\text{m}} \leq 69.3 \frac{\text{kN}}{\text{m}^3} = 77 \frac{\text{kN}}{\text{m}^3} \cdot 0.9$$

The utilisation ratio of the design values is, therefore,

$$\mu = \frac{33.4 \text{ kN/m}}{69.3 \text{ kN/m}} = 0.48$$

The comparison of the utilisation ratio makes it clear that the second assumed failure body is critical in this example. However, the small hydraulic gradient results in an adequate factor of safety against hydraulic heave failure for both failure bodies.

If the conditions are such that a hydraulic heave failure appears probable, then suitable measures to prevent this must be considered at the design stage. In these cases, it is especially important that quay walls be embedded sufficiently deep in the soil in order to keep the hydraulic gradient as small as possible.

When using a very short wall embedment depth, the utilisation ratio for hydraulic heave failure decreases with decreasing embedment depth. This area must be avoided by constructional measures (Odenwald and Herten 2008).

Damage to a quay wall (e. g. declutching in a sheet pile wall) shortens the seepage path for the groundwater flows and, thus, increases the gradient significantly. Therefore, such damage should be evaluated in particular in terms of the risk of hydraulic heave failure.

Loosening the soil or soil replacement measures to improve the drivability of sheet piles changes the geohydraulic conditions in the region of a sheet pile wall. This must be taken into account when calculating the flows beneath the sheet pile wall and when checking for hydraulic heave failure.

3.5 Earth pressure

3.5.1 General

The following recommendations relate to the application of the active earth pressure on the loading side. However, it is always necessary to check whether other forms of earth pressure might be critical. To limit deformations, or when using anchorages with a high prestress, it may be necessary to apply the earth pressure at rest. The compaction pressure can govern when compacting backfill behind a waterfront structure. The silo effect can reduce the active earth pressure where the area of backfill is confined, e.g. between two sheet pile walls. DIN 4085 includes information on determining such active earth pressures.

The characteristic earth pressures can be determined according to the graphical and/or analytical methods given in DIN 4085. Other methods, e.g. numerical ones, are permitted provided that it can be ensured that they result in the same earth pressures as with the methods in that standard. Indices are not added to the soil properties in the following figures and equations unless the characteristic value (index k) or design value (index d) is expressly referred to.

The design of sheet pile walls is dealt with in Section 8.2. Section 8.2.5 contains information on estimating the angle of active earth pressure δ_a.

3.5.2 Considering the cohesion in cohesive soils

The cohesion in cohesive soils may be taken into account when calculating the active and passive earth pressures provided that the following conditions are satisfied:

- The soil must be undisturbed.
- When backfilling with cohesive material, the soil must be placed and compacted without leaving voids.
- The soil should be permanently protected against drying out and freezing.
- The soil should not become very soft when kneaded.

3.5.3 Considering the apparent cohesion (capillary cohesion) in sand

The apparent cohesion c_c (capillary cohesion according to DIN 18137-1) in sand is caused by the surface tension of the pore water. This cohesion is lost when the soil is fully saturated or dries out completely. It is, therefore, not normally included in active or passive earth pressure calculations, instead functions as an inherent reserve factor for stability. Apparent cohesion may be taken into account for temporary conditions during construction provided that it can be ensured that it remains effective throughout the period concerned. Table 3.5 contains characteristic reference values for the apparent cohesion of strata of medium or greater density.

3.5.4 Determining active earth pressure for a steep, paved embankment in a partially sloping waterfront structure

A steep embankment case exists if the inclination of the slope β is greater than the effective angle of internal friction φ' of the in situ soil. The stability of the embankment is only

Table 3.5 Characteristic reference values for the apparent cohesion of an embedding that is at least medium dense (TGL 35983/02 1983).

Soil type	Designation in DIN 4022-1	Apparent cohesion $c_{c,k}$ [kN/m²]
Gravelly sand	G/S	≤ 2
Coarse sand	gS	≤ 3
Medium sand	mS	≤ 5
Fine sand	fS	≤ 9

guaranteed if the soil has a permanently effective cohesion c' and is permanently protected against surface erosion, e. g. by means of dense turf or a revetment.

If the cohesion is insufficient to guarantee the stability of the embankment, it requires protection, e. g. in the form of paving, which must be able to resist downslope forces and be structurally connected to the quay wall. The revetment must be designed in such a way that the resultant of the applied actions always lies within the middle third of its cross-section. The active earth pressure for the embankment area down to the top of the beam supporting the revetment (earth pressure reference line in Figure 3.11) can be determined graphically if the cohesion does not dominate,

$$\frac{c'}{\gamma \cdot h} < 0.1$$

The self-weight of the revetment is not taken into account.

In doing this, any resultant water pressure must be allowed for in addition to the active earth pressure E_a. This is illustrated in Figure 3.11 for impermeable paving. The resultant water pressure is lower in the case of permeable paving. The loading assumptions for bank protection are shown in Figure 3.11b. The reaction force R_R between the bank protection and the wall is given by the polygon of forces shown in Figure 3.11c.

The reaction force R_R must be taken fully into account in the calculations for the wall and its anchorages. In doing so, it should be noted that the active earth pressure load E_{ado} and the dead load of the bank protection are already included in the reaction force R_R and are carried directly by the wall and its anchorages. By way of approximation, the active earth pressure load E_{adu} below the active earth pressure reference line can also be determined with a wall projecting above the active earth pressure reference line by the fictitious height

$$\Delta h = \frac{1}{2} \cdot h_B \cdot \left(1 - \frac{\tan \varphi'}{\tan \beta}\right)$$

with a fictitious slope angle φ' at the same time (Figure 3.12).

3.5.5 Determining the active earth pressure shielding on a wall below a relieving platform with average ground surcharges

A wall can be shielded from the active earth pressure to some extent by building a relieving platform. The shielding effect depends, above all, on the width and position of the platform, but also on the shear strength and compressibility of the soil behind the wall and

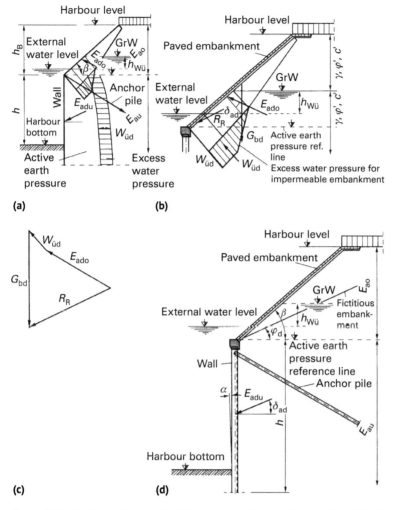

Figure 3.11 Partly sloping bank with steep, paved embankment: (a) sketch of system; (b) forces acting on a steep revetment; (c) polygon of forces for determining the reaction force R_R between embankment paving and quay wall; (d) overall view.

beneath the structure. A relieving platform can have a favourable influence on the critical active earth pressure distribution used for calculating the internal forces. In homogeneous, non-cohesive soils with average ground surcharges (typically 20–40 kN/m² as uniformly distributed load), the active earth pressure shielding can be determined according to Brennecke and Lohmeyer (1930) (Figure 3.13). As can be confirmed by Culmann investigations, the use of the Lohmeyer method works well under the foregoing prerequisites.

If the soil also has a cohesion c', then the shielded active earth pressure can be considered approximately by first calculating the shielded active earth pressure distribution without taking c' into account and then superposing the cohesion on it thus:

$$\Delta e_{ac} = c' \cdot K_{ac}$$

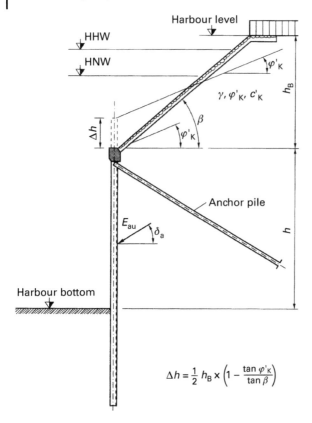

Figure 3.12 Approximate approach for determining E_{au}.

where K_{ac} is the active earth pressure coefficient for cohesion; see DIN 4085. However, this procedure is only permissible when the cohesion forms a small proportion of the total active earth pressure.

The calculations according to Figures 3.13–3.15 may not be used in cases where several relieving platforms are positioned one above the other. Furthermore, irrespective of the shielding, the overall stability of the structure has to be verified for the corresponding limit states according to DIN 1054 – with the full active earth pressure being applied in the critical reference planes.

3.5.6 Earth pressure distribution under limited loads

The active earth pressure due to vertical strip or line loads may be applied to the retaining wall via a simplified limited load figure. The load on the wall is spread over an area limited by the angle φ'_k from the front edge of the load and the angle $\vartheta_{a,k}$ from the back edge of the load (Figure 3.16). The distribution of the load must be selected taking into account any conceivable deformations. If the conditions are such that active earth pressure redistribution can take place (Section 8.2.3.2), then these load components must also be redistributed, in particular to ensure that the load concentration via supports is not underestimated. DG-GT (2012), 3.5 (EB 7), contains further recommendations.

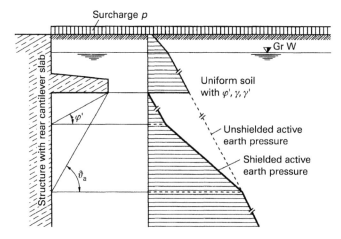

Figure 3.13 Solution according to Lohmeyer for homogeneous soil.

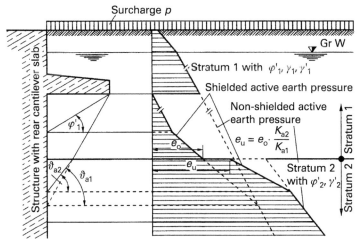

In the example, $\varphi'_2 < \varphi'_1$; $K_{a2} > K_{a1}$; $\vartheta_{a1} < \vartheta_{a1}$; $\gamma'_2 < \gamma'_1$;

Figure 3.14 Solution according to Lohmeyer expanded for stratified soil (option 1).

When it comes to strip or line loads limited in the longitudinal direction of the wall, the load spreading beyond the end may also be taken into account at an angle of 45°. The associated relief on the wall may be taken into account over an area at an angle ±45° from the end of the load (Figure 3.17). If areas overlap in the case of very short-loaded areas, then the load is distributed over the section of the wall bounded by the outer spreading lines.

3.5.7 Determining active earth pressure in saturated, non-consolidated or partially consolidated, soft cohesive soils

In drained conditions, effective stresses and effective shear parameters are to be expected. For undrained conditions, effective stresses and, accordingly, effective shear parameters,

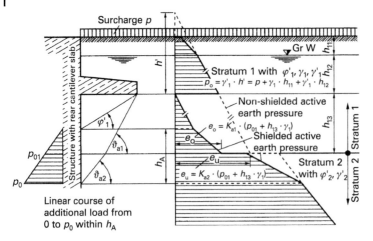

Figure 3.15 Solution according to Lohmeyer expanded for stratified soil (option 2).

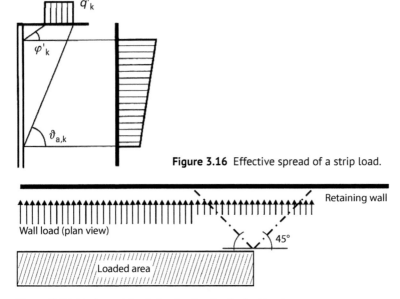

Figure 3.16 Effective spread of a strip load.

Figure 3.17 Horizontal load distribution for limited loads.

should be used in the calculations. Performing calculations with total stresses and, accordingly, undrained shear parameters, is not recommended because c_u depends on:

- The effective stresses in the initial state.
- The overconsolidation ratio (OCR).
- The chronological order of the loading (stress path).
- The rate of load application.

Moreover, these influences can only be fully ascertained by analysing the development of the effective stresses and calculating the active earth pressure with the effective shear parameters.

3.5 Earth pressure

If a surcharge is applied over such a short period that the soil cannot consolidate, then the surcharge is initially absorbed by the pore water, and the effective stress σ' remains unchanged. Only after consolidation begins is the surcharge transferred to the soil skeleton. The following applies for the horizontal load on the wall σ_h directly after applying the load:

$$\sigma_h = e_{ah} + u$$

where

$$e_{ah} = \sigma' \cdot K_{agh} - c' \cdot K_{ach} \quad \text{(with } K_{agh} \text{ and } K_{ach} \text{ to DIN 4085)}$$
$$\Delta p = \Delta u$$
$$u = u_0 + \Delta u$$

e_{ah} horizontal active earth pressure due to effective stresses
σ' effective stress
c' effective cohesion
Δp surcharge applied "quickly"
Δu excess pore water pressure due to surcharge
u_0 hydrostatic water pressure
u total water pressure
σ_h total horizontal load on wall

If the load is applied slowly, however, consolidation begins during the application of the load. Consequently, the excess pore water pressure Δu due to the surcharge Δp is smaller than the load, and the effective stress σ' increases during the application of the load. Following full consolidation, the surcharge Δp is added in full to the effective stress σ' (Figure 3.18b).

Intermediate states can be considered by determining that proportion of Δp for which the soil is already consolidated when calculating the degree of consolidation. This is then added to the above equation for the effective stress; the proportion not yet consolidated (residual excess pore water pressure) is superposed in full on the active earth pressure.

Active earth pressure redistribution (Section 8.2.3.2) is only permitted for active earth pressures due to consolidated soil strata.

In the example shown in Figure 3.18, the active earth pressure distribution is shown for the case where the uniformly distributed surcharge Δp extends unrestrictedly on the land side, i.e. a plane state of deformation applies. In the soft cohesive stratum, the horizontal load on the wall is due to the active earth pressure as a result of the respective effective stress σ' and cohesion c', increased by the excess pore water pressure Δu due to the surcharge Δp and the hydrostatic water pressure u_0.

When it comes to a quick application of the load through hydraulic filling, checks must be performed to see how quickly the sand drains. If this happens immediately, then the surcharge due to the hydraulic fill must be determined using the wet unit weight of the sand. Otherwise, it should be assumed that increased water levels are present in the sand, meaning that part of the fill is still subjected to buoyancy.

3 Geotechnical principles

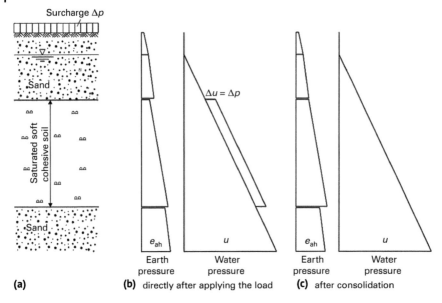

Figure 3.18 Example of calculating the horizontal components of earth pressure distribution for the initial state with the shear parameters of the drained soil.

3.5.8 Effect of water pressure difference beneath beds of watercourses

A difference in water pressure occurs where the harbour bottom or riverbed consists of a cohesive stratum of low permeability lying on a groundwater-bearing, non-cohesive stratum and at the same time the low water level of the body of water lies below the concurrent hydraulic head of the groundwater (Figure 3.19). The effects of this artesian water pressure on the active and passive earth pressures must be taken into account in the design. The artesian water pressure acts on the confining stratum from below and causes a flow through this, which thus reduces its effective unit weight γ'. As a result, the active and passive earth pressures also decrease.

In tidal areas, a difference in water pressure can be caused by the changing tidal water levels, specifically at low water. The magnitude of the difference in water pressure can then be the same as or greater than the dead load of the confining stratum. At low water, the confining stratum is then lifted off the underlying non-cohesive soil due to the effect of artesian water pressure acting from below and slowly begins to float correspondingly to the inflow of pore water. During the subsequent high water, the confining stratum is then pressed back onto the underlying stratum, thus displacing the pore water again.

This process takes place with every tide and is not normally critical under natural conditions. However, if a confining stratum subjected to a difference in water pressure depending on the tide is weakened by dredging during the course of construction work, then local heaving of the confining stratum can occur. Such heaving results in local disturbance of the soil in the area around the heave. At the same time, however, the artesian water pressure is relieved, which means that the process is limited to these local disturbances.

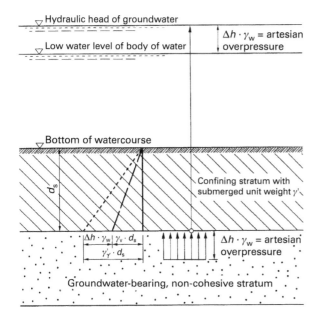

Figure 3.19 Difference in water pressure beneath a watercourse bed with low permeability.

3.5.9 Considering active earth pressure and resultant water pressure, and construction guidance for waterfront structures with soil replacement and a contaminated or disturbed base of excavation

3.5.9.1 General

When soil replacement is used behind waterfront structures in accordance with Section 5.9, it is important to analyse how the contamination in the base of the excavation and slopes to the rear of the excavation can affect active earth pressure and water pressure. This is especially necessary when mud deposits are expected. In addition, in the interests of economic efficiency, soil replacement should be carried out in such a way that intermediate states in which the replacement soil – but primarily any cohesive deposits on the base of the excavation and the rear slope – needs to be considered should not be critical for the design.

3.5.9.2 Approach for determining active earth pressure

Besides the usual design of the structure for the improved soil conditions and ground failure investigations according to DIN 4084, it is necessary to take into account the boundary and disturbing influences arising from a failure plane created by the excavation as per Figure 3.20.

The following factors are critical for the active earth pressure E_a:

1. Length and – if applicable – inclination of the restraining section l_2 of the failure plane determined by the base of the excavation.
2. Thickness, shear strength τ_2 and effective load of intrusive layer on l_2.
3. Any dowelling action for section l_2 due to piles or similar.
4. Thickness of the adjacent soft cohesive soil along the inshore edge of the excavation, also its shear strength together with the form and angle of the rear slope.

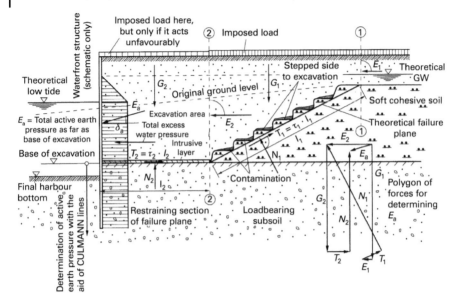

Figure 3.20 Determining the earth pressure E_a acting on a waterfront structure (E_a: total earth pressure down to dredged bottom).

5. Sand load and imposed load, especially on the rear slope.
6. Properties of the backfilling.

The distribution of the active earth pressure E_a as far as the base of the excavation depends on the deformations and the type of waterfront structure.

The active earth pressure below the base of the excavation can be determined, for example, with the aid of the Culmann method. In doing so, the shear forces in section l_2, including any dowelling, must be taken into account.

At all stages of construction, including the original excavation of the harbour bottom or any subsequent deepening of the harbour, the shear stress τ_2 acting at the time in the intrusive layer on section l_2 can be calculated as follows:

$$\tau = (\sigma - \Delta u) \cdot \tan \varphi' \approx \sigma' \cdot \tan \varphi'$$

where σ' is the effective stress due to vertical loads effective at the place and time of the investigation, and φ' is the effective angle of friction of the material in the intrusive layer. The final shear strength at the end of the consolidation process is then

$$\tau = \sigma'_a \cdot \tan \varphi'$$

where σ'_a is the effective stress due to vertical loads in the area investigated in section l_2 at the end of consolidation ($\Delta u = 0$).

Separate calculations are necessary for considering the effects of dowelling section l_2 by means of piles (Brinch Hansen and Lundgren, 1960).

If the rear slope of the excavation has been properly excavated in larger steps, then the critical failure plane passes through the rear corners of the steps and, thus, lies in undisturbed soil (Figure 3.20). In this case, owing to the long consolidation periods of soft cohesive soils, the shear strength used for the failure plane is equal to the initial shear strength of the soil prior to excavation. If the soft cohesive soil is made up of strata with different initial shear strengths, these must be taken into account accordingly.

Where a rear slope in soft soil is severely disturbed, has been excavated in small steps or is unusually contaminated, the shear strength of the disturbed failure plane must be used in the calculations instead of the initial shear strength of the undisturbed soil. This is normally less than the initial shear strength and should, therefore, be determined in laboratory tests.

The consolidation of soft cohesive soils below the soil replacement area can take a long time owing to the long drainage paths. Considering the increase in the shear strength due to consolidation for such soils is generally only permissible when consolidation is accelerated by closely spaced drains.

3.5.9.3 Approaches for determining resultant water pressure

When employing soil replacement, the resultant water pressure should be calculated from the difference in levels between the groundwater in the area of reference line 1-1 (Figure 3.20) and the lowest simultaneous external water level. A drainage system behind the waterfront structure can lower the groundwater level temporarily, but experience shows that such drainage is not effective over the long term.

The resultant water pressure may be applied in the customary, approximate form as the difference between the water pressures acting on both sides of the wall and is assumed to be hydrostatic. A more accurate approach can be derived from an investigation of the flow net around the structure (Section 3.3).

3.5.9.4 Guidance for the design of waterfront structures

Investigations of excavations have revealed that intrusive layers on the base and sloping sides are fully consolidated within the construction period provided that these are no thicker than about 20 cm. Full consolidation during the construction period cannot be assumed for thicker intrusive layers without more detailed investigations. In such cases, the expected shear strength of the intrusive layer must be calculated, e. g. from an estimate of the course of consolidation using the shear parameters of the intrusive layer.

However, in order that the reduced shear strength of an only partially consolidated intrusive layer does not become critical for the design, it may be necessary to schedule certain construction measures, e. g. initial dredging or deepening of the harbour, such that consolidation is completed before the loads associated with these measures become effective.

Anchorage forces are transferred via piles or other structural elements into the load-bearing subsoil beneath the base of the excavation because anchorage forces transferred above the base of the excavation would place additional loads on the sliding mass of soil.

Apart from the length required for structural reasons, the length of section l_2 in Figure 3.20 is to be such that that all structural piles fit within the base of the excavation. This guarantees that the bending stresses in the piles due to settlement of the backfill are kept low.

When mud deposits on the base of the excavation are severe and, therefore, thicker intrusive layers and/or very loosely compacted sand zones cannot be avoided despite taking every care during soil replacement, the result can be high loads on the piles and, in particular, high bending stresses. To avoid brittle fractures due to these actions, the piles must be made from killed steel (see Sections 8.1.2 and 8.1.4.2).

If foundation piles have been installed to create a dowelling effect for the failure plane in section l_2 according to Figure 3.20 and are included when verifying the stability of the overall system according to DIN 4084 (Brinch Hansen and Lundgren, 1960), the maximum principal stress due to axial loads, shear and bending should in no case exceed 85% of the yield stress when checking the stresses in these piles. To calculate the dowelling forces the magnitudes of the pile deformations must be compatible with the other movements of the structure and its components, i.e. generally only a few centimetres. In soft cohesive soils, these deformations are not sufficient to ensure effective dowelling (Figure 3.20). Piles that are stressed up to yield due to settlement of the subsoil or the backfilling may not be considered as contributing to dowelling.

In order to prevent the properties of an intrusive layer on the base or sloping sides of the excavation becoming critical for the design of the structure, efforts must be made to ensure that the base of the excavation is cleaned immediately prior to backfilling with the replacement soil. In addition, section l_2 according to Figure 3.20 must be sufficiently long, and the sloping side to the excavation must be as shallow as possible (see the effects in polygon of forces in Figure 3.20).

Where there is only a thin intrusive layer, the shear resistance of the failure plane in section l_2 can be substantially improved if it is covered with a layer of rubble. If sufficient time is available, closely spaced drains in the soft cohesive soil behind the slope to the back of the excavation can help speed up consolidation and, hence, reduce the active earth pressure. Temporarily reducing the imposed load on the backfill over the slope to the excavation or temporarily lowering the groundwater level as far as a point beyond reference Section 1-1 can help overcome unfavourable initial conditions.

If the restraining section l_2 according to Figure 3.20 is to be omitted in regions with marine clay, i.e. the slope to the excavation extends up to the back of the wall, then the slope must be as shallow as possible. It should certainly be shallower than about 1:4 in this case because then sliding of the replacement soil on the slope does not result in any additional loads on the waterfront structure. In any case, this should be verified by calculation.

3.5.10 Effect of groundwater flow on resultant water pressure and active and passive earth pressures

Where groundwater flows around a structure, the flowing water exerts a pressure on the masses of the sliding wedges of soil for the active and passive earth pressures, thus changing the magnitudes of these forces.

A potential distribution can be used to calculate the total effects of the groundwater flow on the active earth pressure E_a and the passive earth pressure E_p. To do this, the water pressures acting on the sliding body of soil are determined from the hydraulic heads along the boundaries of the sliding body of soil and the ensuing pressures W_1, W_2 and W_3 taken into account in the Coulomb polygons of forces for the active earth pressure (Figure 3.21a) and the passive earth pressure (Figure 3.21b). Figure 3.21 shows the forces that are to be included for the case of a straight failure plane: G_a and G_p are the dead loads of the sliding wedges for the saturated soil, W_1 the resultant water surcharge on the sliding wedges, W_2 the resultant water pressure between the structure and the sliding wedge area and W_3 the resultant water pressure acting at the failure plane. Pressures W_2 and W_3 can be determined from the potential distribution (see, for example, Section 3.3.4.5, Figure 3.7). Forces Q_a and

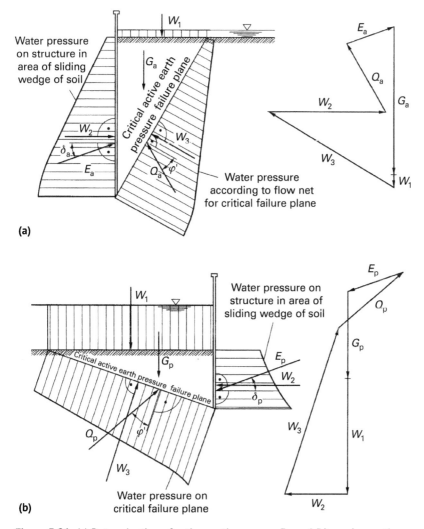

Figure 3.21 (a) Determination of active earth pressure E_a, and (b) passive earth pressure E_p taking into account the effects of flowing groundwater.

Q_p are the soil reactions acting at an angle φ' to the failure plane normal and E_a and E_p, acting at angles δ_a and δ_p, respectively, to the wall normal, are the total active and passive earth pressures, respectively.

The resultant water pressure results from the difference between the water pressure W_2 acting on the structure from inside and outside.

The accuracy of the water pressure, active earth pressure and passive earth pressure figures determined in this way depend on how well the flow net matches the actual conditions.

The above approach has proved useful for waterfront structures in principle. However, more detailed investigations are required for narrow trenches and trench corners in order to take into account the mutual impact of opposing or abutting sides (Ziegler et al. 2009).

The solution according to Figure 3.21 only provides the resultants of E_a and E_p, not the distribution of the active and passive earth pressures over the height of the wall. Separate consideration of the horizontal and vertical flow forces is, therefore, worthwhile for practical applications. In this case, the horizontal effects are added to the resultant water pressure, which is related to the respective failure plane for the active or passive earth pressure. The vertical effects are added to the vertical soil pressures due to the dead load of the soil, i. e. considered in the unit weight of the soil assumed when calculating the active earth pressure. This approach is dealt with in Section 3.3.4.6.

The resultant water pressure on a waterfront structure can generally be determined according to Section 3.3.2 or, for predominantly horizontal flow, in accordance with Section 3.3.4. However, in the event of large differences in water level on the two sides of the structure, the influence of the flow can be so large that it is worthwhile carrying out a more detailed investigation with a flow net.

3.5.11 Determining the amount of displacement required to mobilise passive earth pressure in non-cohesive soils

As a rule, a considerable amount of displacement is required to mobilise the full passive earth pressure in front of a waterfront structure. The amount is chiefly dependent on the embedment depth, the in situ density of the soil and the type of movement. DIN 4085 provides information on the horizontal earth pressure force E'_{pgh} attained depending on the displacement s.

According to research by Weißenbach (1961), good mobilisation is achieved in front of narrow soil resistance compression zones even with minor displacements. The displacement s_p^r required for the maximum three-dimensional passive earth pressure is calculated as follows:

$$s_p^r = 40 \cdot \frac{1}{1+0.5I_D} \cdot \frac{d^2}{\sqrt{b_0}}$$

where

b_0 width of compression zone ($b_0 < d/3$)
I_D relative density

The test results of Weißenbach (1961) are approximated well by the mobilisation function proposed by Horn (1972) in the range of values $0.1 \leq s/s_p^r \leq 1$:

$$E^{r'}_{pgh} = E^r_{pgh} \cdot \frac{s/s_p^r}{0.12 + 0.88 \cdot s/s_p^r}$$

where

$E^{r'}_{pgh}$ mobilised three-dimensional passive earth pressure
E^r_{pgh} maximum three-dimensional passive earth pressure (passive earth pressure in front of narrow soil resistance compression zones)
s_p^r displacement required to achieve E^r_{pgh}

3.5.12 Measures for increasing the passive earth pressure in front of waterfront structures

3.5.12.1 General

The following are examples of suitable underwater measures for increasing the passive earth pressure in front of waterfront structures:

1. Replacement of in situ soft cohesive soil with non-cohesive material (soil replacement).
2. Compaction of loosely bedded, non-cohesive in situ soil, with additional surcharge if necessary.
3. Consolidation of soft cohesive soils with surcharge.
4. Placing of a fill.
5. Consolidation of the in situ soil.
6. A combination of measures 1–5.

These measures differ in terms of their costs; in certain cases, they may also prevent the subsequent deepening of a basin or the upgrading of a waterfront structure, e. g. by driving a new wall in front of the old one. In order to retain these options, the various measures taken in specific cases should also be assessed with respect to the further development of the waterfront facilities.

In principle, all the above measures are common in ground engineering. Special aspects of their use in port and harbour engineering are given below.

3.5.12.2 Soil replacement

When replacing soft cohesive subsoil with non-cohesive material, Section 5.9 must be observed so far as the works themselves are affected. When determining the passive earth pressure, any intrusive layers on the base of the excavation must be taken into account. The remarks in Section 3.5.9 apply accordingly.

The extent and depth of soil replacement in front of a waterfront structure is normally determined according to earth pressure aspects. In order to exploit fully the higher passive earth pressure achieved by the replacement soil deposited, the passive earth pressure sliding wedge must lie completely within the area of the soil replacement.

3.5.12.3 Soil compaction

Loose non-cohesive soils can be compacted by means of vibro compaction. The spacing of the vibration points (grid size) depends on the in situ subsoil and the mean degree of compaction required. The greater the desired improvement in the degree of compaction required and the finer the grains of the soil to be compacted, the closer the spacing must be. An average grid size of 1.80 m is recommended as a guide. In the case of soil replacement, the compaction must encompass all the new soil, i. e. reach as far as the base of the excavation.

Vibro compaction should encompass the entire area of the passive earth pressure sliding wedge in front of the structure and in doing so extend a sufficient distance beyond the critical passive earth pressure failure plane starting from the theoretical base of the sheet piling. In cases of doubt, curved or polygonal failure planes should also be checked to ensure that the area of compaction is sufficiently large.

Deep vibratory compaction causes temporary liquefaction of the soil in the immediate vicinity of the deep vibrator rig, but this soil is then compacted due to the soil surcharge. Therefore, the effect of compaction depends on the soil surcharge, and the soil near the surface (about 2–3 m thick) can only be compacted when a temporary surcharge in the form of fill is introduced during compaction.

Vibro compaction can also be used for subsequent strengthening of waterfront structures. However, when using this method, appropriate procedures should be chosen to guarantee that the stability of the waterfront structure is not adversely affected by the temporary local liquefaction of the soil. Experience has shown that extensive and prolonged liquid states can occur especially in loosely bedded, fine-grained, non-cohesive soils and fine sand.

Soil compaction can reduce the risk of liquefaction effectively in seismic regions.

3.5.12.4 Soil surcharge

Under certain conditions, e.g. stabilising an existing waterfront structure, it may be expedient to improve the support for the structure by introducing a fill with a high unit weight and high angle of internal friction in the region of passive earth pressure. Suitable materials here are metal slags or natural stone. The buoyant unit weight of the materials is critical; metal slags can attain values $\gamma' \geq 18\,\mathrm{kN/m^3}$. The characteristic angle of internal friction may be taken as $\varphi'_k = 42.5°$.

In the case of soft in situ subsoil, care must be taken to ensure that the fill material does not subside by choosing a suitable grading for the fill material, including a filter layer between fill and existing soil or limiting the depth of the fill.

The fill material must be checked constantly for compliance with the conditions and specification. This applies particularly to unit weight.

Sections 3.5.12.2 and 3.5.12.3 apply accordingly with respect to the necessary extent of the works. Additional vertical drains can be installed to accelerate the consolidation of soft strata below fill material.

3.5.12.5 Soil stabilisation

If readily permeable, non-cohesive soils (e.g. gravel, gravelly sand or coarse sand) are present in the passive earth pressure area, then grout injection represents another method of stabilising the soil. In less permeable, non-cohesive soils, stabilisation by means of high-pressure injection is a potential solution. Stabilisation with chemicals is also possible, provided that the chosen stabilising medium can cure properly, allowing for the chemical properties of the pore water. Generally, however, chemical stabilisation is too expensive, which restricts it mainly to localised areas with special boundary conditions and tight construction schedules.

It should be noted that stabilisation of the in situ soil constitutes a barrier to any subsequent deepening of the basin and/or driving a new waterfront wall in front of the old one.

The prerequisite for all types of stabilisation through injection is an adequate surcharge, which must be placed in advance and then possibly removed again.

The dimensions required for the stabilisation area can be stipulated in accordance with Sections 3.5.12.2 and 3.5.12.3. Soil core samples and/or penetrometer tests are always necessary to verify the success of soil stabilisation measures.

3.5.13 Passive earth pressure in front of abrupt changes in ground level in soft cohesive soils with rapid load application on the land side

When determining the passive earth pressure in front of sheet piling with a rapid load application on the land side, the same principles apply as for determining active earth pressure for this load case (Section 3.5.7).

In the load case "excavating in front of wall", the passive earth pressure should be determined with effective shear parameters (c', φ') because excavating does not generally proceed at such a rate that undrained behaviour governs.

In the load case "backfilling behind wall", the load can be applied so rapidly that undrained conditions can occur on the side of the base support, caused by horizontal movement of the wall. A "rapid" surcharge occurs if the rate of load application is significantly greater than the rate of consolidation of the cohesive stratum. In normally consolidated soils, resultant water pressure can occur in the passive earth pressure area, which results in a weakening of the base support. In overconsolidated soils, negative pore water pressure is the result due to the dilatory material behaviour of such soils, but this is generally negligible.

A more accurate analysis of the influence of a rapid loading rate on the passive earth pressure is only possible with numerical modelling. Such an analysis must assess the stress–strain relationship of the soil and the sequence of construction operations as accurately as possible and allow a forecast of the development of the excess pore water pressure over time (Wehnert 2006, AK Numerik 1991).

3.5.14 Waterfront structures in seismic regions

3.5.14.1 General

In practically all countries where earthquakes can be expected, there are various standards, directives and recommendations – for buildings primarily – that normally contain detailed requirements for earthquake-resistant design and construction. Designers of projects in Germany should consult DIN 4149, and when it comes to ports and harbours, PIANC (2001), for example.

The intensity of the earthquakes to be expected in the various regions is generally expressed in such publications by means of the horizontal ground acceleration a_h. Any simultaneous vertical acceleration a_v is generally negligible compared with the acceleration due to gravity g.

The acceleration a_h causes not only immediate loads on the structure but also has an influence on the active and passive earth pressures, the water pressure and, in some cases, the shear strength of the soil beneath the foundation as well. In unfavourable circumstances, this shear strength may temporarily disappear completely (liquefaction).

The additional actions during an earthquake are normally taken into account in the actual structure in such a manner that additional horizontal forces

$$\Delta H = \pm k_h \cdot V$$

acting at the centre of gravity of each accelerated mass are applied simultaneously with the other loads.

In the above equation,

$k_h = a_h/g$ seismic coefficient = ratio of horizontal ground acceleration to gravitational acceleration

V dead load of structural member or sliding wedge of soil considered, including pore water

The magnitude of k_h depends on the intensity of the earthquake, the distance from the epicentre and the in situ subsoil. The first two factors are taken into account in most countries by dividing the endangered regions into earthquake zones with appropriate k_h values (DIN 4149). In cases of doubt, agreement on the magnitude of k_h to be used may need to be reached by consulting a seismic design expert.

In the case of tall, slender structures at risk of resonance, i.e. when the natural frequency of the structure lies within the earthquake frequency spectrum, inertial forces must also be taken into account in the calculations. This is, however, not generally the case with waterfront structures.

Therefore, the principal requirement for the design and construction of an earthquake-resistant waterfront structure is to ensure that it can also accommodate the additional horizontal forces occurring during an earthquake with a reduction in passive earth pressure.

3.5.14.2 Effects of earthquakes on the subsoil

Waterfront structures in earthquake regions must also take account of the soil conditions deeper underground. For example, seismic vibrations are most severe where loose, relatively thin deposits overlie solid rock.

The most sustained effects of an earthquake make themselves felt when the subsoil, especially ground beneath foundations, is liquefied by the earthquake, i.e. loses its shear strength temporarily. This situation (due to settlement flows, liquefaction) arises in loosely bedded, fine-grained, non- or weakly cohesive, saturated and marginally permeable soil (e.g. loose fine sand or coarse silt). The lower the overburden at the depth in question and the greater the intensity and duration of the vibrations, the sooner liquefaction will occur. If the risk of liquefaction cannot be entirely ruled out, it is advisable to investigate the true liquefaction potential (e.g. Idriss and Boulanger, 2004).

Looser soil strata with a liquefaction tendency can be compacted in advance. Cohesive soils cannot be liquefied by seismic action.

3.5.14.3 Determining the effects of earthquakes on active and passive earth pressures

The influence of earthquakes on active and passive earth pressures is also generally determined according to Coulomb – with the additional forces ΔH generated by earthquakes according to Section 3.5.14.1 being considered as well. The resultant of the dead loads of the wedges of soil and the additional horizontal forces is no longer vertical. This is taken into account by relating the angle of the active or passive earth pressure reference plane and angle of the ground surface to the new force direction (Witt, 2018). The result is fictitious angles for the reference plane ($\pm\Delta\alpha$) and the ground surface ($\pm\Delta\beta$):

$$k_h = \tan \Delta\alpha \quad \text{bzw.} \quad k_h = \tan \Delta\beta \qquad \text{(Figure 3.22)}$$

3.5 Earth pressure

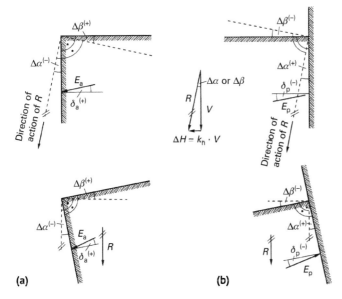

Figure 3.22 Determination of fictitious angles $\Delta\alpha$ and $\Delta\beta$ and diagrams of systems rotated through angles $\Delta\alpha$ and $\Delta\beta$ (signs after Krey): (a) for calculating the active earth pressure; (b) for calculating the passive earth pressure.

The active or passive earth pressure is then calculated on the basis of the imaginary system rotated through an angle $\Delta\alpha$ or $\Delta\beta$, respectively (reference plane and ground surface).

An equivalent procedure considers the angle of rotation according to Figure 3.22 by calculating the active and passive earth pressures with a wall inclination $\alpha \pm \Delta\alpha$ and a ground inclination $\beta \pm \Delta\beta$.

When determining the active earth pressure below groundwater level, it must be realised that the mass of the soil and the mass of the water trapped in the pores of the soil are accelerated during an earthquake. However, the reduction in the unit weight of the submerged soil remains unchanged due to buoyancy. Therefore, it is expedient to use a larger seismic coefficient – the so-called apparent seismic coefficient k'_h – in calculations for the area below groundwater level.

In the section considered in Figure 3.23

$$\sum p_v = p + \gamma_1 \cdot h_1 + \gamma'_2 \cdot h_2$$

and

$$\sum p_h = k_h \cdot \left[p + \gamma_1 \cdot h_1 + (\gamma'_2 + \gamma_w) \cdot h_2 \right]$$

The apparent seismic coefficient k'_h for determining the active earth pressure below groundwater level is, thus,

$$k'_h = \frac{\sum p_h}{\sum p_v} = \frac{p + \gamma_1 \cdot h_1 + (\gamma'_2 + \gamma_w) \cdot h_2}{p + \gamma_1 \cdot h_1 + \gamma'_2 \cdot h_2} \cdot k_h$$

A similar procedure can be employed for the passive earth pressure.

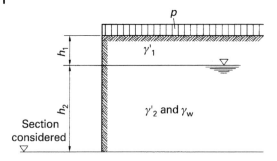

Figure 3.23 Sketch showing the approach for calculating k'_h.

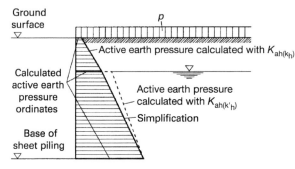

Figure 3.24 Simplified approach for active earth pressure.

For the special case where the groundwater level is at the surface and where there is no ground surcharge, the result for the active earth pressure side with $\gamma_w = 10 \text{ kN/m}^3$ is

$$k'_h = \frac{\gamma' + 10}{\gamma'} \cdot k_h = \frac{\gamma_r}{\gamma_r - 10} \cdot k_h \cong 2k_h$$

where

γ' the buoyant unit weight of soil
γ_r the unit weight of saturated soil

For simplicity, the unfavourable value of k'_h determined in this way for the active earth pressure side is normally also used as the basis for further calculations, even when the groundwater level is lower and for transient loads.

The active earth pressure coefficient K_{ah} determined using k_h and k'_h results in an abrupt change in the active earth pressure at the level of the water table, as shown in Figure 3.24. If a more accurate calculation is deemed unnecessary, the active earth pressure can be applied in simplified form according to Figure 3.24.

In difficult cases for which tabular values are not available for calculating the active and passive earth pressures, it is possible to determine the influences of both the horizontal and – if present – vertical ground accelerations on active and passive earth pressures by applying an extended Culmann method. The forces due to ground accelerations acting on the wedges investigated with the respective seismic coefficient k_h must then be taken into account in the polygons of forces as well. Such an approach is also recommended for larger horizontal accelerations, especially when the soil lies partly below the groundwater level.

3.5.14.4 Application of the resultant water pressure

The resultant water pressure for waterfront structures in the seismic loading case may be approximated as for the normal case, i.e. according to Sections 3.3.2 and 3.3.3, because the effects of the earthquake on the pore water were already taken into account when determining the active earth pressure with the apparent seismic coefficient k'_h according to Section 3.5.14.3. It should be noted, however, that in the seismic loading case, the critical active earth pressure failure plane is inclined at a shallower angle to the horizontal than in the normal case. For this reason, a higher resultant water pressure can act on the failure plane.

3.5.14.5 Transient loads

As the simultaneous occurrence of an earthquake, maximum transient load and maximum wind load is unlikely, it suffices to combine the seismic loads with only half the transient load and half the wind load (see also DIN 4149). Crane wheel loads due to wind and the wind component of line pull should, therefore, be reduced accordingly as well. Loads due to the travel and slewing movements of cranes do not need to be superposed on earthquake effects.

However, loads that in all probability remain constant for a longer period of time, e.g. loads from filled tanks or silos and from bulk cargo storage, may not be reduced.

3.5.14.6 Design situation and partial safety factors

According to DIN EN 1990, earthquake forces may be taken into account for earthquake design situations (DS-E) without applying partial safety factors to actions and resistances.

3.5.14.7 Guidance for considering seismic influences on waterfront structures

Taking into account the foregoing and other EAU recommendations allows waterfront structures to be designed and constructed systematically and with adequate stability for earthquake regions as well. Supplementary information regarding sheet pile structures can be found in Section 8.2.15 and for pile bents/trestles in Section 11.4.

Experience gained from the earthquake in Japan in 1995 is outlined in JSCE (1996). For further implications, the reader is referred to PIANC (2001).

References

AK Numerik (1991). Empfehlungen des AK Numerik in der Geotechnik. Deutsche Gesellschaft für Geotechnik. *Geotechnik* 14: 1–10.

Brennecke, L. and Lohmeyer, E. (1930). *Der Grundbau*, 4th ed., Vol. II. Berlin: Ernst & Sohn.

BAW (2017). Merkblatt Einteilung des Baugrunds in Homogenbereiche, Bundesanstalt für Wasserbau, Karlsruhe.

Bjerrum, L. (1973). Problems of soil mechanics and constructions on soft clays and structurally unstable soils (collapsible, expansive and others). *Proc. 8th ICSMFE*, Moscow, 3: 111–155.

Brinch Hansen, J. (1953). *Earth Pressure Calculations*. Copenhagen: The Danish Technical Press.

Brinch Hansen, J. and Lundgren, H. (1960). *Hauptprobleme der Bodenmechanik*. Berlin: Springer.

Busch, K.-F., Luckner, L. and Tiemer, K. (1993). *Geohydraulik*, 3rd ed. Berlin: Borntraeger.

Busse, M. (2009). Einpressen von Spundwänden – Stand der Verfahrens- und Maschinentechnik, bodenmechanische Voraussetzungen. Tagungsband zum Workshop Spundwände – Profile, Tragverhalten, Bemessung, Einbringung und Wiederverwendung. *Veröffentlichungen des Instituts für Geotechnik und Baubetrieb TU Hamburg Harburg* 19: 27–45.

DGGT (2012). *Empfehlungen des Arbeitskreises "Baugruben", Deutsche Gesellschaft für Geotechnik e. V.*, 5th ed., revised & enlarged. Berlin: Ernst & Sohn.

Gebreselassie, B. (2003). Experimental, analytical and numerical investigations in normally consolidated soft soils. *Schriftenreihe Geotechnik der Universität Kassel* Heft 14.

Gudehus, G. (1981). *Bodenmechanik*. Stuttgart: Ferdinand Enke Verlag.

Horn, A. (1972). Resistance and movement of laterally loaded abutments, *Proc. 5th Intl. Conf. on Soil Mechanics and Foundations*, Madrid.

Idriss, I.M. and Boulanger, R.W. (2004). Semi-empirical procedures for evaluating liquefaction potential during earthquakes. *Joint 11th Intl. Conf. Soil Dyn. Earthq. Eng. (ICSDEE) and 3rd Intl. Conf. Earthq. Geotech. Eng. (ICEGE)*, 7–9 Jan 2004, Berkeley, USA.

Jamiolowski, M., Ladd, C.C., Germaine, J.T. and Lancellotta, R. (1985). New developments in field and laboratory testing of soils. *Proc. 11th Int. Conf. Soil Mech. Found. Eng. in San Franc. (USA)*, Vol. 1, pp. 57–153.

Iai, S., Sugano, T., Ichii, K. et al. (1996). JSCE – Japan Society of Civil Engineering: The 1995 Hyogoken-Nanbu Earthquake, *Investigation into Damage to Civil Engineering Structures – Committee of Earthquake Engineering*, Tokyo, 1996.

Ladd, C.C. and DeGroot, D.J. (2003). Recommended practice for soft ground site characterisation. *Proc. 12th Panam. CSMGE, Arthur Casagrande Lecture*, Cambridge (USA).

Ladd, C.C. and Foott, R. (1974). New design procedure for stability of soft clays. *J. Geotech. Eng. Div., ASCE, GT7* 100 (1): 763–786.

Leinenkugel, H.J. (1976). Deformations- und Festigkeitsverhalten bindiger Erdstoffe. *Veröffentlichungen des Instituts für Bodenmechanik und Felsmechanik der Universität Karlsruhe* Heft 66.

Odenwald, B. and Herten, M. (2008). Hydraulischer Grundbruch: Neue Erkenntnisse. *Bautechnik* 85 (9): 585–595.

Odenwald, B. and Ratz, K. (2017). Berücksichtigung von Grundwasserströmungskräften beim Nachweis von Uferspundwänden. *Bautechnik* 94 (8): 535–541.

Odenwald, B. and Schneider, A. (2015). Wasserdruckansätze zur statischen Berechnung von Uferspundwänden an Kanalstrecken (ohne Hochwasser- und Tideeinfluss). BAWBrief 03/2015, Bundesanstalt für Wasserbau (BAW), Karlsruhe.

Odenwald, B., Hekel, U. and Thormann, H. (2018). Grundwasserströmung – Grundwasserhaltung. *Grundbau-Taschenbuch*, Teil 2: Geotechnischen Verfahren, 8th ed. (ed. by K.J. Witt), pp. 635–819. Berlin: Ernst & Sohn.

PIANC (2001). PIANC-Report "Effect of Earthquakes on Port Structures". Report of MarCom, WG 34.

Randolph, M.F. (2004). Characterisation of soft sediments for offshore applications. *2nd Int. Site Charact. Conf.*, Port, Portugal, Vol. 1, pp. 209–232.

Rollberg, D. (1976). Bestimmung des Verhaltens von Pfählen aus Sondier- und Rammergebnissen. Forschungsberichte aus Bodenmechanik und Grundbau FBG 4, TH Aachen.

Rollberg, D. (1977). Bestimmung der Tragfähigkeit und des Rammwiderstands von Pfählen und Sondierungen. *Veröffentlichungen des Instituts für Grundbau, Bodenmechanik, Felsmechanik und Verkehrswasserbau der TH Aachen* 3: 43–224.

Seah, T.H. and Lai, K.C. (2003). Strength and deformation behavior of soft Bangkok clay. *Geotech. Test. J.* 26 (4), https://doi.org/10.1520/GTJ11260J.

Scherzinger, T. (1991). Materialverhalten von Seetonen – Ergebnisse von Laboruntersuchungen und ihre Bedeutung für das Bauen in weichem Baugrund. *Veröffentlichungen des Institutes für Bodenmechanik und Felsmechanik der Universität Karlsruhe* Heft 122.

Terzaghi, K. and Peck, R.B. (1961). *Die Bodenmechanik in der Baupraxis*. Berlin: Springer.

TGL (1983). TGL 35983/02, Dec 1983 ed., Sicherung von Baugruben und Leitungsgräben, Böschung in Lockergestein, Fachbereichsstandard der Deutschen Demokratischen Republik.

VOB (2016). German Construction Contract Procedures – Part A (DIN 1960:2016-09), Part B (DIN 1961:2016-09), Part C (ATV) (ed. by DIN/DVA) Berlin: Beuth.

Verruijt, A. (1982). *Theory of Groundwater Flow*, 2nd ed. London: Macmillian.

Wehnert, M. (2006). Ein Beitrag zur drainierten und undrainierten Analyse in der Geotechnik. Mitteilung 53. Institut für Geotechnik, Universität Stuttgart.

Weißenbach, A. (1961). Der Erdwiderstand vor schmalen Druckflächen. *Mitteilung des Franzius-Instituts TH Hannover* 19: 220.

Witt, K.J. (ed.) (2018). *Grundbau-Taschenbuch*, 8th ed., Parts 1–3. Berlin: Ernst & Sohn.

Ziegler, M., Aulbach, B., Heller, H. and Kuhlmann, D. (2009). Der Hydraulische Grundbruch – Bemessungsdiagramme zur Ermittlung der erforderlichen Einbindetiefe. *Bautechnik* 86 (9): 529–541.

Standards and regulations

DIN 1054: Subsoil – Verification of the safety of earthworks and foundations – Supplementary rules to DIN EN 1997-1.

DIN 1055-2: Actions on structures – Part 2: Soil properties.

DIN 4020: Geotechnical investigations for civil engineering purposes– Supplementary rules to DIN EN 1997-2.

DIN 4084: Soil – Calculation of embankment failure and overall stability of retaining structures.

DIN 4085: Subsoil – Calculation of earth-pressure.

DIN 4094-1: Subsoil – Field investigations – Part 1: Cone penetration tests.

DIN 4094-2: Subsoil – Field testing – Part 2: Borehole dynamic probing.

DIN 4094-4: Subsoil – Field testing – Part 4: Field vane test.

DIN 4096: Subsoil; Vane Testing; Dimensions of Apparatus, Mode of Operation, Evaluation (replaced by DIN 4094-4).

DIN 4149: Buildings in German earthquake areas – Design loads, analysis and structural design of buildings.

DIN 18134: Soil – Testing procedures and testing equipment – Plate load test.

DIN 18137: Soil, investigation and testing – Determination of shear strength – Part 1: Concepts and general testing conditions.

DIN 18196: Earthworks and foundations – Soil classification for civil engineering purposes.

DIN EN 1990: Eurocode: Basis of structural design.

DIN EN 1997-1: Eurocode 7: Geotechnical design – Part 1: General rules,

DIN EN 1997-2: Eurocode 7: Geotechnical design – Part 2: Ground investigation and testing.

DIN EN ISO 22475-1: Geotechnical investigation and testing – Sampling methods and groundwater measurements – Part 1: Technical principles for the sampling of soil, rock and groundwater (ISO 22475-1:2021).

DIN EN ISO 22476-2: Geotechnical investigation and testing – Field testing – Part 2: Dynamic probing (ISO 22476-2:2005 + Amd 1:2011).

4
Loads on waterfront structures

4.1 Vessel berthing velocities and pressures

4.1.1 Guide values

When vessels approach a berth transversely, it is necessary to determine the berthing velocities that must be assumed for the design of the fenders.

Guide values for sea-going ships that can berth without the help of tugs can be found in ROM (1990).

ROM (1990) also specifies guide values for berthing velocities for vessels that require tug assistance. However, compared with the figures given in PIANC (2002), these velocities are relatively high, which new studies have confirmed; see Hein (2014). Therefore, the recommendation is to apply the PIANC (2002) figures when calculating berthing velocities for vessels that require tug assistance.

The berthing velocities given in Figure 4.1, which are taken from DIN EN 14504:2019-09, can be assumed for inland waterway vessels approaching a berth transversely.

During preliminary design, no exceptional accident impacts need to be taken into consideration, just the typical berthing forces. The magnitude of these berthing forces depends on each ship's dimensions, the berthing speed, the fenders and the deformation of the ship's side and the structure.

In order to provide quays with adequate loading capacity to resist typical berthing forces, but at the same time avoiding unnecessarily large dimensions, it is recommended to design the structural components affected by berthing manoeuvres in such a way that a single compression load equal to the critical line pull force can be applied at any point. For quay walls in seaports, the load should be as per Section 4.9.1 and in inland ports as per Section 4.9.2, but the total load should not exceed the permissible limits.

The single load can be distributed according to the fendering; without fenders, distribution is permissible over an area measuring 0.50×0.50 m. In the case of sheet pile walls without heavyweight superstructures, only walings and waling bolts need to be designed for this compression load.

Berthing forces on dolphins are dealt with in Chapter 12.

If the failure of the waterfront structure as a result of a collision (e. g. ship impact) poses particular risks, e. g. for another facility situated immediately behind it, further delibera-

4 Loads on waterfront structures

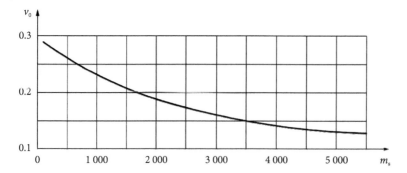

Figure 4.1 Berthing velocities of inland waterway vessels as a function of the vessel mass according to DIN EN 14504:2019-09.

tions may be necessary, and the measures to be taken then agreed upon between designer, client and the authorities.

4.1.2 Loads on waterfront structures due to fender reaction forces

The energy that can be absorbed by the fenders is determined using the deterministic calculation according to Section 7.4. See Section 4.1.1 for the guide values for berthing velocities that should be used in the calculations.

The maximum fender reaction force that can act on the waterfront structure or fender dolphin can be determined using the appropriate diagrams/tables of the manufacturer of the type of fender selected and the calculated energy to be absorbed. This reaction force is to be understood as a characteristic value.

Normally, the reaction force does not result in an additional load on the quay and only the local load derivation has to be investigated, unless special structures have been arranged for fenders, e. g. separate suspended fender panels.

4.2 Vertical imposed loads

In this section, all quantitative loads (actions) are characteristic values.

4.2.1 General

Vertical imposed loads (variable loads within the meaning of DIN EN 1991-2:2010-12 (Eurocode 1)) are essentially loads due to stored goods and means of transport. The changing positions of load influences due to mobile cranes (road or rail types) must be considered separately where these affect the waterfront structure. In the case of waterfront structures in inland ports, loads from mobile cranes generally only apply to those waterfront areas expressly intended for heavy-duty loading/unloading with mobile cranes. In seaports, in addition to rail-mounted quayside cranes, mobile cranes are being increasingly used for general cargo handling, i. e. not just for heavy loads.

With respect to dynamic load influences, a distinction is made between three different basic situations (Table 4.1):

- In basic situation 1, the load-bearing members of the structures are directly loaded by the means of transport and/or stacked goods, e. g. jetties (Table 4.1, basic situation 1).
- In basic situation 2, the means of transport and stacked goods place a load on a layer of a certain thickness, which spreads and transmits the loads to the load-bearing members of the waterfront structure. This form of construction is used, for example, for structures built over embankments, which include a layer of material above the structure to spread the loads (Table 4.1, basic situation 2).
- In basic situation 3, the means of transport and stacked goods only place a load on the body of soil behind the waterfront structure, which, consequently, is only loaded indirectly via an active earth pressure resulting from the imposed loads. Typical examples of this are exclusively sheet pile walls or a partly sloping bank (Table 4.1, basic situation 3).

There are also intermediate cases between the three basic situations, e. g. a pile trestle supporting a short slab.

Provided that a complete and reliable basis for calculations is available, the magnitudes of the imposed loads should be assumed to be those expected in normal circumstances. The higher the proportion of dead loads and the better the distribution of loads within the structure, the easier it is to accommodate any increases in the imposed loads that may be necessary at a later date within the scope of the permissible limits. Structural systems complying with basic situation 2 and, in particular, basic situation 3 offer advantages in this respect.

Table 4.1 Vertical imposed loads (from EAU 2012).

Basic situation	Traffic loads[a]		
	Rail	Road Vehicles	Mobile cranes
a) Basic situation 1	Loading assumptions to German Federal Railways guideline RIL 804 and/or DIN EN 1991-2:2010-12 Dynamic coefficient: amounts > 1.0 can be reduced to half their value	Loading assumptions to DIN EN 1991-2: 2010–12	Forklift truck loads to DIN EN 1991-1-4: 2010–12; outrigger loads for mobile cranes to Sections 4.2.2 and 4.10
b) Basic situation 2	As for basic situation 1 but a further reduction in the dynamic coefficient of up to 1.0 for a layer depth of 1.00 m. For a layer depth $h \geq 1.50$ m, use a uniformly distributed surface load of 20 kN/m²		
c) Basic situation 3	Loads as for basic situation 2 with a layer depth > 1.50 m		

a) Crane loads in accordance with Section 4.10.

See Section 1.1.4 for the allocation of the respective loads to load cases (DS-P, DS-T, DS-A, DS-E).

4.2.2 Basic situation 1

The railway traffic loads correspond to loading diagram 71 of DIN EN 1991-2: 2010-12 (Eurocode 1). With respect to road traffic, the loading assumptions according to DIN EN 1991-2: 2010-12 (Eurocode 1) should be used, generally load model 1. In the dynamic coefficients specified for railway bridges, used to factor the traffic loads, it is generally possible to reduce amounts > 1.0 to half their value on account of the slow speeds. For road bridges, load model 1 already presumes a slow vehicle travelling speed (congestion situation), and so no reduction is permitted. For jetties in seaports, forklift truck loads to DIN EN 1991-2: 2010-12 (Eurocode 1) and mobile crane outrigger loads of 1950 kN should be assumed with an outrigger size of 5.5 × 1.3 m, unless higher figures are necessary in particular cases (see Tables 4.11 and 4.12 in Sections 4.10.3 and 4.2.5).

4.2.3 Basic situation 2

This is essentially the same as basic situation 1. However, depending on the depth of the layer, the dynamic coefficients for railway bridges can be further reduced in a linear manner and eventually completely disregarded if the layer depth is at least 1.00 m (calculated from the top of the rail in the case of tracks embedded in paving). The loading must be considered bay by bay, however.

If the layer is at least 1.50 m deep, the total rail traffic load can be replaced by a uniformly distributed load of 20 kN/m².

4.2.4 Basic situation 3

Loads as for basic situation 2 with a layer depth > 1.50 m.

4.2.5 Loading assumptions for quay surfaces

When heavy road cranes or similar heavy vehicles and heavy construction equipment, e.g. crawler excavators etc., operate just behind the front edge of the waterfront structure, the following loads should be assumed as a minimum for designing the waterfront structure and any upper anchorage required:

a) Imposed load = 60 kN/m² from rear edge of top of wall landwards over a width of 2.0 m.
b) Imposed load = 40 kN/m² from rear edge of top of wall landwards over a width of 3.50 m.

Both a) and b) include the influences of an outrigger load $P = 500$ kN provided that the distance between the axis of the waterfront structure and the axis of the outrigger is at least 2 m. The reader is referred to Sections 4.2.2 and 4.10 in cases when higher outrigger loads need to be considered.

In accordance with PIANC (1987), the following imposed loads are used as a basis outside traffic zones. The container loads take into account gross loads of 300 kN for 40 ft containers and 240 kN for 20 ft containers.

- General cargo at least 20 kN/m^2
- Containers:
  - empty, stored in four tiers 15 kN/m^2
  - full, stored in two tiers 35 kN/m^2
  - full, stored in four tiers 65 kN/m^2
- RoRo load 30–50 kN/m^2
- Multi-purpose installations 50 kN/m^2
- Offshore supply bases see Section 7.5.

Further details of the material parameters of bulk and stacked goods can be found in DIN EN 1991-4:2010-12 (Eurocode 1) and the tables in ROM (1990).

When it comes to calculating active earth pressures on retaining structures, the different loads in traffic and container areas can generally be combined into an average uniformly distributed load of 30–50 kN/m^2.

Higher loads can occur at facilities serving the offshore wind industry, e.g. for building and handling offshore wind turbines, their foundations and transformer platforms. These loads must be agreed with the operator in each individual case.

4.3 Sea state and wave pressure

4.3.1 General

The wave loads on maritime and port structures are essentially due to the wind-generated sea state, the significance of which for the design has to be checked with reference to the local boundary conditions. In coastal areas, it is generally not just the sea state generated locally that is relevant to design, since fetch distances and shallow depths limit the wave energy component. Instead, the local sea state (wind-driven waves) combined with the sea state of the open sea beyond the project area and the sea state approaching the coast (groundswell) should be considered together.

The following descriptions are limited to fundamental processes and simplified approaches to the determination of hydraulic boundary conditions and loads on structures. Detailed information on this can be found, for example, in the coastal protection recommendations (EAK 2002/2015).

It is recommended that an institute or consulting engineers experienced in coastal engineering be consulted when it comes to investigating the wave conditions in the project area and the specific loads on structures. The need to carry out further physical or numerical studies should be looked at carefully prior to starting detailed design work.

4.3.2 Description of the sea state

The natural sea state can basically be described as an irregular chronological sequence of waves of varying height (or amplitude), period (or frequency) and direction. It represents the superposition in space and time (superposition principle) of various short and long-period sea state components. The direct influence of the wind generates an irregular, short-period (short-crested) sea state, also known as wind-driven waves. A long-period, irregular sea state arises through the superposition of wave components with the same direction, with the waves sorted by various interactions and the sea state no longer influenced directly by the wind.

The prevailing natural, irregular sea state in a project area is comprised of the local, short-period sea state (wind-driven waves) and the long-period sea state (groundswell) originally generated as wind-driven waves outside the project area.

In order to take into account the sea state relevant for design that is actually present as a load variable in the existing design methods, it is first necessary to parameterise the irregular sea state. This is because, in general, only individual, characteristic sea state parameters (see Section 4.3.3) can be included in the calculations. This parameterisation of the irregular sea state may be carried out both (EAK 2002/2015)

1. chronologically (direct short-term statistical evaluation of the time series) by determining and presenting characteristic wave parameters (wave heights and wave periods) as arithmetic mean values, and
2. in terms of frequency (Fourier analysis) by determining and presenting the results as a wave spectrum, with the energy content of the sea state being determined as a function of wave frequency.

Owing to the evaluation, the parameterisation of the sea state relevant for design results in the loss of all the information on the wave time series, its statistics and the wave spectrum.

In the design and detailing of maritime and port structures, the results of the sea state studies from the chronological and frequency analyses must be taken into consideration, depending on the location of the project area. In most cases, the parameterisation of the sea state relevant for design and the characterisation by way of individual wave height, wave period and wave direction parameters are normally sufficient. Where complex wind and wave conditions prevail and, in particular, in shallow water zones with breakwaters, it may be necessary to determine further local parameters characterising the sea state in order to reliably design the load variables required in the design process (EAK 2002/2015).

4.3.3 Determining the sea state parameters

4.3.3.1 General

Sea state parameters are characteristic values that describe and quantify certain properties of the irregular sea state varying with space and time. Depending on the evaluation process (see Section 4.3.2), these are:

1. In the time domain, the mean values of individual parameters, e.g. wave heights or wave periods, or combinations thereof.
2. In the frequency domain, prominent frequencies or integrals from the spectral density of the sea state spectrum.

The wave conditions in the project area must be analysed on the basis of measurements or observations over a sufficiently long period with respect to their theoretical probability of occurrence. To this end, and depending on the particular task in hand, the significant wave parameters resulting from short-term statistical analysis – such as wave heights, wave periods and wave approach directions – must be determined with respect to their seasonal frequencies or maximum long-term values in order to be able to derive data relevant for design. If such measurements are not available, empirical-theoretical or numerical methods must be used to determine the wave parameters from wind data (hindcasting), and these must be verified using any measured wave values available.

The parameterisation of the natural sea state is carried out based on the fact that there are statistical correlations between the heights of the individual waves of a natural sea state ascertained by measurements. These correlations can be described with a Rayleigh distribution according to Longuet-Higgens (1952) assuming a narrow-band wave spectrum and a large number of different waves (see Section 3.7.4 in EAK 2002/2015 and CEM 2001).

In deep water ($d \geq L/2$), wave height distributions based on measured data agree very well with the Rayleigh distribution even for wider wave band spectrums.

In shallow water ($d \leq L/20$), various effects (see Section 4.3.5) influence the waves, and so there are greater discrepancies between the measured wave height distribution and the theoretical Rayleigh distribution. The wave spectrum in the shallow water zone is no longer a narrow band, and the associated wave height distribution may differ substantially from the Rayleigh distribution due to the breaking waves.

Deviations from the Rayleigh distribution increase with larger wave heights and decrease with the narrowing of the spectral bandwidth. The Rayleigh distribution tends to overestimate large wave heights in all water depths.

The determination of the sea state parameters in the time and frequency domains and their relationships is explained below.

4.3.3.2 Sea state parameters in the time domain

When evaluated in the time domain, the wave heights and wave periods recorded during an observation period are described by stochastic variables of the frequency distribution. Where the wave heights are concerned, the Rayleigh function provides a good approximation of the probability $P(H)$ of the occurrence of a wave of height H (individual probability) and the probability $P(H)$ of the occurrence of a number of waves up to height H (total probability); see also Figure 4.2 after Oumeraci (2001):

$$P(H) = 1 - e^{-\frac{\pi}{4}\left(\frac{H}{H_m}\right)^2}$$

with

N	frequency of wave heights H in observation period, expressed as a percentage
H_m	mean value of all wave heights from sea state records
H_d	most frequent wave height
$H_{1/3}$	mean value of 33% highest waves
$H_{1/10}$	mean value of 10% highest waves
$H_{1/100}$	mean value of 1% highest waves
H_{max}	the maximum wave height
H_{rms}	the measurement of mean wave energy; it corresponds roughly to $1.13H_m$
H_S	the significant wave height (see Section 4.3.3.4)

4 Loads on waterfront structures

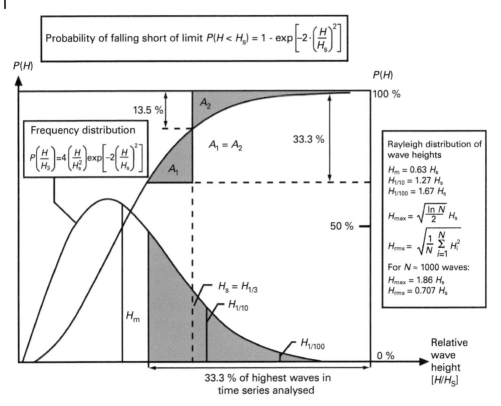

Figure 4.2 Rayleigh distribution of the wave heights of a natural sea state (schematic).

According to Longuet-Higgins (1952) and Schüttrumpf (1973), the frequency distribution of the wave heights results, approximately, in the following relationships, assuming a theoretical wave height distribution of the sea state corresponding to a Rayleigh distribution:

$$H_m = 0.63 \cdot H_{1/3}$$
$$H_{1/10} = 1.27 \cdot H_{1/3}$$
$$H_{1/100} = 1.67 \cdot H_{1/3}$$

These theoretical ratios agree well with ratios determined from sea state measurements despite the wave spectrum possibly having a greater bandwidth than that presumed by the Rayleigh distribution.

The maximum wave height H_{max} depends, in principle, on the number of waves recorded during the measuring period available. According to Longuet-Higgins (1952), if we assume

$$H_{max} = 0.707 \cdot \sqrt{\ln(n)} \cdot H_{1/3}$$

for $n = 1000$ waves, the result is a maximum wave height $H_{max} = 1.86 \cdot H_{1/3}$. As far as practical engineering is concerned, it is sufficient to estimate the maximum wave height as

$$H_{max} = 2 \cdot H_{1/3}$$

A further sea state parameter commonly used in practice is the wave height H_{rms} (rms = root-mean-square). A sea state with Rayleigh distribution results in the relationship $H_{rms} =$

$0.7 \cdot H_{1/3}$. The value H_{rms}, as a measurement of the mean wave energy, gives greater weight to the higher waves in the wave spectrum than the simple mean H_m.

Similarly to the ratios of the wave heights, the wave periods in the time domain can be estimated based on measurements from nature (EAK 2002/2015).

The actual ratios of the wave heights and periods depend on the actual wave height distribution, the specific shape of the sea state spectrum, the duration of measurements, etc., and can deviate from the above theoretical values, especially in shallow water, due to the true distribution of the waves and their asymmetry. Short measurement periods, e.g. 5 or 10 min, can lead to considerable errors when determining ratios, which is why EAK (2002/2015) proposes a period of at least 30 min for measuring and evaluating sea state measurements in order to rule out statistical irregularities.

4.3.3.3 Sea state parameters in the frequency domain

When parameterising an irregular sea state in the frequency domain, superposing the individual wave components converts the time series from the sea state records into an energy density spectrum and the associated wave phases into a corresponding phase spectrum (Fourier transformation, Figure 4.3). Presenting the sea state spectrum jointly for all wave directions is known as a "one-dimensional spectrum"; a separate presentation for different wave directions is a "directional spectrum".

The sea state spectrum enables the following and other characteristic sea state parameters to be specified as a function of the frequency f [Hz] taking into account spectral moments of the nth order:

$$m_n = \int S(f) \cdot f^n \, df \quad \text{where} \quad n = 0, 1, 2 \ldots$$

and also as a function of the wave approach direction (Figure 4.3):

H_{m_0} characteristic wave height, $H_{m_0} = (4m_0)^{0.5}$, where m_0 is the area of the wave spectrum

T_{01} mean period $= m_0/m_1$

T_{02} mean period $= (m_0/m_2)^{0.5}$

T_p peak period, i.e. wave period at maximum energy density

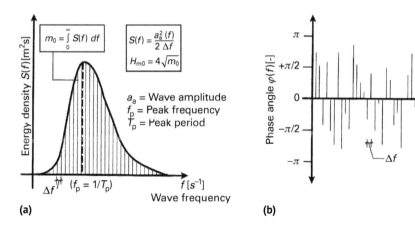

Figure 4.3 Wave spectrum parameters – definition sketches after Oumeraci (2001): (a) energy density spectrum; (b) phase spectrum.

The wave periods T_{01} and T_{02} have a fixed ratio, which depends on the shape of the wave spectrum, and describe the bandwidth of the spectrum. The wave spectrum allows the identification of long-period wave components in particular, e.g. groundswell waves, wave components transformed by the structure or changes in the spectrum caused by shallow water. These can be significant when defining the hydraulic boundary conditions for the design of maritime and port structures.

4.3.3.4 Relationships between sea state parameters in the time and frequency domains

The "significant wave height" H_s was introduced to characterise the irregular sea state for practical engineering applications (PIANC 1986). Further, presuming a Rayleigh distribution of the sea state, it is assumed for practical applications that H_s can be determined from the wave height $H_{1/3}$ in the time domain or the wave height H_{m_0} in the frequency domain:

$$H_s = H_{1/3} = H_{m_0}$$

Moreover, theoretically, the wave periods T_m (time domain) and T_{02} (frequency domain) can be equated. As regards further interrelationships between sea state parameters in the time and frequency domains, which can always vary depending on the respective wave spectrum, the reader is referred to the EAK (2002/2015).

4.3.4 Design concepts and specification of design parameters

The design sea state is understood to be the sea state event that leads to the critical load on a structure or part thereof or describes its effect in characteristic terms and is the result of a critical combination of different influencing variables.

With respect to the design of structures, a distinction is made (see Section 3.7 in EAK (2002/2015)) between

- the structural design, i.e. verification of stability for an extreme event, and
- the functional design, which deals with the effect and influence of the structure on its surroundings.

The irregularity of the sea state and its description as an input variable in the corresponding design procedure is crucial for determining the load variables that actually occur. Depending on the structure to be designed and the design method being used, the design sea state can be

- A characteristic individual wave, which is used to determine a specific load (deterministic method; possible application: load on a flood defence wall).
- Considered as a characteristic wave time series, the result of which is a time series of the loads occurring on the structure, which can be evaluated statistically and assessed with respect to the maximum and total loading (stochastic method; possible application: structures resolved in three dimensions, e.g. offshore platforms).
- Integrated into a fully statistical distribution, which enables a failure probability to be determined for the structure taking into account various limit states of the structure (probabilistic method).

Deterministic design methods are predominantly used in practical engineering applications. These will be looked at in more detail below. The EAK (2002/2015) contains advice

Table 4.2 Recommendation for stipulating design wave height.

Structure	$H_d/H_{1/3}$
Breakwaters	1.0–1.5
Sloped moles	1.5–1.8
Vertical moles	1.8–2.0
Flood defence walls	1.8–2.0
Quay walls with wave chambers	1.8–2.0
Excavation enclosures	1.5–2.0

on how the actual irregularity of the sea state can be taken into account in studies, calculations and design work on the basis of regular waves.

Since both wave and wind measurements rarely encompass the planned duration of use or the return periods associated with extreme sea state situations, the wave data available should be extrapolated to a longer period (frequently 50 or 100 years) by using a suitable theoretical distribution (e.g. Weibull). The extrapolation should not exceed three times the measurement period. The theoretical return period, and hence the parameters of the design wave height H_d, must be specified taking into account the potential damage or the permissible risk of flooding or destruction of the structure type (type of failure), also the database and other aspects (structural planning).

As far as functional planning is concerned, considerably shorter return periods sometimes have to be used in order to be able to estimate the anticipated restrictions on use and hazard situations as averages.

In the case of high safety requirements, the ratio of the design wave height H_d to the significant wave height $H_{1/3}$ should be taken as 2.0 (see Table 4.2). However, in order to draw up safe and economic solutions, hydraulic modelling is advisable to achieve a more precise analysis of the actual loads and the stability properties of the structure.

The design wave height is the maximum wave height to be used when designing a structure or structural component. The action effect resulting from the design wave height must be multiplied by the partial safety factors for the critical design situation, which results in the internal forces of the design.

If the frequency distribution used is based on long observation periods or corresponding extrapolation (approximately 50 years) or corresponding theoretical or numerical studies, the ensuing design wave may be classified as a rare wave as described in Section 1.1.4.3 for design situation DS-A. With shorter observation/study periods, the ensuing design wave should be defined as a frequent wave and classified in design situation DS-T according to Section 1.1.4.2.

4.3.5 Conversion of the sea state

Only in exceptional cases are the wave conditions in the immediate vicinity of the proposed structures actually known. As a rule, therefore, the deep-water sea state has to be converted for the project area on the coast. Where waves enter shallow water or encounter obstacles, various effects come into play:

1. Shoaling
   When a wave touches bottom, its velocity, and hence wavelength, decreases. Following a local, insignificant reduction, the wave height, therefore, gradually increases as the wave enters shallow water – for reasons of energy equilibrium (up to breaking point). This effect is called shoaling (Wiegel 1964).
2. Bottom friction and percolation
   Frictional losses and exchange processes at the bottom reduce the wave height. This effect is normally negligible for design purposes (Walden and Schäfer 1969).
3. Refraction
   Refraction occurs where the bottom depth varies and the waves approach the coast obliquely (or, more accurately, obliquely to the contour lines of the bottom). The waves tend to turn parallel to the shoreline as a result of local variations in the shoaling effect, so that – depending on the shape of the coastline – the effective wave energy is reduced or even increased (e. g. by focusing wave energy on a promontory).
4. Breaking waves
   Generally, waves can break if either the limit steepness is exceeded (parameter H/L) or the wave height reaches a certain dimension with respect to the water depth (parameter H/d). Once the associated limit water depth is exceeded, the height of deep-water waves entering shallow water is limited by the breaking process. The ratio of breaker height H_b to limit water depth d_b is normally $0.8 < H_b/d_b < 1.0$ (breaker criterion), although higher values have also been observed in special cases (Siefert 1974). Owing to the different wave heights in a sea state spectrum, the breaking of the waves generally takes place over what is known as a surf zone, the location and extent of which is determined by the underwater topography, the tides and other factors.
   To be precise, the ratio of breaker height H_b to water depth d_b is a function of the beach slope α and the steepness of the deep-water wave H_0/L_0. These parameters are combined in the breaker index ξ, which specifies the breaker type of regular waves (i. e. surging, plunging or spilling breakers) approximately. Further details can be found in Battjes (1975), Siefert (1974), Galvin (1972) and EAK (2002/2015).
   The breaker index ξ can be related to both the deep-water wave height $H_0(\xi_0)$ and the wave height at the breaking point $H_b(\xi_b)$ (Table 4.3):

   $$\xi = \frac{\tan \alpha}{\sqrt{H/L_0}}$$

   where

   α \qquad angle of inclination of bottom [°]
   H/L_0 \quad wave steepness
   H \qquad local wave height
   L_0 \qquad wavelength in deep water

   Highly reflective structures and influences due to foreshore geometry can have a considerable effect on the breaking process. Corresponding breaker criteria are then required (see Section 4.3.6, for example).

Table 4.3 Specification of breaker types (the values are based on studies with slope gradients from 1:5 to 1:20) (EAK 2002/2015).

Breaker type	ξ_0	ξ_b
Surging breaker	> 3.3	> 2.0
Plunging breaker	0.5–3.3	0.4–2.0
Spilling breaker	< 0.5	< 0.4

5. Diffraction
   Diffraction occurs when waves encounter obstacles (structures, but also features such as islands lying off the coast). Following diversion around the obstacle, the waves run into the lee of the structure, which results in energy being transported along the crest of the wave and, hence, normally a reduction in the wave height. At certain points beyond the wave shadow, the superposition of diffraction waves from closely spaced obstacles can result in higher waves (SPM 1984).
6. Reflections from the structure
   Waves approaching the shore or structures are reflected to a certain extent, the degree of which depends essentially on the properties of the reflecting edges (angle, roughness, porosity, etc.) and the water depth in front of the structure. Non-breaking waves are almost fully reflected where they strike a vertical structure perpendicularly such that, theoretically, a standing wave twice the height of the incoming wave is formed. Moreover, the reflection coefficient of sloping structures depends largely on the steepness of the wave and, thus, varies for the waves contained in the wave spectrum.
   Since the aforementioned influences depend on many factors, including factors specific to the structure and/or location, a general stipulation is not possible. For more details, see HTG (1996), EAK (2002/2015) and Oumeraci (2001).

4.3.6 Wave pressure on vertical quay walls in coastal areas

4.3.6.1 General

The wave pressure, or rather, the wave motion on the front of a waterfront structure is to be taken into account for:

- Block-type walls in the uplift pressure beneath the base and pressure in the joints.
- Structures above embankments with a non-backfilled front wall, taking into account the effective excess water pressure on both sides of the wall.
- Non-backfilled sheet pile walls.
- Flood defence walls.
- Loads during construction.
- Backfilled structures in general, also because of the lowered outer water level in the wave trough.

In addition, quay walls are loaded by line pull forces, vessel impacts and fender pressures resulting from ship movements caused by waves.

4 Loads on waterfront structures

When taking into account the wave pressure on vertical quay walls, a distinction must be made between three types of loading:

1. Walls loaded by non-breaking waves.
2. Walls loaded by waves breaking on the structure.
3. Walls loaded by waves breaking before reaching the structure.

Which of these three types of loading applies depends on the water depth, the sea state and the morphological and topographical conditions in the area of the structure.

Load assumptions are explained in the sections below for the various loading types. In addition, reference is made to the fact that the loads resulting from the standing, breaking or already broken waves of a natural sea state can be determined in accordance with Goda (2000) and EAK (2002/2015). The dynamic pressure increase coefficient determined empirically according to Takahashi (1996) can be used to calculate dynamic pressure loads. The disadvantage of this method is that only landward-facing load components are covered.

4.3.6.2 Loads due to non-breaking waves

A structure with a vertical or approximately vertical front wall in water of such a depth that the highest incoming waves do not break is loaded on the water side by the excess water pressure (which is higher due to reflection) at the crest, or on the land side by the higher excess water pressure at the wave trough.

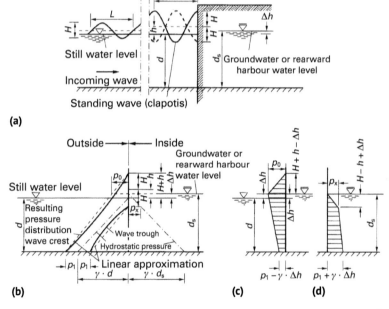

Figure 4.4 Dynamic pressure distribution on a vertical wall for total wave reflection after Sainflou (1928) and excess water pressures at wave crest and trough. (a) Explanation of calculation approach; (b) application of hydrostatic and dynamic pressures; (c) maximum excess water pressure from outside; (d) maximum excess water pressure from inside.

Standing waves are formed when incoming waves are superimposed on the backwash. In reality, true standing waves never occur; the irregularity of the waves creates certain wave impact loads, but these are generally negligible compared with the following load assumptions, and so are considered to be practically static. The wave height doubles as a result of reflection when the waves strike a vertical or approximately vertical wall and if no losses are incurred (coefficient of reflection $\kappa_R = 1.0$). A reduction in the wave height due to partial reflection ($\kappa_R < 0.9$) at vertical walls should only be assumed when verified by large-scale model tests. Otherwise, use the coefficients of reflection listed in EAK (2002/2015).

The method according to Sainflou (1928) as per Figure 4.4 is recommended for calculations when waves impact at 90° to the structure. However, this method supplies loads that are marginally too large if the waves are steep, whereas the loads from very long-period, shallow waves are underestimated. Further details and other design procedures, e.g. according to Miche-Lundgren, can be found in CEM (2001) and EAK (2002/2015).

Notation for Figure 4.4:

H height of incoming wave
L length of incoming wave
h height difference between still and mean water levels in the reflection area in front of wall:

$$h = \frac{\pi H^2}{L} \coth \frac{2\pi d}{L}$$

Δh difference between still water level in front of wall and groundwater or inner water level
d_s depth of groundwater or inner water level
d_s depth of groundwater or inner water level
γ unit weight of water
p_1 pressure increase (wave crest) or decrease (wave trough) at base of structure due to wave effect:

$$p_1 = \gamma H / \cosh \frac{2\pi d}{L}$$

p_0 maximum excess water pressure ordinate at land-side water level according to Figure 4.4c:

$$P_0 = (p_1 + \gamma \cdot d) \cdot \frac{H + h - \Delta h}{H + h + d}$$

p_x excess water pressure ordinate at level of wave trough according to Figure 4.4d:

$$p_x = \gamma \cdot (H - h + \Delta h)$$

Assigning this method to the case of an oblique wave approach is dealt with in Hager (1975). Accordingly, the assumptions for right-angled wave approaches should also be used for acute-angled wave approaches, especially in the case of long structures.

4.3.6.3 Loads due to waves breaking on a structure

Waves breaking on a structure can exert extreme impact pressures of 10 000 kN/m² and more. These pressure peaks are, however, very localised and of very brief duration (1/100–1/1000 s).

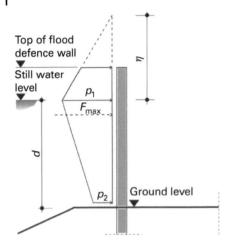

Figure 4.5 Loads due to plunging breakers (HTG 1996, Kortenhaus and Oumeraci 1997).

Owing to the huge pressure impulses and dynamic loads that occur, the structure should be suitably arranged and designed to ensure that, as far as possible, high waves do not break directly on the structure. If this is not possible, model studies at the largest possible scale are recommended for the final design. For further information regarding design for pressure impacts, the reader is referred to EAK (2002/2015, Section 4.3.2.3), and CEM (2001).

The following method of calculation may be used for simple geometries:

Tests on a large-scale hydraulic model of a caisson structure on a rip-rap foundation resulted in the following approximation for the impact pressure load on vertical walls (HTG 1996, Kortenhaus and Oumeraci 1997)

According to Figure 4.5, the maximum static horizontal force F_{max} on a quay wall is

$$F_{max} = \varphi \cdot 8.0 \cdot \rho \cdot g \cdot H_b^2 \quad [kN/m]$$

The point of application of this force lies just below still water level. An approximation for reducing the load as a result of overwash is explained in HTG (1996).

- Breaking wave height H_b: the wave steepness-related breaker criterion developed for relatively steep embankments (HTG 1996, Kortenhaus and Oumeraci 1997) results in the following equation:

$$H_b = L_b \cdot \left[0.1025 + \frac{0.0217(1 - \chi_R)}{(1 - \chi_R)}\right] \tanh(2\pi d_b / L_b)$$

where

χ_R coefficient of reflection of quay wall
d_b water depth at breaking point
L_b wavelength of breaking wave, $L_b = L_0 \tanh(2\pi d_b / L_b)$

- With a coefficient of reflection of 0.9 and assuming that the water depth d_b and the wavelength L_b are approximately equal to the corresponding values on the foreshore (water depth d at wall and wavelength L_d), the result is

$$H_b \approx 0.1 \cdot L_0 \cdot [\tanh(2\pi d / L_d)]^2$$

where

L_0 wavelength in deep water, $L_0 = 1.56 \cdot T_p^2$
T_p peak period in wave spectrum
L_d wavelength in water depth $d \approx L_0[\tanh(2\pi d/L_0)^{3/4}]^{2/3}$

- Impact factor φ: the impact coefficients given below were derived from calculations for the dynamic interaction of impulse-like, wave pressure impact loads varying over time with the stress and deformation conditions of the structure and the subsoil (HTG 1996). The impact coefficient φ depends on the level of the section being checked and is $\varphi = M_{dyn}/M_{stat}$ (wall moment for impact-type load/wall moment for quasi-static load) (HTG 1996, Kortenhaus and Oumeraci 1997, Heil et al. 1997).
- Walls with a yielding support in the subsoil, e.g. non-anchored vertical cantilever walls (see Figure 4.5, for example) or walls supported deeper than 1.50 m below ground level (BHFU 2013):

    $\varphi = 1.2$ for all analyses < 1.50 m below ground level.
    $\varphi = 0.8$ for all analyses > 1.50 m below ground level.

- Walls with rigid support (e.g. concrete walls on quay structures) or walls supported at a level > 1.50 m below ground level:

    $\varphi = 1.4$ for all analyses < 1.50 m below ground level.
    $\varphi = 1.0$ for all analyses > 1.50 m below ground level.

- Pressure ordinate p_1 at still water level:

$$p_1 = \frac{F_{max}}{0{,}625 \cdot d_b + 0{,}65 \cdot H_b}$$

- where η is the level of the pressure figure (the difference in height between the top of wave pressure load and the still water level):

$$\eta = 1.3 \cdot H_b$$

- Pressure ordinate p_2 at ground level:

$$p_2 = 0.25 \cdot p_1$$

4.3.6.4 Loads due to broken waves

SPM (1984) provides an approximate calculation of the loads from waves that have already broken. In this calculation, it is assumed that the broken wave continues with the same height and velocity after breaking, although this overestimates the actual loads. For a more accurate determination of the actual loads, EAK (2002/2015), therefore, proposes a correction to the characteristic values based on Camfield's method (Camfield 1991), which is not shown here.

4.3.6.5 Additional loads caused by waves

If a structure supported on a permeable foundation does not have a watertight face, e.g. in the form of an impervious diaphragm, the effect of the waves on the uplift pressure beneath the base must also be taken into account along with the water pressure on the wall surfaces. This also applies to the water pressure in joints between blocks.

4.4 Effects of waves due to ship movements

4.4.1 General

Waves of different types are generated by the bow and stern of every moving vessel. Depending on the local circumstances, these cause different loads on waterfronts and/or protective features (Figures 4.6 and 4.7). Initially, an accumulation of water is established ahead of the bow, the size of which may be several ship lengths viewed in the direction of travel. A further local accumulation occurs directly in front of the bow (bow wave). Alongside the ship, the water level dips as a result of the backwash (squat effect), and this tracks the vessel. The ensuing depression extends the length of the ship and extends across the full width of canals and narrow channels. On open water, the depression reduces with the distance from the ship; in a first approximation its width corresponds to 1.5–2.0 times the length of the ship (BAW 2010).

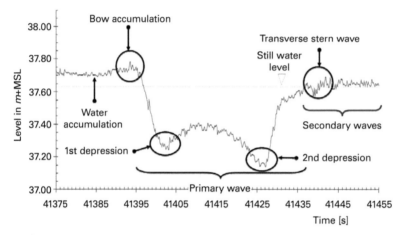

Figure 4.6 Change in water level for a ship travelling along a confined watercourse.

Figure 4.7 Large motorised vessel travelling at close to its critical speed along a canal; the water accumulation at the bow, the depression, the breaking stern wave and the secondary wave system of transverse stern waves are all clearly visible (BAW 2010).

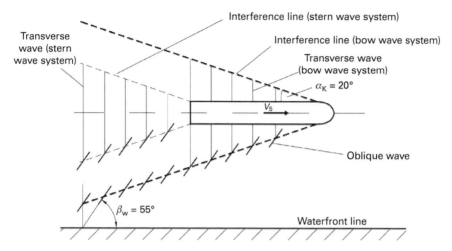

Figure 4.8 Wave pattern, schematic, after BAW (2010).

Secondary waves are superposed on this primary wave system and spread out from the ship in a fixed pattern (Figure 4.8). Of special significance with regard to the loads imposed on waterfronts are oblique waves, the crests of which are at an angle of 55° to the ship's axis.

4.4.2 Wave heights

Particular allowance must be made for the effects of the ship wave system on embankments and waterfront revetments in confined shipping channels. The critical design loads result from the increase and decrease in pressure in the area of the depression in the primary wave system and the breaking of the waves from the primary (stern wave) and secondary (oblique waves and transverse stern waves) wave systems at the transition to the shallow water area on the banks – and depending on the direction of the waves. The stone size required for revetments is usually determined by the load exerted by breaking transverse stern waves or the superposition of these waves on the secondary bow wave. Owing to their hydrostatic pressure changes, the bow wave and depression alongside the ship affect the pore water pressures underground and result – via temporary excess pressure in the subsoil – in a destabilisation of the waterfront protection. They, thus, generally determine the required thickness of a revetment (BAW 2010).

Where reflections are possible, e.g. in short branches with a perpendicular termination (lock basins) or along upright waterfront protection, the size of the accumulation or depression can double. More precise values can be determined in model tests.

The chronological sequence of the accumulation or depression might need to be taken into account for permeable waterfront structures, together with its effect on groundwater movement. Attention is drawn to the possible effects on automatically operating closures, e.g. dyke sluices (opening and closing of gates as a result of sudden pressure changes), and on lock gates.

The accumulation in front of a ship can be regarded as a "solitary wave". The height of the accumulation is generally small and rarely exceeds 0.2 m above the still water level.

The height of the waves on the interference line of oblique bow and stern waves may be estimated in accordance with BAW (2010) as follows:

$$H_{sec} = A_W \frac{v_S^{8/3}}{g^{4/3}(u')^{1/3}} f_{cr}$$

where

A_W the wave height factor [–], depending on shape and dimensions of ship, loaded draught and depth of water
       $A_W = 0.25$ for conventional inland waterway vessels and tugs
       $A_W = 0.35$ for empty pushed lighter convoys
       $A_W = 0.80$ for fully laden pushed lighter convoys
F_{cr} speed factor [–] ($f_{cr} = 1$ applies for $v_S/v_{crit} < 0.8$)
g acceleration due to gravity [m/s²]
H_{sec} secondary wave height [m]
u' distance from ship's side to waterfront line [m]
v_S speed of ship through water [m/s]

The water level depression corresponds to the backwash below and alongside the submerged body of the vessel, and its form and size depend on the vessel's shape, means of propulsion, travelling speed and the conditions of the waterway (ratio n of the watercourse cross-section to the submerged main frame cross-section of the vessel, proximity and shape of the waterfront). The maximum depression seldom exceeds approximately 15% of the depth of the water, even when the vessel attains its critical speed. Figures 4.9 and 4.10 enable the designer to make a safe estimate of the depression Δh as a function of n and the ship's critical speed. Depression and wave height change with the distance from the ship, chiefly in the vicinity of the waterfront; see BAW (2010), for instance. The maximum values determined can be used as design values.

Notation for Figure 4.9:

b_{WS} width at water level [m]
h maximum depth of water [m] in profile
h_m mean water depth [m]
Δh_{crit} mean water level depression at v_{crit}
n cross-section ratio [–]
v_{crit} critical speed of vessel [m/s]; maximum feasible hydraulic velocity of vessel travelling in displacement mode.
       The critical speed of the vessel in shallow water or a navigable channel where the water displaced by the vessel can no longer be fully displaced rearwards in the flowing state opposite to the direction of travel, the transition from subcritical to supercritical flow (Froude number in narrowest cross-section alongside ship = 1). Generally speaking, v_{crit} cannot be exceeded by displacement vessels.
x_{crit}, y_{crit} non-dimensional variables [–]

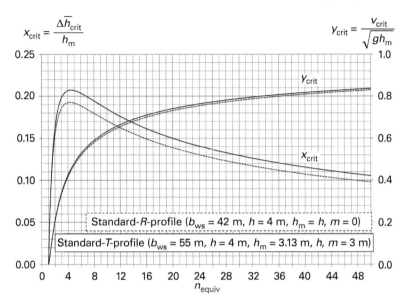

Figure 4.9 Critical vessel speed v_{crit} in relation to mean depth of water h_m, acceleration due to gravity g and cross-section ratio n for watercourses with rectangular (R) and trapezoidal (T) profiles ($n = n_{equiv}$ for typical ship and watercourse dimensions) (BAW 2010).

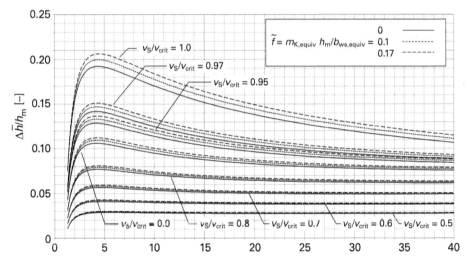

Figure 4.10 Mean drop in water level $\Delta \bar{h}$ in relation to mean depth of water h_m and relative (i.e. v_{crit} related) vessel speed v_s for different cross-section ratios n ($n = n_{equiv}$ for typical ship and channel dimensions) (BAW 2010).

4.5 Choosing a greater design depth (allowance for scouring)

Opting for a greater design depth (allowance for scouring) ensures that scour does not endanger the stability of the waterfront structure down to the depth of the scouring allowance. In conjunction with regular soundings, the allowance for scouring allows the effects on the bed associated with berthing and deberthing procedures to be monitored without endangering the stability of the facility. Decisions can then be made on the basis of this experience as to whether the effects of port operations make it necessary to protect the bed in the long term.

The size of the allowance for scouring depends on the local conditions. Experience of facilities with similar environmental conditions is the best guide in this respect.

Further indications as to the size of the scouring allowance can also be gained by estimating the maximum likely scouring depth expected with particular types of vessel or particular currents. However, the empirical approaches available should be applied with care, and the calculated scouring depths can vary significantly. Drewes et al. (1995) describe model experiments that allow an estimate of the scouring depth arising solely from vessel manoeuvres. See DWA (2020) for further information.

Where the allowance for scouring exceeds 10–20% of the height of the waterfront structure, the recommendation is to verify whether it is more economical to cover the bed in front of the waterfront structure protection against scouring.

4.6 Loads arising from surging and receding waves due to the inflow or outflow of water

4.6.1 General

Surging and receding waves occur in bodies of water as a result of a temporary or temporarily increased inflow or outflow of water. However, surging and receding waves essentially only manifest themselves with wetted watercourse cross-sections that are small in comparison with the inflow/outflow volume per second. Therefore, surging and receding waves and their effects on waterfront structures are generally only significant for navigation channels. In these cases, the effects of changes in water levels on embankments, the linings to watercourses, revetments and other facilities should be taken into account.

4.6.2 Determining wave values

Surging and receding waves are shallow-water waves in the range

$$\frac{d}{L} < 0.05$$

The wavelength depends on the duration of the water inflow or outflow. The wave propagation velocity can be roughly calculated with

$$c = \sqrt{g \cdot (d \pm 1.5H)} \quad [m/s] \begin{cases} + \text{ for surging} \\ - \text{ for receding} \end{cases}$$

where

g   acceleration due to gravity
d   depth of water
H   rise for surging or fall for receding compared with still water level

If the H/d ratio is small, then

$$c = \sqrt{g \cdot d}$$

can be used.

The rise or fall in water level is roughly

$$H = \pm \frac{Q}{c \cdot B}$$

where

Q   volume of water inflow or outflow per second
B   mean width at water level

The wave height can increase or decrease as a result of reflections or subsequent surging or receding waves. With uniform canal cross-sections and smooth canal linings in particular, the wave attenuation is small, and so the waves can move back and forth several times, especially with short reaches.

In navigation channels, the most frequent cause of surging and receding phenomena is the inflow or outflow of lockage water. In order to prevent extreme surging and receding phenomena, the lockage volume is generally restricted to 70–90 m³/s.

Lockages at intervals of the reflection time or a multiple thereof, particularly in canal stretches, can result in a superposition of the waves and, hence, in an increase in the degree of surging and receding.

4.6.3 Loading assumptions

Loading assumptions for waterfront structures must take account of the hydrostatic load due to the height of the surging or receding wave and its possible superposition by reflected or subsequent waves, also any potential simultaneous fluctuations in the water level, e. g. from the raising of the water level by wind and ship waves, in the least favourable configuration each time. Owing to the long-period nature of surging and receding waves, the ensuing effect on the flow gradient of the groundwater must also be checked in the case of permeable revetments.

The dynamic effects of surging and receding waves can be ignored owing to the mostly low flow rates caused by such waves.

The loads calculated in this way are characteristic values.

4.7 Wave pressure on piled structures

4.7.1 General

When designing piled structures, the loads originating from wave motion are to be taken into account with respect to the loads on both a single pile and the entire structure, in so far as this is necessitated by local circumstances. Superstructures should be located above the crest of the design wave, if possible. Otherwise, large horizontal and vertical loads from direct wave actions can affect superstructures; determining such actions does not fall within the remit of this recommendation since reliable values for such cases can only be obtained from model studies. The level of the crest of the design wave should be determined taking into account the simultaneous occurrence of the highest still water level and, where applicable, also allowing for the raising of the water level by wind, the influence of the tides and the rising and steepening of the waves in shallow water.

The superposition method according to Morison et al. (1950) is suitable for slender structural components, whereas the calculation method for wider structures is based on the diffraction theory of MacCamy and Fuchs (1954).

The subject of this recommendation is the superposition method of Morison et al. (1950), which applies to non-breaking waves. Owing to a lack of accurate calculation methods for breaking waves, a makeshift method is proposed in Section 4.7.5.

The Morison method provides useful values provided that the following is satisfied for a single pile:

$$\frac{D}{L} \leq 0.05$$

where

D pile diameter or, for non-circular piles, characteristic width of structural component (width transverse to direction of wave propagation)
L length of "design wave" in accordance with Section 4.3 in conjunction with Table 4.4

This criterion is generally satisfied.
   Definition of symbols used in Table 4.4:

$$\vartheta = \frac{2\pi \cdot x}{L} - \frac{2\pi \cdot t}{T} = kx - \omega t \text{ (phase angle)}$$

$$k = \frac{2\pi}{L}; \quad \omega = \frac{2\pi}{T}, \quad c = \frac{\omega}{k}$$

t duration
T wave period
c wave velocity
k wave number
ω wave angular frequency

For the determination of the wave loads, the reader should refer to Hafner (1977) and SPM (1984), which contain helpful tables and diagrams. The diagrams in SPM (1984) are based

Table 4.4 Linear wave theory: physical correlations (Wiegel 1964).

	Shallow water $\frac{d}{L} \leq \frac{1}{20}$	Transition area $\frac{1}{20} < \frac{d}{L} < \frac{1}{2}$	Deep water $\frac{d}{L} \geq \frac{1}{2}$
1. Profile of free surface	General equation $\eta = \frac{H}{2} \cdot \cos\vartheta$		
2. Wave velocity	$c = \frac{L}{T} = \frac{g}{\omega} kd = \sqrt{gd}$	$c = \frac{L}{T} = \frac{g}{\omega}\tanh(kd) = \sqrt{\frac{g}{k}\tanh(kd)}$	$c = \frac{L}{T} = \frac{g}{\omega} = \sqrt{\frac{g}{k}}$
3. Wavelength	$L = c \cdot T = \frac{g}{\omega} kdT = \sqrt{gd} \cdot T$	$L = c \cdot T = \frac{g}{\omega}\tanh(kd) \cdot T = \sqrt{\frac{g}{k}\tanh(kd)} \cdot T$	$L = c \cdot T = \frac{g}{\omega} \cdot T = \sqrt{\frac{g}{k}} \cdot T$
4. Velocity of water particles			
a) horizontal	$u = \frac{H}{2} \cdot \sqrt{\frac{g}{d}} \cdot \cos\vartheta$	$u = \frac{H}{2} \cdot \omega \cdot \frac{\cosh[k(z+d)]}{\sinh(kd)} \cdot \cos\vartheta$	$u = \frac{H}{2} \cdot \omega \cdot e^{kz} \cdot \cos\vartheta$
b) vertical	$w = \frac{H}{2} \cdot \omega \cdot \left(1 + \frac{z}{d}\right) \sin\vartheta$	$w = \frac{H}{2} \cdot \omega \cdot \frac{\sinh[k(z+d)]}{\sinh(kd)} \cdot \sin\vartheta$	$w = \frac{H}{2} \cdot \omega \cdot e^{kz} \cdot \sin\vartheta$
5. Acceleration of water particles			
a) horizontal	$\frac{\partial u}{\partial t} = \frac{H}{2} \cdot \omega \cdot \sqrt{\frac{g}{d}} \cdot \sin\vartheta$	$\frac{\partial u}{\partial t} = \frac{H}{2} \cdot \omega^2 \cdot \frac{\cosh[k(z+d)]}{\sinh(kd)} \cdot \sin\vartheta$	$\frac{\partial u}{\partial t} = \frac{H}{2} \cdot \omega^2 \cdot e^{kz} \cdot \sin\vartheta$
b) vertical	$\frac{\partial w}{\partial t} = -\frac{H}{2} \cdot \omega^2 \cdot \left(1 + \frac{z}{d}\right) \cos\vartheta$	$\frac{\partial w}{\partial t} = \frac{H}{2} \cdot \omega^2 \cdot \frac{\sinh[k(z+d)]}{\sinh(kd)} \cdot \cos\vartheta$	$\frac{\partial w}{\partial t} = \frac{H}{2} \cdot \omega^2 \cdot e^{kz} \cdot \cos\vartheta$

on stream function theory and may be applied to waves of varying steepness up to the breaking limit, whereas the diagrams in Hafner (1977) are only applicable under the prerequisites of linear wave theory.

To determine the upward wave load, the reader is referred to Section 4.7.9.

4.7.2 Method of calculation according to Morison et al. (1950)

The wave load on a single pile is made up of the components of flow force and acceleration force (inertial force), which must be determined separately and superposed to suit the phases.

According to Hafner (1977), Streeter (1961) and SPM (1984), the total horizontal load per unit length on a vertical pile is

$$p = p_D + p_M = C_D \cdot \frac{1}{2} \cdot \frac{\gamma_w}{g} \cdot D \cdot u \cdot |u| + C_M \cdot \frac{\gamma_w}{g} \cdot F \cdot \frac{\partial u}{\partial t}$$

For a pile with a circular cross-section:

$$p = C_D \cdot \frac{1}{2} \cdot \frac{\gamma_w}{g} \cdot D \cdot u \cdot |u| + C_M \cdot \frac{\gamma_w}{g} \cdot \frac{D^2 \cdot \pi}{4} \cdot \frac{\partial u}{\partial t}$$

where

p_D	flow force caused by flow resistance per unit length of pile
p_M	inertial force due to unsteady wave motion per unit length of pile
p	total load per unit length of pile
C_D	drag coefficient for flow pressure
C_M	drag coefficient for flow acceleration
g	acceleration due to gravity
γ_w	unit weight of water
u	horizontal component of velocity of water particles at pile location under consideration
$\frac{\partial u}{\partial t} \approx \frac{du}{dt}$	horizontal component of acceleration of water particles at pile location under consideration
D	pile diameter or, for non-circular piles, characteristic width of structural component
F	cross-sectional area of the pile within the flow in the direction of flow in area under consideration

The velocity and acceleration of the water particles are taken from the wave equations, which may be based on various wave theories. For linear wave theory, the correlations required have been compiled in Table 4.4. SPM (1984), Kokkinowrachos (1980) and EAK (2002/2015) should be consulted regarding the application of theories of a higher order.

4.7.3 Determining the wave loads on a single vertical pile

Since the velocities and, accordingly, accelerations of the water particles are a function of, among other factors, the distance of the location considered from the still water level, the wave load diagram of Figure 4.11 results from the calculation of the wave pressure load for various values of z.

The coordinate's zero point lies at still water level, but can be chosen to be at any point on the axis. The symbols used in Figure 4.11 are defined as follows:

- z ordinate of point investigated ($z = 0$ = still water level).
- x x coordinate of point investigated.
- η level of water table varying over time, related to still water level (water surface displacement).
- d water depth below still water level.
- D pile diameter.
- H wave height.
- L wavelength.

The phase displacement of the components of the wave load max. p_D and max. p_M should not be ignored. The calculation must, therefore, be performed for different phase angles and the maximum load determined by a phase-adjusted superposition of the components due to flow velocity and flow acceleration. For example, when applying linear wave theory,

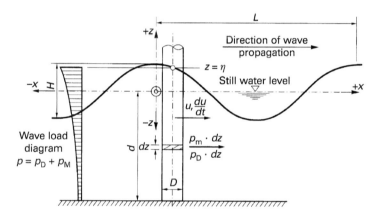

Figure 4.11 Wave action on a vertical pile according to CEM (2011).

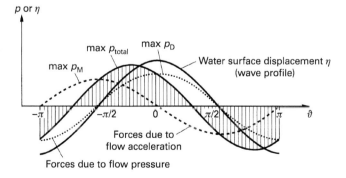

Figure 4.12 Variation in the forces due to flow pressure and flow acceleration over one wave period.

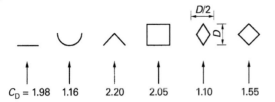

Figure 4.13 C_D values of pile cross-sections with stable separation points (Hafner 1978).

the phase of the acceleration force is displaced by 90° ($\pi/2$) with respect to the flow force, which lies in the same phase as the wave profile (Figure 4.12).

4.7.4 Coefficients C_D and C_M

4.7.4.1 Drag coefficient for flow pressure C_D

The drag coefficient for the flow pressure C_D is determined from measurements. It depends on the shape of the body within the flow, the Reynolds number Re, the surface roughness of the pile and the initial degree of turbulence of the current (Hafner 1978, Streeter 1961, Burkhardt 1967). The location of the separation point of the boundary layer is critical for the flow force. In the case of piles where the separation point is defined by corners or flow separation edges, the C_D value is practically constant (Figure 4.13).

On the other hand, in the case of piles without a stable separation point, e. g. circular cylindrical piles, a distinction is made between a subcritical range of the Reynolds number with a laminar boundary layer and a supercritical range with a turbulent boundary layer.

Since, however, generally speaking, high Reynolds numbers prevail in nature, in the case of smooth surfaces the recommendation is to assume a uniform value $C_D = 0.7$ (SPM 1984, Hafner 1978). Further information can be found in Sparboom (1986).

Larger C_D values should be expected with rough surfaces; see Det Norske Veritas (1991), for example.

4.7.4.2 Drag coefficient C_M for flow acceleration

Using potential flow theory results in a value $C_M = 2.0$ for a circular cylindrical pile, although C_M values of up to 2.5 have been ascertained from tests involving circular cross-sections (Dietze 1964).

Normally, the theoretical value $C_M = 2.0$ can be used. Otherwise, the designer should consult SPM (1984), Det Norske Veritas (1991) and Sparboom (1986).

4.7.5 Forces from breaking waves

Currently, there is no usable calculation method available for determining the forces from breaking waves correctly. The Morison formula is, therefore, again used for this range of waves, but under the assumption that the wave acts on the pile as a water mass with high velocity but no acceleration. In this regard, the inertia coefficient C_M is set to 0, while the flow pressure coefficient C_D is increased to 1.75 (SPM 1984).

4.7.6 Wave load on a group of piles

When determining the wave load on a group of piles, the phase angle ϑ critical for the respective pile location must be taken into account.

Using the designations according to Figure 4.14, the total horizontal load on a piled structure consisting of N piles is

$$P_{tot} = \sum_{n=1}^{N} P_n(\vartheta_n)$$

where

N number of piles
$P_n(\vartheta_n)$ wave load on a single pile n, taking into account the phase angle $\vartheta = k \cdot x_n - \omega \cdot t$
x_n distance of pile n from y–z plane

It should be noted that for piles situated closer together than about four pile diameters, there is an increase in load on piles situated side by side in the direction of the wave, but a decrease in load on piles situated one behind the other.

In this situation the correction factors given in Table 4.5 are proposed for such loads (Dietze, 1964).

4.7.7 Raking piles

The phase angle ϑ for the local coordinates x_0, y_0, z_0 differs for the individual pile segments d_s of a raking pile, and this must be taken into account.

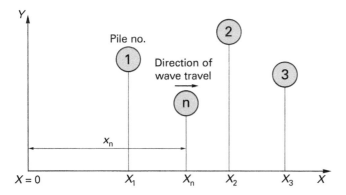

Figure 4.14 Details for a pile group (plan view) according to SPM (1984).

Table 4.5 Correction factors for closely spaced piles (Dietze, 1964).

Pile centre-to-centre distance e / Pile diameter D	2	3	≥ 4
For piles in rows parallel with crest of wave	1.5	1.25	1.0
For piles in rows perpendicular to crest of wave	0.7[a]	0.8[a]	1.0

a) The reduction does not apply to the front pile directly exposed to the wave action.

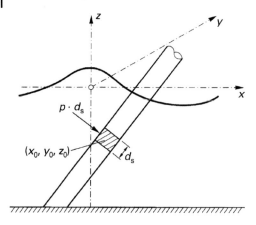

Figure 4.15 Calculating wave forces on a raking pile (SPM, 1984).

Consequently, the pressure on the pile at the location under consideration should be determined using the coordinates x_0, y_0 and z_0 in accordance with Figure 4.15.

According to SPM (1984), the local force $p \cdot d_s$ due to the flow and acceleration of the water particles towards the pile element d_s ($p = f[x_0, y_0, z_0]$) can be equated with the horizontal force on an equivalent vertical pile at position (x_0, y_0, z_0). However, when the pile slopes at a steeper angle, it is necessary to check whether determining the load taking into account the components of the resulting velocity acting vertical to the pile axis

$$v = \sqrt{u^2 + w^2}$$

and the resultant acceleration

$$\frac{\partial v}{\partial t} = \sqrt{\left(\frac{\partial u}{\partial t}\right)^2 + \left(\frac{\partial w}{\partial t}\right)^2}$$

supply less favourable values.

4.7.8 Safety factors

Designing piled structures to resist wave action is highly dependent on the choice of "design wave" (Section 4.3 in conjunction with Table 4.4). The wave theory used, and the values of coefficients C_D and C_M, remain influential.

This applies to piled structures in shallow water in particular. In order to allow for such uncertainties, the recommendation is to multiply the calculated loads by increased partial safety factors according to SPM (1984).

Consequently, when the "design wave" occurs only rarely, i.e. in the normal case with deep water conditions, the resulting wave load on piles must be increased by a partial safety factor $\gamma_d = 1.5$. When the "design wave" occurs frequently, which is usually the case in shallow water conditions, a partial safety factor $\gamma_d = 2.0$ is recommended.

Readers should refer to Sparboom (1986), EAK (2002/2015) regarding the possibility of using coefficients C_D and C_M depending on the Reynolds and Keulegan–Carpenter numbers and a corresponding reduction in the partial safety factor.

Critical vibrations can occur occasionally in piled structures, especially when separation eddies act transverse to the inflow direction, or the eigenfrequency of the structure is close to the wave period, resulting in resonance phenomena. In this case, regular waves lower than the "design wave" can have a more unfavourable effect, which calls for special investigations.

4.7.9 Vertical wave load ("wave slamming")

Horizontal structural components located in the vicinity of the water level may experience substantial vertical upward loads if the waves reach the structural component. This process is known as "wave slamming" and induces high impact loads. As these loads are considerable, the general aim is to try to avoid wave slamming completely by raising the structural components or positioning them at an angle.

4.7.9.1 Determining the necessary level of the structural components

In order to avoid wave slamming on horizontal slabs, e. g. the decks of jetties or offshore platforms, the level that is usually chosen for the underside of the deck structure is approximately 1.5 m above the crest of the design wave ("air gap approach"). The crest position is determined for the maximum wave height H_{max} of the storm on which the design is based (normally 1.6–2.0 H_s) while taking account of the design water level. According to Muttray (2000), the crest position predicted by Rienecker and Fenton (1981) in accordance with Fourier wave theory can be estimated thus:

$$h_{cr} = h_{DWL} + H_{max}\left(\frac{1}{2} + \frac{1}{3}\frac{\Pi}{3\Pi + \frac{1}{2}} + \frac{1}{6}\frac{\Pi^2}{\Pi^2 + \frac{1}{30}}\right)$$

$$\text{with } \Pi = \frac{H_{max}}{L}\coth^3\left(\frac{2\pi d}{L}\right) \tag{4.1}$$

where

h_{cr} crest [m]
h_{DWL} design water level [m]
H_{max} maximum wave height [m]
Π non-linearity parameter [–]
L wavelength [m] (according to linear wave theory)
d water depth [m]

4.7.9.2 Estimating the load exerted on structural components due to wave slamming

For functional or economic reasons, e. g. with very high design waves, it can be difficult to arrange the structural components high enough above the crest. In these cases, the structural components affected must be designed for vertical wave loads. The calculation methods presented below for the vertical loads due to wave slamming are suitable for preliminary designs. However, model tests at the largest scale possible are recommended for validating these methods or for detailed design work. As with investigations into breaking waves, on account of the reproduction of the proportion of air in the water, the wave height in the model should be at least 0.5 m.

Vertical wave forces acting from below on horizontal surfaces are comparable with horizontal impact loads from breaking waves on vertical walls or piles. As a rule, the action is made up of an impact-type, dynamic load component of very brief duration and a periodic, quasi-static load component. The first contact between wave and structural component leads to high pressure peaks, which, however, act only briefly and over a relatively small area. The subsequent periodic wave pressure affects a larger area, the extent of which changes as the wave passes.

4.7.9.3 Design approach for horizontal cylindrical structural components

A simple approach for estimating the vertical wave forces was developed for horizontal cylindrical structural components with truss-type offshore platforms. The vertical force F_s is described here as a line load:

$$F_S = \frac{1}{2}\rho_w C_S D w |w| \tag{4.2}$$

where

F_s vertical force per unit length [kN/m]
ρ_w unit weight of water [t/m³]
C_s slamming coefficient [–]
D diameter of cylindrical structural component [m]
w vertical velocity of surface of water [m/s]

It is assumed here that the pressure peak is reduced by the three-dimensional movement of the surface of the water in the natural sea state and by air pockets and the entry of air.

With respect to circular cylinders, slamming coefficients C_s in the range 0.5–1.7 times the theoretical value $C_s = \pi$ were established in model tests. A slamming coefficient $C_s \geq 3.0$ is recommended for static design. The dynamic (i.e. transient) load peaks can be estimated using a coefficient in the order of magnitude $C_s = 4.5$–6.0. Further information regarding the magnitude of slamming coefficients can be found in British Standard 6349-1 (2000). and Det Norske Veritas (1991). Surface water movements can be estimated with various (linear or non-linear) wave theories (see also Section 4.7 and Eq. (4.7) below).

4.7.9.4 Design approach for horizontal slabs

There is no generally recognised design approach for the vertical wave forces due to wave slamming on horizontal slabs. With respect to the preliminary design, the following simple analytical approach according to Tanimoto and Takahashi (1978) and Ridderbos (1999) can be used for the approximate determination of vertical wave forces on a deck. The dynamic wave pressure (impact pressure) is then essentially influenced by the contact angle β between water level and deck (see Figure 4.16 for initial contact between wave and deck). The total wave is the sum of the quasi-static and dynamic pressure components:

$$p = p_0 + p_S \tag{4.3}$$

$$p_0 = C_0 \rho_w g (\eta - R_c) \tag{4.4}$$

$$p_S = \frac{1}{2}\rho_w C_S w^2 \tag{4.5}$$

$$C_S = \min\left\{1 + \left(\frac{\pi}{2}\cot\beta\right)^2 ; 300\right\} \tag{4.6}$$

$$w = c\sin\beta \tag{4.7}$$

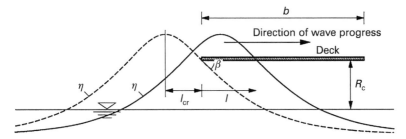

Figure 4.16 Definitions for slamming.

where

p wave pressure (total) [kN/m²]
p_0 quasi-static pressure component [kN/m²]
p_s dynamic pressure component [kN/m²]
η water surface displacement [m]
R_c freeboard of deck (above design water level) [m]
ρ_w density of water [kg/m³]
C_0 quasi-static pressure coefficient [–] ($C_0 = 1$)
w velocity component of vertical, upward water particles at wave surface
C_s slamming coefficient [–] ($C_s \leq 300$)
β contact angle between front of wave and deck [°]
c velocity of wave [m/s]; $c = L/T$ with wave period T [s] and wavelength L [m] (based on linear wave theory)

The maximum quasi-static pressure load occurs beneath the crest (Eq. (4.4)).

The resulting quasi-static wave force (per unit width) is derived from the integration of the wave pressure p_0 (Eq. (4.4)) along the contact area between the crest of the wave and the deck. The length of the contact area l varies during the passage of the wave ($l = ct$) and is limited by the geometry of the wave profile or the deck (with deck length b). The water surface displacement $\eta(x, t)$ can be taken from Table 4.6 or Figure 4.17. A coefficient $C_0 = 1$ is recommended for design.

The maximum value of the dynamic load component can be determined using Eq. (4.7). A rapid rise in pressure and an exponential pressure drop are typical. The maximum pressure load migrates along the deck structure; the spatial extent of the peak pressure in this regard is, however, very limited. The dynamic peak pressure p_s, therefore, only has to be taken into consideration when designing (small volume) structural components beneath the deck.

The resulting dynamic wave force (per unit width) F_s on the deck can be approximated according to Kaplan (1992) and Broughton/Horn (1987) as follows:

$$F_S = \frac{\pi}{4}\rho_w wcl'; \quad l' = \min\{l, l_{cr}, b\} \tag{4.8}$$

where

l length of contact area ($l = ct$)
b length of deck (in direction of wave progress)
l_{cr} horizontal distance between crest of wave and original contact point between wave and deck (see Figure 4.16)

Table 4.6 Wave profile (left) and contact angle (right) according to Fourier wave theory.

Distance x/L [−]	Non-linearity parameter Π [−]						Level η/H [−]	Non-linearity parameter Π [−]					
	0.1	0.2	0.3	0.4	0.5	0.6		0.1	0.2	0.3	0.4	0.5	0.6
	Relative water surface displacement η/H [−]							Contact angle β/H [°]					
−0.5	−0.40	−0.31	−0.25	−0.21	−0.18	−0.16	0.8			0.0	8.4	8.7	
−0.4	−0.36	−0.29	−0.24	−0.20	−0.18	−0.16	0.7		0.0	11.9	15.6	17.2	17.5
−0.3	−0.21	−0.22	−0.20	−0.18	−0.17	−0.16	0.6	0.0	10.7	16.5	18.8	19.9	19.6
−0.2	0.06	−0.03	−0.08	−0.10	−0.11	−0.12	0.5	6.0	13.1	17.0	18.3	19.0	19.2
−0.1	0.41	0.35	0.27	0.21	0.17	0.14	0.4	7.5	13.4	16.2	16.4	17.1	16.0
0	0.60	0.69	0.75	0.79	0.82	0.84	0.3	8.1	12.9	14.5	14.7	14.3	13.7
0.1	0.41	0.35	0.27	0.21	0.17	0.14	0.2	8.1	11.8	13.0	13.5	12.4	11.2
0.2	0.06	−0.03	−0.08	−0.10	−0.11	−0.12	0.1	7.6	10.4	10.9	10.6	10.5	9.6
0.3	−0.21	−0.22	−0.20	−0.18	−0.17	−0.16	0.0	7.0	8.4	8.3	6.8	6.5	5.9
0.4	−0.36	−0.29	−0.24	−0.20	−0.18	−0.16	−0.1	6.0	6.2	5.3	4.3	3.7	2.9
0.5	−0.40	−0.31	−0.25	−0.21	−0.18	−0.16	−0.2	4.9	3.9	1.9	0.3		
							−0.3	3.4	0.8				
							−0.4	0.5					

Once the crest of the wave has passed the front of the deck, a dynamic pressure load no longer occurs. Therefore, l' is always smaller than or equal to l_{cr}.

The wave profile and the contact angle (according to Fourier wave theory, see Rienecker/Fenton (1981)) can be determined using the non-dimensional figures in Table 4.6 and Figure 4.17. At the same time, the standardised wave profile varies with the non-linearity parameter Π; the actual wave profile is obtained by multiplying the values in the table by the actual wavelength L. The non-linearity parameter can be calculated according to linear wave theory using the wavelength (Eq. 4.1).

4.7.9.5 Construction details
Vertical wave forces due to wave slamming can be reduced by choosing an open, permeable deck structure (e.g. gratings). Furthermore, damage caused by wave slamming can be limited by a deck structure with loose cover plates, which are simply lifted and displaced by large waves.

4.8 Moored ships and their influence on the design of mooring equipment and fenders

4.8.1 General

This recommendation should be regarded as a supplement to the proposals and advice for the planning, design and detailing of fenders and mooring equipment, especially Sections 4.9 and 12.2.3.

The loads for mooring installations – such as bollards or quick-release hooks with their associated anchorages, foundations, support structures, etc. – ensuing according to this recommendation only replace the load variables given in Section 4.9 when the influence

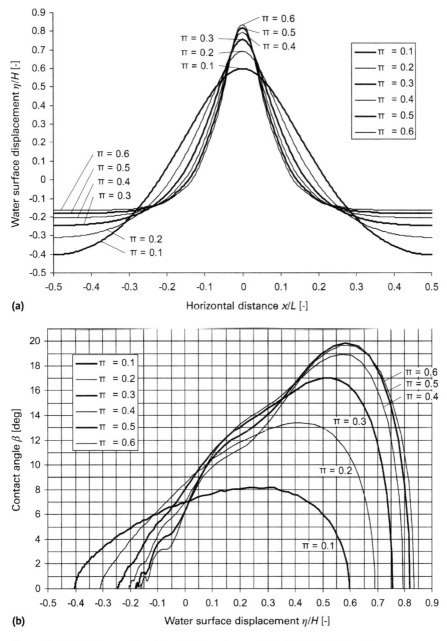

Figure 4.17 (a) Wave profile; (b) contact angle according to Fourier wave theory.

of swell, waves and currents on a ship's moorings can be neglected. Otherwise, the latter must be specifically verified and taken into consideration.

Section 4.1 is not affected by this recommendation. When determining the "normal berthing loads" dealt with in that section, the reference to Section 4.9 thus remains applicable without restrictions.

4 Loads on waterfront structures

Table 4.7 Conversion factors for mean wind speeds for a 10 min wind according to Thoresen (2018).

Mean wind for a period of ...	Conversion factor for 10 min wind
3 s	1.35
10 s	1.30
15 s	1.27
30 s	1.21
1 min	1.15
10 min	1.00
60 min	0.89

4.8.2 Critical wind speed

Owing to the inertia of vessels, it is not the short-term peak gusts (in the order of magnitude of seconds) that are critical when determining line pull forces; instead it is the mean wind speed over a period T, which can be assumed to be 10 min.

Conversion factors for various mean wind speeds according to Thoresen (2018) are given in Table 4.7.

It is recommended that wind measurements be used to determine the critical wind speed. If wind data from the immediate vicinity are not available, then wind measurements from more remote measuring stations can be used by means of interpolation or numerical methods of calculation, taking into account the orography. The time series of the wind measurements should be used to produce statistics of extreme values. A return period of 50 years is recommended for the design value.

If no other specific data on the wind conditions are available for the area of the ship's berth, the values given in DIN EN 1991-1-4:2010-12 can be used as critical wind speeds v for all wind directions.

This basic value can be differentiated according to wind direction if more detailed data are available.

4.8.3 Wind loads on moored vessels

The loads quoted are characteristic values.
Wind load components:

$$W_t = (1 + 3.1 \sin \alpha) \cdot k_t \cdot A_W \cdot v^2 \cdot \varphi$$
$$W_1 = (1 + 3.1 \sin \alpha) \cdot k_1 \cdot A_W \cdot v^2 \cdot \varphi$$

Equivalent loads for $W_t = W_{tb} + W_{th}$:

$$W_{tb} = W_t \cdot (0.50 + k_e)$$
$$W_{th} = W_t \cdot (0.50 - k_e)$$

(see Figure 4.18 for force diagram)

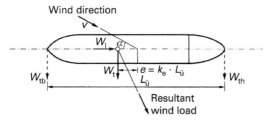

Figure 4.18 Wind loads acting on a moored vessel.

where

A_W area exposed to wind
v critical wind speed
W_t, W_l wind load components
k_t, k_l wind load coefficients
k_e coefficient of eccentricity
φ factor for dynamic and other non-ascertainable influences, generally $\varphi = 1.25$

The area exposed to the wind is derived from the most unfavourable load condition observed for each case, including any cargo on deck that may be present.

In line with international experience, the load and eccentricity coefficients given in Tables 4.8 and 4.9 can be used for vessels up to 300 m long, and lie on the safe side. Alternatively, a dynamic mooring analysis can be carried out to obtain an accurate figure.

Table 4.8 Load and eccentricity coefficients for ships $\leq 50\,000$ DWT.

α [°]	k_t [kN s²/m⁴]	k_e [−]	k_l [kN s²/m⁴]
0	0	0	$9.1 \cdot 10^{-5}$
30	$12.1 \cdot 10^{-5}$	0.14	$3.0 \cdot 10^{-5}$
60	$16.1 \cdot 10^{-5}$	0.08	$2.0 \cdot 10^{-5}$
90	$18.1 \cdot 10^{-5}$	0	0
120	$15.1 \cdot 10^{-5}$	−0.07	$-2.0 \cdot 10^{-5}$
150	$12.1 \cdot 10^{-5}$	−0.15	$-4.1 \cdot 10^{-5}$
180	0	0	$-8.1 \cdot 10^{-5}$

Table 4.9 Load and eccentricity coefficients for ships $> 50\,000$ DWT.

α [°]	k_t [kN s²/m⁴]	k_e [−]	k_l [kN s²/m⁴]
0	0	0	$9.1 \cdot 10^{-5}$
30	$11.1 \cdot 10^{-5}$	0.13	$3.0 \cdot 10^{-5}$
60	$14.1 \cdot 10^{-5}$	0.07	$2.0 \cdot 10^{-5}$
90	$16.1 \cdot 10^{-5}$	0	0
120	$14.1 \cdot 10^{-5}$	−0.08	$-2.0 \cdot 10^{-5}$
150	$11.1 \cdot 10^{-5}$	−0.16	$-4.0 \cdot 10^{-5}$
180	0	0	$-8.1 \cdot 10^{-5}$

The reader is referred to the tables in PIANC (2002) for more precise data on different types of ships.

4.8.4 Loads on mooring equipment and fenders

A structural system, consisting of vessel, hawsers and mooring and fender structures, must be introduced in order to determine the mooring and fender forces. The elasticity of the hawsers, which depends on their material, cross-section and length, is just as important as the inclination of the hawsers in the horizontal and vertical directions during varying loading and water level conditions. The elasticity of the mooring and fender structures must be ascertained for all support and bearing points of the structural system. Anchored sheet piling and structures with raking pile foundations may be considered as rigid elements in this context. Please note that the structural system can alter if individual lines fall slack, or fenders remain unloaded under specific load situations.

The wind-shielding effect of structures and facilities may be taken into account to a reasonable extent.

4.9 Loads on bollards

The term "bollard" is used to cover all forms of bollard, e. g. edge bollards, recessed bollards, dolphin bollards, mooring hooks, mooring rings, etc.

Bollards can be in the form of single or double bollards and can accommodate several hawsers simultaneously. They should be designed in such a way that they can be easily repaired or replaced.

All bollards must be clearly marked with their maximum permissible line pull force.

DIN 19703:2014-06 applies to locks and lock basins on inland waterways. Quick-release hooks (QRH), which are frequently used at berths for bulk cargos and especially at LNG or oil and gas terminals, are not covered in this section. Dynamic mooring analyses should always be carried out for such mooring equipment (see also Section 4.8). The arrangement of bollards is covered in Section 7.2.6.

4.9.1 Loads on bollards for sea-going vessels

A mooring system consists of three main components (Figure 4.19):

- The winch on the vessel
- The mooring line
- The bollard on land.

In an optimum mooring system, the braking effect of the winch gives way first, then the line breaks and, finally, the infrastructure on the land or the on-board equipment fails.

In recent years, when designing vessels, attention has focused increasingly on secure mooring.

The principal documents for the design and mooring of vessels are the publications of the OCIMF (2018) and the IACS (2016). The maximum bollard load and, hence, the safe work-

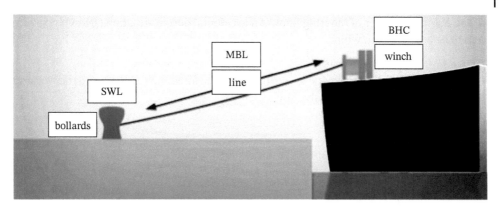

Figure 4.19 The components of a mooring system.

ing load (SWL) required for a bollard mounted on the land is determined by the following parameters in particular:

MBL minimum breaking load:
the minimum load at which a line fails.
MBL_{SD} Ship design MBL:
a theoretical value describing the minimum breaking load of new, dry mooring lines designed for the mooring system of a vessel.
WLL Working load limit:
the nominal load-bearing capacity or maximum working load of lines during normal operations, and according to OCIMF (2018), equivalent to 50% of MBL_{SD} for synthetic lines.
BHC Brake holding capacity:
the characteristic braking effect of a winch, for which the following reference values are available:
- For bulk carriers, container ships, cruise liners and special vessels with tall superstructures: 80% of MBL_{SD}.
- For tankers: 60% of MBL_{SD}.

SWL Safe working load:
the permissible characteristic working load, usually specified in tonnes on the bollard.

Mooring systems in which the line forces are calculated to exceed the WLL should be avoided for normal operations. If the loads on the mooring system increase, e. g. as a result of exceptional circumstances, then the braking effect of the most heavily loaded winch should give way first. This leads to a redistribution of the forces and to a stabilisation of the system. Only when more than one winch gives way simultaneously might the vessel break loose from its moorings in certain circumstances.

The following load cases should be examined:

In the permanent design situation (DS-P), the load on one bollard is:

$$F_{k,1} = BHC + (N-1) \cdot WLL$$

where N is the number of mooring lines. An accidental design situation (DS-A) is considered to be a situation in which

- Case 1: all winches give way.
- Case 2: a maximum of two winches are blocked (line forces up to MBL), and the working load limit (WLL) acts on the other lines.

$$F_{k,2} = \text{MAX} \begin{bmatrix} F_{k,2,1} \\ F_{k,2,2} \end{bmatrix}$$

where

$$F_{k,2,1} = N \cdot \text{BHC}$$
$$F_{k,2,2} = \text{MBL for} \quad N = 1$$
$$F_{k,2,2} = 2 \cdot \text{MBL} + (N - 2) \cdot \text{WLL for} \quad N > 1$$

Typical values for the large container ships in use these days (as of 2019) are: BHC = 100 t and WLL = 75 t (MBL = 150 t). Owing to recent amendments to the regulations and the use of ever larger vessels, the need for mooring lines with much higher MBL values can be expected in the future. This means that an SWL of at least 300 t is recommended for bollards at future container ship berths.

4.9.2 Loads on bollards for inland waterway vessels

The line pull loads acting on bollards are in the first place dependent on the size of the vessel, the speed of and distance from passing vessels, the flow rate of the water at the berth and the quotient of the water cross-section to the submerged ship's cross-section.

For new-build measures, the loads according to Table 4.10, which are based on BAW (2012), are recommended for the line pull loads acting on bollards at port basins and berths alongside inland waterways in non-exposed positions (no currents, no ice states).

According to BAW (2012), when checking the structural integrity of existing facilities, the line pull loads for inland waterway classes III and IV may be reduced to 180 kN.

Bollards must not be used to brake ships. Therefore, such a braking load is ignored in the load assumptions.

Table 4.10 Characteristic line pull loads on bollards E_k.

Inland waterway class	Characteristic line pull load on bollard E_k [KN]
I	100
II	150
III/IV	200
Va/Vb	200
VI or higher	300 or as calculated for individual case

For waterway classes VI or higher, unless 300 kN is assumed, and for especially exposed locations, e.g. due to currents or wind conditions, then dynamic mooring analyses, for example, must be carried out or the actual line pull forces measured by suitable means.

4.9.3 Direction of line pull load

A line pull load acting on a bollard can act at any angle in the direction of the water side. Line pull loads acting towards the land side are not considered, unless the bollard is also required for a waterfront structure situated behind it or it is a corner bollard that has to satisfy special requirements. In the calculations for a waterfront structure, a line pull load is normally assumed to act horizontally.

When designing the bollard itself and its connections to a waterfront structure, then corresponding line pull loads inclined upwards at up to 45° must also be considered.

4.9.4 Design for line pull loads

The line pull loads acting on bollards given in this section are characteristic values and should be allocated to design situation DS-P.

When checking the fixings between a bollard and a structural element and the structural element (tie bars) itself, 1.5 times the characteristic line pull loads acting on the bollard according to Sections 4.9.1 and 4.9.2 should be taken as the characteristic actions.

4.10 Quay loads from cranes and other transhipment equipment

The following loads are characteristic values that must be multiplied by the partial safety factors for the relevant load cases (see Section 1.1.4) according to DIN 1054:2010-12.

4.10.1 Conventional general cargo cranes

General cargo cranes are only used at existing, older port facilities. The corresponding loads must be taken from the as-built documents. Mobile cranes are often used to supplement old cranes at ports. The corresponding loads are generally specified on the datasheets of the crane manufacturers.

4.10.2 Container cranes

Container cranes are constructed as gantries with cantilevering booms and trolleys (ship-to-shore cranes), and usually run on up to 32 wheels. The centre-to-centre spacing of crane rails at existing container terminals usually lies between 15.24 m (50 ft) and 30.48 m (100 ft), but in some cases can be as much as 35.00 m.

Depending on the operating specification (e.g. Twinlift, Double-Forty, Double-Hoist), a safe working load of between 85 and max. 135 t can be assumed for the ropes of large container cranes. The maximum corner load is very much dependent on the type of crane and the length of the boom, which varies depending on the type of container ship handled

114 | 4 Loads on waterfront structures

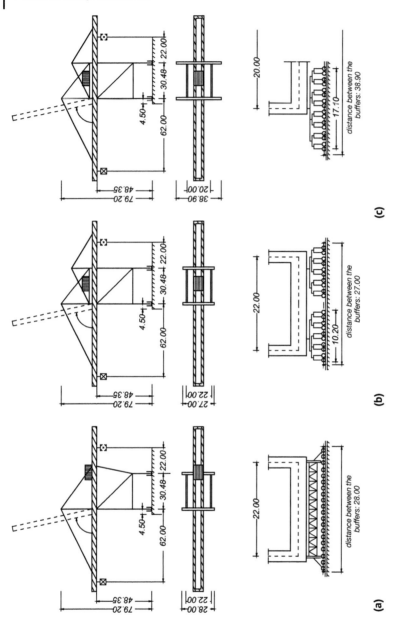

Figure 4.20 Dimensions of various container cranes currently in use: (a) 20 running wheels in one group; (b) 20 running wheels in two groups; (c) 32 running wheels in two groups.

at the terminal. Booms up to 80 m long can be necessary for ultra large container ships (ULCS) (see Figure 4.20).

The trend in container crane development is towards booms that can handle 24–26 rows of containers on the ships. That results in booms cantilevering up to 80 m beyond the waterside rail. The recommendation is to obtain accurate design data from the terminal operator because the large number of potential solutions does not allow data to be specified in greater detail. It is especially important to consider developments in which more than one container can be transported simultaneously, e. g. Twinlift cranes, which are already in widespread use.

4.10.3 Load specifications for port cranes

The support structure always consists of a portal-type substructure, either with slewing and height-adjustable jib or with a rigid cantilever beam that can be raised when not in operation. The portal usually stands on four corner points, each with several wheels arranged in swing-arm bogies, depending on the magnitude of the corner load. The corner load is distributed as evenly as possible over all the wheels of the corner point. Further to the remarks in Sections 4.10.1 and 4.10.2, general load and dimensional data can be found in Tables 4.11 and 4.12. These data assist in the general preliminary design of waterfront structures. Final geotechnical and structural calculations must be performed using the specific load specifications of the port cranes envisaged for handling the goods.

Table 4.11 Dimensions and characteristic loads of slewing and container cranes.

	Slewing cranes	Container cranes and other goods-handling equipment
Lifting capacity [t]	7–50	10–110
Self-weight [t]	180–350	200–3000[a]
Portal span [m]	6–19	9–45
Portal clearance [m]	5–7	5–13
Max. vertical corner load [kN]	800–3000	1200–20 000
Max. vertical wheel load [kN/m]	250–600	250–1150
Horizontal wheel load		
Transverse to direction of rail	up to approximately 10% of vertical load	
In direction of rail	up to approximately. 15% of vertical load of braked wheels	

a) Due to recent developments, the intrinsic weights may have increased even further.

Table 4.12 Dimensions and characteristic loads of mobile cranes.

	Mobile cranes				
Max. lifting capacity [t]	42	64	84	104	140
Self-weight [t]	130	170	250	420	460
Associated jib length [m]	12	12	15	22	20
Static outrigger load [kN]	920	1250	1660	2600	3250
Dynamic outrigger load [kN]	1080	1450	1950	3050	3650
Outrigger size [m]	5.5 × 0.8	5.5 × 0.8	5.5 × 1.3	5.5 × 1.8	5.5 × 1.8

4.10.4 Notes

Further information on port cranes can be found in the recommendations and reports of the Committee for Port Handling Equipment (AHU der HTG), the recommendations and reports of the Technical Committee for Inland Ports (ETAB 1981) and VDI Technical Rule 3576 (2011).

4.11 Impact and pressure of ice on waterfront structures, fenders and dolphins in coastal areas

4.11.1 General

Loads on hydraulic engineering structures due to the effects of ice can occur in various ways:

a) As ice impact from collisions with ice floes moved by the current or by the wind.
b) As ice pressure acting through ice thrusting against an ice layer adjacent to the structure or through vessel movements.
c) As ice pressure acting on the structure through an unbroken ice layer as a result of thermal expansion.
d) As imposed ice loads when ice forms on the structure or as surcharges or uplift loads when water levels fluctuate.

Among other factors, the magnitude of possible actions depends on:

- The shape, size, surface finishes and elasticity of the obstacle with which the ice mass collides.
- The size, shape and rate of advance of the ice masses.
- The nature of the ice and the ice formation.
- The salt content of the ice and the ice strength dependent on this.
- The angle of incidence.
- The critical strength of the ice (compressive, bending, shear).
- The rate of application of the load.
- The temperature of the ice.

The calculated ice loads are characteristic values. Consequently, according to Section 1.1.4, a partial safety factor of 1.0 can generally be used.

The calculations presented below apply to sea ice with a salinity (salt content) $S_B \geq 5\%$. Section 4.12 applies for lower salinities and freshwater ice. Further, the recommendations mentioned represent rough assumptions for ice loads on structures; they do not apply to extreme ice conditions, as are encountered in Arctic regions, for example. Where possible, the recommendation is to check the load values for waterfront structures, including piled structures, against the assumptions for completed installations that have performed well, or against ice pressures measured in situ or in the laboratory.

With respect to protected areas (bays, harbour basins, etc.) significantly reduced load assumptions can apply where the size of the ice field is small, in seaports with clear tidal effects and a substantial amount of maritime traffic and when adopting measures designed to reduce the ice load, e. g.:

- The effect of the current.
- The use of air bubble systems.
- Heating or other thermal transfer systems.

In any given case where a precise stipulation of the ice loads is required, experts should be called in and, if necessary, model tests carried out.

When positioning port entrances and orienting harbour basins, particular consideration must be given to the wind direction, current and the shear zone formation of the ice when it comes to determining ice loads and ice formation processes.

If the ice loads on dolphins are substantially greater than the loads resulting from vessel impact or line pull, a check should be carried out to determine whether such dolphins should be designed for the higher ice loads or whether, for economic reasons, occasional overloads can be accepted.

The reader is referred to the explanatory notes in Hager (1996). Advice taken from other international regulations (of the United States, Canada, Russia, etc.) can be found in Hager (2002). Other methods for determining loads due to ice pressure can be found in Thoresen (2018). The values to be used for structures in the Port of Hamburg are specified in ZTV-TB (HPA) (2020).

4.11.2 Determining the compressive strength of ice

The mean compressive strength of ice σ_0 essentially depends on its temperature, salt content and specific rate of expansion, i.e. the ice drift speed. In addition, ice has noticeably anisotropic properties, i.e. the maximum compressive strength depends on the direction of pressure. If no detailed studies of the material properties of the ice are available, then the following assumptions apply for the north German coasts:

- Linear temperature distribution over the thickness of the ice, with a temperature on the underside of approx. $-2.0\,°C$ for the German North Sea coast and approximately $-1.0\,°C$ for the German Baltic Sea coast (varies according to the salt content of the water) and the top surface of the ice at air temperature.
- The salinity of the ice in the North Sea and the Baltic Sea in accordance with Table 4.13.
- A specific rate of expansion $\varepsilon = 0.001\ s^{-1}$ (the compressive strength of the ice depends on the rate of expansion and attains the maximum value in the range between ductile and brittle failure at $\varepsilon \approx 0.001\ s^{-1}$).

Based on the salt content and temperature of the ice, the porosity φ_B, i.e. the quantity of salt crystals and air pockets in the ice, according to Kovacs (1996) is as follows:

$$\varphi_B = 19.37 + 36.18 S_R^{0.91} \cdot |\vartheta_m|^{-0.69}$$

where

φ_B porosity [%]
S_B salinity [%]
ϑ_m mean ice temperature, $(\vartheta_o + \vartheta_u)/2$ [°C]
ϑ_u temperature on underside of ice ($\vartheta_u = -1\,°C$ in the German region of the Baltic Sea, $\vartheta_u = -2\,°C$ in the German region of the North Sea) [°C]
ϑ_o temperature on top surface of ice (corresponds to air temperature) [°C]

Table 4.13 Guidance values for salt content (salinity S_B) of the seawater and sea ice along the German North Sea and Baltic Sea coasts according to Kovacs (1996).

North Sea	Salinity of water [%]	Salinity of ice [%]	Baltic Sea	Salinity of water [%]	Salinity of ice [%]
German	32–35	14–18	Belt Sea	15–20	10–12
Bight Estuaries	25–30	12–14	Bay of Kiel	15	8–10
			Bay of Mecklenburg	15	8–10
			Arkona Basin and Bornholm Sea	8–10	5–7
			Gotland Sea	5–7	a)
			Gulf of Finland and Gulf of Bothnia	1–5	a)

a) The compressive strength of ice for salinity < 5% is determined in accordance with Section 4.12.

The horizontal uniaxial compressive strength of the ice σ_0 according to Germanische Lloyd (2005) and Kovacs (1996) is derived from the material properties specified above or, in an ideal situation, by means of the material properties determined in situ or through experiments:

$$\sigma_0 = 2700 \dot{\varepsilon}^{1/3} \cdot \varphi_B^{-1}$$

where

σ_0 horizontal uniaxial compressive strength of the ice [MN/m²]
$\dot{\varepsilon}$ specific rate of expansion ($\dot{\varepsilon} = 0.001$) [s⁻¹]
φ_B porosity [%]

If more precise ice strength studies are not available, the flexural strength σ_B can be assumed to be roughly $1/3\sigma_0$ and the shear strength τ roughly $1/6\sigma_0$.

4.11.3 Ice loads on waterfront structures and other structures of greater extent

4.11.3.1 Mechanical ice pressure

The following design approaches can be used to determine the horizontal ice loads on vertical planar structures on the north German coasts for ice thicknesses d in the order of magnitude of $0.25\,\text{m} \leq d \leq 0.75\,\text{m}$ plus compressive strengths as per Section 4.11.2:

a) A horizontal mean line load p_0 acting at the most unfavourable height of the water levels under consideration. It is assumed here that the maximum load calculated from the uniaxial compressive strength of the ice σ_0 is only effective, on average, over one-third of the length of the structure (contact coefficient $k = 0.33$). The mean line load is:

$$p_0 = k \cdot h \cdot \sigma_0$$

where

p_0 mean line load [MN/m]
k contact coefficient (0.33) [–]
h thickness of ice [m]
σ_0 compressive strength of the ice [MN/m²]

b) A local load per unit area p acting over the thickness of the ice must be taken into account in local analyses. This results in the following equation:

$$p = \sigma_0$$

where

p local load per unit area [MN/m²]
σ_0 uniaxial compressive strength of the ice [MN/m²]

c) A reduced horizontal mean line load p'_0 acting at the most unfavourable respective height of the water levels under consideration in the case of platforms and revetments in tidal areas when the layer of ice has broken as a result of fluctuating water levels. According to Hager (1996), this results in the following equation:

$$p'_0 = 0{,}40 \cdot p_0$$

where

p'_0 reduced line load [MN/m]
p_0 mean line load [MN/m]

It is not necessary to consider the simultaneous effect of ice coupled with wave load and/or vessel impact.

If no maximum ice thickness values specific to the location are available, the maximum values specified in Table 4.14 can be assumed, which are based on ice observations conducted over many years.

4.11.3.2 Thermal ice pressure

Thermal ice pressure, as a further form of static loading on waterfront structures and other planar structures in the water, is caused by rapid changes in temperature and simultaneous restraint to expansion. In confined, iced-up port basins or similar configurations, considerable loads can occur as a result of thermal expansion. A precise determination of the thermal ice pressure on waterfront structures is complex, since the calculation approaches available frequently require numerous input parameters such as air temperature, wind speed, solar radiation, snow covering, etc., which can only be determined with a considerable degree of uncertainty.

With respect to the thermal ice pressure of sea ice as a function of the rate of temperature change (°C/h), ISO/FDIS 19 906 (2010) specifies the values given in Figure 4.21 for various ice temperatures and thicknesses.

Table 4.14 Measured maximum ice thicknesses as guidance values for design (BSH 2001).

North Sea	max h [cm]	Baltic Sea	max h [cm]
Helgoland	30–50	"Nord-Ostsee" Canal	60
Wilhelmshaven	40	Flensburg (outer fjord)	32
"Hohe Weg" lighthouse	60	Flensburg (inner fjord)	40
Büsum	45	Schleimünde	35
Meldorf (harbour)	60	Kappeln	50
Tönning	80	Eckernförde	50
Husum	37	Kiel (harbour)	55
Wittdün harbour	60	Bay of Lübeck	50
		Wismar harbour	50
		Bay of Wismar	60
		Rostock-Warnemünde	40
		Stralsund–Palmer Ort	65
		Saßnitz harbour	40
		Koserow–Usedom	50

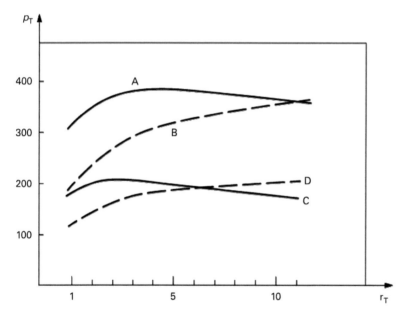

Figure 4.21 Thermal ice pressure as a function of ice temperature and thickness (according to ISO/FDIS 19906 2010). A: $\vartheta_m = -30\,°C$, $h = 1.0\,m$; B: $\vartheta_m = -30\,°C$, $h = 0.5\,m$; C: $\vartheta_m = -20\,°C$, $h = 1.0\,m$; D: $\vartheta_m = -20\,°C$, $h = 0.5\,m$; p_T: thermal ice pressure [kN/m]; r_T: temperature change rate [°C/h]; h: ice thickness; ϑ_m: mean ice temperature.

DIN 19704-1:2014-11 recommends considering a load of 0.25 MN/m² uniformly distributed over the ice thickness as a calculation parameter for German coasts and ice thicknesses $h \leq 0.8\,m$. The reader should refer to (HTG 2010) for further information and calculation approaches.

4.11.4 Ice loads on vertical piles

The ice loads acting on piles depend on the shape, rake and arrangement of the piles, as well as the critical compressive, bending or shear strengths of the ice responsible for breaking the ice. In addition, the magnitude of the load depends on the type of load (static or impact load due to ice floes) and the associated rates of expansion and deformation.

In the case of structural components up to 2 m wide with a ratio of structural component width to ice thickness $d/h \leq 12$ and where the ice is flat, the horizontal ice load on vertical piles with a rake steeper than $6:1$ ($\beta \geq 80°$) is derived as follows according to Schwarz et al. (1974):

$$P_p = k \cdot \sigma_0 \cdot d^{0,5} \cdot h^{1,1}$$

where

P_p ice load [MN]
σ_0 uniaxial compressive strength of the ice at a specific rate of expansion, $\dot{\varepsilon} = 0.001$ s^{-1} [MN/m²]
D width of single pile [m]
h thickness of ice [m]
k empirical contact coefficient from Table 4.15 [m$^{0.4}$]

If no studies specific to the location are available, then the ice thickness h can be estimated using the guidance values of Table 4.14 for the German North Sea and Baltic Sea coasts. In the event of ice ridges or ramparts, i. e. banked or humped, compacted sea ice with ice floes on top of each other, the ice loads determined must be doubled.

Along the German North Sea coast, the ice load on free-standing piles is frequently taken to be 0.5–1.5 m above mean high tide, depending on local conditions.

With respect to ice loads on tapering and inclined structural components, the guidance and calculation methods of Germanischer Lloyd (2005) should be taken into account.

4.11.5 Horizontal ice load on a group of piles

The ice load on a group of piles is derived from the sum of the ice loads on the individual piles. Generally, assuming the sum of the ice loads acting on the piles facing the drifting ice will suffice.

Table 4.15 Empirical contact coefficients k after Hirayama et al. (1974).

Load impact	k
Ice movement with layer of ice not tight up against structural component (drifting ice)	0.564
Ice movement with layer of ice tight up against structural component (icebound structural component)	0.793

4.11.6 Ice surcharges

Ice surcharges must be taken into account depending on local conditions. Without more detailed data, a minimum ice surcharge of 0.9 kN/m² can be regarded as sufficient (Germanischer Lloyd 1976). In addition to the ice surcharge, an estimate of the typical snow load to DIN EN 1991-3-1:2010-12 should also be considered. Conversely, variable loads that are absent during thicker ice formation do not normally need to be considered to act simultaneously.

4.11.7 Vertical loads with rising or falling water levels

Vertical ice forces, which occur when a covering of ice around the pile freezes solid and the water level subsequently changes, are limited by the flexural strength $\sigma_B(\sigma_f)$ of the ice. A prerequisite for the transmission of vertical ice forces to piles is a bond (adhesion) between the surface of the pile and the ice.

When determining upward and downward vertical ice loads on piles, the following approach based on the Russian standard (SNiP 1995) is recommended according to Kohlhase et al. (2006) and Weichbrodt (2008):

$$A_V = \left(0.6 + \frac{0.15D}{h}\right) \cdot 0.4 \cdot \sigma_0 h^2$$

where

A_V vertical ice load [kN]
h thickness of ice cover [m]
D pile diameter [m]
σ_0 compressive strength of ice cover [kN/m²]

The vertical ice loads determined using the recommended approach apply to individual vertical piles. If the spacing of the piles or the distance from piles to fixed structures is smaller than the extent of the deformation of the layer of ice under a vertical load (the characteristic length of ice layer ℓ_c), the vertical ice loads per pile are reduced.

The vertical ice load on groups of piles or piles close to fixed structures can be calculated by multiplying the load by a "geometric factor" f_g from the vertical ice load for a single pile.

This factor f_g is calculated from the relationship between the area of deformation, which might be limited at the pile location as a result of neighbouring piles or nearby structures, and the possible area of deformation when assuming an unlimited layer of ice.

The possible deformation area of an unlimited layer of ice is taken to be the area of a circle around the pile with a radius equal to the characteristic length of the ice layer ℓ_c, which as an approximation can be assumed to be 17 times the thickness of the ice. The limited deformation area is taken to be the mean of the areas of four circles whose radius r in each case is half the distance to the next structure fixed to the ground. The distances are determined in four directions at 90° to each other in accordance with the orientation of the structure. The radii r may not be greater than the characteristic length of the ice layer ℓ_c. The geometric factor f_g is determined as follows in accordance with Edil et al. (1988):

$$f_g = \frac{r_1^2 + r_2^2 + r_3^2 + r_4^2}{4\ell_c^2}$$

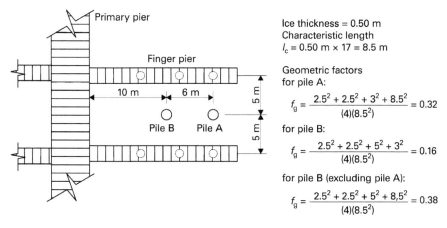

Figure 4.22 Examples of calculations for determining the geometric factor f_g.

where

f_g geometric factor [–]
r_{1-4} half the distance from pile to next solid structure [m]
ℓ_c characteristic length of ice layer [m]

Figure 4.22 explains the stipulation of the radii r_1 to r_4.

4.12 Impact and pressure of ice on waterfront structures, piers and dolphins at inland facilities

4.12.1 General

Most of the data provided in Section 4.11 apply to inland facilities as well. This is true for both the general statements and the loading assumptions because these depend on the respective dimensions of the structure, the thickness of the ice and the strength properties of the ice.

In accordance with Section 4.11.1, the ice loads determined are characteristic values to which, generally speaking, a partial safety factor of 1.0 should be applied.

The supplementary information provided in Section 4.11.1 must be observed.

4.12.2 Ice thickness

According to Korzhavin (1962), the ice thickness can be deduced from the sum of degrees of cold occurring on a daily basis during an ice period – the "cold sum". Hence, for example, according to Bydin (1959):

$$h = \sqrt{\sum |t_L|}$$

where h is the ice thickness in cm and $\sum |t_L|$ the sum of the absolute values of the mean daily sub-zero air temperatures in °C.

If no more detailed statistics or measurements are available, a theoretical ice thickness $h \leq 30$ cm can generally be assumed provided that the conditions given in Section 4.12.1 apply.

4.12.3 Compressive strength of the ice

The compressive strength of the ice depends on the mean ice temperature ϑ_m, corresponding to half the ice temperature at the surface, as 0 °C is always reached on the underside.

The uniaxial compressive strength of the ice vertical to the direction of ice formation can be determined approximately according to Schwarz (1970) as follows:

$$\sigma_0 = 1.10 + 0.35|\vartheta_m| \quad \text{for} \quad 0° < \vartheta_m < -5\,°C$$
$$\sigma_0 = 2.85 + 0.45|\vartheta_m + 5| \quad \text{for} \quad \vartheta_m < -5\,°C$$

where ϑ_m is the mean temperature of the ice [°C].

4.12.4 Ice loads on waterfront structures and other structures of greater extent

4.12.4.1 Mechanical ice pressure

Generally speaking, Section 4.11.3 applies to the force resulting from the pressure of the ice acting as a line load:

$$p_0 = k \cdot h \cdot \sigma_0$$

where

p_0 mean line load [MN/m]
k contact coefficient (0.33) [–]
h thickness of the ice [m]
σ_0 uniaxial compressive strength of the ice [MN/m²]

With respect to local structural components, a local load per unit area p acting over the thickness of the ice can be taken as

$$p = \sigma_0$$

where

p local load per unit area [MN/m²]
σ_0 uniaxial compressive strength of the ice [MN/m²]

On sloping surfaces, the horizontal force due to the pressure of the ice according to Korzhavin (1962) can be taken as

$$p'_0 = k \cdot h \cdot \sigma_B \cdot \tan \beta$$

where

p'_0 maximum line load [MN/m]
k contact coefficient (0.33) [–]
σ_B flexural strength of the ice, $1/3\,\sigma_0$ [MN/m²]
$\tan\beta$ angle of slope [°]

4.12.4.2 Thermal ice pressure

Generally speaking, the remarks in Section 4.11.3.2 apply. A load per unit area of 0.15 MN/m² distributed evenly over the ice thickness should be taken as the reference value for the thermal ice pressure resulting from thermal expansion and for ice thicknesses $0.30\,\text{m} \leq h \leq 0.50\,\text{m}$ (HTG 2010).

4.12.5 Ice loads on narrow structures (piles, dolphins, bridge and weir piers and ice deflectors)

The approaches regarding vertical piles according to Section 4.11 are also valid for inland areas taking into consideration the critical ice strengths. They also apply to pier structures and ice deflectors, allowing for their cross-sectional and surface form as well as the angle.

4.12.6 Ice loads on groups of structures

The information in Section 4.11 applies.

In the case of installations such as bridge piers in watercourses, avoiding ice rafting and the ensuing reduced discharge capacity, the risk of ice rafting can be estimated as shown in Figure 4.23.

The structures should be chosen such that the conditions of Figure 4.23 are complied with. If this is not possible, additional studies, e. g. physical model tests, are recommended. However, ice rafting does not necessarily increase the ice load in every situation, e. g. when the failure conditions of the banking ice are critical. Additionally, changes to the

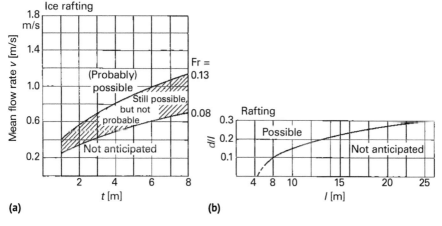

Figure 4.23 Risk of possible ice rafting according to Hager (1996) in terms of (a) water depth t and flow rate v; (b) the distance between piers l and the ratio of pier width d to pier spacing l.

load distribution and height of load application must be taken into account. Likewise, additional loads resulting from accumulated water and changes to currents as a result of cross-sectional restrictions must also be taken into account.

4.12.7 Vertical loads with rising or falling water levels

The information in Section 4.11.7 applies.

References

AHU der HTG. *Empfehlungen und Berichte des Ausschusses für Hafenumschlagtechnik (AHU) der Hafenbautechnischen Gesellschaft e. V.*, Hamburg.

Battjes, J.A. (1975). Surf similarity. *Proc. 14th Int. Conf. Coastal Eng.* Copenhagen 1974, Vol. I.

BAW (2010). *BAW-Merkblatt: Grundlagen zur Bemessung von Böschungs- und Sohlensicherungen an Binnenwasserstraßen*. Federal Waterways Engineering & Research Institute.

BAW (2012). *Empfehlungen zum Pollerzug (Trossenzugansatz) für Binnenschiffsschleusen. R&D final report A39510210106*. Karlsruhe: Federal Waterways Engineering & Research Institute (ed.).

BHFU (2013). *Berechnungsgrundsätze für Hochwasserschutzwände, Flutschutzanlagen und Uferbauwerke im Bereich der Tideelbe der Freien und Hansestadt Hamburg, Landesbetrieb Straßen, Brücken und Gewässer*, Hamburg: Hamburg Port Authority, April 2013.

Broughton, P. and Horn, E. (1987). Ekofisk platform 2/4C: Re-analysis due to subsidence. *Proc. Inst. Civ. Eng.*

BSH (2001). *Eiskarten der deutschen Nord- und Ostseeküste (seit 1879)*. Federal Maritime & Hydrographic Agency.

Burkhardt, O. (1967). Über den Wellendruck auf senkrechte Kreiszylinder. *Mitteilungen des Franzius-Instituts Hannover* 29.

Bydin, F.I. (1959). Development of certain questions in area of river's winter regime. *3rd Hydrologic Congress*, Leningrad.

Camfield, F.E. (1991). Wave forces on a wall. *J. Waterway Port Coast. Ocean Eng.* 117 (1): 76–79.

CEM (2001). *Costal Engineering Manual Part VI. Design of Coastal Projects Elements*. Washington: US Army Corps of Engineers.

Dietze, W. (1964). Seegangskräfte nichtbrechender Wellen auf senkrechte Pfähle. *Bauingenieur* 39 (9): 354.

Det Norske Veritas (1991). Environmental conditions and environmental loads. *Classification Notes* No. 30.5.

Drewes, U., Römisch, K. and Schmidt, E. (1995). Propellerstrahlbedingte Erosionen im Hafenbau und Möglichkeiten zum Schutz für den Ausbau des Burchardkais im Hafen Hamburg. *Mitteilungen des Leichtweiß-Instituts für Wasserbau der Technischen Universität Braunschweig* 134: 285.

EAK (2002/2015). *Empfehlungen für die Ausführung von Küstenschutzbauwerken durch den Ausschuss für Küstenschutzwerke der DGGT und HTG. Die Küste 65. Korrigierte Ausgabe 2007*. Karlsruhe: Federal Waterways Engineering & Research Institute (BAW), (unrevised reprint, 2015).

EAU (2015). *Recommendations of the Committee for Waterfront Structures, Harbours and Waterways EAU 2012*. (ed. by Deutsche Gesellschaft für Geotechnik e.V. and Hafentechnische Gesellschaft e.V.). Berlin: Ernst & Sohn.

Edil, T.B., Roblee, C.J. and Wortley, C.A. (1988). Design approach for piles subject to ice jacking. *J. Cold Reg. Eng.* 2 (2): 65–86.

ETAB (1981). *Empfehlungen und Berichte des Technischen Ausschusses Binnenhäfen*. Neuss: Bundesverband öffentlicher Binnenhäfen e. V.

Galvin, C.H. Ir. (1972). Wave breaking in shallow water. In: *Waves on Beaches*, (ed. by R.E. Meyer). New York: Academic Press.

Germanischer Lloyd (1976). *Vorschriften für Konstruktion und Prüfung von Meerestechnischen Einrichtungen, Band I – Meerestechnische Einheiten – (Seebauwerke)*. Hamburg: Eigenverlag des Germanischen Lloyd.

Germanischer Lloyd (2005). *Guideline for the Construction of Fixed Offshore Installations in Ice Infested Waters, Rules and Guidelines IV-Industrial Services*, Teil 6, Kap. 7. Hamburg: Eigenverlag des Germanischen Lloyd.

Goda, Y. (2000). Random seas and design of maritime structures. In: *Advanced Series of Ocean Engineering*, Vol. 15, 2nd, amended edn. University of Tokyo Press, World Scientific Singapore.

Hafner, E. (1977). Kraftwirkung der Wellen auf Pfähle. *Wasserwirtschaft* 67 (12): 385.

Hafner, E. (1978). Bemessungsdiagramme zur Bestimmung von Wellenkräften auf vertikale Kreiszylinder. *Wasserwirtschaft* 68 (7/8): 227.

Hager, M. (1975). Untersuchungen über Mach-Reflexion an senkrechter Wand. *Mitteilungen des Franzius-Instituts für Wasserbau und Küsteningenieurwesen der Technischen Universität Hannover* 42.

Hager, M. (1996). Eisdruck. In: *Geotechnical Engineering Handbook*, Pt. 1, Chap. 1.14, 5th edn. Berlin: Ernst & Sohn.

Hager, M. (2002). Ice loading actions. In: *Geotechnical Engineering Handbook*, Vol. 1: Fundamentals. Berlin: Ernst & Sohn.

Heil, H., Kruppe, J. and Möller, B. (1997). Berechnungsansätze für HWS-Wände und Uferbauwerke. *Hansa* 134 (5): 77 ff.

Hein, C. (2014). Anlegegeschwindigkeiten von Großcontainerschiffen, *33rd Int. Schifffahrtskongr.*, 1–5 June 2014, San Francisco, USA, German presentations, pp. 1–5.

Hirayama, K. I., Schwarz, J. and Wu, H.C. (1974). An investigation of ice forces on vertical structures. Iowa Institute of Hydraulic Research (IIHR), Report No. 158.

HTG (1996). Hochwasserschutz in Häfen – Neue Bemessungsansätze. *Tagungsband zum HTG-Sprechtag Oktober 1996*. Hamburg: Hafenbautechnische Gesellschaft (HTG) e. V.

HTG (2010). Empfehlungen des Arbeitsausschusses Sportboothäfen und wassertouristische Anlagen. Handlungsempfehlungen für Planung, Bau und Betrieb von Sportboothäfen und wassertouristischen Anlagen. Draft, May 2010, in preparation.

IACS (2016). Anchoring, Mooring and Towing Equipment, rec. 10, rev. 3. *International Association of Classification Societies*.

Kaplan, P. (1992). Wave impact forces on offshore structures: re-examination and new Interpretations. *Offshore Technol. Conf.*, OTC 6814, Houston, USA.

Kohlhase, S., Dede, C., Weichbrodt, F. and Radomski, J. (2006). *Empfehlungen zur Bemessung der Einbindelänge von Holzpfählen im Buhnenbau, Ergebnisse des BMBF-Forschungsvorhabens Buhnenbau*. University of Rostock.

Kokkinowrachos, K. (1980). Hydromechanik der Seebauwerke. In: *Handbuch der Werften*, Vol. 15. Hamburg: Schifffahrtsverlag Hansa.

Kortenhaus, A. and Oumeraci, H. (1997). Lastansätze für Wellendruck. *Hansa* 134 (5): 77 ff.

Korzhavin, K.N. (1962). Action of ice on engineering structures. English translation, US Cold Region Research and Engineering Laboratory. *Trans. T. L.* 260.

Kovacs, A. (1996). Sea-ice part II. Estimating the full-scale tensile, flexural, and compressive strength of first-year ice, US Army Corps of Engineers, CRREL Report 96-11.

Longuet-Higgins, M.S. (1952). On the statistical distribution of the heights of sea waves. *J. Mar. Res. XI* (3).

MacCamy, R.C. and Fuchs, R.A. (1954). Wave Forces on Piles: A Diffraction Theory. Techn. Memorandum 69, US Army, Corps of Engineers, Beach Erosion Board, Washington, Dec 1954.

Morison, J.R., O'Brien, M.P., Johnson, J.W. and Schaaf, S. A. (1950). The force exerted by surface waves on piles. *Pet. Trans.* 189: 149–154.

Muttray, M. (2000). Wellenbewegung an und in einem geschütteten Wellenbrecher. Dissertation. Braunschweig TU.

OCIMF (2018). *Mooring Equipment Guidelines (MEG4)*, 4th edn. Issued by Oil Companies International Marine Forum, ISBN: 978-1-85609-771-0.

Oumeraci, H. (2001). Küsteningenieurwesen. In: *Taschenbuch der Wasserwirtschaft*, (ed. by K. Lecher et al.), 8th edn., 657–743. Berlin: Paul Parey.

Oumeraci, H. and Kortenhaus, A. (1997). Anforderungen an ein Bemessungskonzept. *Hansa* 134: 751 ff.

PIANC (1986). IAHR/ PIANC: Int. Ass. For Hydr. Research/Permanent Int. Ass. of Navigation Congresses. List of Sea State Parameters. Supplement to Bulletin No. 52, Brussels.

PIANC (1987). Report of Pianc Working Group II-9 "Development of modern Marine Terminals". Supplement to Bulletin No 56, Brussels.

PIANC (2002). PIANC Report "Guidelines for the Design of Fender Systems: 2002". Report of MarCom, WG 33, Brussels.

Ridderbos, N.L. (1999). Risicoanalyse met behulp van foutenboom en golfbelasting ten gevolge van „slamming" op horizontale constructie. Master thesis. TU Delft, Faculty of Civil Engineering & Geoscience, Delft.

Rienecker, M.M. and Fenton, J.D. (1981). A Fourier approximation method for steady water waves. *J. Fluid Mech.* 104.

ROM (1990). Recomendaciones para Obras Maritimas. (English Version), Maritime Works Recommendations (MWR): Actions in the Design of Maritime and Harbor Works (ROM 0.2-90), Ministerio de Obras Publicas y Transportes, Madrid.

Sainflou, M. (1928). Essai sur les digues maritimes verticales. *Annales des Ponts et Chaussées* tome 98 II. Translation: Treatise on vertical breakwaters. US Army, Corps of Engineers.

Schüttrumpf, R. (1973). Über die Bestimmung von Bemessungswellen für den Seebau am Beispiel der südlichen Nordsee. *Mitteilungen des Franzius-Instituts für Wasserbau und Küsteningenieurwesen der Technischen Universität Hannover* 39.

Schwarz, J. (1970). Treibeisdruck auf Pfähle. *Mitteilung des Franzius-Instituts für Grund- und Wasserbau der Technischen Universität Hannover* 34.

Schwarz, J., Hirayama, K. and Wu, H.C. (1974). Effect of Ice Thickness on Ice Forces. *Proc. Of 6th Annual Offshore Technol. Conf.*, Houston, USA.

Siefert, W. (1974). Über den Seegang in Flachwassergebieten. *Mitteilungen des Leichtweiß-Instituts für Wasserbau der Technischen Universität Braunschweig* 40.

Sparboom, U. (1986). Über die Seegangsbelastung lotrechter zylindrischer Pfähle im Flachwasserbereich. *Mitteilungen des Leichtweiß-Instituts der TU Braunschweig* 93: 1.

SPM (1984). *Shore Protection Manual*. Vicksburg: US Army Corps of Engineers, Coastal Engineering Research Center.
Streeter, V.L. (1961). *Handbook of Fluid Dynamics*. New York: McGraw-Hill.
Takahashi, S. (1996). *Design of Breakwaters*. Yokosuka: Port & Harbour Research Institute.
Tanimoto, K. and Takahashi, S. (1978). Wave forces on horizontal platforms. *Proc. Of 5th Intl. Ocean Dev. Conf.*
Thoresen, C.A. (2018). *Port Designer's Handbook*, 4th edn. London: ICE Publishing.
Walden, H. and Schäfer, P.J. (1969). *Die winderzeugten Meereswellen, Teil II, Flachwasserwellen*. Heft 1 und 2, Einzelveröffentlichungen des Deutschen Wetterdienstes. Seewetteramt Hamburg.
Weichbrodt, F. (2008). Entwicklung eines Bemessungsverfahrens für Holzpfahlbuhnen im Küstenwasserbau. Published dissertation. Rostock.
Wiegel, R.L. (1964). *Oceanographical Engineering. Prentice Hall Series in Fluid Mechanics*. Englewood Cliffs: Prentice-Hall.
ZTV-TB (HPA) (2020). Zusätzliche technische Vertragsbedingungen – Technische Bearbeitung, Hamburg Port Authority, Statische Prüfstelle Hafen (SPH), Hamburg, Jan 2020.

Standards and Regulations
BS 6349-1:2000: Maritime Structures, Part 1: Code of practice for general criteria, Section 5.
DIN 1054:2010-12: Subsoil – Verification of the safety of earthworks and foundations – Supplementary rules to DIN EN 1997-1.
DIN 19703:2014-06: Locks for waterways for inland navigation – Principles for dimensioning and equipment.
DIN 19704-1:2014-11: Hydraulic steel structures – Part 1: Criteria for design and calculation.
DIN EN 1991-1-1:2010-12: Eurocode 1: Actions on structures – Part 1-1: General actions – Densities, self-weight, imposed loads for buildings.
DIN EN 1991-1-2:2010-12: Eurocode 1: Actions on structures – Part 1-2: General actions – Actions on structures exposed to fire.
DIN EN 1991-1-3:2010-12: Eurocode 1: Actions on structures – Part 1-3: General actions – Snow loads.
DIN EN 1991-1-4:2010-12: Eurocode 1: Actions on structures – Part 1-4: General actions – Wind actions.
DIN EN 14504:2019-09: Inland navigation vessels – Floating landing stages and floating bridges on inland waters – Requirements, tests.
DWA(2020) Auskolkungen an pfahlartigen Bauwerksgründungen. Merkblatt DWA-M 529. Vorlage für Weißdruck, Deutsche Vereinigung für Wasserwirtschaft, Abwasser und Abfall e. V., Hennef, Feb 2020.
ISO/FDIS 19906 (2010). (E) Petroleum and natural gas industries. Arctic offshore structures.
SNiP (1995). SNiP 2.06.04-82: Loads and Effects on Hydrotechnical Facilities (caused by Wave, Ice, Vessels). Federal Registry of National Building Codes & Standards, Ministry of Construction of Russia, Moscow.
VDI Technical Rule 3576 (2011). Rails for crane systems – Rail connections, rail beddings, rail fastenings, tolerances for crane tracks. Verein Deutscher Ingenieure, Düsseldorf.

5
Earthworks and dredging

5.1 Dredging in front of quay walls in seaports

This section deals with the technical options and conditions to be taken into account when planning and executing dredging work in front of quay walls in ports and harbours.

A distinction must always be made between new dredging and maintenance dredging when it comes to the necessary authorisations and approvals, etc.

Dredging down to the design depth according to Section 5.1 should be carried out by grab dredgers, bucket-ladder dredgers, cutter-suction dredgers, cutter-wheel suction dredgers, plain suction dredgers or hopper suction dredgers. Harrows and water-jetting machines can be employed as well.

When using cutter-suction dredgers, plain suction dredgers or hopper suction dredgers in areas of passive earth resistance in front of quay walls, the dredgers must be equipped with devices that ensure exact adherence to the planned dredging depth. Cutter-suction dredgers with high capacity and high suction force as well as plain suction dedgers are less suitable because of the danger of dredging too deep and reducing the strength of the underlying soil.

It should also be noted that, even under favourable conditions and with careful working, bucket-ladder, cutter-suction and hopper suction dredgers cannot achieve the exact theoretical nominal depth when making the final dredging cut, because a wedge of soil 3–5 m wide remains unless the soil is able to slide down. The need to remove this residual wedge depends on the type of fender on the quay wall and on the block coefficients of the ships that will berth there. Only grab dredgers can remove any residual wedge of soil. Additionally, the troughs of the sheet piles must be flushed free of cohesive soil in certain circumstances.

When dredging a port or harbour with a floating equipment, the work is usually divided into dredging cuts of between 2 and 5 m, depending on the type and capacity of the equipment. The intended use of the excavated soil can also be relevant to the choice of dredging plant in the case of varying soil types.

To ensure early detection of any structural deformation, it is advisable to survey the front face of the quay prior to and at intervals throughout the dredging operations, and not only afterwards.

Inspections by divers are required to detect any damage to sheet pile interlocks that might have occurred in wall areas exposed by the dredging (Section 5.4.4).

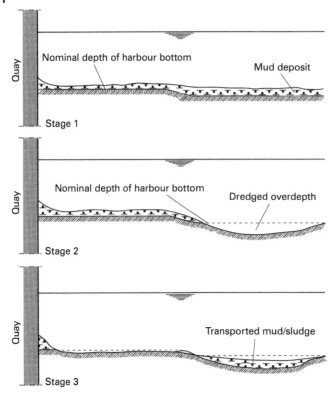

Figure 5.1 Dredging work in front of vertical sheet pile walls in seaports. Stage 1: existing situation; Stage 2: situation after dredging; Stage 3: situation after using harrow (hedgehog) or grab dredger.

An approach such as that depicted in Figure 5.1 can be economical and less disruptive to port operations.

Following overdepth dredging (stage 2), the mud/sludge lying in front of the quay wall is moved into the overdepth area of the harbour bottom with grab dredgers or a harrow (stage 3). Wherever possible, the overdepth area should be created during the new dredging works.

It is essential to check the stability of the quay wall before every dredging operation that might fully exploit the theoretical total depth beneath the nominal depth of the harbour bottom. In addition, the behaviour of the quay wall must be monitored before, during and after dredging.

5.2 Dredging and hydraulic fill tolerances

5.2.1 General

The specified dredging depths and hydraulic fill heights are to be achieved within clearly defined permissible tolerances. If the difference in level between individual points of the base of an excavation or a filled area exceed those tolerances, then supplementary measures

Achim Hettler, Karl-Eugen Kurrer

Earth Pressure

- Collection of working instructions for daily and unusual tasks
- Commentary to DIN 4085
- Technological history of the development of earth pressure theory
- Comprehensive and unique handling of the subject without competition

After describing the development of earth pressure theory, the book concentrates on the current basis for calculations. It offers a collection of working instructions for foundation and structural engineers in construction companies, consultants and in building supervision.

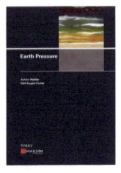

414 pages · 259 figures · 24 tables

Hardcover
ISBN 978-3-433-03223-7 € 89*

eBundle (Print + PDF)
ISBN 978-3-433-03207-4 € 149*

ORDER
+49 (0)30 470 31-236
marketing@ernst-und-sohn.de
www.ernst-und-sohn.de/en/3223

* All book prices inclusive VAT.

will be necessary. Specifying excessively tight tolerances can lead to disproportionately high extra costs for dredging works. Stipulating tolerances for dredging and hydraulic filling work should, therefore, be based on technical but, first and foremost, economic aspects. The client must consider how important it is to achieve a certain accuracy for the base of an excavation or a height of fill.

In addition to deviations from the intended level (vertical tolerances), horizontal tolerances must also be specified when trenches have to be dug for soil replacement measures, inverted siphons and tunnels, for instance. Here, too, it is almost always necessary to reach an optimum balance between the extra costs involved in more extensive dredging and filling quantities with more generous tolerances, and the extra costs due to reduced performance of the equipment because of more precise and hence slower working plus the costs of any potential additional measures.

The achievable accuracy of dredging work for inland waterways is generally better than that in waterways for ocean-going vessels, where tides, waves, shoaling by sand, and/or sludge play a major role. For nautical reasons, a minimum depth is normally required in waterways.

Where dredging work is followed by construction measures such as profiling in front of embankments, installing revetments or concreting bottom slabs, dredging tolerances must be tailored to meet the requirements of subsequent operations.

5.2.2 Dredging tolerances

Dredging tolerances are to be specified taking into account the following factors:

a) Quality demands regarding the accuracy of the depths to be achieved resulting from the objective of the dredging work, e. g.:
   - Regular, recurring maintenance dredging to eliminate all sediments and maintain navigability.
   - The creation or deepening of a berth in front of a quay wall or a navigable channel to improve navigability.
   - The creation of a watercourse bottom to accept a structure (inverted siphon or tunnel structures, bottom protection measures, etc.).
   - Dredging to remove non-load-bearing subsoils within the scope of soil replacement.
   - Dredging to remove contaminated soil sediments.

   Each of these objectives calls for specific and distinctly different dredging accuracies. This affects the choice of plant and so has an influence on the cost of the dredging work.

b) In addition, structural boundary conditions, the extent of dredging and the properties of the soil to be dredged must all be taken into account, e. g.:
   - The stability of nearby underwater slopes, breakwaters, quay walls, etc.
   - The depth below the planned bottom level in which disturbance of the subsoil may be accepted.
   - The horizontal or vertical dimensions of the dredging work, the length and width of the excavation, the thickness of the stratum to be dredged and the overall volume of dredging works.
   - The soil types and properties, particle sizes and distribution and shear strengths of the soils to be dredged.
   - The use of dredged soil.

- Any possible contamination, with specific requirements for dealing with the dredged soil.

This last point in particular is becoming increasingly significant due to the normally extremely high cost of treating and disposing of contaminated soils, and can, therefore, make it necessary to work to very tight tolerances.

c) Local circumstances continue to play a decisive role, influencing the use and control of dredging equipment and the general strategy for the dredging works. Examples of this are:
- The depth of the water.
- Accessibility for the dredging equipment.
- Tide-related changes in water levels with changing currents.
- Any seawater/freshwater changes.
- Weather conditions (wind and current conditions).
- Waves, sea conditions, swell.
- Interruptions to the dredging works by waterborne traffic.
- Constraints due to the proximity of berths or ships at anchor.
- The extent of regularly recurring sediments (sand or silt), possibly even during the dredging works.

d) Finally, the dredging equipment itself and its equipment also play their part in terms of:
- The size and type of equipment as well as its dredging accuracy depending on soil and depth.
- The instrumentation on board the dredging plant (positioning instruments, depth measurement, performance measurement, quality of monitoring and logging technology for the entire dredging process).
- The experience and skills of dredging crews.
- The magnitude of the specific drop in performance of dredging equipment owing to the tolerances that have to be maintained, and the ensuing costs.

It must also be taken into account that the outcome of dredging work and the accuracy achieved is generally checked and recorded by sounding. The actual results determined through sounding are always affected by the accuracy of the sounding equipment itself. Therefore, when specifying dredging tolerances, the resolution of the sounding system employed to check the work must also be considered.

The International Association of Dredging Companies has published figures for dredging accuracy, which are reproduced in Table 5.1 to serve as a guide. The figures should be understood as both positive and negative, i.e. the excavating tolerance for a bucket-ladder dredger excavating in gravel under normal conditions, for instance, is ±20 to ±30 cm, i.e. as much as 60 cm. The surcharges for water depths, currents and unprotected watercourses increase the tolerance range in both directions.

The figures are, on the whole, generous. Modern methods for determining positions can enable tighter tolerances, which are especially necessary when only small depths of subsoil have to be excavated. It might be worthwhile asking tenderers not only to quote prices for the accuracy called for in the tender but also request them to quote prices for the dredging tolerances they themselves suggest and can guarantee.

Horizontal tolerances have deliberately been omitted because in the case of slopes these tolerances depend on the vertical tolerance due to the required angle of the slope (Sec-

Table 5.1 Guide values for vertical dredging tolerances in cm (SBRCURnet 2014).

Dredger	Non-cohesive soils			Cohesive soils		Surcharge for		
	Sand	Gravel	Rock	Silt/mud	Clay	water depth 10–20 m	current 0.5– 1.0 m/s	Unprotected watercourse
Cable grab dredger	40–50	40–50	—	30–45	50–60	10	10	20
Bucket-ladder dredger	20–30	20–30	—	20–30	20–30	5	10	10
Cutter-suction dredger	30–40	30–40	40–50	25–40	30–40	5	10	10
Cutter-wheel suction dredger	30–40	30–40	40–50	25–40	30–40	5	10	10
Suction dredger with reduced turbidity	10–20	—	—	10–20	—	5	5	—
Backhoe dredger	25–50	25–50	40–60	20–40	35–50	10	10	10
Hopper suction dredger	40–50	40–50	—	30–40	50–60	10	10	10

These figures should be understood as both positive and negative, e. g. 40 means +40 and −40.

tion 5.7). Furthermore, the horizontal tolerances should be specified for each individual case in conjunction with the specific requirements and the equipment available for determining the position of the dredger. Soundings should be carried out with instruments that measure the actual depth to the bottom and do not give a false reading to the top of an overlying suspended layer.

5.3 Hydraulic filling of port areas for planned waterfront structures

5.3.1 General

Section 5.4 covers the direct backfilling of waterfront structures.

Port areas with a good bearing capacity behind waterfront structures can be created by hydraulic filling, provided well-graded sand is available. In hydraulic filling over water, a greater degree of compaction is achieved without additional measures than when filling underwater for the same conditions (Section 5.7).

In all hydraulic filling work, but especially in tidal areas, measures must be taken to ensure good run-off of the filling water, as well as the water flowing in with the tide.

The fill sand should contain as few fine particles as possible (silt and clay < 0.06 mm). The allowable volume of fine particles in any specific case depends not only on the load-bearing capacity of the intended waterfront structure of the planned port area, but also on the magnitude of the anticipated residual settlement or when port operations are due to start.

The upper 2 m of hydraulic fill must be easy to compact and exhibit a load-bearing capacity suitable for setting up manoeuvring and storage areas.

If the port area will include settlement-sensitive facilities, silt and clay deposits must be avoided and the proportion of fine particles in the hydraulic fill in load-bearing areas should be < 10%. For economic reasons, it is frequently necessary to obtain the fill material locally or to use material obtained, for example, from dredging work in the port itself. The dredged soil is often loosened with cutter-suction or plain suction dredgers and pumped via pipelines directly to the planned port area.

In such cases, meaningful soil investigations beforehand at the source are indispensable to establish the types of soil and their properties, especially the proportion of fine particles. Continuous sampling of the sand obtained using core samples is expedient because these also reveal thin inclusions of cohesive material. Employed in conjunction with supplementary cone penetration tests, these investigations provide worthwhile knowledge of the proportion of fine particles in the sand intended for the fill and how it is distributed vertically and horizontally over the source site.

If the fill sand is pumped via pipelines directly into the port area from the area from which it is being sourced, the content of fine particles is completely restored at the filling site, possibly even concentrated in layers around the outlets. Where hopper suction dredging is used or the fill sand is transported to the filling site in barges, some of the fine particles are washed out with the overflow. In this respect, this type of backfill for sand deposits with a larger silt and clay content is more advantageous than direct pumping. However, turbidity in the watercourse resulting from the overflow of fines must be acceptable.

If mud or clay has been deposited locally on the surface of the hydraulic fill, e.g. around the outlet, then this material must be removed to a depth of 1.5–2.0 m below the future ground level and replaced by sand (see also Section 5.5).

Silt or clay deposited in layers in the hydraulic fill area prevents rapid run-off of the hydraulic fill water, which means that the sand fill is compacted to a lesser degree than when the water can drain away quickly. Appropriate maintenance of the hydraulic fill area and controlling the flow of hydraulic fill material can ensure that silt or clay is removed from the filling area.

If cohesive inclusions still cannot be prevented, the consolidation of these layers can be promoted and accelerated by using vertical drains, for example (Section 5.11).

Embankments of hydraulic fill material with a medium sand have slopes of 1 : 3 to 1 : 4 above the water table, in some cases up to 1 : 2 in depths of 2 m or more below the water table. Such embankments are unstable, and currents can flatten the slopes.

5.3.2 Hydraulic filling of port above the water table

Hydraulic filling of port areas in the dry is shown schematically in Figure 5.2. Clearly defining the width and length of the area of fill and the positions of the outlets can facilitate the removal of fine particles with the hydraulic fill flow.

The width and length of the filling area and the outlets must be specified in such a way that the water carrying suspended solids and fine particles can drain away as quickly as possible and, in particular, that no eddies ensue. In addition, hydraulic filling operations must continue without interruption if at all possible.

Where interruptions cannot be avoided (e.g. weekends), checks must be carried out after every interruption to establish whether layers of fine particles have settled anywhere. Any such layers must be removed before filling resumes.

If the water containing suspended particles and fine material is to be returned to the watercourse, settlement basins may need to be included to ensure compliance with the corre-

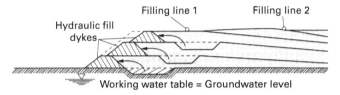

Figure 5.2 Hydraulic filling of port areas above the water table (schematic).

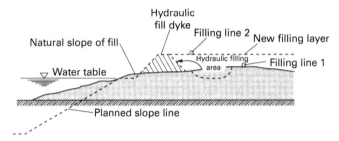

Figure 5.3 Hydraulic filling of port areas on a surface below the water table.

sponding regulations of the authorities with regard to water turbidity and the introduction of suspended particles. Sediments separated off in this way must be disposed of separately

When the hydraulic fill dyke is to form the subsequent boundary to a port or harbour, e. g. the waterfront to a watercourse, it is recommended to construct this as a sand dyke with a covering of plastic sheeting. In order that sand with the coarsest possible grain size settles in the area of the dyke, the filling lines are laid on the hydraulic fill dyke or at the base of the dyke on the backfill side. The coarse sand deposited in front of the hydraulic fill dyke can then be used to raise the dyke further (Figure 5.2). This increase in height must be limited, so that the bearing capacity of the subsoil is not exceeded.

5.3.3 Hydraulic filling of port areas below the water table

5.3.3.1 Filling with coarse-grained sand

Coarse-grained sand can be used as hydraulic fill without any further measures (Figure 5.3). The angle of the natural hydraulic fill slope depends on the coarseness of the fill sand and on the currents. The fill material deposited outside the intended underwater embankment will be dredged away later (Section 5.7).

The sand deposited as fill in the first stage should reach a level of about 0.50 m above the relevant working water level for coarse sand and at least 1.0 m for coarse to medium sand. Above this, work continues between hydraulic fill dykes according to Section 5.3.2. Filling in tidal areas may need to be carried out to suit the tides.

5.3.3.2 Filling with fine-grained sand

Fine-grained fill sand is placed underwater by pumping or dumping between hydraulic fill dykes of rock fill material (Figure 5.4). This method can also be recommended when, for instance, waterborne traffic requirements do not leave sufficient space for a natural hydraulic fill slope.

The fill material used for the hydraulic fill dykes underwater should create a stable filter with respect to the fill sand.

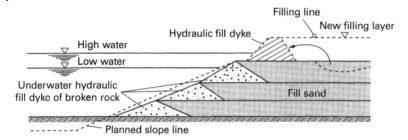

Figure 5.4 Underwater hydraulic fill dyke of broken rock with fine-grained sand deposited by pumping or dumping.

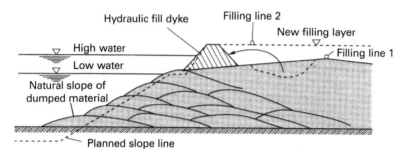

Figure 5.5 Underwater construction of dykes of coarse sand by dumping.

If the hydraulic fill dykes form the boundary embankments to a port or harbour area, they must be able to withstand the effects of currents and waves. If necessary, protection must be provided in the form of revetments. Hydraulic fill dykes of rock fill material are problematic where pile driving works are to be carried out later.

It is also possible to build up the shore in advance with dumped sand (Figure 5.5) and then backfill behind this. The coarsest sand should be used for this method in order to avoid drifting caused by strong currents. Excess dumped sand outside the theoretical underwater slope line is dredged away later (Section 5.7). An alternative to dumped sand is the so-called rainbow method. In this method the filling material is deposited through a jet. This method is only possible if the fill material is not carried away by currents.

5.3.3.3 Hydraulic filling of port areas above soft sediment deposits

If existing idle harbour or port basins are to be filled as part of restructuring work for new uses, it can be economical for existing soft cohesive sediments to be left on the bottom and covered by fill material. For structural reasons, however, it is generally necessary to replace any sediments in front of and behind waterfront structures and embankments.

In the basin, the sediment remains on the bottom and is covered using the hydraulic filling method. The fill sand must be placed in layers, the thickness of which is limited by the bearing capacity of the in situ sediments. Where sediments have very little strength, in order to rule out local deformations and ground failure, the first layer of hydraulic fill material can often be only a few decimetres thick. Only the rainbow method is suitable for placing such thin layers. The equipment customarily available these days enables the placing of layers 10 cm thick (Möbius et al. 2002). Maintaining the permissible thickness of the layers of hydraulic fill material must be constantly monitored in order to avoid the fill sand collapsing into the soft in situ sediments. If efforts to do this fail, large differential settlement can occur later in the finished surface.

The in situ sediments consolidate under the load of the newly deposited fill sand. Thus, the second layer of sand can be correspondingly thicker – to suit the increased strength of the sediments. The required duration of the consolidation process can be estimated beforehand with settlement analyses. Obtaining a reliable figure for the required consolidation time is, however, only possible by using geotechnical measurements during the work (see Section 5.12.5.1).

Hydraulic filling of port areas above soft sediment deposits, therefore, requires careful prior soil mechanics investigations of the strength and consolidation characteristics of the soft in situ sediments, plus extensive monitoring of the building measures by way of measurements during the work, e. g. settlement and the development of excess pore water pressure, as well as shear strength.

Once the hydraulic fill is raised to such a height that the surface is suitable for vehicles, consolidation can be accelerated using vertical drains, possibly also in conjunction with preloading (Section 5.11).

When developing port areas by covering soft sediments below bodies of water, the area is first enclosed with a dyke or pile wall. Within this polder it is then possible to backfill over in situ soft sediments up to the intended level.

5.3.3.4 Camber during hydraulic filling to account for settlement

The amount of camber during hydraulic filling is largely dependent on the accuracy with which the settlement of the subsoil and the settlement and subsidence of the fill material can be predicted. An appropriate estimate of the settlement requires a sufficient number of good quality field and laboratory soil mechanics investigations. It will certainly be necessary, however, to carry out profiling work to establish the target levels. Moreover, the fill tolerances are determined in relation to the depth of the fill.

When greater settlement is expected, the amount of settlement – and the camber derived from that – should be indicated in the tender and taken into account in the hydraulic fill specification. If this is not possible, then separate remuneration for the additional filling volume due to settlement can be agreed upon.

5.4 Backfilling of waterfront structures

5.4.1 General

Where waterfront structures are to be erected and subsequently backfilled in open water, it can be useful to remove soft in situ sediments from areas where pile driving will take place and from regions of active and passive earth pressure prior to driving sheet pile walls and other piles and to replace the material with good, compactable soil. Subsequent backfill can, thus, be supported directly on subsoil with sufficient bearing capacity. This avoids construction stages during which the quay wall is affected by the surcharge due to backfill on top of soft cohesive layers that have not been consolidated for these loads. These measures prevent actions due to active earth pressure in the unconsolidated state, and the full passive earth pressure in front of the wall can be assumed from the beginning. In addition, differential settlement between waterfront structure and backfilling can be minimised.

5.4.2 Backfilling in the dry

Waterfront structures erected in the dry should also be backfilled in the dry wherever possible. The backfill must be placed in horizontal layers in depths to suit the compaction equipment used and then well compacted. Sand or gravel should be used as the backfill material wherever possible. Non-cohesive backfill must have a minimum in situ density $D \geq 0.5$, particularly near the top of the backfill. Otherwise, maintenance of roads, rail tracks, etc. can be expected.

The in situ density of the backfill can be checked with cone penetration tests. In a backfilling of inhomogeneous sand in which the content of fines < 0.06 mm dia. is < 10% by wt., the toe resistance q_c should be $> 6\,MN/m^2$. Following compaction, q_c can reach $> 10\,MN/m^2$ at depths below approximately 0.6 m.

For backfilling in the dry, cohesive soil types such as boulder clay, sandy loam, loamy sand and, in exceptional cases, even firm or silty clays may be suitable. Cohesive backfill must be placed in thin layers and well distributed and compacted to achieve a maximum density and rule out voids in the backfilling. A sufficiently deep layer of sand must be added over a cohesive backfill to protect it from the direct effects of vehicular traffic.

5.4.3 Backfilling underwater

Only sand and gravel or other suitable, non-cohesive soil may be used as fill underwater. A moderate in situ density ($0.3 < D < 0.5$) can usually be achieved if inhomogeneous sand is hydraulically pumped in such a manner that it is deposited without segregation. When using homogenous sand as fill, hydraulic pumping alone can generally achieve a loose in situ density only ($D < 0.3$). Higher in situ densities can only be achieved through additional compaction by means of vibro-flotation. When using vibro-flotation methods in an area directly affecting the quay wall, always ensure that the structure can accommodate the vibrations generated and the pressure from the local liquefied soil.

When the sand for hydraulic fill contains fine particles or mud is to be expected, then backfilling must proceed in such a manner that no continuous sediment horizon is created in the body of soil behind the quay wall. Such horizons can form failure planes with reduced strength in which a higher active earth pressure and/or lower passive earth pressure might have to be assumed.

If during backfilling the upper layers are deposited by hydraulic methods above the water table, the hydraulic filling is to be controlled such that the water can drain away quickly. Otherwise, a higher excess water pressure must be assumed for the quay wall. Any existing drainage system for a waterfront structure must not be used for draining the hydraulic fill water because it could become irreparably clogged and, therefore, ineffective.

As a rule, subsoil settlement beneath a backfill surcharge and settlement of the backfill itself do not abate before dredging works are carried out in front of the waterfront structure and the area becomes operational. Therefore, residual settlement can occur in the first years of port operation.

5.4.4 Additional remarks

Sheet pile interlocks are occasionally damaged during driving (declutching), which results in a considerable flow of water at these points when there is a difference in water pressure. Backfill can then be washed out locally and the watercourse bottom possibly eroded in front of the sheet piling.

Voids ensue in the backfill, which owing to the arching effect can remain stable for some time. However, sooner or later, the voids become visible at the surface in the form of subsidence. Washout at the watercourse bottom can be identified using divers and/or sounding methods. Where declutching results in only a very small opening, subsidence often only occurs after many years of port operation but can then cause considerable risks for persons and property.

To prevent this, immediately after dredging, divers should inspect the structure for driving damage between the water level and the base of the excavation or watercourse bottom.

5.5 In situ density of hydraulically filled non-cohesive soils

5.5.1 General

The bearing capacity of port areas with hydraulic fill is essentially determined by the in situ density and strength of the uppermost 1.5–2.0 m of the hydraulic fill. The in situ density of hydraulically filled ground depends primarily on the following factors:

- The granulometric composition, especially the silt content of hydraulic fill material. To achieve the highest in situ density, it is important to limit the proportion of fine particles <0.06 mm to maximum 10%. This can be guaranteed through, for example, correct barge loading (Section 5.3).
- The type of extraction and further processing of the fill material.
- The shape and setup of the hydraulic fill site.
- The positioning and type of drainage for hydraulic fill water.

During hydraulic filling above water, a higher in situ density is generally achieved without additional measures than is the case under water.

The influence of tides and waves often compacts the hydraulically filled sand within a short time, which, therefore, results in a very high in situ density.

5.5.2 Empirical values for in situ density

Experience has shown that hydraulic filling underwater results in the following in situ densities D:

- Fine sand with different uniformity coefficients and a mean grain size $d_{50} < 0.15$ mm:

    $D = 0.35 - 0.55$

- Medium sand with different uniformity coefficients and a mean grain size $d_{50} = 0.25-0.50$ mm:

    $D = 0.15 - 0.35$

Table 5.2 Use-related in situ density D of non-cohesive soils for port areas.

Type of use	In situ density D	
	Fine sand $d_{50} < 0.15$ mm	Medium sand $d_{50} = 0.25–0.50$ mm
Storage areas	0.35–0.45	0.20–0.35
Traffic areas	0.45–0.55	0.25–0.45
Structure areas	0.55–0.75	0.45–0.65

As the granulometric composition and silt content of the material do not remain constant during hydraulic filling work, the aforementioned empirical values represent only a rough guide to the actual in situ density that can be achieved.

5.5.3 In situ density required for port areas

The upper 1.5–2.0 m of a port area should exhibit the following in situ densities D depending on the respective use and particle sizes in the fill (Table 5.2). Therefore, for the same loads, fine sand always requires a higher in situ density than medium sand.

5.5.4 Checking the in situ density

The in situ density in the upper part of hydraulic fill can be determined with the customary tests for density determination to DIN EN 22475-1, normally using equivalent methods, and by plate bearing tests to DIN 18134 or radiometric penetration sounding equipment. However, these can only measure the in situ density or load-carrying capacity of the upper areas (max. 1 m). At greater depths, the in situ density can be determined through cone or dynamic penetration tests to DIN EN 22476-2 or with a radiometric depth sounder.

The cone penetration test (CPT) is ideal for checking the in situ density of hydraulically filled sands. However, the heavy dynamic penetration test (DPH) is also suitable when, for instance, surfaces are inaccessible for the cone penetration test. The light dynamic penetration test (DPL) represents a further option for depths < 3 m. The values given in Table 5.3 are empirical values for the correlation between the respective test findings and the in situ density. They only apply from about 1.0 m below the application point of the test.

5.6 In situ density of dumped non-cohesive soils

5.6.1 General

This information essentially supplements Sections 5.3–5.5.

The dumping of non-cohesive soils generally leads to a more or less pronounced segregation of the material. The result is that the composition of the backfill changes considerably, particularly as currents can also wash the fine constituents out of the soil. Embankments formed by dumping non-cohesive soils are initially relatively steep. They are not stable, however, and repeated slope failures cause them to become shallower. Such movements al-

Table 5.3 Correlation between in situ density D, toe resistance q_c of cone penetration test and dynamic penetration test for number of blows N_{10} in hydraulically filled sands (empirical values for non-uniform fine sand and uniform medium sand).

		Type of use		
		Storage areas	Traffic areas	Structure areas
In situ density D	Fine sand	0.35–0.45	0.45–0.55	0.55–0.75
	Medium sand	0.20–0.35	0.25–0.45	0.45–0.65
Cone penetration test CPT 15 m q_c [MN/m²]	Fine sand	2–5	5–10	10–15
	Medium sand	3–6	6–10	> 15
Heavy dynamic penetration test DPH, N_{10}	Fine sand	2–5	5–10	10–15
	Medium sand	3–6	6–15	> 15
Light dynamic penetration test DPL, N_{10}	Fine sand	6–15	15–30	30–45
	Medium sand	9–18	18–45	> 45
Light dynamic penetration test DPL-5, N_{10}	Fine sand	4–10	10–20	20–30
	Medium sand	6–12	12–30	> 35

so loosen the soil. Embankments with a slope of 1 : 5 or less are stable over a longer period of time.

Dumped non-cohesive soils can also be further compacted by tide and wave effects.

5.6.2 Influences on the achievable in situ density

The in situ density of dumped non-cohesive soils depends primarily on the following factors:

a) Generally, a non-uniform granulometric composition produces a higher in situ density than a uniform one. The silt content should not exceed 10%.
b) Segregation increases with the depth of the water, especially for non-cohesive soils with a uniformity coefficient $U > 5$. This changes the particle size distribution, with the coarse grained fractions reaching a higher in situ density than the fine grained ones. On the whole, this results in a body of soil with an inhomogeneous in situ density.
c) The greater the flow, the greater the segregation and the more irregular the settlement of the soil.
d) In general, a higher in situ density is achieved with split-hopper barges than with bottom-dump barges.

Owing to the fact that these influences sometimes have opposing effects on the in situ density, the density of dumped non-cohesive soils can vary considerably.

The earth surcharge has only a minor influence on the in situ density of dumped non-cohesive soils. Even with increasing overburden pressure, there is generally scarcely any change to the in situ density of the dumped sand. Therefore, only a loose in situ density should be assumed in dumped sand without additional compaction.

5.7 Dredging underwater slopes

5.7.1 General

In many cases, underwater slopes are constructed as steep as stability requirements will allow. With respect to the long-term stability of dredged underwater slopes, the influences of wave impact and currents are taken into account along with the effects of the dredging work itself. Waterborne traffic must also be evaluated. Experience has shown that slope failures frequently take place during and shortly after the dredging work.

The high costs of restoring failed slopes justify careful soil surveys and soil mechanics investigations in advance as a basis for specifying the intended angle of the dredged embankment and the tender for the dredging work.

In general, the recommendation is to withdraw groundwater by means of wells installed immediately behind the slope and, thus, create a flow gradient towards the fill. This results in the flow gradient during dredging work – which would otherwise be away from the fill – no longer being critical for the stability of the slope.

5.7.2 Dredging underwater slopes in loose sand

Special problems can arise when dredging underwater slopes in loose sand. Effects such as vibrations and local stress changes in the soil due to dredging operations can mobilise large quantities of sand. The sand then temporarily behaves like a heavy liquid, and so this type of slope failure is also referred to as a "flow failure". The latent flow sensitivity of the in situ soil must be investigated prior to dredging in order to initiate countermeasures such as compacting the soil. Failing this, the slope must be kept shallow from the beginning, so that a flow failure is ruled out. However, this is often not possible in practice because even a thin layer of loosely bedded sand in the mass of soil to be dredged can trigger a flow failure.

5.7.3 Dredging equipment

In principle, underwater slopes can be excavated by all common dredgers. The dredger must be selected to suit the operating conditions.

Slopes down to a depth of approximately 30 m can be successfully excavated with large cutter-suction dredgers and cutter-wheel suction dredgers, approximately 35 m with large bucket-ladder dredgers. Backhoe dredgers can currently excavate maximum 20 m deep.

Backhoe dredgers are preferably employed for dredging in heavy soils. Grab dredgers can be used when dredging only small quantities, or when dredging is to be carried out in accordance with Section 5.1.

Plain suction dredgers are only advisable for forming slopes with low requirements regarding accuracy. Their operating mode means that they are apt to cause slope failures. They are, therefore, not generally considered for specifically dredging underwater slopes.

Undercutting must be avoided at all costs when forming underwater slopes.

5.7.4 Execution of dredging work

5.7.4.1 Rough dredging

Preliminary dredging is carried out from above to just below the water level, where, for example, the profile of this part of the slope can be properly formed with a grab dredger. Before dredging the remainder of the underwater slope, dredging is carried out at such a distance from the slope that the dredger can operate as closely as possible to full capacity without causing any risk of slope failure in the future slope.

Indications as to the safe distance to be maintained between the dredger and the planned embankment are gained by observing the in situ soil during the preliminary dredging work.

After concluding the rough dredging work, a strip of soil remains at the top of the underwater slope. This must be removed by a suitable method chosen such that the stability of the slope is not at risk (Figures 5.6 and 5.7).

5.7.4.2 Slope dredging

Dredging to form underwater slopes must be carried out carefully so that slope failures are kept within bounds and under control. The following types of equipment are acceptable. Their suitability in each case must be researched and assessed.

5.7.4.2.1 Bucket-ladder dredger

In the past, bucket-ladder and grab dredgers were employed without exception for both the rough dredging and slope dredging work. Small bucket-ladder dredgers can be employed for dredging from approximately 3 m below the water level.

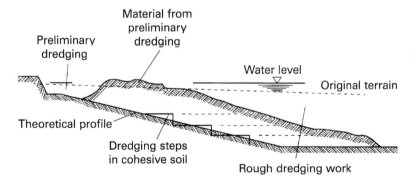

Figure 5.6 Dredging an underwater slope with a bucket-ladder dredger.

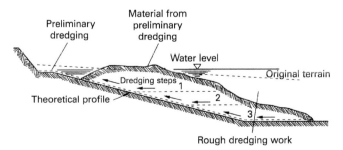

Figure 5.7 Dredging an underwater slope with a cutter-suction or cutter-wheel suction dredger.

For practical reasons, the bucket-ladder dredger operates parallel to the slope, generally dredging layer by layer. Full or semi-automatic control of movements of the dredger ladder is possible and is to be recommended.

The slope is dredged in steps. The type of soil determines the extent to which the steps may intrude into the theoretical slope line (Figure 5.6).

In cohesive soils, the steps are generally dredged in the intended slope line. In non-cohesive soils, however, intruding into the intended slope line is not permitted. The potential removal of the excess soil depends on the tolerances stipulated contingent on the soil conditions and the boundary conditions listed in Section 5.2.

The height of the steps depends on the soil conditions and other factors and is generally between 1.0 and 2.5 m. The planned slope angle, the type of soil and the capabilities and experience of the dredger crew are just some of the factors affecting the precision with which slopes can be formed in this manner.

With slopes of 1 : 3–1 : 4 in cohesive soils, it is possible to achieve an accuracy of ±50 cm measured perpendicularly to the theoretical slope line. The tolerance in non-cohesive soils should be +25 to +75 cm, depending on the dredging depth.

5.7.4.2.2 Cutter-suction or cutter-wheel suction dredgers

For economic reasons, cutter-suction and cutter-wheel suction dredgers are also used these days for forming underwater slopes.

When dredging, the cutter-suction dredger preferably moves parallel to the slope, dredging layer by layer like the bucket-ladder dredger. Computerised control of the dredger and dredger ladder is recommended.

Figure 5.7 shows how the cutter-wheel works upwards, parallel to the theoretical slope line, after having made a horizontal cut. Highly accurate underwater slopes can be produced in this way. In the case of computerised dredger control, tolerances $T_h = +25$ cm measured transversely to the slope can be achieved with small cutter- suction dredgers, +50 cm for larger models. If dredging is carried out without special control, the same tolerances apply as for bucket-ladder dredging, provided that flow failures can be ruled out.

5.8 Subsidence of non-cohesive soils

Subsidence, in a narrower geotechnical sense, is a non-loading-related deformation of a non-cohesive soil that leads to a higher in situ density. Subsidence is caused by the cessation of capillary cohesion (apparent cohesion) due to saturation or dynamic actions. Subsidence also occurs in cemented soils if the bonds between soil aggregates disintegrate, or cohesion breaks down as a result of chemical bonds caused by external influences.

Figure 5.8 shows subsidence as a sudden occurrence of settlement when an earth-moist sand is saturated during a compression test.

Subsidence is especially likely in the first few years following earthworks. It is most likely in soils with a low in situ density; in loosely bedded sands, subsidence can amount to 10% of the stratum thickness. In densely bedded soils, subsidence can reach an order of magnitude of up to 0.5% of the stratum thickness. However, subsidence in densely bedded soils is only triggered by particular actions, e. g. vibrations due to pile driving or other operations.

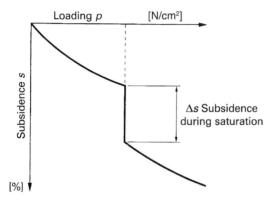

Figure 5.8 Load-settlement diagram of an earth-moist sand during saturation.

In coarse-grained, and, therefore, sufficiently permeable, soil, the possibility of subsidence is reduced by adding water during compaction (the "wet branch" of the Proctor curve). In fine-grained soils, i. e. with low permeability, adding water can prevent compaction.

Subsidence can occur in rock fill when it is dammed for the first time. In this case, subsidence is mainly triggered by weaknesses in the rock structure due to water absorption and the resulting material fracture at very highly stressed contact points.

In general, subsidence is greater in round-grained than in angular-grained soils. Uniform sands show a greater degree of subsidence than non-uniform sands. The difference is, however, only discernible in loose and very loose deposits.

In a broader sense, subsidence can also occur as a result of removing material in deeper strata. This can be caused by hydraulic material transport (suffosion, contact erosion, etc.), damaged areas in the supporting walls, leaching (sinkholes) or organic decomposition processes.

5.9 Soil replacement along a line of piles for a waterfront structure

5.9.1 General

Soil replacement can be useful where soils presenting difficult driving conditions and/or obstacles are present along a line of piles for a waterfront structure. In this case, there is a risk that driving damage cannot be avoided, especially with combined sheet piling. The cost of the soil replacement is compensated for by the greater certainty for design and construction and by the fact that damage is much less likely.

Soil replacement is also helpful where low-strength cohesive soils are encountered in thicker layers along the line of the waterfront structure. In such a case, soil replacement is the prerequisite for guaranteeing the stability of the quay wall during all stages of construction and in its final state.

An economic prerequisite for soil replacement is that the new soil, normally a sand that can be readily installed and compacted, can be obtained inexpensively.

The design basis for soil replacement is a meaningful soil investigation during the preliminary design phase. The results should allow exact determination of the replacement area and enable an optimised plan to be drawn up for the necessary equipment.

To optimise equipment usage, it can be advisable to excavate a trial pit prior to extensive dredging. Moreover, observing the pit allows the slope stability to be assessed under the effect of waves, currents and mud deposits.

5.9.2 Dredging

5.9.2.1 Dredger selection

Only bucket-ladder dredgers, cutter-suction dredgers, cutter-wheel suction dredgers or dipper dredgers can be used for excavating cohesive soil. If soil containing obstacles has to be dredged (e. g. soil interlaced with rubble), the risk when using a suction dredger is that rubble not picked up by the dredger remains on the base of the excavation and forms a layer that is almost impossible to penetrate during later driving work. In such instances, soil must be replaced to such a depth that obstacles brought up by dredging lie beneath the embedment depth of all pile sections.

When dredging cohesive soils underwater, it is impossible to avoid soil from the bucket becoming deposited on the base of the excavation as a layer of sludge. Owing to overfilled buckets, incomplete emptying of overflowing buckets and barges, large deposits of sediment must be reckoned with when excavating with bucket-ladder dredgers in particular (Figure 5.9). The ensuing layer of sediment, possibly in conjunction with sludge from the watercourse, only has a very low strength and should, therefore, be removed prior to filling. To ensure that the layer is removed as thoroughly as possible, a shallower cut must be used when reaching the base of the excavation (Figures 5.9 and 5.10).

Using cutter-suction and cutter-wheel suction dredgers results in an undulating base to the excavation (see Figure 5.10), and the layer of remoulded sediments is thicker than that produced by a bucket-ladder dredger.

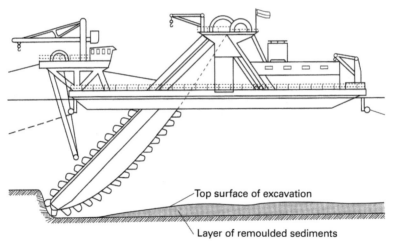

Figure 5.9 Formation of layer of remoulded sediments when dredging with a bucket-ladder dredger.

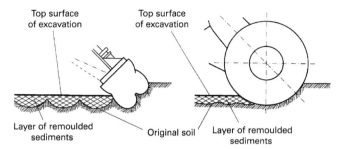

Figure 5.10 Formation of layer of remoulded sediments when excavating with a cutter-suction or cutter-wheel suction dredger.

The thickness of the layer of remoulded sediments can be reduced by using a special cutting head shape, slower rotation, short thrusts and a slow cutting speed.

5.9.2.2 Executing and checking dredging work

Excavation is performed in steps corresponding to the mean profile angle at the edge of the base of the excavation. The height of the steps depends on the type and size of the equipment and the type of soil. Strict control of cut widths must be maintained because cuts that are too wide can cause excessively steep slopes in places, possibly resulting in slope failures.

Proper dredging progress can be monitored by modern surveying methods (e. g. depth sounders in combination with the global positioning system, GPS). It is also possible to detect any profile changes, possibly caused by underwater slope failures, in good time. Merely marking the dredger cutting width solely on the side lines of the dredger is not adequate for checking dredging works. Measurements with inclinometers around the edge of the excavation have also proven successful for monitoring underwater slope failures.

The last sounding should be taken immediately before the sand fill is placed. In order to obtain information on the characteristics of the base of the excavation, soil samples should be taken for testing. A hinged sounding tube (sediment core drill) with a minimum diameter of 100 mm and a gripping attachment (core catcher) has proved effective. Depending on requirements, this tube is driven 0.5–1.0 m or even deeper into the base of the excavation. After extracting and opening the tube, the sample inside permits an assessment of the soil strata at the base of the excavation.

See Section 5.3 for information on the quality of the fill sand.

5.9.3 Cleaning the base of the excavation before filling it with sand

A sediment-free base to the excavation is especially important in the filling area in front of a quay wall because a layer of soft sediment remaining on the base of the excavation could severely reduce the passive earth pressure here. Therefore, any deposits of cohesive soils – possibly the result of layers of mud building up between dredging and sand fill operations – must be removed from the base of the excavation prior to placing the sand fill. However, if a period of several days or longer has elapsed between the finishing of dredging work and the start of mud removal, the deposits may already be so solid that suction is impossible, and another cleaning cut may be necessary. Water injection dredging has also proved to be effective here. The cleanliness of the base of the excavation must be checked regularly.

The sounding tube described in Section 5.9.2.2 can be used for this. A suitably designed grab (hand grab) can be used for taking the samples in soft deposits. A combination of silt soundings at discrete positions and depth soundings (with differing frequencies to ensure reliable detection of the base of the excavation) is a good evaluation option.

If it cannot be guaranteed that no mud deposits remain on the base of the excavation, then it must be ensured that the replacement soil interlocks with the in situ, load-bearing soil beneath the base of the excavation once it has been installed, so that the layer of mud cannot form a continuous sliding surface. For example, the first layer of the fill can be of crushed rock, which displaces the sediment and, therefore, interlocks with the soil beneath the base of the excavation. The thickness of the layer of crushed rock must be chosen such that the sediment can be accommodated in the pores of the crushed rock without the mineral contact between the individual stones being lost. If it can be ensured that there will be no pile driving in nor deepening of the watercourse bottom at a later date, keying between in situ soil and fill sand can be achieved with coarse rubble rather than crushed rock. With non-cohesive soils and a shallow excavation, keying between the fill material and the subsoil can also be achieved by the "dowelling" effect achieved with vibro-flotation with a unit of two to four vibrators.

5.9.4 Placing the sand fill

The sand filling operation must be carried out in such a way that no continuous layers of sludge can build up. In the case of severe deposits, this can be achieved through continuous, efficient operations, which is not even interrupted at weekends. For all other cases, see the details for placing sand underwater given in Section 5.3.3.

The use of water-jetting on the surfaces of layers of sand already deposited during the filling operations has proved to be effective. Any deposits are then turned into suspensions again and layers of mud are prevented from the very outset. Sand losses due to this water-jetting must be taken into account when determining the quantities.

In order to avoid causing active earth pressure on the waterfront structure in excess of the design load, the excavation must be filled in such a way that silted-up slopes occurring during the filling have an inclination opposite to that of the failure plane of the active earth pressure wedge that will later act on the waterfront structure. The same applies similarly to soil replacement on the passive earth pressure side.

For additional information, please refer to PIANC (1980).

5.9.5 Checking the sand fill

Soundings should be taken constantly during the sand filling operation and the results logged. This enables the filling processes itself and the effects of the currents to be checked to a certain extent. At the same time, these records show clearly how long surfaces have remained stable and whether sediments have accumulated.

The taking of samples from the fill area can be dispensed with only when there is fast, uninterrupted hydraulic filling and/or dumping. If sand filling was interrupted, however, the surface must be checked for silt deposits, as described above, before further filling can take place.

After completing the filling work, samples of the new soil must be taken and tested randomly, but systematically, by means of core samples and cone penetration tests. Sampling and testing are to be carried out as far as the soil beneath the base of the excavation.

An acceptance certificate forms the basis for the final design of the waterfront structure and any measures that may be required to adapt the design to the conditions at the site.

5.10 Dynamic compaction of the soil

Effective compaction with heavy weights dropped from a height is primarily suitable for soils with good water permeability. This method can also be used on weakly cohesive and non-saturated cohesive soils because the impacts increase the pressure in the compressible water and air in the pores to such an extent that it tears open the soil structure. The decrease in the pressure in the pores leads to a relatively rapid consolidation rate.

The use of very heavy weights and high drop heights enables even saturated cohesive soils to be compacted with this method.

In order to assess the success of dynamic compaction measures reliably, the soil to be compacted must be examined for its suitability beforehand using soil mechanics methods. It is also necessary to define any areas where dynamic compaction cannot be used; soil replacement is then necessary in such areas.

The drop height required, the size of the weight and the number of compaction passes should be determined in advance in tests. This also permits an assessment of the side-effects of compaction such as noise and the vibration of neighbouring structures. The compaction achieved can be checked using the methods outlined in Section 5.5.

5.11 Vertical drains to accelerate the consolidation of soft cohesive soils

5.11.1 General

Vertical drains accelerate consolidation because, during the consolidation process, they enable radial drainage of the strata instead of drainage just upwards and downwards. The radial drainage path is half the spacing of the drains and, in contrast to upward drainage (= stratum thickness) or upwards and downwards (= half the stratum thickness), it can be influenced. Consolidation settlement (primary settlement) of soft cohesive, relatively impermeable, strata can be considerably speeded up with vertical drains. At the same time, the settlement due to consolidation, especially in soft cohesive strata, is greater with vertical drains than without them. The reason for this is that with shorter drainage paths, the stagnation pore water pressure (at which the consolidation stops) is lower than when using longer drainage paths.

Secondary settlement (creep settlement) is, however, not influenced by vertical drains. Such settlement can be relatively large and continues for some time in soft cohesive and organic soils in particular.

Vertical drains accelerate consolidation primarily in horizontal strata of soft cohesive soils such as marine clay because they exploit the horizontal permeability of these soils. This is also true for stratified soils with alternating permeability (e. g. layers of marine clay and mudflats sand). The strata of low permeability are drained via the adjacent layers of higher permeability, which feed the water into the vertical drains.

5.11.2 Applications

Vertical drains are used where bulk materials, dykes, dams or fill are dumped on soft cohesive soils. The period of consolidation is shortened, and the in situ soil attains the bearing capacity required for the intended use at an earlier date. Vertical drains are also used to stabilise embankments or terraces and when it is necessary to limit lateral flow movements from fill.

There are limits to the use of vertical drains, e. g. when contamination in the soil could be mobilised unacceptably.

5.11.3 Design

When designing a vertical drain system, take the following factors into account:

- The consolidation of a soil requires a surcharge in every case. Vertical drains can speed up consolidation. In an ideal situation, consolidation is finished before the structure goes into operation. However, secondary settlement still takes place, the course of which cannot be influenced by vertical drains, and in any event occurs during the operational phase.
- The application of a surcharge greater than the total of all intended loads can compensate for a part of the secondary settlement. If the surcharge in excess of the intended load is removed again, the soil is overconsolidated.
- The settlement to be expected from consolidation can be estimated using Terzaghi's consolidation theory. Approaches for estimating the consolidation and secondary settlement simultaneously can be found in Koppejan (1948). However, as those methods were developed for very much simplified boundary conditions, the results they provide might have limited applicability within the scope of detailed design work. In any case, settlement should always be calculated for the probable bandwidths of the soil mechanics parameters, so that the inhomogeneity of the in situ soils is also taken into account.
- Owing to the different consolidation theories, the chronological sequence of consolidation cannot be derived for certain; at best, the theories provide rough estimates. However, the input parameters for the calculations can be calibrated in conjunction with settlement measurements.
- Normally, the settlement due to consolidation with vertical drains is greater than that without vertical drains for the same conditions because the stagnation gradient of the consolidation is lower with vertical drains than it is without them.
- If there is a confined water table below the stratum to be consolidated, the drains should terminate about 1 m above the lower stratum. Otherwise, groundwater is forced upwards through the vertical drains.
- The structure of the subsoil must be very carefully investigated beforehand in order to specify the optimum drain spacing. It is primarily the permeability of the in situ soil types that can often only be determined reliably by means of trial pumping.

When setting up vertical drains it is important to avoid contaminating the surface in contact with the soil, which might increase the resistance to water entering the drain to such an extent that drainage of the soil is prevented. Moreover, it must be ensured that mechanical overstresses (e. g. due to a local ground failure) do not limit or even inhibit the effect of the vertical drains.

Vertical drains are covered with a layer of sand or gravel in which the water escaping from the soil is held and which, in turn, discharges the water into an outfall.

Vertical drains suffer from the same settlements as the soil strata in which they are installed. Large settlements can bend plastic drains, which can seriously impair their function. Nowadays, vertical drains are mainly made from plastic.

5.11.4 Design of plastic drains

The aim of the design is to determine the spacing of the drains so that the consolidation is completed within a period specified in the construction schedule (usually less than 2 years), where this is called for. The time t needed to achieve a degree of consolidation U_h is calculated on the basis of the one-dimensional consolidation theory according to Kjellmann (1948) and Barron (1948). Hansbo (1976) simplified Kjellmann's approach by assuming uniform soil deformation and undisturbed soil conditions:

$$t = \frac{D_e^2 \alpha}{8 c_h} \cdot \ln\left(\frac{1}{1 - U_h}\right)$$

where

$$\alpha = \frac{n^2}{n^2 - 1} \cdot \left[\ln(n) - \frac{3}{4} + \frac{1}{n^2} \cdot \left(1 - \frac{1}{4 \cdot n^2}\right)\right]$$

$$n = \frac{D_e}{d_w}$$

t time available for consolidation [s]
D_e diameter of drained soil cylinder [m], area of influence of the drain
c_h horizontal consolidation coefficient [m$_2$/s]
d_w equivalent diameter of vertical drain (drain circumference/π) [m]
U_h average degree of consolidation [–]

With a symmetrical arrangement of the drain starting points and partial overlap of the drained soil cylinders, the distance s between each drain is as follows:

- For equilateral triangle grids: $s = \frac{D_e}{1.05}$.
- For square grids: $s = \frac{D_e}{1.13}$.

For the design, the drain spacing s is chosen first and then D_e is calculated from this. Thereafter, n is calculated as the ratio of the diameter of the drained soil cylinder D_e to the equivalent diameter d_w of the selected drain and the coefficient α. Finally, the duration of consolidation t can be calculated with these values and the target value of the degree of consolidation U_h. If t is too long, the drain spacing s must be reduced, and the calculation repeated.

The design can also be carried out with nomograms and computer programs, which are available from the drain manufacturers. Further calculation approaches can be found in Hansbo (1981) and Sondermann and Kirsch (2009).

Since the required service life of plastic vertical drains is usually less than 2 years, there are no special requirements in terms of durability. Only in the case of large settlements is it necessary to prove that the drains remain functional even with kinks. Guidelines for the assessment of water flow in kinked drains can be found in the Dutch recommendations BRL 1120 (1997).

5.11.5 Installation

Sand drains have diameters of approximately 25–35 cm and are installed using driving, drilling or water-jetting methods. Thin layers of fine cohesive sediments at the contact surface between drain and soil can only be reliably avoided by using driven vertical drains.

Plastic drains are supplied in widths of approximately 10 cm. Their walls are 5–10 mm thick, and their water drainage capacity lies between 0.01 and 0.05 l/s with a ground pressure of 350 kN/m². This capacity is, therefore, relatively large in comparison to the volume to be drained from the soil.

Plastic vertical drains are pressed or vibrated into the ground using special equipment; however, in many cases, a flat working area must be set up for this equipment in advance. The material for this must be at least 0.5 m deep, but an excessively thick working level can make installation of the drains more difficult.

In the event of contaminated subsoil, avoid installation procedures that produce drilling or jetting debris. Moreover, any existing sealing stratum should not be penetrated. In order to gain time for consolidation, it is advisable to install the drains at an early stage and initiate consolidation by means of preloading.

5.12 Consolidation of soft cohesive soils by preloading

5.12.1 General

Frequently, only areas with soft cohesive soil types of inadequate bearing capacity are available for extensions to ports and harbours. However, in many cases, the bearing capacity of these areas can be improved by preloading. At the same time, settlement is pre-empted such that residual settlement due to port operations does not exceed the permissible tolerances.

Apart from preloading by applying fill material, the bearing capacity can also be improved by vacuum consolidation.

5.12.2 Applications

The aim of preloading is to pre-empt the settlement of a soft soil stratum that would otherwise take place during the later utilisation of the area (Figure 5.11). The time needed for this depends on the thickness of the soft strata, their permeability and the magnitude of the preload. However, preloading is only effective when there is sufficient time for the consolidation to take place. It is advisable to use the maximum possible preload and to apply

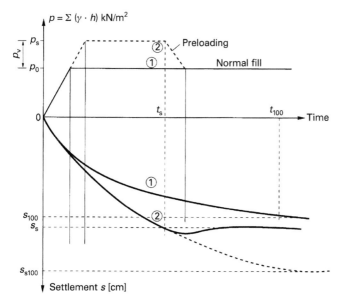

Figure 5.11 Relationship between settlement, time and surcharge (principle).

this at such an early stage that all primary settlement is completed before commencing any construction work for the actual waterfront structure.

Preloading in accordance with Figure 5.11 is divided into:

a) That part of the preload fill constructed in the form of earthworks (permanent filling). It generates the surcharge stress p_0.
b) The excess preload fill, which temporarily acts as an additional surcharge with the preload stress p_v.
c) The sum of both fills (total fill), which results in the overall stress $p_s = p_0 + p_v$.

5.12.3 Bearing capacity of in situ soil

The magnitude of the initial preload is limited by the bearing capacity of the in situ subsoil. The maximum depth of a preload fill can be estimated with

$$h = \frac{4c_u}{\gamma}$$

The shear strength c_u – and, thus, the permissible depth h of further preload fill – increases with the consolidation.

When filling underwater, it must be expected that there is a soft layer of sediment on the bottom of the watercourse. If this remains in the soil, it affects the settlement behaviour of the future port area. Where this is unacceptable, it must be removed (Section 5.9.3).

Occasionally, attempts are made to displace the layer of soft soils by pushing the fill material ahead in one single thick layer. Experience has shown that this method is only successful for very soft sediments. It is much more often the case that the displacement is not completed or is unsuccessful. Sediment layer residue then remains in the subsoil, with the risk of differential settlement.

5.12.4 Fill material

The permanent fill material must constitute a stable filter with respect to the soft subsoil. If applicable, filter layers or geotextiles should be laid before the permanent fill is applied. Otherwise, the required quality of the permanent fill material is governed by the intended use.

5.12.5 Determining the depth of preload fill

5.12.5.1 Soil mechanics principles

The requirements with respect to the preload fill depth are essentially due to the consolidation time available. The dimensioning is based on the consolidation coefficient c_v. If c_v values are derived from the time settlement curves of compression tests, experience shows that only a rough estimate of the consolidation time is possible. Therefore, such values should be used for preliminary appraisals only. This also applies to consolidation coefficients, which can be calculated using

$$c_v = \frac{k \cdot E_s}{\gamma_w}$$

from which the stiffness modulus E_s and the permeability k can be derived. It should be remembered that considerable scatter can affect the permeability coefficient k.

A reliable estimate of the course of settlement is possible, however, when the consolidation coefficient c_v is determined in advance from a trial fill. The settlement and, if possible, the pore water pressure should also be measured. Using Figure 5.12, c_v is then calculated with the following equation:

$$c_v = \frac{H^2 \cdot T_v}{t}$$

where

T_v specific consolidation time [–]
t consolidation time for trial fill [s]
H drainage length of soft soil stratum [m]

For a degree of consolidation $U = 0.95$, i.e. virtually complete consolidation, the specific consolidation time T_v = approximately 1.0. With a consolidation time t_{100} corresponding to the degree of consolidation $U = 0.95$, the consolidation coefficient can, thus, be calculated:

$$c_v = \frac{H^2}{t_{100}}$$

5.12.5.2 Determining the preload fill

The drainage length of the soft soil stratum and the c_v value must be known to determine the preload fill. Additionally, the consolidation time t_S (Figure 5.11) must be specified (construction schedule). The figure $t_S/t_{100} = T_v$ is determined with $t_{100} = H^2/c_v$, and the required degree of consolidation $U = s/s_{100}$ under preloading is determined with the aid of

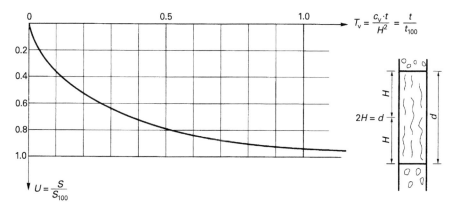

Figure 5.12 Relationship between time factor T_V and degree of consolidation U.

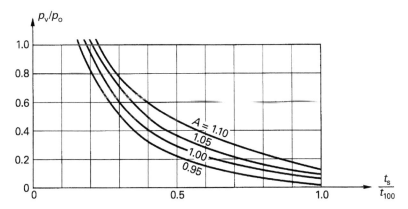

Figure 5.13 Determination of preload p_v as a function of time t_s.

Figure 5.12. The 100% settlement s_{100} of the permanent fill p_0 is determined with the aid of a settlement calculation in accordance with DIN 4019.

The magnitude of the preload p_v (Figure 5.11) is then derived in accordance with Horn (1984) as

$$p_v = p_0 \cdot \left(\frac{A}{U} - 1\right)$$

where A is the ratio of the settlement s_s after removing the preload to the settlement s_{100} of the permanent fill: $A = s_s/s_{100}$. The ratio A must be 1.0 if complete elimination of the settlement is to be achieved (Figure 5.13).

Thus, for example, with a stratum drained on both sides, a depth $d = 2H = 5\,\mathrm{m}$, $c_v = 3\,\mathrm{m}^2/\mathrm{year}$ and a given consolidation time $t_s = 1\,\mathrm{year}$, the time necessary for complete consolidation of this stratum at $t_{100} = (2.5\,\mathrm{m})^2/(3\,\mathrm{m}^2/\mathrm{year}) = 2.08\,\mathrm{years}$. Thus, $t_s/t_{100} = 1\,\mathrm{year}/2.08\,\mathrm{years} = 0.48$. The degree of consolidation as shown in Figure 5.12 after 1 year $U =$ approx. 0.78, i.e. 80% of the consolidation settlement has taken place.

Should the settlement s_s due to the preload be approximately 5% greater than the settlement from the later permanent load s_{100}, then $A = 1.05$, and, therefore, the magnitude of the preload can be estimated as

$$p_v = p_0 \left(\frac{A}{U} - 1\right) = p_0 \left(\frac{1.05}{0.78} - 1\right) = 0.35 p_0$$

As stated above, p_v is limited by the bearing capacity of the in situ soil. Thus, it may be necessary to apply preload fill in several stages.

5.12.6 Minimum extent of preload fill

In order to save preload material, the soil stratum to be stabilised is generally preloaded in sections. The sequence of preloading is governed by the construction schedule. To maximise the even distribution of stress in the subsoil, the area of the preload fill should not be too small. As a guide, the smallest side length of the preload fill area should be two to three times the sum of the depths of the soft stratum and the permanent fill.

5.12.7 Soil improvement through vacuum consolidation with vertical drains

With conventional preloading, the increase in the shear strength is achieved through an additional surcharge (total stresses). In contrast to this, the vacuum method achieves the consolidation by reducing the pore water pressure – the total stresses remain unchanged.

First of all, a layer of sand at least approximately 0.8 m deep must be placed on the soil to be consolidated to serve as a working platform for installing the vertical drains and also to function as a drainage layer. The strength of very soft cohesive soils is usually insufficient for applying the drainage layer in the necessary thickness in one operation.

The water within the drainage layer is collected and drained away with the help of a horizontal drainage system. The drainage layer is covered with a layer of plastic waterproofing material.

Special pumps that can pump, or rather extract, water and air simultaneously are connected to the drainage layer; these generally achieve a maximum vacuum of 75% of atmospheric pressure (approximately 0.75 bar). The pumping capacity $n < 1.0$ denotes the ratio of applied vacuum to atmospheric pressure.

The stresses in the soil result from the atmospheric pressure P_a, the density γ of the moist soil in the drainage layer, the density γ_r of the in situ soil under the drainage layer and the density γ_w of the water, as well as the depth z and thickness h of the drainage layer in the following relationships:

total stress: $\quad \sigma = z \cdot \gamma_r + h \cdot \gamma + P_a$
pore water pressure: $u = z \cdot \gamma_w + P_a$
effective stress: $\quad \sigma' = \sigma - u = z \cdot \gamma' + h \cdot \gamma$

Once the vacuum pumps are in operation, the atmospheric pressure decreases by P_a, which increases the effective stress by this amount to

$$\sigma'_{vacuum} = \sigma' + n \cdot P_a$$

From the consolidation with the additional stress $\Delta\sigma = nP_a$, it follows that the degree of consolidation U_t leads to an increase in the shear strength $\Delta\tau$ amounting to

$$\Delta\tau = U_t \cdot (\tan\varphi' \cdot \Delta\sigma)$$

The final settlement and the progression of the settlement can be calculated using the consolidation theory of Terzaghi and Barron. The shear strength after consolidation corresponds to that due to an equivalent sand surcharge. The temporary increase in strength of non-cohesive soils due to the negative pressure is again lost once the vacuum pump is switched off.

5.12.8 Execution of soil improvement through vacuum consolidation with vertical drains

Firstly, a working platform (thickness > approx. 0.8 m) is installed. Vertical drains (a size equivalent to 5 cm diameter is usual, but flat and round drains can be used in combination) are then installed from here extending down to about 0.5–1.0 m above the underlying non-cohesive strata to prevent hydraulic contact with the strata below reliably. Therefore, in inhomogeneous subsoil conditions, indirect investigations (e. g. penetration tests) during the work are necessary in order to satisfy this condition reliably and establish the depth quickly beforehand.

In subsoil conditions with major variations, test installations without drains can be carried out on a 10 × 10 m grid, for example. The test installations enable the equipment operator to detect increased penetration resistance of intermediate sand strata or the topside of in situ sands.

The success of soil improvement using vacuum consolidation depends on the covering of plastic waterproofing material maintaining a seal during the consolidation. Any defects in this sheeting are difficult to locate and repair. It should, therefore, not be covered with stony, angular material. Flooding with water can be used to protect the sheeting.

In special cases, the in situ soft stratum can itself function as the sealing layer maintaining the vacuum. To do this, the vertical drains are interconnected hydraulically by a horizontal drain approximately 1.0 m below the top of the soft stratum.

Owing to the limited depth of vertical drains, the vacuum process is currently limited to layer thicknesses of approximately 40 m of soil to be improved.

5.12.9 Checking the consolidation

As settlement and degree of consolidation are generally difficult to predict accurately in soil improvement measures involving soft cohesive soils, soil improvement using vacuum consolidation must be monitored through observations of settlement and pore water pressure. The results can be used to calibrate the calculations.

At the edges of the fill, inclinometers can measure whether the bearing capacity of the subsoil has been exceeded.

A limiting value should be specified for the rate of settlement at which the preload fill can be removed, e. g. in mm/day or cm/month.

5.12.10 Secondary settlement

It should be noted that preloading can pre-empt secondary settlement, which is independent of consolidation, to only a very limited extent (in highly plastic clays, for example). If secondary settlement is expected on a fairly large scale, special supplementary investigations are necessary.

5.13 Improving the bearing capacity of soft cohesive soils with vertical elements

5.13.1 General

Grouted or non-grouted ballast or sand columns, supported on low-lying load-bearing strata, are frequently constructed for the foundations to earthworks on soft cohesive soils.

The columns carry the vertical loads and transfer them to the loadbearing subsoil, and in doing so they are supported by the surrounding soil. Therefore, the soil must exhibit a strength $c_u > 15 \text{ kN/m}^2$ at least. Grouted vibro-replacement stone columns only need such support during installation.

Very soft, organic cohesive soils cannot guarantee the necessary degree of support, and, therefore, such foundation systems can be built in this soil type only when the lateral support is achieved by other means, e. g. geotextile jackets.

Ballast and sand columns exhibit drainage properties similar to those of vertical drains (see Section 5.11) and, hence, also increase the shear strength of the in situ soil through consolidation.

5.13.2 Methods

Vibroflotation is a method of improving the subsoil with vertical elements, which has been used and accepted for a long time. Using vibrations, the soil is compacted in the sphere of influence of the vertical element. The resulting loss of volume is compensated for by adding soil. Positioning the compaction points on a triangular or square grid improves the bearing capacity over a wide area. However, the use of this method in non-cohesive, compactable soils is limited. Even just small proportions of silt can prevent compaction by vibration because the fine soil particles cannot be separated from each other by vibration.

Vibro-displacement compaction was developed for these soils. Soil improvement by this method is achieved, on the one hand, by displacing the in situ soil and, on the other, by introducing columns of compacted, coarse-grained material into the in situ soil. However, even vibro-displacement compaction requires the support of the surrounding soil. As a guide to the applicability of this method, the undrained shear strength should generally be $c_u > 15 \text{ kN/m}^2$.

In cases of very soft cohesive soils with $c_u < 15 \text{ kN/m}^2$, a geotextile sleeve must be used to compensate for the lack of lateral support from the soil.

When designing geotextile-encased columns, the following factors must be taken into account:

- The geotextile-encased column is a flexible load-bearing element that can adapt to horizontal deformations.
- Practical experience on site and in the laboratory has shown that there is no punching shear risk in low-strength strata beneath the columns because the settlement of the geotextile-encased column is equal to that of the surrounding soil.

- Residual settlement can be pre-empted by applying an overload greater than the sum of all the later loads (excess fill).
- The geotextile-encased columns are arranged on a uniform grid (usually triangular). The spacing between the columns is approximately 1.5–2.0 m, depending on the grid.
- The calculation of the respective tensile force in the geotextile is based on the compression, or rather settlement, of an individual column as a result of the effective stress concentration above the column and the horizontal support of the segment of surrounding soft soil. This results in a volume-based bulging of the column over the depth, which in turn causes stretching of the geotextile. Further details and methods of calculation can be found in Raithel (1999) and Kempfert (1996).

Measurements of the pore water pressure and stresses in and above the soft strata during and after construction must be carried out to check the true state of consolidation. In terms of consolidation, geotextile-encased columns behave similarly to a large-diameter vertical drain.

The load-bearing elements arranged on a grid are embedded in the low-lying load-bearing soil strata. The load transfer into the load-bearing elements takes place via a layer of sand above the elements. This concentrates the stress over the load-bearing elements and relieves the surrounding cohesive soil.

The load-bearing effect is enhanced by including geotextile reinforcement above the load-bearing elements, spanning across the soft strata like a membrane.

When designing geotextile-reinforced earthworks on pile-type foundation elements, the following boundary conditions and dimensions must be taken into account (see Figure 5.14):

- The diameter of the pile-type load-bearing elements is generally 0.6–0.8 m, and the spacing between these, arranged on a regular grid, is approximately 1.0–2.5 m.
- The geotextile reinforcement is generally placed 0.2–0.5 m above the tops of the columns. Additional layers of reinforcement are placed at a distance of 0.2–0.3 m above the first layer. The sand between the reinforcement layers prevents a failure in the form of one geotextile sliding on another.
- Analysing the stability for the construction phases and the final condition can be carried out according to DIN 4084 using curved failure planes or rigid body failure mechanisms. In doing so, the three-dimensional system should be converted into an equivalent planar system with wall-like plates but maintaining the area ratios. Resistances due to the "truncated" piles and the geotextile reinforcement may be taken into account.
- The allocation of the loads to the pile elements and the surrounding, settlement-sensitive soft soil is expressed by the load redistribution E in the load-bearing layer. The redistribution E is the force F_P that must be carried by one pile related to the area of influence F_{AE}:

$$E = \frac{F_P}{F_{AE}} = 1 - \frac{\sigma_{zo} \cdot (A_E - A_P)}{(\gamma \cdot h) \cdot A_E}$$

The load redistribution E, and, hence, the stress σ_{zo} acting on the soil between the columns, can be determined numerically with the help of a vaulting model in which it is assumed that a redistribution of stress takes place only within a limited zone above the soft stratum (Zaeske 2001).

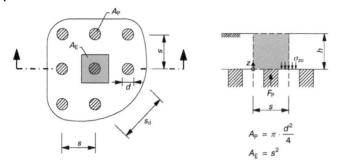

Figure 5.14 Loading on the piles (Zaeske 2001).

The load redistribution in the load-bearing layer is directly dependent on the shear strength of the material of the load-bearing layer. This is valid for:

- Rectangular grid with axis dimensions s_x and s_y:

$$F_p = E \cdot (\gamma \cdot h + p) \cdot s_x \cdot s_y$$

- Triangular grid with axis dimensions s_x and s_y:

$$F_p = E \cdot (\gamma \cdot h + p) \cdot \frac{1}{2} s_x \cdot s_y$$

The membrane effect of the geotextile reinforcement is improved by laying the reinforcement as closely as possible over the almost rigid piles. To prevent shearing of the geotextile, provide a levelling layer over the heads of the piles, so that the material does not rest directly on the piles.

The membrane effect of the geotextile layer can also relieve the soft strata even further. The reader is referred to Zaeske (2001) and Kempfert et al. (1997) for more details of the calculations.

5.13.3 Construction of pile-type load-bearing elements

To construct the pile elements, a working platform with adequate bearing capacity for the necessary equipment is essential. This can be achieved through improving the soil near the surface or by replacing the topmost layer of soil.

Vibro-displacement stone columns are produced with a bottom-feed vibrator. After reaching the design depth or the load-bearing subsoil, the bottom-feed vibrator is raised by a few decimetres, and coarse-grained material is driven out of a chamber in the vibrator by compressed air or water. Afterwards, the vibrator is lowered again, which compacts the material just added. The in situ soil is thus displaced and also compacted. This process is repeated in several stages until a compacted pile element is created from the bottom upwards.

To prevent the pile material escaping into the surrounding soil, vibro-displacement stone columns are only recommended for cohesive soils with an undrained shear strength $c_u \geq 15 \, \text{kN/m}^2$.

Grouted or partly grouted vibrated stone columns make use of the same method as the vibro-displacement stone column. However, the coarse-grained material is mixed with a cement suspension as it is installed; alternatively, the column can be entirely of concrete. The same rule applies here: an undrained shear strength of cohesive strata $c_u > 15\,\text{kN/m}^2$ is recommended. The shear strength may be lower in intermediate strata less than approximately 1.0 m thick but must be $c_u > 8\,\text{kN/m}^2$.

Soil displacement using vibration gives rise to an excess pore water pressure, which in the case of vibro-displacement stone columns is dispersed relatively quickly owing to the drainage effect of the load-bearing elements. This excess pore water pressure is, therefore, not relevant for the design. In grouted or partly grouted vibrated stone columns or concrete columns, the drainage effect is at best severely limited. In these cases, it must be verified that the excess pore water pressure does not affect the load-carrying capacity or serviceability.

Soil displacement or soil excavation within a casing is used to install geotextile-encased columns.

During installation via soil excavation, an open casing down to the level of the load-bearing subsoil is introduced by vibration. The soil is then excavated from within the casing, and a prefabricated geotextile jacket is suspended in the casing and filled with sand or gravel. Afterwards, the casing is extracted while being vibrated, and the geotextile jacket that is filled with sand or gravel remains in the ground.

With the displacement method, flaps at the bottom of the drive pipe are closed, and the geotextile sleeve is driven to its final depth with the pipe, while filling simultaneously. Afterwards, the pipe is extracted.

The prefabricated geotextile jacket is either woven directly as a tube or factory-sewn to form a tube.

The displacement method is more economical than excavating. In addition, soil displacement can already increase the initial shear strength of the soft soil during the installation of the geotextile-encased column. This may need to be checked by measurements before and after installation of the columns.

A brief increase in the excess pore water pressure after constructing the column is to be expected when using soil displacement. However, this is quickly dispersed by the drainage effect of the column. Soil displacement causes a temporary lifting of the top level of the soft stratum.

References

Barron, R.A. (1948). Consolidation of fine-grained soils by drain wells. *Trans. ASCE* 113: Paper No. 2346.

BRL 1120 (1997). *Nationale Beoordelingsrichtlijn, Geokunststoffe: Geprefabriceerde verticale drains*. Rijswijk, Kiwa.

Hansbo, S. (1976). *Consolidation of clay by band-shaped prefabricated drains*. Ground Engineering, Foundation Publications Ltd.

Hansbo, S. (1981). Consolidation of fine-grained soils by prefabricated Drains, *10th Int. Conf. on Soil Mech. & Found. Eng.*, Stockholm.

Horn, A. (1984). Vorbelastung als Mittel zur schnellen Konsolidierung weicher Böden. *Geotechnik* 3: 189.

Kempfert, H.-G. (1996). Embankment foundation on geotextile-coated sand columns in soft ground. *Proc. of 1st Euro. Geosynth. Conf. EurGeo 1*, Maastricht, The Netherlands.

Kempfert, H.-G., Stadel, M. and Zaeske, D. (1997). Berechnung von geokunststoffbewehrten Tragschichten über Pfahlelementen. *Bautechnik* 74 (12): 818–825.

Kjellmann, W. (1948). Accelerating consolidation of fine-grained soils by means of card-board wicks. *2nd Int. Conf. on Soil Mech. & Found. Eng.*

Koppejan, A.W. (1948). A Formular combining the Terzaghi Load-compression relationship and the Buisman secular time effect. *Proc. 2nd Int. Conf. on Soil Mech. & Found. Eng.*

Möbius, W., Wallis, P., Raithel, M., Kempfert, H.-G. and Geduhn, M. (2002). Deichgründung auf geokunststoffummantelten Sandsäulen. *Hansa* 139 (12): 49–53.

PIANC (1980). Report of the 3rd International Wave Commission. Supplement to Bulletin No. 36, Brussels.

Raithel, M. (1999). Zum Trag- und Verformungsverhalten von geokunststoffummantelten Sandsäulen. In: *Schriftenreihe Geotechnik*. Universität Kassel 6.

SBRCURnet (2014). Construction and Survey Accuracies for the execution of rock works. Best practices from the "Maasvlakte 2" Port Expansion Project, Rotterdam, The Netherlands.

Sondermann, W. and Kirsch, K. (2009). Baugrundverbesserung, In: *Grundbau-Taschenbuch*, Kap. 2.1, (ed. by K.J. Witt), https://doi.org/10.1002/9783433600559.ch2.

Zaeske, D. (2001). Zur Wirkungsweise von unbewehrten und bewehrten mineralischen Tragschichten über pfahlartigen Gründungselementen. In: *Schriftenreihe Geotechnik*. Universität Kassel 10.

Standards and Regulations

DIN 4019: Soil – Analysis of settlement.

DIN 4084: Soil – Calculation of embankment failure and overall stability of retaining structures.

DIN 4094-1: Subsoil – Field investigations – Part 1: Cone penetration tests (withdrawn).

DIN 18134: Soil – Testing procedures and testing equipment – Plate load test.

DIN EN ISO 22475-1: Geotechnical investigation and testing – Sampling methods and groundwater measurements – Part 1: Technical principles for the sampling of soil, rock and groundwater (ISO 22475-1:2021).

DIN EN ISO 22476-2: Geotechnical investigation and testing – Field testing – Part 2: Dynamic probing (ISO 22476-2:2005 + Amd 1:2011).

6 Protection and stabilisation structures

6.1 Bank and bottom protection

6.1.1 Embankment stabilisation on inland waterways

6.1.1.1 General

In the presence of dynamic hydraulic loads, soil banks are only permanently stable at very shallow angles (1 : 8–1 : 15). Steeper slopes require armour that can ensure stability against hydraulic loads and adequate overall bank stability.

When choosing the slope angle, the technical benefits of a shallower bank should be compared with the disadvantage of the larger armoured area and the greater amount of land consumed. The construction and maintenance costs should, therefore, not be disproportionate to the economic and ecological benefits.

With greater hydraulic loads (lock entrances/exits, berths), the bottoms of waterways can also be protected by revetments. Where the bottom of a waterway comprises an impervious lining, a suitable revetment will be required to protect it.

The following conditions apply for the design:

- The angle of the embankment stabilisation should be as steep as possible without compromising stability.
- It should be possible to install the armour with mechanical equipment wherever possible.

The following information refers primarily to inland waterways but, in principle, also applies in other situations. Further guidance on the design and construction of revetments was published by the Federal Waterways Engineering & Research Institute (MAR 2008).

6.1.1.2 Loads on inland waterways

The banks of manmade inland waterways (canals) are essentially subjected to hydraulic actions caused by waterborne traffic, and the banks of impounded or free-flowing rivers are subjected to natural currents as well.

The hydraulic loads caused by waterborne traffic can be divided into propeller wash, return current and the resulting drawdown alongside a vessel, a transverse stern wave with slope supply flow at the level of the ship's stern and the secondary wave system. Every action results in a separate load on the waterfront. The return current primarily affects the bank below the still water level, whereas the transverse stern wave and the secondary wave system essentially affect the zone around the still water level. The critical hydraulic actions for designing revetments on inland waterways are generally the transverse stern wave with the slope supply flow and the drawdown. Information about wave loads due to the inflow and outflow of water, as well as waterborne traffic can be found in Sections 4.4 and 4.6.

Waterfront revetments must be designed so that they can withstand the hydraulic shear and flow forces. Which loading component is critical for the design depends on the types of vessel expected (propulsion power and cross-section) and the cross-section of the waterway.

Water level differences arising from waterborne traffic and tides or those occurring naturally also lead to loads on the waterfront. A distinction must be made here between the upward water pressure below the revetment, which increases as the permeability of the revetment decreases, and the hydraulic gradient in the filter layers and the subsoil. With permeable revetments, the open water interacts with the groundwater. With a restricted navigable cross-section, the drop in the water table as a vessel passes takes place faster than the corresponding pressure drop in the pore water but is dependent on the permeability of the subsoil. The result is an excess pore water pressure in the subsoil, the decrease in which in the direction of the surface can be represented with good approximation as an exponential function (Köhler and Schulz 1986). Excess pore water pressure in the soil reduces the shear strength of the soil and, thus, leads to a loss of stability that must be taken into account when designing the bank and its revetment. Soils with low permeability and no or only very little cohesion, e. g. silty fine sands, are particularly at risk.

6.1.1.3 Construction of bank protection

Riprap revetments are the most common form of bank protection. The various components used, which depend on the specific requirements, are – from top (external) to bottom (internal) – as follows (Figure 6.1):

- An armour layer
- A filter/separating layer
- An impervious lining.

In order to create a smooth subgrade, a levelling layer may also be needed below the revetment. Likewise, a cushioning layer can be used to protect individual components of the revetment against loads, e. g. an impervious lining or geotextile filter.

Other parts of a revetment are the toe protection and, where applicable, connections to other elements.

6.1.1.3.1 Armour layer

The armour layer is the uppermost, erosion-resistant layer of bank protection measures and can be either permeable or impermeable. It is designed to resist hydraulic and geotechnical

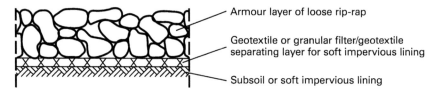

Figure 6.1 Construction of a revetment.

aspects. Currents, wave attack and the drop in the water table must not cause displacement of parts of the construction or the subsoil, and pressure impacts must be absorbed without damage. At the same time, wave energy should be dissipated. The guidelines for scour protection given in Section 6.1.4 also apply, in principle, to armour layers of bank protection.

The armour layer is often in the form of loose armourstones (rip-rap). The armourstones must be stable as well as lightfast and frost and weather-resistant (DIN EN 13383-1). They should also have the highest possible density to ensure their stability over the long term. This type of construction is very flexible and easily adapts to deformations in the subsoil. Rip-rap is relatively simple to install and also easy to repair in the event of damage. The stones are mostly natural stone but can also be made industrially during metal production (slags). For loose armour layers, hydraulic loads would normally cause the stones to move, but this can be kept to a minimum through suitable design. Fundamental research backed up by experiments has led to the development of design approaches for such revetments in recent years (PIANC 1987a,b 1992a; GBB 2010, CIRIA, CUR, CETMEF 2007).

In order to increase the resistance to hydraulic loading, rip-rap can be either partially or fully grouted with a hydraulically bonded mortar. This form of construction is recommended for heavily loaded areas, e.g. manoeuvring areas such as berths or outer basins, in particular. Partial grouting must ensure that the armour layer retains sufficient flexibility and permeability. Ideally, partial grouting produces conglomerates that are securely interlocked with each other and possess the necessary adaptability and erosion resistance of large individual stones. Owing to the high hydraulic resistance of partially grouted armour layers, it is in some cases possible to use much smaller armourstones than would be the case with loose rip-rap.

A similar effect can be achieved using precast concrete blocks laid like paving, provided that the armour layer is sufficiently permeable, and the blocks are linked together. The weight per unit area that can be achieved with concrete blocks is, however, limited.

Impermeable (fully grouted) armour layers act both as lining and protective layer, and exhibit better resistance to erosion and other mechanical damage. In general, they can be thinner than permeable revetments. They are, however, inflexible and ecologically controversial.

6.1.1.3.2 Filter

In principle, any bank protection scheme must be built to prevent materials being washed out from the subsoil (geometrically closed). To this end, filter layers should be placed between the subsoil and the armour layer as required, and must have mutually compatible properties. All layers must constitute a stable filter with respect to the adjacent layer. Where a special clay lining is integrated into the revetment, a geotextile or mineral filter should be laid between the armour layer and the lining as a separating layer without filter function (Figure 6.1).

The filter should be designed according to geohydraulic aspects (i.e. pore water flows and their interaction with the granular structure). Besides granular filters, geotextile filters are also suitable (Section 6.1.3). Water permeability and filter stability (filter and separating functions) are typical design requirements for both types of construction.

Owing to the turbulent inflows and alternating throughflows, both granular and geotextile filters in bank and bottom stabilisation are subjected to high unsteady hydraulic loads in contrast to their use in drainage, where they are exposed to flows in one direction only. The properties of the filters should, therefore, be chosen to suit the subsoil, the armour layer and the loading conditions.

6.1.1.3.3 Impervious lining

Impervious lining is required for reasons of water management or – in the case of embankment dams – to improve stability. In principle, a distinction should be made between watertight revetments (surfacing or full grouting with asphalt or hydraulically bonded mortar) that act as both lining and protection and those revetments with a separate impervious lining such as natural clay, earth composites or geosynthetic clay liners (GCL, bentonite mat). Section 6.1.6 contains information on mineral linings.

Unlike a permeable revetment, an excess water pressure can act on the underside of a watertight revetment, which should be taken into account when designing the revetment. The magnitude of the excess water pressure depends on the magnitude of the changes to the water level on the bank and the simultaneous groundwater levels behind the revetment. The excess water pressure reduces the potential friction force between the revetment and the soil below it.

Asphalt is a viscous material. Roots and rhizomes can, thus, penetrate revetments with asphalt grouting, and the revetment can creep. Revetments made from asphalt or those with full asphalt grouting should be regarded as rigid solutions if the not insignificant ground deformations below the revetment caused by erosion, undermining or subsidence take place more quickly than the creep process in the asphalt, which proceeds very slowly. Detailed guidelines on asphalt linings can be found in EAAW (2008).

It is a requirement for all load cases that the self-weight component of the revetment perpendicular to the bank always be greater than the maximum water pressure arising directly underneath it, so that the lining is never subjected to uplift.

6.1.1.3.4 Levelling layer

Levelling layers are used to create a smooth subgrade where this cannot be achieved through excavation. A levelling layer can be designed as a filter layer where there is very inhomogeneous soil underneath. In this case, the particle distribution of the levelling layer is adjusted to suit the coarser in situ soil and acts as a filter with respect to the finer material. This solution avoids having to construct filters differently in different areas. A levelling layer is also an alternative to soil replacement, e. g. where an embankment cannot be shaped properly because the in situ soil is unstable.

6.1.1.3.5 Cushioning layer

In certain cases, cushioning layers are used for protection against very heavy loads on the underlying layers (e. g. particularly large rip-rap stones on a geotextile filter). A cushioning layer must comply with the filter criteria of the adjacent layers.

6.1.1.4 Toe protection

On steep banks, the downslope forces from the bank protection cannot be fully resisted by the friction between revetment and subsoil. The portion of the downslope forces exceeding the friction must be resisted by toe protection. The requirement to resist the downslope forces entirely through friction would lead to revetments that are too thick or the inclination of the bank is too shallow, both of which are uneconomical solutions.

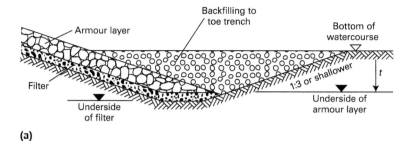

(a)

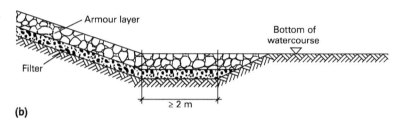

(b)

Figure 6.2 Toe protection by means of (a) an embedded toe or (b) a toe apron (after MAR 2008).

In the case of slopes that continue down to the bottom of the watercourse, the revetment normally continues at the same angle into the subsoil (embedded toe, Figure 6.2a). For subsoil at risk of erosion (sands and non-cohesive sand/silt mixtures), the embedment depth should not be less than 1.5 m (e. g. MAR 2008, GBB 2010). Alternatively, the revetment can also continue horizontally across the bottom (toe apron, Figure 6.2b). However, such a toe apron should not be used if there are soils in the bottom at risk of erosion (sands and finer materials). Where there is a revetment on the bottom as well, the revetment is supported by this.

The toe of a revetment can also be supported by a sheet pile wall at the toe of the bank, provided that the soil is suitable for driving. In terms of its design and construction, this solution corresponds to supporting revetments on partial slopes behind sheet piling (see Sections 7.1.4 and 7.1.5).

6.1.1.5 Junctions

Junctions with structures and covering materials or the subsoil require particular attention. In practice, many cases of damage can be traced back to design and/or construction errors at these junctions. When connecting a revetment to a sheet pile wall or other component, care should be taken to ensure that there is a good transfer of forces and stable filter action, and that the joint is protected against erosion.

It can often make sense to specify full grouting for a strip of the loose rip-rap 0.5–1.0 m wide at the junction with a rigid structure and partial grouting adjacent to this with a decreasing quantity of grout.

The connections between filter layers or impervious linings and structures should also be designed and built with particular care and attention.

6 Protection and stabilisation structures

6.1.1.6 Design of revetments
The following aspects must be considered when designing revetments:

- Stability of individual stones with respect to hydraulic attack.
  The size of stone required is essentially determined by the height of the transverse stern wave and the velocities of the return current and the slope supply flow, as well as – disproportionately – by the density of the stone.
- Sufficient revetment weight to avoid bank failure where there is a rapid drop in water level:
  The weight of revetment required is provided by the thickness of the armour layer. The key influencing factors are the density of the stone, the magnitude and speed of the drop in water level, the type of filter and the permeability of the subsoil.

For inland waterways, the design can be carried out in accordance with GBB (2010). Under certain boundary conditions, standard forms of construction for revetments according to MAR (2008) can be applied without the need for numerical verification. Stones of class $LMB_{5/40}$ with a minimum density of 2650 kg/m³ are recommended for class V waterways with modern waterborne traffic. Typical armour layer thicknesses lie between 60 and 80 cm and depend on the subsoil and the type of filter.

6.1.2 Slopes in seaports and tidal inland ports

6.1.2.1 General
The waterfronts in ports where bulk cargoes are handled, at berths and at port/harbour entrances and turning basins can be constructed as permanently stable sloped banks even where the tidal range is large, and the water level undergoes wide fluctuations. Certain design principles should be followed to avoid the need for more extensive maintenance. Large sea-going vessels generally enter ports under their own power with tug assistance. Embankments can suffer considerable damage from propeller wash during berthing and deberthing. In addition, the propeller wash as well as the bow and stern waves of large tugboats, inland vessels, small sea-going ships and coastal vessels can also strike the slope over a depth of about 6–7 m below the respective water level (Section 6.1.4). Bow and stern thrusters or Azipod propulsion units (azimuth thrusters) cause certain actions that require specially tailored solutions in each case (e.g. ferry ports).

6.1.2.2 Examples of permeable revetments
Figure 6.3 shows a solution employed in Bremen. The transition from the protected to the unprotected area of the bank is in the form of a 3.00 m wide horizontal berm covered with rip-rap. Above this berm, the stone revetment has a gradient of 1:3. A concrete beam, 0.5 m wide × 0.6 m deep, located on the level of the port operations area, serves as the upper boundary to the revetment. The revetment consists of heavy rip-rap dumped in a layer approximately 0.7 m thick just above mean low tide level. On the slope above this, the armour layer consists of an approximately 0.5 m thick layer of rip-rap tightly packed together. During laying, it should be ensured that the stones are properly fixed and that there is sufficient mutual support between stones to prevent individual stones being washed away by waves.

For the maintenance of the revetment and to provide access to berths, a 3.00 m wide maintenance road suitable for heavy vehicles should be included 2.50 m behind the upper con-

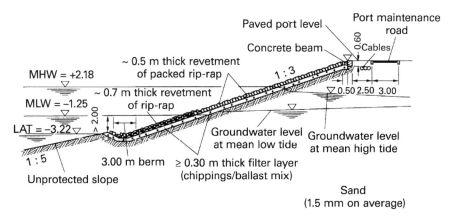

Figure 6.3 Port embankment in Bremen with a permeable revetment (example).

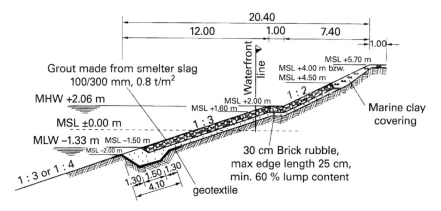

Figure 6.4 Port embankment in Hamburg with a permeable revetment (example).

crete beam (Figure 6.3). Electricity cables for the port facilities and the port's navigation lights, as well as communication cables are laid in a strip between the concrete beam and the maintenance road.

Figure 6.4 shows a section through an embankment in the Port of Hamburg. In this solution, an abutment made from brick rubble, approximately 3.5 m³/m, is built on top of geotextile at the toe of the revetment (see Figure 6.4). Above this, a uniform double-layer revetment covers the majority of the slope.

Taking into account the in situ soil conditions, bank protection generally only continues down to 0.7 m below mean tide low water level. Where washout occurs below this level, this is easily corrected by dumping additional brick rubble. In this form of construction, the rip-rap can be torn away when ice forms, although the cost for supplementing the rip-rap is seen as relatively low in Hamburg.

To assure that banks blend into the natural landscape as much as possible, the banks with rip-rap covering in Hamburg include pockets for planting (Figure 6.5). In the standard cross-section as per Figure 6.4, a horizontal strip approximately 8.00–12.00 m wide, depending on the space available, is included in the region of AMSL +0.4 m. This strip is filled with marine clay 0.4–0.5 m thick to create a zone where vegetation can grow. The

6 Protection and stabilisation structures

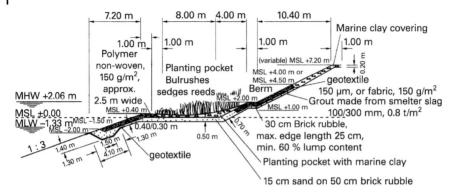

Figure 6.5 Port embankment in Hamburg with a permeable revetment and planting pocket (example).

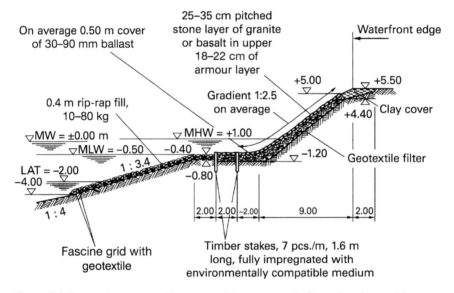

Figure 6.6 Port embankment with a permeable revetment in Rotterdam (example).

brick rubble substructure is thickened to 0.5 m and there is a 0.15 m thick layer of Elbe sand acting as a ventilation zone between the marine clay and the brick rubble.

Bulrushes (*Schoenoplectus tabernaemontani*), sedges (*Carex gracilis*) and reeds (*Phragmitis australis*) are planted in this pocket, staggered according to the steepness of the location. Willow cuttings are used above the berm at AMSL +2.0 m, i. e. also within the standard riprap revetment. Planting (in NW Europe) should take place in April/May due to the better growing conditions.

However, the special shape of the planting pocket can only be used on banks with sufficient space and little surge from waterborne traffic and wave impacts.

Figure 6.6 shows a solution with a permeable revetment in the Port of Rotterdam. Apart from the deck layer itself, its construction is largely similar to the Rotterdam solution with an impermeable revetment. Further design details can, therefore, be found in Section 6.1.2.3. Figure 6.7 shows another solution for a permeable port embankment.

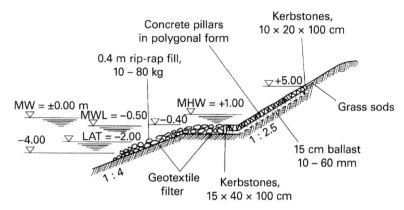

Figure 6.7 Port embankment with a permeable revetment near Rotterdam (example).

6.1.2.3 Examples of impermeable revetments

Figure 6.8 shows a revetment developed and tested in Rotterdam. It has an "open toe" to reduce the excess water pressure. This toe consists of coarse gravel fill $d_{50} \geq 30$ mm secured by two rows of closely spaced timber stakes (2.00 m long × 0.2 m thick) that are fully impregnated with an environmentally compatible medium. At the bottom end of the asphalt-grouted quarry stone covering, the coarse gravel is covered by a permeable layer of large (25–30 cm) granite or basalt stones.

There is a geotextile filter beneath this layer of coarse gravel, which also extends below a significant part of the impermeable armour layer.

There is a 2.0 m wide berm next to the "open toe", below that underwater stabilisation in the form of a timber mattress with a – in sand – 1:4 gradient down to about 3.5 m below mean low tide level. The bundles in the fascine grid are laid horizontally and in the direction of the embankment. On top of that there is a 0.30–0.50 m deep fill of quarry stones because the load from the armour layer is intended to be about 3–5 kN/m^2.

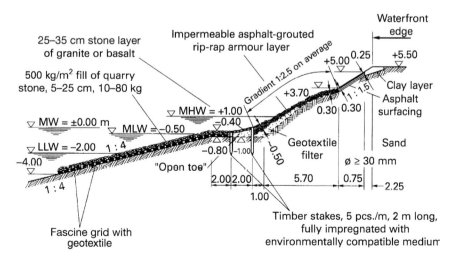

Figure 6.8 Port embankment with impermeable revetment in Rotterdam (example).

The asphalt-grouted quarry stone revetment extends from the land-side row of timber stakes up to about 3.7 m above mean low tide level and has an average gradient of 1 : 2.5. Its weight per unit area (thickness) must be designed for each situation depending on the magnitude of the critical water pressure acting on its underside. In the standard case, the thickness reduces from approximately 0.5 m to approximately 0.3 m from bottom to top.

The individual quarry stones weigh between 10 and 80 kg.

Adjoining the revetment at a level of 1.3 m is a 0.25–0.30 m thick asphalt surfacing layer with a gradient of 1 : 1.5, and above this at the same gradient there is a clay cover at a level of 0.5 m. This is designed to allow pipes and cables to be laid subsequently without having to disturb the bank protection.

6.1.3 Use of geotextile filters in bank and bottom protection

6.1.3.1 General

Geotextiles in the form of wovens, nonwovens and composites are used for bank and bottom protection.

To date, the following polymers have proved to be effective rot-resistant materials for geotextile filters: acrylic, polyamide, polyester, polyvinyl alcohol, polyethylene and polypropylene. Details of the properties of these materials can be found in PIANC (1987a,b).

Where geosynthetics are to be used in bank and bottom protection, their properties must comply with DIN EN 13253. Threshold values for these properties depend on the specific application. Examples of threshold values can be found in PIANC (1992a) and TLG (2003).

The benefit of geotextile filters over mineral filters is the factory prefabrication, which results in very consistent properties. Geotextile filters can also be used underwater as long as certain installation rules and product requirements are met. The geotextile filters themselves weigh very little. A thicker armour layer can, therefore, sometimes be required compared with mineral filters. For non-cohesive fine-grained soils, the actions of waves in the zone between high and low water levels and below can give rise to a risk of liquefaction and a shifting of the soil below the revetment. To prevent this, the filter must meet geometrical filter criteria and the revetment must be sufficiently heavy (MAG 1993).

6.1.3.2 Design principles

Geotextile filters in bank and bottom protection can be designed in terms of the mechanical and hydraulic effectiveness of the filter, the in situ loads, such as punching and tensile forces, and their durability with respect to abrasion in non-bonded armour layers in accordance with the rules set out in PIANC (1987a,b), MAG (1993) and DWA (2017). PIANC (1987a,b) and DWA (2017) contain design rules for unsteady hydraulic loads which are based on past experience with steady loads. MAG (1993) contains design rules based on throughflow tests ("soil-type method") designed for unsteady hydraulic loads. Both methods are essentially based on German domestic experience. International experience and design principles can be found, for example, in Veldhuyzen van Zanten (1994), Koerner (2005) and CFEM (2006).

Besides the mechanical and hydraulic filter effectiveness, geotextile filters must be mainly designed to resist in situ loads. In this respect, relatively thick ($d \geq 4.5$ mm) or heavy ($g \geq 500$ g/m^2) geotextiles have proved effective for installation underwater while waterborne traffic operations continue.

6.1.3.3 Requirements

The tensile strength of the geotextile filters at breaking point should be at least 1200 N/10 cm in the longitudinal and transverse directions when laid wet.

It is necessary to verify the perforation resistance of armour layers made from dumped stone material (RPG 1994).

The abrasive resistance of the geotextile should be verified if abrasive wear movements of the armour layer stones can occur as a result of wave or current loads (RPG 1994).

6.1.3.4 Additional measures

Where required, the properties of the geotextile can be improved through the use of additional measures. Examples of such are set out below. By including coarser additional layers on the underside of the geotextile matched to the subgrade grain sizes, it is possible to achieve an interlock with the subsoil, thus stabilising the boundary layer between subsoil and filter. However, in most cases, it is better to increase the load on the geotextile filter, especially as such additional layers can also give rise to negative effects (e.g. lifting of the geotextile, unintended drainage effect). A fascine grid on the top of a geotextile can prevent creases forming in the geotextile during installation and enhance the stability of the fill material on the geotextile. Such grids have long been used successfully in the construction of fascine mattresses. A combination of woven fabric and needle-punched nonwovens can be used to increase the friction between the woven geotextile and the subsoil and to improve the filter action.

A factory-produced mineral fill consisting of sand or other granulates ("sand mat") increases the weight per unit area of the geotextile, which improves stability during installation. Furthermore, the filling reduces the risk of creases during laying.

6.1.3.5 General installation guidelines

Prior to laying geotextile filters, it is important to verify that the geotextile supplied complies with the contract and the relevant terms of supply, e.g. (RPG 1994) and TLG (2003). Once on site, the geotextile should be carefully stored and protected against UV radiation, the weather and other detrimental effects.

In order to prevent functional defects, care should be taken to ensure that multi-layer geotextile filters (composites) with filter layers graded according to pores are laid the right way up (the top and the bottom are different).

Geotextiles must be laid without wrinkles or folds in order to avoid creating any water channels and, thus, the possibility of soil particle migration.

Nailing to the subsoil at the top of the embankment is only permitted when this does not cause any restraint stresses in the geotextile as construction progresses. A better alternative to rigid fixing by nailing is to embed the geotextile filter in a trench at the top of the embankment. This allows the geotextile to give way in a controlled manner if there are any high loads during subsequent stages of construction. As geotextiles float during wet installation, they must be held in position by installing either the armour layer or a cushioning layer on top directly after being laid. Geotextiles should not be laid at temperatures below +5 °C.

Careful stitching or overlapping when joining together individual sheets of geotextile filter material is particularly important for their soil retention capacity. With stitching, the strength of the stitches should meet the required minimum strength for that particular geotextile. With installation in dry conditions on an embankment at a gradient of 1 : 3 or shallower, the planned overlaps must be at least 0.5 m. In wet conditions and for all steeper embankments, the overlap should be at least 1 m wide. Where the subsoil is soft, a check should be carried out to determine whether larger overlaps are required, so that any movement of the geotextile sheets does not result in uncovered areas when installing rip-rap.

In principle, site stitching and overlapping should always follow the slope of the bank. If in exceptional cases overlapping in the longitudinal direction cannot be avoided, the sheet further down the slope must overlap the sheet further up in order to avoid downslope erosion of the bank through the overlapping.

When laying above water, care should be taken to avoid the relatively light geotextile from becoming displaced by the wind.

The following points should be taken into consideration when installing geotextile filters underwater in areas with waterborne traffic in order to ensure that the geotextile filter is laid on the subgrade with sufficient overlap and without creases, folds, gaps or distortion:

- Warning signs must be set up to indicate the construction site in such a way that all vessels are warned to pass by at slow speed only.
- The subgrade must be carefully prepared and cleared of all stones.
- The installation equipment must be positioned such that laying is not impaired by currents and the drawdown of passing vessels, and so that the geotextile is not subjected to any inadmissible forces (equipment mounted on stilts is preferable for installation).
- The risk of the geotextile sheets floating should be countered by using appropriate installation methods. It is advantageous to press the geotextile onto the subsoil when laying. There should only be a short interval between laying the geotextile and dumping rip-rap, and the dumping height should be kept small.
- Mechanisms for fixing the geotextile sheets to the installation equipment must be released upon dumping rip-rap.
- Installing rip-rap on slopes with geotextiles must proceed from bottom to top.
- Laying underwater is only permitted when the contractor can prove that all requirements can be met.
- Diver inspections are essential.

6.1.4 Scour and protection against scour in front of waterfront structures

6.1.4.1 General

Waterfront structures, especially vertical ones, divert and concentrate currents, which can result in bed material being eroded, a phenomenon known as "scouring". Its causes are essentially twofold:

1. Natural currents carrying material away from the base of the waterfront structure. This occurs, for example, around pierheads at entrances to seaports and inland ports, which are subjected to strong cross-currents, or on the outer banks of river bends, where port facilities are often located due to the greater depth of water.
2. Wash from ship propellers and other manoeuvring aids such as bow thrusters or tugboats. This carries material away from the bed of the waterway in front of the (in most cases) vertical waterfront structure.

Scouring is a process, which means that scour does not appear immediately, but develops over time, as the current carries away a certain amount of subsoil from the bed only gradually. This is particularly important for scour caused by vessel manoeuvres because the flow forces they cause are high but act for a short time only, whereas natural currents, while often relatively small forces, are long-term, ongoing effects.

The two causes of scour can be superimposed. However, it is often also the case that natural sedimentation processes can cause the mooring basins in front of quay facilities to silt up, e.g. in basins with no through-currents. The localised scouring resulting from vessels berthing and deberthing, as well as vessels moored at a berth once again carries off the accretions and thus reduces maintenance requirements.

When designing waterfront structures, it is necessary to resolve the question of whether scouring is likely in that particular case. Experience of other waterfront structures in a similar location can be very helpful when making an assessment. In addition, the following boundary conditions should also be given consideration:

- Features of the particular body of water, e.g. the strength of the natural current along the waterfront structure and the concentration of sediments in the water.
- The type and properties of the in situ soil on the bed. Non-cohesive, fine-grained soils are particularly prone to scouring, whereas cohesive soils with a semi-firm to firm consistency are mostly resistant to erosion and, hence, less prone to scouring.
- The manner in which vessels berth and deberth. How are manoeuvring aids such as tugs or bow thrusters used and also the vessel's main engines? Do strong currents or winds make berthing or deberthing more difficult?
- Type of vessel: Ro-Ro ships and ferries generally always dock at the same place and so cause scouring at the same place. Container quays are subjected to different conditions; vessels varying significantly in size dock at different locations each time, so the effects from different berthing and deberthing procedures are superimposed. This can cause the removal of soil from new scouring to fill older scouring, partially or completely.

The risk of scouring can be counteracted either by deepening the basin or watercourse or by protecting the bottom. Section 4.5 deals with the choice of a deeper design depth (scour allowance).

6.1.4.2 Covering the bottom (scour protection)

To protect the bottom of a waterway against scour, the following measures should be considered:

1. Covering the bottom with loose stone fill.
2. Covering the bottom with a grouted (stable) stone fill.
3. Covering the bottom with a flexible composite system.
4. An underwater concrete bottom (e.g. in ferry berths).
5. Designing the quay wall to deflect wash, including covering the bottom where applicable.

The details of these five approaches are given below:

6 Protection and stabilisation structures

6.1.4.2.1 Covering the bottom with loose stone fill

Loose stone fill (natural stones and waste material such as slag) is one of the mostly commonly used protection systems. The requirements for this solution are as follows:

- Adequate stability against damage caused by propeller wash.
- Proper covering of the subsoil, i.e. two or three layers of stones.
- A filter-type installation, i.e. on a granular or geotextile filter tailored to the particular subsoil (MAG 1993 and MAK 2013).
- Current, and hence erosion-proof connection to the waterfront structure, especially in the case of sheet pile quay walls.

Section 6.1.1 contains further guidance on design.

The thickness of the layer of stones, or the weight of an individual stone, depends on the current loads and/or velocities at the bottom. For currents induced by propeller wash, Section 6.1.4.3 provides a method of calculating the current velocity near the bed. The formula for this is shown below.

The mean stone diameter required for loose stone fill according to Römisch (1994) is as follows:

$$d_{reqd} \geq \frac{v_{bed}^2}{B^2 \cdot g \cdot \Delta'}$$

d_{reqd} mean stone diameter required for protection [m] (armour layer)
v_{bed} bed velocity according to Section 6.1.4.3 [m/s]
B stability coefficient [–] after Römisch (1994)
 = 0.90 for stern thrusters without central rudder
 = 1.25 for stern thrusters with central rudder
 = 1.20 for bow thrusters
g = 9.81 (acceleration due to gravity) [m/s²]
Δ' relative density of bed material underwater [–]
 = $(\rho_s - \rho_0)/\rho_0$
ρ, ρ_0 density of fill material and water respectively [t/m]

The particle diameter d_{50} is normally used for d_{reqd}, sometimes also the particle diameter d_{75} of a stone size class. The stability coefficient B determined experimentally takes into account the different turbulence intensity (erosive effect of current) that can arise due to the various configurations of propellers and rudders.

For bottom velocities > 3 m/s, loose stone fill becomes increasingly uneconomic as the associated mean diameters become greater than about 0.5 m, making the armour layer disproportionately deep. For higher bottom velocities, grouted stone fills, special forms of construction with flexible covers or underwater concrete are necessary.

6.1.4.2.2 Covering the bottom with a grouted (stable) stone fill

A distinction is made between partial and full grouting for grouted stone fills or revetments. With full grouting, all the voids in the stone fill are filled with grout, which results in an armour layer similar to a plain concrete bottom. Normally, the grout is applied in such a way that the tips of the stones still protrude and contribute to dissipating the energy of the current. With partial grouting, only as much grout is added to the stone structure as is needed to fix the individual stones in position; the stone fill still has sufficient permeability to prevent excess water pressure below the armour layer. Section 6.1.1 contains further guidance.

Owing to the interlocking effect, partially grouted stone fills remain stable up to bed velocities of 6–8 m/s (see Römisch 2000). Propeller wash does not usually cause faster current velocities at the bottom.

6.1.4.2.3 Covering the bottom with a flexible composite system

Composite systems are designed to create a planar protective system by coupling together individual elements. An important principle is that the coupling should be flexible enough to adapt well to edge scour and, thus, stabilise it. The following technical coupling forms are known:

- Concrete elements coupled together with ropes or chains.
- Interlocking precast concrete blocks.
- Wire mesh containers filled with quarry stone (stone or gravel mattresses or gabions).
- Mortar-filled geotextile matrasses.
- Geotextile mats with permanently connected concrete blocks.
- Mats made from fabric-reinforced heavy rubber.
- Sandbags or sand-filled geotextile nonwoven bags.

These systems have excellent stabilising properties provided that they are adequately dimensioned. Owing to the wide variety of systems on offer, a universal flow-mechanics design approach is only available for special cases, see Römisch (1993). Dimensions are often, therefore, based on the empirical values of suppliers.

Where the coupling is sufficiently flexible, these systems can themselves stabilise edge scour and, thus, prevent any regressive erosion. The wire mesh of stone mattresses is, however, prone to corrosion, sand abrasion and mechanical damage despite its good stabilising and protective properties regarding edge scour. If the wire mesh is damaged, the gabions lose their mechanical stability. Mattresses or gabions must be joined together with tension-resistant connections.

6.1.4.2.4 Underwater concrete scour protection (e.g. at ferry berths)

An underwater concrete scour protection, which can be constructed to a far higher degree of precision in terms of its thickness compared with stone fill, constitutes very effective erosion protection in certain situations (e.g. ferry berths). Owing to the homogeneous structure of the concrete, the thrust transmitted locally onto the bed by ship propellers is distributed over a wide area, so a bed protected by a carefully cast concrete slab remains stable even with very severe wash actions.

One disadvantage is that differential settlement can fracture the rigid concrete slab. Further, on its own it is not capable of stabilising edge scour, which calls for special solutions. Cut-off walls have proved to be effective around the edges of underwater concrete slabs. Such slabs are cast in thicknesses of 0.3–1.0 m, depending on the flow forces and the installation method.

6 Protection and stabilisation structures

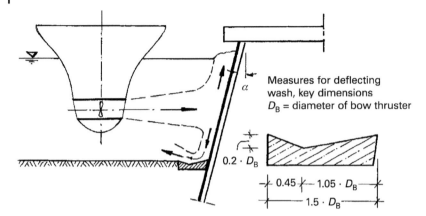

Figure 6.9 Measures for deflecting wash at a quay wall to reduce scouring (Römisch 2001), minimum dimensions.

Underwater concrete should only be installed by a specialist contractor who is able to demonstrate the necessary expertise in this field of work. As the layers are generally relatively thin, it is not possible to place the concrete with a tremie pipe. Instead, it is necessary to use erosion-resistant underwater concrete, which does not segregate as it falls through the water.

6.1.4.2.5 Designing the quay wall to deflect wash, including covering the bottom where applicable

Inclining the sheet pile wall as per Figure 6.9 to create a cushion of water between the front edge of the quay and the side of a vessel, possibly in combination with measures to deflect the wash, can be an effective way of avoiding or minimising loads on the bottom. Such approaches are particularly suitable for reducing scouring due to wash erosion caused by bow or stern thrusters.

Where the measures for deflecting the wash are adequate, e. g. sloping the wall at $\alpha = 10°$ and placing a concrete apron on the bed as per Figure 6.9, the scouring effect of the propeller wash is reduced to such an extent that no further measures are needed to protect the bottom; see Römisch (2001).

6.1.4.3 Current velocity at revetment due to propeller wash

Designing a loose stone fill to act as a revetment protecting against scour from propeller wash requires an estimate of the wash-induced current velocity near the bottom. Certain types of structure also require this approach. A procedure for estimating this velocity is set out below.

6.1.4.3.1 Wash caused by stern thrusters

According to Römisch (1994), the wash velocity caused by a rotating propeller, i. e. the induced wash velocity occurring directly behind the propeller, can be calculated as follows (see Figure 6.10):

$$v_0 = 1.6 \cdot n \cdot D \cdot \sqrt{k_T}$$

where

n rotational speed of propeller [1/s]
D propeller diameter [m]
k_T thrust coefficient of propeller [−], $k_T = 0.25$–0.50

The thrust coefficient takes into account the different types of propeller, the number of propeller blades and their pitch, which all depend on the type of ship. A simplified way of arriving at a mean value for the thrust coefficient is to use the following equation:

$$v_0 = 0.95 \cdot n \cdot D$$

The wash velocity is, therefore, essentially the product of the rotational speed of the propeller n and the propeller diameter D. The rotational speed of the propeller when berthing and deberthing is crucial when designing measures to protect against scour at a waterfront structure. Practical experience indicates that the rotational speed of the propeller when manoeuvring is between 30 and 50% of the rated speed; ship speeds "dead slow ahead" (30%), "slow ahead" (40%) and "half ahead" (50%, for small vessels only) according to Bruderreck et al. (2011). As the required diameter of the stones for a loose stone fill is calculated from the square of the rotational speed (see Section 6.1.4.2), the estimate of this value is very important. The data in the literature varies considerably.

The rated speed of the propeller and its diameter are key design features of a vessel's propulsion. The larger the propeller, the lower its rotational speed must be to avoid cavitation on the tips of the propellers. Table 6.1 lists customary dimensions. It shows that the product of the rated speed and the propeller diameter are relatively constant for a wide range of sizes and types of vessel.

The wash continues to spread out in a conical shape due to turbulent exchange and mixing processes (Figure 6.10), and its velocity decreases as the distance increases.

According to Römisch (1994), the maximum wash velocity found in the region of the bottom – and which is primarily responsible for scouring – can be calculated as follows:

$$\frac{\max v_{\text{bed}}}{v_0} = E \cdot \left(\frac{h_P}{D}\right)^a$$

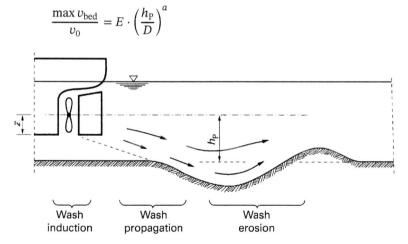

Figure 6.10 Wash caused by stern thrusters.

Table 6.1 Typical values for propeller diameter and rated propeller speed.

Type of vessel	Propeller diameter D [m]	Rated rotational speed n [min^{-1}]	Peripheral propeller speed $n \cdot D$ [m/s]
Container ship			
800 TEU	5.2	135	12
2 500 TEU	7.2	105	13
8 000 TEU	9.2	100	15
19 000 TEU	10.5	79	14
Multi-purpose cargo vessel			
5 000 DWT	3.4	200	11
12 000 DWT	5.2	150	13
25 000 DWT	6.1	120	12
Bulk carrier			
20 000 DWT	4.8	140	12
50 000 DWT	6.3	115	12
75 000 DWT	6.8	105	12
180 000 DWT	8.1	82	11
Tanker			
10 000 DWT	4.4	180	13
20 000 DWT	5.2	140	12
44 000 DWT	6.4	115	12
120 000 DWT	7.8	90	12
300 000 DWT	9.6	75	12

E 0.71 for single-propeller ship with central rudder
0.42 for single-propeller ship without central rudder
0.42 for twin-propeller ship with central rudder, valid for $0.9 < h_P/D < 3.0$
0.52 for twin-propeller ship with two rudders behind the propellers, valid for $0.9 < h_P/D < 3.0$

a −1.00 for single-propeller ships
−0.28 for twin-propeller ships

h_P height of propeller axis above bed [m] (Figure 6.10)
D propeller diameter [m]

6.1.4.3.2 Wash caused by bow thrusters

A bow thruster is a propeller that operates in a pipe placed transversely to the ship's longitudinal axis. It is designed to perform manoeuvres from a standstill and is, therefore, installed near the bow, less commonly near the stern. When using the bow thruster in the vicinity of a quay, the wash it produces strikes the quay wall directly and is deflected in all directions. Crucial for the quay wall is the proportion of the wash directed towards the bed, which can cause scouring directly adjacent to the wall when it strikes the bed (Figure 6.11).

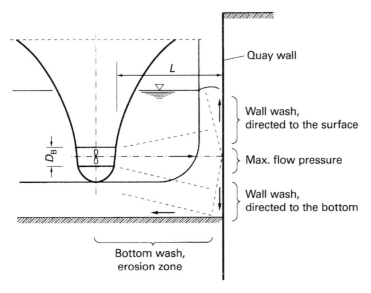

Figure 6.11 Wash loads on watercourse bottom caused by bow thrusters.

According to Römisch (1994), the wash velocity $v_{0,B}$ at the bow thruster exit can be calculated as follows:

$$v_{0,B} = 1.04 \cdot \left(\frac{P_B}{\rho_0 \cdot D_B^2} \right)^{1/3}$$

where

P_B power of bow thruster [kW]
D_B inside diameter of bow thruster opening [m]
ρ_0 density of water [t/m³]

The bow thrusters of large container ships ($P_B = 2500$ kW, $D_B = 3.00$ m) are likely to produce wash velocities of 6.5–7.0 m/s.

The part of the wash velocity responsible for bed erosion max v_{bed} is calculated as follows:

$$\frac{\max v_{bed}}{v_{0,B}} = 2.0 \cdot \left(\frac{L}{D_B} \right)^{-1.0}$$

where L is the distance [m] between the bow thruster opening and the quay wall (Figure 6.11).

A bow or stern thruster is normally operated at full power.

6.1.4.4 Designing scour protection

Determining the dimensions of any scour protection should take account of flow mechanics factors, ensuring that the wash velocities near the edges of the protection are reduced to such an extent that there is no risk of the protection being undermined by edge scour. However, this requirement can lead to the dimensions of the bottom protection being very large, which can increase costs significantly.

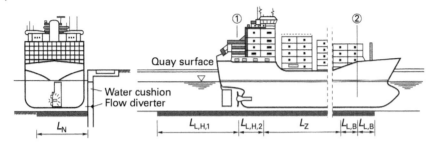

Figure 6.12 Dimensions of protected areas in front of a quay wall: (1) maximum forward position of the stern; (2) maximum rearward position of the bow.

For economic reasons and following the principle that the structure (quay wall or similar) and not the bottom should be protected, the dimensions of the scour protection should ensure that at least the intensive current loading is withstood. Furthermore, the minimum dimensions of the protection should ensure that the area of the structurally effective passive earth pressure wedge at the base of the quay wall is protected against edge scour.

The values shown in Figure 6.12 are recommended as an initial approximation for the minimum dimensions. It should be remembered that about 70–80% of the maximum bottom velocity is still present at the edges of the scour protection when using these dimensions for scour protection measures. For soils on the bottom that are prone to erosion, the protection should be suitably designed along the edges so that it can adapt flexibly to edge scour and, thus, stabilise it.

The recommended minimum dimensions of scour protection for single-propeller ships are as follows:

- Perpendicular to the quay:

$$L_N = (3 \cdots 4 \cdot D) + \Delta EP$$

- Parallel to the quay:

$$L_{L,H,1} = (6 \cdots 8 \cdot D) + \Delta EP$$
$$L_{L,H,2} = 3 \cdot D + \Delta EP$$
$$L_{L,B} = (3 \cdots 4 \cdot D_B) + \Delta EP$$

where

D propeller diameter
ΔEP allowance for edge protection, approximately 3–5 m

For twin-propeller ships, the dimensions given above should be doubled. The total extent of the protection along the quay depends on the anticipated variation in berthing positions. For berths with precisely defined ship positions, the intermediate length L_Z can be left unprotected.

For berths in frequent use, e. g. ferry docks, and for quay structures particularly susceptible to settlement, the extent of the protection should be investigated in greater detail – going beyond the scope of the above recommendation (minimum dimensions) – by analysing the reach of the propeller wash.

6.1.5 Scour protection at piers and dolphins

In principle, scour protection at piers and dolphins should be built like a revetment (Section 6.1.1). The armour layer should withstand the maximum loading and a filter layer should guarantee the long-term stability of the scour protection (preventing erosion of the soil through the voids in the revetment). The installation of scour protection is made more difficult by currents and waves. Solutions that combine the necessary filter function with sufficient weight to withstand the hydraulic loading are, therefore, needed. Geosynthetic containers offer such properties. Geosynthetic containers can also be laid to form a filter layer in those situations where mineral or geotextile filters can no longer be reliably installed because of fast currents. With a suitable choice of size and filling, geosynthetic containers can even be laid at high current velocities. Experiments involving a barrier of stacked geocontainers (three layers up to a height of 1.8 m) in a hydraulic test channel resulted in stability with a maximum flow velocity of approximately 4 m/s and average of 1.5–2 m/s perpendicular to the barrier (Pilarczyk and Zeidler 1996). Adequate stability is also to be expected at significantly higher velocities, where containers are placed over a larger area.

To ensure that the filter layer functions properly, there should be no gaps between the elements. Therefore, two layers of containers are normally installed. Furthermore, the filling should not exceed 80% of the theoretical volume, as fully filled geocontainers cannot adapt to the subsoil, structures or adjacent geocontainers; and with lower fill rates, the geotextile can become displaced by oscillating motions (flapping) caused by the current, which can lead to fatigue failure of the geotextile material.

Owing to their good extensibility, there is a very low risk of geotextiles sustaining mechanical damage during installation. The nonwoven containment is very flexible, and so the container can absorb the high impact loads that can occur when the container hits the bottom or when rip-rap is dumped on top. A minimum weight per unit area of 500 g/m^2 for the nonwoven geotextile and a minimum tensile strength of 25 kN/m are recommended to ensure sufficient robustness during installation and usage. As the friction angle of a nonwoven geotextile is greater than that of a woven fabric, geotextile containers with nonwoven containment are also suitable for protecting relatively steep embankments.

An armour layer of rip-rap (loose or partially grouted) is normally laid on top of a filter layer of geotextile containers. However, the containers themselves can also be used as a permanent armour layer. Ultraviolet radiation limits the design life of geotextile containers above water unless additional measures are taken. The UV radiation is of only minor importance underwater, so containers can be used, for example, as permanent protection against scouring around bridge piers and dolphins. Sufficient resistance to abrasion is required, which can be verified, for example, by corresponding RPG tests (RPG 1994).

6.1.6 Installation of mineral impervious linings underwater and their connection to waterfront structures

6.1.6.1 Concept

A mineral impervious lining consists of a natural, fine-grained soil whose composition or pretreatment gives it a very low permeability without the need for additional materials to achieve the sealing effect, or which achieves the necessary properties through suitable additives. (Impervious linings in the form of fully grouted rip-rap revetments are covered in Section 6.1.1.3.)

6.1.6.2 Installation in dry conditions

Mineral linings installed in dry conditions are covered in detail in DWA (2012). Geosynthetic clay liners are dealt with in EAG-GTD (2002) and EAO (2002).

6.1.6.3 Installation in wet conditions

6.1.6.3.1 General

When deepening or extending lined basins or waterways, impervious linings often have to be installed underwater, sometimes with waterborne traffic still operating. In this situation, it is inevitable that part of the bed is temporarily without a lining. The resulting effects in terms of the water levels to be assumed and the quality of the groundwater should be considered during the detailed design work. Depending on the installation procedure, the sealing material will have to satisfy certain requirements.

6.1.6.3.2 Requirements

Mineral lining materials installed underwater cannot be compacted mechanically, or at best to a limited degree only. They should, therefore, be homogenised beforehand and installed in a consistency that provides a uniform sealing effect from the outset, ensures that the material used adapts to any unevenness in the subgrade without splitting, can withstand the erosion forces of waterborne traffic during installation and can guarantee that junctions with waterfront structures are sealed, even if these structures deform.

Impervious linings installed on slopes must be strong enough to ensure that the sealing material remains stable on the slope.

When proposing a mineral impervious lining, it should be verified that it has adequate resistance to:

- The risk of the newly installed lining material disintegrating underwater.
- Erosion due to the return currents of waterborne traffic adjusting to the conditions imposed by the building works.
- The lining developing cracks or holes in the form of narrow channels on coarse-grained subsoil (piping).
- Sliding on embankments with a gradient of up to 1 : 3.
- Loads due to dumping filter materials and rip-rap on the lining.

Impervious linings made from natural soils without additives generally meet these requirements provided that the sealing material fulfils the following criteria (geotextile clay liners must be considered as a special case):

- Proportion of sand ($d \geq 0.063$ mm) < 20%.
- Proportion of clay ($d \leq 0.002$ mm) > 30%.
- Permeability $k \leq 10^{-9}$ m/s.
- Undrained shear strength $15 \text{ kN/m}^2 \leq c_u \leq 25 \text{ kN/m}^2$.
- Thickness (for a water depth of 4 m) $d \geq 0.20$ m.

When specifying mixes that include certain additives plus a proportion of cement and that solidify after installation, it is important that the flexibility of the lining in its final state is not compromised. This must be verified through testing, e.g. according to Henne (1989).

Special investigations are required when installing mineral impervious linings at greater depths, in gravelly soil, where the subsoil has large pores, on embankments steeper than 1:3 and when designing the lining material in terms of its self-healing properties at any cracks and the sealing effect at butt joints (Schulz 1987a,b).

See ZTV-W 210 (2015) for suitability and monitoring tests.

Several methods are now available (some patented) for single-layer soft mineral impervious linings (EAO 2002). Installation equipment mounted on stilts is recommended for all methods.

6.1.6.4 Connections

Mineral impervious linings are normally connected to structures via butt joints. The sealing material is generally pressed on with equipment that ensures that the lining adapts to the shape of the joint (e.g. a sheet pile section). An adequate quantity of sealant is applied to the line of the joint beforehand using suitable equipment. As the sealing effect comes from the perpendicular contact stress between lining material and joint (Schulz 1987a,b), the pressing procedure should be carried out with great care.

For lining materials with an undrained shear strength $c_u < 25\,\text{kN/m}^2$, the contact length between a mineral lining and a sheet pile wall or component should be at least 0.5 m, and at least 0.8 m for a higher strength. The shear strength of a mineral lining in the area of the wall connection should not exceed $c_u = 50\,\text{kN/m}^2$. A geosynthetic clay liner is connected via a sealing wedge made from suitable sealing material with a contact length to the lining material of at least 0.8 m.

6.2 Flood defence walls in seaports

6.2.1 General

Flood defence walls protect land against flooding. They require considerably less space than dykes and are, therefore, often used in and around ports and harbours. The particular requirements to be met by such walls are described in the following sections.

6.2.2 Critical water levels

6.2.2.1 Critical water levels for flooding
6.2.2.1.1 Outer water level and nominal level

The nominal level of a flood defence wall is derived from the critical still water level (design water level corresponding to highest expected storm tide level, highest astronomical tide) plus freeboard allowances for the localised sea state effects (waves, Section 4.3.6) and surge where applicable.

Owing to the larger run-up of waves on walls, the top of a flood defence wall is set higher than for dykes, unless brief overflow over the walls is acceptable. In this case, it should, therefore, be ensured that the floodwater causes no scouring behind the wall and can drain away without causing any damage (Section 6.2.7.1).

The following values for the permitted wave overtopping rate are recommended in Table A 4.2.3 of EAK (2002):

- $q_T < 0.5 \, l/(s \, m)$ for flat terrain with stationary traffic.
- $q_T < 5 \, to \, 10 \, l/(s \, m)$ for paved but empty areas.

Where applicable, a different value should be assumed for areas around ports and harbour facilities, depending on the potential for damage.

The wave overflow can be effectively reduced by building an overflow deflector along the top edge of the wall (Hamburg 2013).

6.2.2.1.2 Inner water level

The inner water level (groundwater) should normally be taken as the level of the top of the terrain directly behind the wall (Figure 6.13).

Design situation DS-T (transient) can be used to verify stability for flooding, taking the mean wave height into account. If the maximum wave pressure or a special load according to Section 6.2.5 is considered, verification can be carried out for DS-A (accidental). For extremely rare load combinations, the extreme case can be applied.

6.2.2.2 Critical water levels for low water

6.2.2.2.1 Outer water level

Mean low tide should be considered as the standard low water level for DS-P (permanent).

Exceptionally low outer water levels occurring only once a year should be allocated to DS-T (transient).

The lowest astronomical tide ever measured or lowest outer water level expected in the future should be categorised in DS-A (accidental).

6.2.2.2.2 Inner water level

Generally, the inner water level should be assumed to be the top of the terrain, unless a lower water level can be demonstrated through more precise studies of the flows or can be permanently ensured through construction measures such as drainage. There should nevertheless still be a safety factor ≥ 1.0 for the event of failure of the drainage (extreme case). The critical inner water level for an individual case can also be determined by observing the groundwater levels – presuming detailed knowledge of the local conditions.

6.2.2.2.3 Flood run-off

Flood run-off can give rise to water level differences that correspond to the low water situation (excess pressure on land side) but also lead to a higher loading on the wall, e. g. with a water level above the terrain on the inner side.

6.2.3 Excess water pressure and unit weight of soil

The progression of the excess water pressure coordinates can be calculated with the help of a flow net in accordance with Section 3.3.4. The change in the effective unit weight due to the flowing groundwater can be taken into account according to Section 3.3.4.

If a gap forms between the wall and a less permeable stratum due to deflection of the wall, this stratum is to be considered as fully permeable. Further information on calculations for walls in flowing groundwater can be found in Section 3.3.4.

6.2.4 Minimum embedment depths for flood defence walls

The minimum depth of embedment for a flood defence wall can be derived from the structural calculations and the required verification of safety against slope failure. The reduction in the unit weight as a result of the upward vertical flow through the soil in the passive earth pressure zone must be taken into account (see also Section 3.3.4). The risk to the subsoil and the operations on site due to potential leaks (declutching) should also be considered, and it should be noted that:

- Just one flaw in a flood defence wall can lead to failure of the entire structure.
- It is not possible to carry out a suitability test for the flooding design load case.
- The driving depth allowance, taking into account that an embankment could potentially act in an unfavourable manner, should be determined in accordance with Section 8.2.9.

Values for the flow path in the soil should, therefore, not drop below the following for the flooding load case:

- For homogeneous soils with a relatively permeable soil structure and where a gap has formed due to deflection of the wall: four times the difference between the design water level and the top of the terrain on the land side (regardless of the actual inner water level).
- Stratified soils with permeability differences exceeding two powers of 10: three times the difference between the design water level and the top of the terrain on the land side (regardless of the actual inner water level); horizontal seepage paths due to, for example, settlement beneath a superstructure slab cannot be included.

6.2.5 Special loads on flood defence walls

Apart from the normal imposed loads, it is also necessary to allow for other loads (min. 30 kN) caused by impacts from floating objects (including vessels) during flooding and from the collision of land-based vehicles (Figure 6.13). A reasonable line of action of the load must be specified. However, a much higher impact load should be assumed where the location is vulnerable due to unfavourable current and wind conditions or is easily accessible.

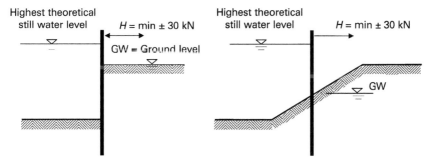

Figure 6.13 Critical water levels for flooding.

Using suitable constructional measures to distribute the loads is permitted provided that the functionality of the flood defence wall is not impaired.

Special loads also include ice pressure. For special loads, verification of stability may be conducted according to DS-A (accidental).

6.2.6 Guidance on designing flood defence walls in slopes

The low water levels of the outer water are generally critical when designing flood defence walls in or in the vicinity of slopes.

The most unfavourable resulting water pressure should be used for the design. In the case of a quay wall in a groundwater flow, it is necessary to consider the effects on water pressure and active and passive earth pressures (Section 6.2.3). The different water levels often also lead to less stability against slope failure. It is, therefore, recommended that when determining the seepage flow, only half of the horizontal seepage path be assumed.

In stratified subsoils (cohesive intermediate strata) that are not secured with sufficiently long sheet piling, then in addition to the normal analysis of slope failure, the stability of the wedge of soil in front of the flood defence wall should also be checked (slip circle failure, safety against sliding).

The outer slope should be protected against scouring using rocky material or similar measures. Stability against slope failure should be verified for DS-T (transient) at least in accordance with DIN 4084. Such slopes should be inspected regularly.

6.2.7 Structural measures

6.2.7.1 Surface protection on the land side of a flood defence wall

In order to avoid land-side scouring caused by overflows during flooding, the surface should be protected. The width of the protection should be at least equal to the exposed land-side height of the wall.

6.2.7.2 Flood defence road

It is recommended that an asphalted road be built near the flood defence wall. It should be a minimum of 2.50 m wide and at the same time protect the surface in accordance with Section 6.2.7.1.

6.2.7.3 Pressure relief filter

A 0.3–0.5 m wide pressure relief filter should be placed directly adjacent to the flood defence wall on the land side to prevent greater water pressure building up beneath the flood defence road.

For sheet piling structures, it is sufficient to fill the land-side troughs with an appropriate filter material (e.g. 35/55 metal slag).

6.2.7.4 Imperviousness of the sheet pile wall

Sections of sheet pile wall projecting above ground level are generally provided with synthetic interlock seals as set out in Section 8.1.3.

6.2.8 Buried utilities in the region of flood defence walls

6.2.8.1 General

For various reasons, buried utilities in the region of flood defence walls can represent weak spots. The main reasons are:

- Leaks from pipes carrying liquids reduce, through washout, the seepage paths in the soil that would otherwise exist.
- Digging trenches to replace damaged/faulty pipes or cables reduces the supporting effect of the passive earth pressure and again shortens seepage paths.
- Decommissioned pipes can leave behind uncontrolled cavities; such pipes should, therefore, be removed, but at the very least filled.

Wherever possible, work on buried pipes or cables should be avoided at times when storm tides are likely. Where this is unavoidable, the construction work should take account of potential flood situations.

6.2.8.2 Buried utilities parallel to a flood defence wall

Pipes or cables parallel to a flood defence wall must not be laid within an adequately wide (> 15 m) safety strip on both sides of the flood defence wall. Existing pipes/cables should be repositioned or decommissioned. Any ensuing voids must be properly backfilled.

Any pipes/cables remaining in the safety strip should be paid particular attention. It must be possible to close off pipes that transport liquids with suitable shut-off valves at the points where they enter and leave the safety strip.

6.2.8.3 Buried utilities crossing a flood defence wall

Pipes or cables passing through a flood defence wall are also potential weak spots and should, therefore, be avoided wherever possible. As such,

- Utilities, especially high-pressure pipes or high-voltage cables, should, therefore, be routed over the flood defence wall wherever possible.
- Individual pipes/cables located in the subsoil outside the safety strip should be combined and routed through the safety strip and the flood defence wall as a single pipe/cable or bundle of pipes/cables.
- Utilities should cross the wall at 90° wherever possible.

The different settlement behaviour of buried utilities and flood defence walls should be taken into account by way of constructional measures (flexible penetrations, articulated pipe joints). Rigid penetrations are not permitted.

The design of pipe/cable intersections depends on the type of buried utility and should take account of any relevant regulations.

6.3 Rouble mound moles and breakwaters

6.3.1 General

Moles differ from breakwaters primarily in the way they are used; it is possible to drive or at least walk along moles. They are, therefore, generally higher than breakwaters, which in some cases do not even project above still water level. Furthermore, breakwaters are not always connected to the land.

Besides careful determination of the wind and wave conditions, currents and any potential sand drift, it is essential to have accurate information about the subsoil when constructing moles and breakwaters.

The positions and cross-sections of large, rouble mound breakwaters are determined not just by their intended purpose, but also by their constructability.

Guidance on geotechnical design can be found in EAK (2002). The design of breakwaters themselves is described in detail in CIRIA, CUR, CETMEF (2007). Below is just a brief overview.

6.3.2 Stability analyses, settlement and subsidence and guidance on construction

Loosely bedded non-cohesive soils below the footprint of an intended mole or breakwater must first be compacted; soils with a low bearing capacity must be replaced.

It is also possible to displace soft cohesive layers by deliberately exceeding their bearing capacity so that the dumped material is embedded in the subsoil. Blasting below the dumped material is also possible so that the subsoil is displaced. However, both procedures result in the finished structure experiencing greater differential settlement because the displacement achieved in this way is never uniform.

Silt strata are displaced by pushing them ahead of the dumped material. The ensuing build-up of silt should be removed because otherwise it can infiltrate the dumped material and have a long-term negative effect on its properties.

In the case of dumped breakwaters, safety against ground and slope failure should be checked. Here, the effect of waves is taken into account with the characteristic value of the design wave. In earthquake zones, the risk of soil liquefaction should be assessed.

The total settlement due to the load of the dumped material, subsidence under the effect of waves and the dumped material becoming embedded in the subsoil, or vice versa, can amount to several metres and must be compensated for by specifying a greater height.

Dumped breakwaters are permeable and are, thus, subjected to throughflows. Where their construction is inhomogeneous, the stability of the filter action of adjacent strata must be guaranteed.

6.3.3 Specifying the geometry of the structure

The key input parameters for determining the cross-section of a breakwater are:

- Design water levels.
- Significant wave heights, wave periods (frequencies), approach direction of waves.
- Subsoil conditions.
- Availability of construction materials.

The crest height is set so that once settlement has ceased, any wave overtopping is kept to a minimum.

The recommendation is to calculate the overall height of a breakwater as follows:

$$R_c = 1.2H_S + S$$

where

R_c freeboard height (overall height above still water level) [m]
H_S significant wave height $H_{1/3}$ of design sea state [m]
S total expected final settlement, subsidence and embedment [m]

When $R_c < H_S$, significant wave overtopping can be expected.
When $R_c = 1.5 H_S$, wave overtopping is negligible.
When specifying the still water level, it might be necessary to take into account any higher water levels that can occur, e. g. during storm surges, if they occur simultaneously with the wave events that govern the design.

Wave overtopping q can be calculated as set out by Van der Meer Janssen (1994) or Owen (1980).

The stone size for the armour layer is determined based on tried-and-tested empirical equations, e. g. after Hudson. The approach used by Hudson is described in the following section.

The sizes of blocks that can be obtained economically from quarries are frequently inadequate for the armour layer. Precast concrete blocks can be used instead, e. g. the standard precast blocks listed in Table 6.2 and illustrated in Figure 6.14.

Reference values for the seaward embankment are given in Table 6.2.
The minimum width of the crest is calculated from

$$W_{min} = (3 \text{ to } 4) D_m$$

$$D_m = \sqrt[3]{\frac{W}{\rho_s}} \quad D_m = \sqrt[3]{\frac{W_{50}}{\rho_s}}$$

where

W_{min} minimum width of breakwater crest [m]
D_m mean diameter of individual stone or block in armour layer [m]
W, W_{50}, ρ_s see Section 6.3.4

As fine-grained material is in most cases considerably less expensive than the coarse armour layer material, most breakwaters have the traditional structure as shown in Figure 6.15:

- Core
- Filter layer
- Armour layer.

However, it is possible that the difference in cost of the core and armour layer materials can sometimes vanish in the case of large transport distances, for example. The breakwater can then be constructed from uniform block sizes, especially when being built with offshore equipment.

Installing a toe filter should be given particular consideration where coarse-grained cores are used.

The stones forming mole crests are often covered with concrete to make them suitable for vehicular traffic.

6 Protection and stabilisation structures

Walls on top of dumped breakwaters are very often used to repel wave overtopping and spray, as well as for access on moles. However, a wall constitutes a "foreign body" that reveals the considerable settlement and differential settlement. Tilting of and cracks in such walls are, therefore, not uncommon issues.

6.3.4 Designing the armour layer

With given wave conditions, the stability of the armour layer depends on the size, weight and form of the constructional elements, as well as the gradient of the armour layer.

Following a series of tests over many years, Hudson developed the following equation for the required block weight (SPM 1984, PIANC 1992a, Bruun 1985), which has proved

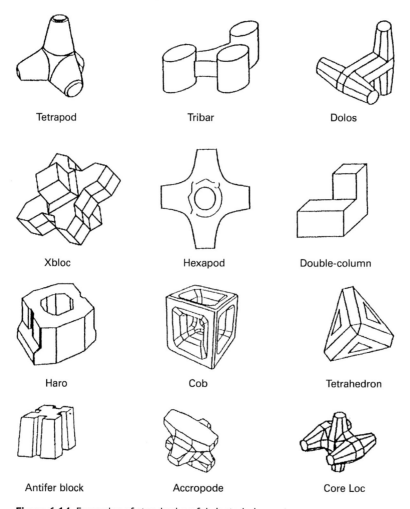

Figure 6.14 Examples of standard prefabricated elements.

Table 6.2 Recommended K_D values for designing the armour layer for a permissible destruction of up to 5 % and only negligible wave overtopping.

Armour layer elements (examples)	Number of layers	Type of arrangement	Breakwater side K_D [a]		Breakwater crest K_D		Gradient
			Breaking waves [e]	Non-breaking waves [e]	Breaking waves	Non-breaking waves	
Smooth, rounded natural stone	2	Random	1.2	2.4	1.1	1.9	1 : 1.5 to 1 : 3
	3	Random	1.6	3.2	1.4	2.3	1 : 1.5 to 1 : 3
Angular quarry stone	2	Random	2.0	4.0	1.9	3.2	1 : 1.5
	3	Random	2.2	4,5	1.6	2.8	1 : 2
	2	Special arrangement [b]	5.8	7.0	2.1	2.3	1 : 3
					1.3	4.2	1 : 1.5 to 1 : 3
					5.3	6.4	1 : 1.5 to 1 : 3
Tetrapod	2	Random	7.0	8.0	5.0	6.0	1 : 1.5
					4.5	5.5	1 : 2
					3.5	4.0	1 : 3
Antifer block	2	Random	8.0	—	—	—	1 : 2
Accropode	1		12.0	15.0	9.5	11.5	Up to 1 : 1.33
Core Loc	1		16.0	16.0	13.0	13.0	Up to 1 : 1.33
Tribar	2	Random	9.0	10.0	8.3	9.0	1 : 1.5
					7.8	8.5	1 : 2
					6.0	6.5	1 : 3
Tribar	1	Arranged uniformly	12.0	15.0	7.5	9.5	1 : 1.5 to 1 : 3
Dolos	2	Random	15.8 [c]	31.8 [c]	8.0	16.0	1 : 2 [d]
					7.0	14.0	1 : 3
Xbloc	1	Random	16	16	13	13	Up to 1 : 1.33

a) For gradients of 1 : 1.5–1 : 5.
b) Longitudinal axis of stones perpendicular to surface.
c) K_D values only confirmed experimentally for a 1 : 2 gradient; for higher requirements (destruction < 2%), K_D values should be halved.
d) Gradients steeper than 1 : 2 are not recommended.
e) Breaking waves appear increasingly where the still water depth is less than the wave height in front of the breakwater.

Source: (extract from SPM 1984)

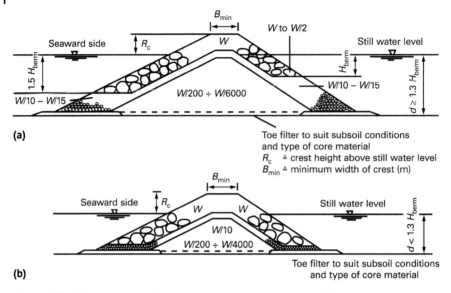

Figure 6.15 Filter structure of breakwater in three gradations: (a) for non-breaking waves and (b) for breaking waves.

effective in practice:

$$W = \frac{\rho_s \cdot H_{des}^3}{K_D \cdot \left(\frac{\rho_s}{\rho_w} - 1\right)^3 \cdot \cot \alpha}$$

where

W	block weight [t]
ρ_s	density of block material [t/m³]
ρ_w	density of water [t/m³]
H_{des}	characteristic height of "design wave" multiplied by partial safety factor [m]
α	gradient of armour layer [°]
K_D	shape and stability coefficient [–]

This equation applies to an armour layer built from stones with a roughly uniform weight. The most common form and stability coefficients K_D for quarry stones and moulded blocks for inclined breakwater armour layers according to SPM (1984) are summarised in Table 6.2. Figure 6.14 shows examples of standard prefabricated elements.

When selecting elements for the armour layer, it should be remembered that, depending on the form of the element, additional tensile, bending, shear and torsion loads can occur with the possible settlement or subsidence movements in accordance with Section 6.3.2. Owing to the high sudden loading, the K_D values should be halved for larger Dolos elements.

According to SPM (1984), the following amended equation is recommended for designing an armour layer of graded natural stone sizes with design wave heights of up to about 1.5 m:

$$W_{50} = \frac{\rho_s \cdot H_{des}^3}{K_{RR} \cdot \left(\frac{\rho_s}{\rho_w} - 1\right)^3 \cdot \cot\alpha}$$

where

W_{50} weight of average-size stone [t]
K_{RR} shape and stability coefficient [–]
 = 2.2 for breaking waves
 = 2.5 for non-breaking waves

Here, the weight of the largest stones should be $3.5W_{50}$ and the smallest $0.22W_{50}$. According to [SPM (1984)], owing to the complex processes involved, the block weight should, in general, not be reduced where waves approach the structure at an angle.

The significance of the design wave for the structure can be seen in the fact that the required weight of the individual blocks W increases to the power of 3 in proportion to the wave height. PIANC (1992a) recommends using $H_{des} = H_s$ as the "design wave" in the Hudson equation for all wave heights. This value can generally be extrapolated with the help of the statistics of extreme values to cover a longer period (e.g. 100-year return). For extrapolation to be reliable, sufficient data on wave measurements must be available (see also Section 4.3).

Partial safety factors are not applied here because the design level of safety is very much dependent on the choice of the return period and the statistics of extreme values. In addition, the structure does not fail if the design sea state is exceeded; instead, damage to the structure gradually increases.

When planning a rouble mound breakwater, economic considerations can lead to different criteria for the lowest possible destruction rate where extreme sea state loads occur only very rarely or at the land end of the breakwater where silting-up occurs on the seaward side to such extent that the armour layer is no longer needed. The more economical options should be chosen if the capitalised repair costs and the likely costs of rectifying any other damage in the port area are lower than the increased capital expenditure when designing the block weight for a particularly high design wave that occurs only rarely. The feasibility of carrying out general maintenance in situ as well as the likely duration of the work should be considered separately in each case.

Further calculation methods are given in PIANC (1973, 1976) and PIANC (1992a). Abromeit (1997) contains fundamental information on how the size, installed thickness and dry bulk density of the rip-rap used influence the stability of a grouted armour layer with respect to current and wave loads as well as suggestions for calculating technically equivalent armour layers.

In addition to the Hudson formula for designing the armour layers of dumped breakwaters, the report by PIANC Working Group 12 of the Permanent Technical Committee II for Coastal and Ocean Waterways (PIANC (1992b) as well as PIANC (1992a)) includes, in particular, the Van der Meer formula.

These equations take into account the breaking form of the waves (plunging and surging breakers), which is calculated based on the height and period (frequency) of the wave according to the Iribarren number. However, they also take into account the duration of the storm, the degree of damage and the porosity of the breakwater. The formulas were derived from model experiments with waves that corresponded to the natural wave spectrum in terms of wave height and wavelength distribution. Hudson (1958, 1959), on the other hand, used only regular waves in his experiments. Van der Meer's method of calculation presumes advance knowledge of many detailed relationships, as can be found in PIANC (1992a). According to Hudson and Van der Meer, the results of calculations for armour layer sizes vary significantly for extreme cases. For large mole or breakwater structures, the recommendation is, therefore, to appoint an approved hydraulic engineering institute to investigate the chosen cross-section as a whole with the help of hydraulic models. Such models can also reveal how a crest wall might influence the overall stability of the breakwater.

6.3.5 Construction of breakwaters

According to the recommendations of SPM (1984), breakwaters in three-layer gradations according to Figure 6.15 have proved effective in practice.

Notation for Figure 6.15:

W weight of individual block [t].
H_{des} height of "design wave" [m].

A single-layer structure made from quarry stones should not be used. The general recommendation for the slope inclination on the seaward side is that it be no steeper than 1 : 1.5.

Particular attention should be paid to the toe support of the armour layer, especially when it does not reach as far down as the base of the seaward slope. Stability requirements might indicate that the slope requires a suitable berm (Figure 6.16).

Filter principles should be observed with respect to the subsoil, too. This can be achieved primarily by including a special filter layer (granular filter, geotextile, sand mat, geotextile containers with filter function) especially underneath block-type outer layers at the base because the installation of these layers is more reliable.

6.3.6 Construction and use of equipment

6.3.6.1 General

The construction of dumped moles and breakwaters often calls for large quantities of materials to be installed in a relatively short time and under difficult local conditions caused by weather, tides, sea states and currents. The mutual dependency of individual operations in

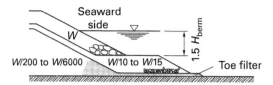

Figure 6.16 Seaward toe stabilisation for a breakwater.

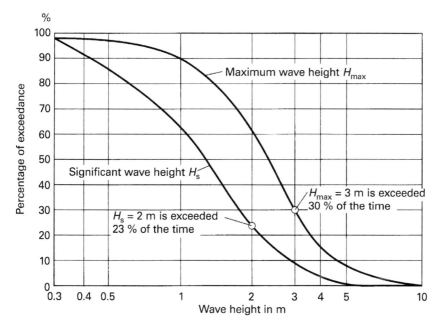

Figure 6.17 Wave height duration curve: period of time over which a particular wave height is exceeded in 1 year, e.g. $H_s = 2$ m; $H_{max} = 3$ m.

such construction conditions requires very careful planning of the construction schedule and the use of the equipment.

The design engineer and the contractor should find out about the wave heights expected during the construction period. To do this they require information about the prevailing sea state during operations, as well as very rare wave events. The duration of wave height H_s and H_{max} occurring in 1 year, for example, can be estimated in accordance with Figure 6.17. Observations must be made over significant periods of time in order to obtain a reliable description of the wave climate by the way of a wave height duration curve.

Breakwaters must be designed so that serious damage, even in the case of a sudden storm, can be avoided, e.g. by using a layered construction with few gradations.

When specifying the productivity of the construction site, or when choosing the size of equipment, realistic approaches must be applied that take account of work being interrupted due to bad weather.

Depending on the particular construction, dumping works are carried out

1. With floating equipment.
2. With land-based equipment building ahead of itself.
3. With fixed equipment, jack-up platforms, etc.,
4. With cableways.
5. Any combination of these methods.

For particularly exposed sites seriously affected by winds, tides, sea states and currents, building methods with fixed equipment, jack-up platforms, etc. are preferred. This is especially the case if there is no sheltered harbour at or near the construction site.

6.3.6.2 Provision of fill and other construction materials
The provision of fill and construction materials requires careful planning, depending on the options available for their acquisition and transportation. Procuring coarse materials is often the main problem.

6.3.6.3 Installing materials with floating equipment
When building with floating equipment, the cross-section of the breakwater must be adapted to the equipment. Split hopper barges always require a sufficient depth of water. Side stone dumping vessels can be moved sideways even in low water depths. By exploiting today's computer-controlled positioning procedures, floating equipment can achieve the accuracy that in the past was only possible with land-based equipment.

6.3.6.4 Installing fill material with land-based equipment
The working level of land-based equipment should normally be above the effects of the normal sea state and surf. The minimum width of this working level should be adjusted to suit the requirements of the equipment being used.

Land-based equipment works by gradually extending the construction ahead of itself, the materials being delivered to site by dump trucks. This construction method, therefore, generally requires a core protruding out of the water with an extra-wide crest. The core serves as a road and, therefore, must be removed again to certain depth before the armour layers can be placed on top so to ensure sufficient interlocking and restore the hydraulic homogeneity.

With a narrow working area it is often advantageous to use a gantry crane for the installation work, as the materials for the ongoing work can be transported underneath it.

Stones to be installed by the crane are mostly delivered in transport containers on flatbed trailers, trucks with a special loading area or low-loaders. Where the roadway is narrow, it will be necessary to use trailers that can be driven backwards without needing to turn around. Large stones and precast concrete elements are placed using orange peel or other, special, grabs. Electronic monitors in the crane cabin make it easier to install materials in accordance with the design profile, even underwater.

Covering the core should follow quickly after dumping just a short length, especially when building the structure progressively ahead of the equipment. This avoids unprotected core material from being washed away. Further information can be found in CIRIA/CUR (1991).

6.3.6.5 Installing material from fixed scaffolds, jack-up platforms, etc.
Installing from fixed scaffolds, jack-up platforms, etc., also with a cableway, is primarily considered for bridging a zone with continuous, powerful surf.

When using a jack-up platform, construction progress generally depends on the capacity of the installation crane. The crane chosen should, therefore, have a large safe working load for the required reach.

The design should indicate clearly which parts of the breakwater cross-section must be installed when the sea is calm and which ones may still be constructed in certain wave conditions. This applies to both the core material and the precast concrete elements of the armour layer. Even in low swell, precast concrete elements can still suffer impacts underwater due to their great weight, which can lead to cracks and fractures.

6.3.7 Settlement and subsidence

Consistent and minor settlement of dumped breakwaters can be allowed for by building higher. Once the settlement has abated, which can always be monitored by taking settlement levels, the concrete on top of the crest should be cast in sections that are not overlong.

Where large differential settlement is expected, walls on the crest should be avoided because their settlement can subsequently lead to a visually unappealing overall appearance of the breakwater even though neither function nor stability are at risk.

6.3.8 Invoicing for installed quantities

As the settlement and subsidence behaviour of such structures is very difficult to predict, the recommendation is to specify realistic tolerances (±) from the outset for the purpose of invoicing based on the drawings. The tolerances should take into account the form of the mole and the installation layers in order to compensate for the settlement and, where applicable, embedded volumes and the technical procedures selected.

The tender should certainly specify whether invoicing is based on measured or actually installed quantities of materials. Where actually installed quantities are to be invoiced, settlement levels should be included in the tender.

If soil investigations reveal that it is likely to prove particularly difficult to invoice actually installed quantities, then the recommendation is to base the invoice on weight CIRIA/CUR (1991) if no other solution specific to the given subsoil conditions is possible. The measurement procedure (highest points of a stone layer or the use of a sphere/hemisphere at the bottom of a measuring stick) should be specified.

References

Abromeit, H.-U. (1997). Ermittlung technisch gleichwertiger Deckwerke an Wasserstraßen und im Küstenbereich in Abhängigkeit von der Trockenrohdichte der verwendeten Wasserbausteine. *Mitteilungsblatt der Bundesanstalt für Wasserbau Karlsruhe* 75, 45.

Bruderreck, L., Römisch, K. and Schmidt, E. (2011). Kritische Propellerdrehzahl bei Hafenmanövern als Basis zur Bemessung von Sohlsicherungen. *Hansa* 148 (5): 85–88.

Bruun, P. (1985). *Design and Construction of Mounds for Breakwaters and Coastal Protection*, Vol. 37, 1st edn., Elsevier Science, eBook ISBN: 9780444600455.

CFEM (2006). *Canadian Foundation Engineering Manual*, 4th edn. Canadian Geotechnical Society, ISBN 0-920505-28-7.

CIRIA/CUR (1991). Manual on the use of rock in coastal and shoreline engineering. In: *CIRIA Special Publication 83, CUR Report 154*. Rotterdam: A.A. Balkema.

CIRIA, CUR, CETMEF (2007). *The Rock Manual. The Use of Rock in Hydraulic Engineering*, 2nd edn. London: C683, CIRIA, ISBN 978-0-86017-683-1.

DWA (Deutsche Vereinigung für Wasserwirtschaft, Abwasser und Abfall e.V.) (2012). Dichtungssysteme im Wasserbau – Teil 1: Erdbauwerke. Merkblatt M 512-1.

DWA (Deutsche Vereinigung für Wasserwirtschaft, Abwasser und Abfall e.V.) (2017). Filtern mit Geokunststoffen. Merkblatt M 511.

EAAW (2008). *Empfehlungen für die Ausführung von Asphaltarbeiten im Wasserbau*, 5th edn. Deutsche Gesellschaft für Geotechnik, http://www.dggt.de/images/PDF-Dokumente/eaaw2008.pdf, last access: 8.6.2023.

EAG-GTD (2002). *Empfehlungen zur Anwendung geosynthetischer Tondichtungsbahnen*, (ed. by Deutsche Gesellschaft für Geotechnik). Berlin: Ernst & Sohn.

EAK (2002). *Empfehlungen für die Ausführung von Küstenschutzwerken durch den Ausschuss für Küstenschutzwerke*. 3. korrigierte Ausgabe 2020. Karlsruhe: Bundesanstalt für Wasserbau (BAW).

EAO (2002). Empfehlungen zur Anwendung von Oberflächendichtungen an Sohle und Böschung von Wasserstraßen. *Mitteilungsblatt der Bundesanstalt für Wasserbau Karlsruhe Nr. 85.*

GBB (2010). *Grundlagen zur Bemessung von Böschungs- und Sohlensicherungen an Binnenwasserstraßen*. Karlsruhe: Federal Waterways Engineering & Research Institute, www.baw.de.

Hamburg (2013). *Freie und Hansestadt Hamburg – Berechnungsgrundsätze für Hochwasserschutzwände, Flutschutzanlagen und Uferbauwerke im Bereich der Freien und Hansestadt Hamburg*. Hamburg Port Authority.

Henne, J. (1989). Versuchsgerät zur Ermittlung der Biegezugfestigkeit von bindigen Böden. *Geotechnik* 12 (2): 96–99.

Hudson, R.Y. (1958). Design of quarry stone cover layers for rubble mound breakwaters, Research Report No. 2.2, WES, Vicksburg.

Hudson, R.Y. (1959). Laboratory investigation of rubble mound breakwaters. *Proc. ASCE* 85: WW3.

Koerner, R.M. (2005). *Designing with Geosynthetics*. Englewood Cliffs: Prentice-Hall.

Köhler, H.-J. and Schulz, H. (1986). Bemessung von Deckwerken unter Berücksichtigung von Geotextilien. *3rd Int. Conf. on Geotext. Viennam*, 1986. Rotterdam: Balkema.

MAG (1993). Merkblatt "Anwendung von geotextilen Filtern an Wasserstraßen", 1993 ed., Federal Waterways Engineering & Research Institute, Karlsruhe.

MAK (2013). Merkblatt "Anwendung von geotextilen Filtern an Wasserstraßen", 2013 ed., Federal Waterways Engineering & Research Institute, Karlsruhe.

MAR (2008). Merkblatt "Anwendung von Regelbauweisen für Böschungs- und Sohlensicherungen an Binnenwasserstraßen". Federal Waterways Engineering & Research Institute, Karlsruhe, www.baw.de.

Owen, M.W. (1980). Design of seawalls allowing for wave overtopping. Hydraulic Research Station, Wallingford, Report No. Ex 924, Wallingford

PIANC (1973). Report of the International Waves Commission, PIANC-Bulletin No. 15, Brussels.

PIANC (1976). Report of the International Waves Commission, PIANC-Bulletin No. 25, Brussels.

PIANC (1987a) Report of PIANC Working Group II-9 "Development of modern Marine Terminals". Supplement to PIANC-Bulletin No. 56, Brussels.

PIANC (1987b) Report of Working Group I-4 "Guidelines for the design and construction of flexible revetments incorporating geotextiles for inland waterways". Supplement to PIANC-Bulletin No. 57, Brussels.

PIANC (1992a) Report of Working Group II-21 "Guidelines for the design and construction of flexible revetments incorporating geotextiles in marine environment". Supplement to PIANC-Bulletin No. 78/79, Brussels.

PIANC (1992b) Analysis of rubble mound breakwaters, PIANC MarCom WG 12.

Pilarczyk, K. and Zeidler, R. (1996). *Offshore Breakwaters and Shore Evolution Control*. Rotterdam: Balkema.
Römisch, K. (1993). Propellerstrahlinduzierte Erosionserscheinungen in Häfen. *Hansa* 130 (8): 62–68.
Römisch, K. (1994). Propellerstrahlinduzierte Erosionserscheinungen – Spezielle Probleme. *Hansa* 131 (9).
Römisch, K. (2000). Strömungsstabilität vergossener Steinschüttungen. *Wasserwirtschaft* 90 (7/8): 356–361.
Römisch, K. (2001). Scouring in Front of Quay Walls Caused by Bow Thruster and New Measures for its Reduction. *5th Int. Sem. on Renovation and Improvements to Existing Quay Structures*, Gdansk TU, 28–30 May 2001.
RPG (1994). *Richtlinien für die Prüfung von geotextilen Filtern im Verkehrswasserbau*. Karlsruhe: Federal Waterways Engineering & Research Institute.
Schulz, H. (1987a) Mineralische Dichtungen für Wasserstraßen. Seminar "Dichtungswände und Dichtsohlen", June 1987 in Braunschweig. *Mitteilungen des Instituts für Grundbau und Bodenmechanik, Braunschweig TU* 23.
Schulz, H. (1987b) Conditions for day sealings at joints. *Proc. of 9th Eur. Conf. on Soil Mech. & Found. Eng.*, Dublin.
SPM (1984). *Shore Protection Manual*. Vicksburg: US Army Corps of Engineers, Coastal Engineering Research Center.
TLG (2003). Technische Lieferbedingungen für Geotextilien und geotextilverwandte Produkte an Wasserstraßen, 2003 ed., Federal Ministry for Transport, Building & Housing. Verkehrsblatt 2003, No. 18.
Van der Meer, J.W. and Janssen, J.P.F.M. (1994). Wave run-up and waver overtopping at dikes and revetments. In: *Delft Hydraulics Publication No. 485*.
Veldhuyzen van Zanten, R. (1994). *Geotextiles and Geomembranes in Civil Engineering*. Rotterdam, Boston: Balkema.
ZTV-W LB 210 (2015). Zusätzliche Technische Vertragsbedingungen – Wasserbau (ZTV-W) für Böschungs- und Sohlensicherungen (Leistungsbereich 210). Federal Ministry for Transport, Building & Urban Development.

Standards and regulations
DIN 4084: Soil – Calculation of embankment failure and overall stability of retaining structures.
DIN EN 13253: Geotextiles and geotextile-related products – Characteristics required for use in erosion control works (coastal protection, bank revetments).
DIN EN 13383-1: Armourstone – Part 1: Characteristics.

Konrad Bergmeister, Frank Fingerloos,
Johann-Dietrich Wörner (Hrsg.)

Beton-Kalender 2023

Schwerpunkte: Wasserundurchlässiger Beton; Brückenbau

- das aktuelle Regelwerk für die Planung und Herstellung wasserundurchlässiger Betonbauwerke
- Entwurf, Bemessung, Konstruktion und Monitoring von Betonbrücken nach den Regeln des Eurocode 2 in Deutschland
- Autor:innen aus Praxis, Normung und Forschung

Der Beton-Kalender 2023 thematisiert die Herstellung wasserundurchlässiger Betonbauwerke auf der Grundlage der aktuellen ÖBV-Richtlinien und der WU-Richtlinie des DAfStb. Im zweiten Teil widmet er sich dem Entwurf und der Konstruktion von Brücken – einschließlich Fragestellungen der Bauwerksdiagnostik und des Schwingungsschutzes.

Teile 1 + 2, 2022 · 988 Seiten · 909 Abbildungen · 185 Tabellen

Hardcover
ISBN 978-3-433-03375-3 € 184*

Fortsetzungspreis € 164*

eBundle (Print + ePDF)
ISBN 978-3-433-03376-0 € 234*

Fortsetzungspreis eBundle € 194*

BESTELLEN
+49 (0)30 470 31-236
marketing@ernst-und-sohn.de
www.ernst-und-sohn.de/3375

* Der €-Preis gilt ausschließlich für Deutschland. Inkl. MwSt.

7
Configuration of cross-sections and equipment for waterfront structures

7.1 Configuration of cross-sections

7.1.1 Standard cross-sectional dimensions for waterfront structures in seaports

7.1.1.1 Standard cross-sections

When building new cargo-handling facilities and extending existing installations, the standard cross-section of Figure 7.1 is recommended, which takes into account all relevant influences.

The distance of 1.75 m between crane rail and edge of quay should be understood as a minimum. In the case of new construction and watercourse deepening works, it is better to allow 2.50 m, as the outboard crane bogies can then be built as wide as required. (The crane bogies of modern port cranes are often approx. 0.60–1.20 m wide, see Figure 7.1.) In addition, it is easier to comply with the health and safety regulations for mooring operations and for embarkation/disembarkation via gangways.

A crane safety clearance must be maintained for railway operations. The centreline of the first track must be at least 3.00 m from the front crane rail. However, these days, railway tracks are only constructed on quay walls in exceptional circumstances.

7.1.1.2 Walkways (towpaths)

The walkway space (towpath) in front of the outboard crane rail is necessary to provide space for the installation of bollards and storing gangways, to serve as a path and working space for line handlers, to allow access to berths and to accommodate the outboard portion of the crane gantry. Consequently, this clearance is of special importance for port operations.

Health and safety regulations must also be considered when selecting the walkway width. Furthermore, account must be taken of the fact that ship superstructures often project beyond the hulls of moored vessels and that cargo handling operations must still be possible on listing ships.

For the reasons outlined above, the walkway must be wide enough so that the outermost edges of the cargo-handling facilities are at least 1.65 m, but preferably 1.80 m, behind the front edge of the quay wall or face of timber fender, fender pile or fender system (Figure 7.1).

Recommendations of the Committee for Waterfront Structures Harbours and Waterways – EAU 2020, 10th Edition. Issued by the Committee for Waterfront Structures of the German Port Technology Association and the German Geotechnical Society.
© 2024 Ernst & Sohn GmbH. Published 2024 by Ernst & Sohn GmbH.

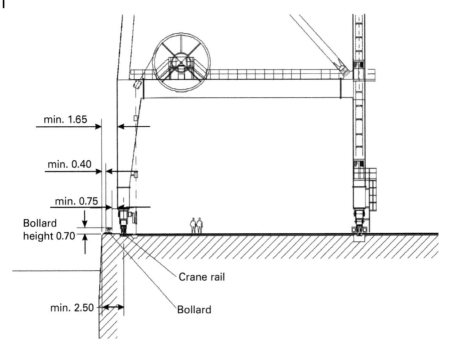

Figure 7.1 Standard cross-section for waterfront structures in seaports (service ducts not illustrated).

7.1.1.3 Railings, rubbing strips and edge protection

Railings are not required at the edges of quays for mooring and cargo-handling operations. However, the edges of such quays should be provided with adequate edge protection according to Section 7.2.13. The edges of quays with public access and those not used for mooring or cargo handling should be provided with railings.

7.1.1.4 Edge bollards

The front edge of an edge bollard must lie 0.15 m behind the front face of the quay wall, because otherwise it is difficult to attach and remove the hawsers of ships moored tight up against the quay wall. The width of the head of a bollard should be taken as 0.50 m.

7.1.1.5 Arrangement of tops of quay walls at container terminals

Owing to the safety requirements and high productivity demands at container terminals, a greater clearance between the front edge of the quay and the axis of the outboard crane rail is recommended, as shown in Figure 7.1. It should be possible to store gangways parallel to the ship between the front edge of the quay and the cranes, also to park service and delivery vehicles and, thus, separate the zones for cargo-handling and ship service traffic. It is accepted that container cranes will require longer jibs.

If containers are automatically transported between crane bridge and storage area, it is essential to separate ship service traffic and container-handling areas for safety reasons. In such cases, the distance of the front edge of the quay from the outboard crane rail should be such that all service vehicle lanes can be accommodated in this strip. Alternatively, one lane can be placed in this strip and another adjacent to the outboard rail beneath the portal of the crane, separated from the container-handling area by a fence.

7.1.2 Top edges of waterfront structures in seaports

7.1.2.1 General

The level of the top edge of a waterfront structure at a seaport is determined by the operations level of the port. When specifying the port operations level, the following factors must be observed:

1. Water levels and their fluctuations, especially heights and frequencies of possible storm tides, wind-induced water build-up, tidal waves, the possible effects of flows from high water levels, and maybe other factors mentioned in Section 7.1.2.2.2.
2. Mean groundwater level, with frequency and magnitude of level fluctuations.
3. Shipping operations, port installations and cargo-handling operations, imposed loads.
4. Ground conditions, subsoil, availability of fill material and ground management.
5. Constructional options for waterfront structures.
6. Environmental concerns.

The factors listed above must be assessed in terms of technical and economic aspects according to the requirements of the port or harbour in order to arrive at the optimum level for the port operations area.

7.1.2.2 Level of port operations area with regard to water levels

When it comes to the level of the port operations area, a fundamental distinction must be made between wet docks and open harbours with or without tides.

7.1.2.2.1 Wet docks

In wet docks protected against flooding, the height of the port operations level above the mean operational water level must be such that it is not flooded at the highest possible operational water level in the port. At the same time, it must lie above the highest groundwater level and be suitable for the envisaged cargo-handling operations.

The port operations area should generally be 2.00–2.50 m, but at least 1.50 m, above the mean operational water level in the port.

7.1.2.2.2 Open harbours

The height and frequency of high tides are critical when it comes to selecting a suitable operations level for an open harbour.

As far as possible, planning work should make use of frequency lines to indicate levels above mean high tide. In addition to the main influencing factors stated in Section 7.1.2.1 (1), the following influences must be taken into consideration:

- Wind-induced water build-up in the harbour basin.
- Oscillating movements of the harbour water due to atmospheric influences (seiching).
- Wave run-up along the shore (Mach effect).
- Resonance of the water level in the harbour basin.
- Secular rises in the water level.
- Long-term coastal uplift or subsidence.

If there are no (meaningful) data records available for the above influences, as many measurements as possible must be taken in situ within the scope of the design work and these linked to high tide levels and wind actions.

7.1.2.3 Effects of (changing) groundwater levels on the terrain and the level of the port operations area

The mean groundwater level and its local seasonal and other changes, as well as their frequency and magnitude, must be taken into consideration when establishing the level of the port operations area, especially with respect to proposed pipes, cables, roads, railways, imposed loads, etc., in conjunction with subsurface conditions. Owing to the need to drain precipitation, the course of the groundwater level with respect to the harbour water must also be given attention.

7.1.2.4 Level of port operations area depending on cargo handling

1. General cargo and container-handling
   In general, an operations level not liable to flooding is essential for general cargo- and container-handling. Exceptions should only be permitted in special cases.
2. Bulk cargo-handling
   Owing to the diversity of cargo-handling methods and types of storage, as well as the sensitivity of the goods and vulnerability of handling gear, it is not possible to give a general recommendation regarding the level of the port operations area for facilities handling bulk cargo. An effort should nevertheless be made to provide an area not liable to flooding, particularly in view of the environmental problems involved.
3. Special cargo-handling
   For ships with side doors for truck-to-truck operations, bow or stern doors for roll-on/roll-off operations or other special types of equipment, the top level of the waterfront structure must be compatible with the type of vessel and equipped with either fixed or movable loading/unloading ramps. Such operations do not necessarily require the top of the waterfront structure to be at the same level as the ground. In tidal areas, it may be necessary to adjust the levels in the vicinity of ramps. Floating pontoons or similar arrangements may even be necessary. In any case, the requirements of the types of vessel that will use such port facilities must be taken into account.
4. Cargo-handling with ship's lifting gear
   In order to achieve adequate working clearances under crane hooks, even for low-lying vessels, the level of a quay where on-board lifting gear is used for handling must generally be lower than the one where cargo is handled by quayside cranes.

7.1.3 Standard cross-sections for waterfront structures in inland ports

7.1.3.1 Port operations level

The operations level of an inland port should normally be arranged above the highest water level. However, in the case of flowing waters with large water level fluctuations, this is frequently only possible at considerable expense. Occasional flooding can be accepted provided that cargos are not damaged and there is no risk of the watercourse being contaminated by any water lying temporarily on the port operations area.

In ports on inland canals, the operations level should be at least 2.00 m above the normal canal water level.

7.1.3.2 Waterfront

As far as possible, waterfronts in inland ports should be straight and have a front face as smooth as possible (Section 7.1.5). Sheet piles are ideal for securing a waterfront apart from a few exceptions (Section 7.2.4).

It is important to ensure that the outermost structural parts of cranes do not protrude as far as or beyond the front edge of the waterfront structure. A crane leg width of 0.60–1.00 m should be assumed.

7.1.3.3 Clearance profile

When positioning crane tracks and designing cargo-handling cranes, care must be taken to comply with the side and overhead safety clearances required by the relevant specifications (in Germany: EBO, BOA, UVV Eisenbahnen); see Figure 7.2.

As far as roads under crane portals are concerned, in Germany the recommendations given in Figure 7.3 by the Association of German Public Inland Ports (EBO, EBA, UVV) apply.

7.1.3.4 Position of outboard crane rail

The aim is to place the outboard crane rail as close as possible to the waterfront edge in order to reduce the length of the crane jib to a minimum and save valuable storage space near

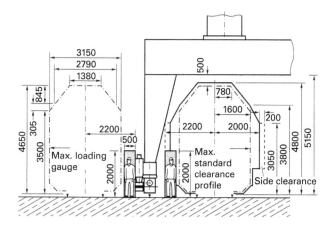

Figure 7.2 Side and overhead safety clearances for railways.

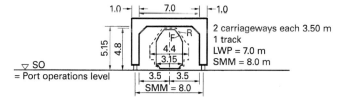

Figure 7.3 Recommended track gauge (SMM) and clear width (LWP) for crane portals over roads and covered tracks.

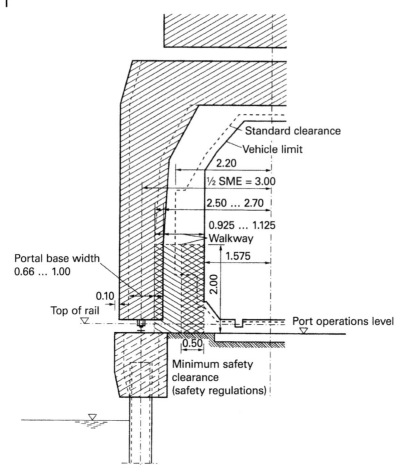

Figure 7.4 Standard cross-sectional dimensions for sheet pile structures in inland ports (crane rail on the centreline of sheet piles).

the waterfront. The walkway is then positioned inboard of the crane portal leg (Figure 7.2). Where a walkway is provided between crane portal and waterfront, it must be at least 0.80 m wide.

Depending on the circumstances, the waterfront structure can be in the form of a reinforced concrete wall, a sheet pile wall or a combination of these two forms of construction – a reinforced concrete wall with a facing of sheet piles on bored piles (Figure 7.6). The advantage of reinforced concrete retaining walls is that they are easily built with smooth wall surfaces.

In the case of sheet piles, the crane rail should be aligned with the centreline of the sheet pile wall (Figure 7.4). However, the necessary geometrical requirements (Section 7.1.3.2) can make it necessary to support the crane rail off-centre (Figure 7.5).

A combined solution (reinforced concrete wall plus facing of sheet piles on bored pile foundations as shown in Figure 7.6) has the advantage that the crane loads are transferred via the bored piles to the subsoil. In addition, the crane rail can be routed close to the edge without vessel impacts having any influence on the crane. Ladders for access to vessels can

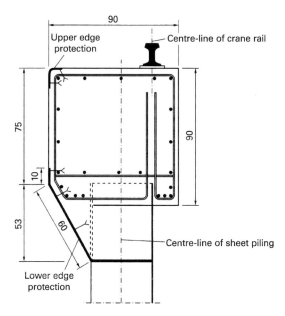

Figure 7.5 Crane rail eccentric to the centreline of sheet piles (example).

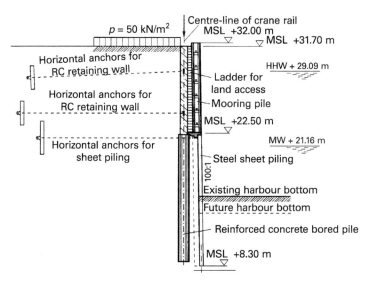

Figure 7.6 Anchored sheet pile wall with mooring piles/reinforced concrete wall (example).

be placed in ideal positions behind mooring piles (Figure 7.6 and recommendation R 42 of the Association of German Public Inland Ports).

7.1.3.5 Mooring equipment

Sufficient mooring equipment for vessels must be installed on the water side of the waterfront structure (Section 4.9).

7.1.4 Upgrading partially sloped waterfronts in inland ports with large water level fluctuations

7.1.4.1 Reasons for partially sloped upgrades

The berthing, mooring, lying and casting-off of vessels in inland ports must be possible at every water level without the use of anchors, and port and operations personnel should have safe access to vessels at all water levels. This is only possible with vertical waterfronts.

If in inland ports with fluctuating water levels a port operations area free from flood water is also required, this can result in very high quay walls. In these cases, it is appropriate to design cargo-handling facilities with a vertical quay wall and an adjoining upper slope (Figures 7.7 and 7.8).

7.1.4.2 Design principles

For cargo-handling operations on a partially sloping waterfront, the level of transition from the vertical wall to the sloping bank should be such that it is not below water for more than 60 days (long-term mean).

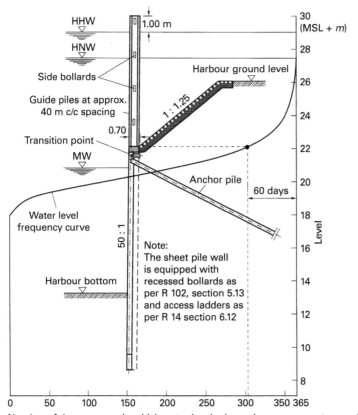

Figure 7.7 Partially sloped bank for berths, particularly for pushed lighters where the port operations level is subject to flooding. Note: the sheet pile wall is equipped with recessed bollards as per Section 7.2.6 and access ladders as per Section 7.2.2.

On the Lower Rhine, for example, this corresponds to a transition level about 1.00 m above mean water level (Figure 7.7). The level of this vertical/sloping transition should remain the same throughout the port basin.

For waterfronts with a port operations area at a very high level, the transition from the quay wall to the bank should be positioned so that – for operating and structural reasons – the bank is a maximum of 6.00 m high (Figure 7.8).

Guide piles about 40 m apart are advisable along the vertical bank section at berths and push-tow coupling quays for unmanned vessels without cargo-handling operations. These piles serve for marking, safe mooring and protecting the bank. They should extend 1.00 m above highest high water but not project over the water (Figure 7.7).

The vertical waterfront is generally constructed of sheet piles considered as fixed at the base and with a single row of anchors at or near the top.

The top of the piles should be finished with a 0.70 m wide steel or reinforced concrete capping beam, which also can also function as a safe walkway between the guide piles (Figures 7.7 and 7.8). It must be possible to pass behind the guide piles.

Ensuring that ships are moored correctly will prevent them grounding in the event of falling water levels.

The outboard edge of a reinforced concrete capping beam is to be protected against damage by a steel plate as described in Section 7.2.13.

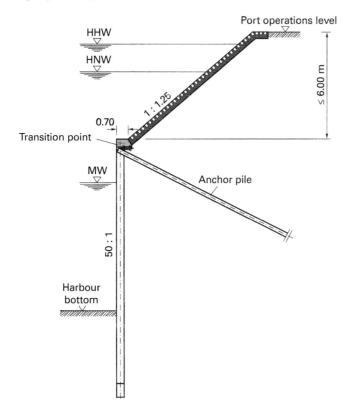

Figure 7.8 Partially sloped bank with flood-free port operations level. Note: the sheet pile wall is equipped with recessed bollards as per Section 7.2.6 and access ladders as per Section 7.2.2.

To ensure safe access via steps, the slope above the wall must not be steeper than 1 : 1.25. Slopes of 1.25–1 : 1.5 are chiefly used.

Bollards should be in accordance with Section 7.2.6.

7.1.5 Design of waterfront areas in inland ports according to operational aspects

7.1.5.1 Requirements

Designs for embankment cross-sections at inland ports are primarily influenced by economic and operational considerations. When handling cargos with cranes, a well-planned facility is important.

Trouble-free shipping operations are ensured when vessels can tie-up and cast-off easily at the waterfront, and passing ships or push-tow convoys do not create disadvantageous ship movements. Mooring cables and ropes must be able to slacken as the water level changes.

The waterfront structure should also act as a guide when berthing ships.

Pushed lighters have relatively large masses and are box-shaped with sharp corners and edges. Therefore, waterfront structures for cargo-handling operations involving pushed lighters should be as flat as possible.

To ensure fast and safe operations, vessels should move as little as possible during loading/unloading. On the other hand, if necessary, it should be possible to warp the vessel without difficulty.

Regarding the carriage of passengers between land and ship, it must be possible to transfer them directly or use a gangway safely (R 42, ETAB).

7.1.5.2 Design principles

In principle, long straight waterfronts are preferable. If changes in direction are unavoidable, they should be designed in the form of angled, not rounded, turns. Distances between changes of direction should be such that the intermediate straight stretches match the lengths of the ships or push-tow convoys using the facility. Trouble-free shipping operations are easiest to achieve with smooth quay walls without recesses or protruding structures. The front faces of the waterfront walls can be sloped, partially sloped or vertical.

7.1.5.3 Waterfront cross-sections

1. Embankments

   Embankment surfaces should be designed so as to be as flat as possible. Intermediate landings should be avoided if feasible. Steps and stairs should be installed in the direction of the slope, i. e. at a right-angle to the shoreline. Bollards and mooring rings must not protrude above the embankment surface. If intermediate berms are unavoidable on high embankments, they should not be located in the area of frequent water level fluctuations. Instead, they should be situated in the high water zone. See Section 7.1.4.2 for details of the transition from the sloping to the vertical waterfront.

   On sloping and partially sloping waterfronts, safe guiding of vessels can only be guaranteed in conjunction with closely spaced mooring dolphins.

2. Vertical waterfronts

   Vertical or only slightly inclined waterfront structures in concrete or masonry must be built with a smooth front face. This condition is easily fulfilled when constructing a new structure in a dry excavation. If built in the form of a diaphragm or bored cast-in-place pile wall, the side of the wall on the water side usually requires further work to meet the operational requirements for a flat wall.

Using sheet piles for vertical waterfront structures represents a proven, economical solution. Should the shipping operations require it, sheet pile walls can be reinforced as described in Section 7.2.4 in order to achieve a flat outer surface.

7.1.6 Nominal depth and design depth of the harbour bottom

7.1.6.1 Nominal depth in seaports

The nominal depth is the water depth below a defined reference level. When stipulating the nominal depth of the harbour bottom in front of a quay wall, the following factors should be considered:

1. The draught of the largest, fully laden ship berthing in the port. The draught must be calculated taking into account the salinity of the water and the listing of the ship.
2. A safety clearance between ship's keel and nominal depth. The safety clearance depends on the regulations of the local harbour authorities, but should be at least 0.50 m.

The reference level for the nominal depth is generally a statistically based low water level.

In regions without tides, e.g. the Baltic Sea, the low water level is derived from data collected over many years.

When determining the reference level and hence the nominal depth in front of waterfront structures in tidal regions, appropriate consideration of the tide-related changes to the water level is necessary so that an adequate water depth is available with adequate statistical frequency. In this case, the chart datum (CD) is frequently chosen as the reference level. Up until the end of 2004, the chart datum in Germany was derived from the mean low water spring level. Since 2005, the chart datum has been defined as the lowest astronomical tide (LAT). This is a standardised reference level, which is also common internationally, for all countries bordering the North Sea.

LAT defines the lowest possible water level caused by astronomical influences. For the German North Sea, LAT is about 0.50 m below mean low water spring.

The reference level has to be fixed with respect to local requirements and can also be different from the chart datum LAT when a lower water level due to exceptional meteorological or astronomical conditions with a higher statistical frequency is acceptable. Therefore, the reference level must be agreed unanimously by all those involved prior to fixing the nominal depth for the harbour bottom.

7.1.6.2 Nominal depth of harbour bottom for inland ports

The nominal depth of the harbour bottom in inland ports and harbour entrances should be selected so that ships can reach their destinations with the greatest possible loaded draught. In inland ports on rivers, the water depth should generally be 0.30 m deeper than that of the adjoining waterway in order to rule out any dangers for ships in ports at low water levels.

7.1.6.3 Design depth in front of quay wall

If dredging is to be carried out in front of a quay wall because of silt, sand, gravel or rubble deposits, the dredging must be deeper than the intended nominal depth of the harbour bottom stipulated in Sections 7.1.6.1 and 7.1.6.2 (Figure 7.9).

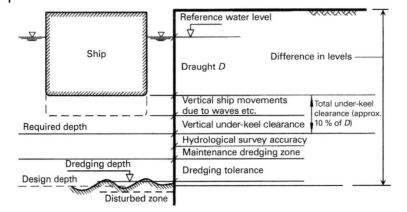

Figure 7.9 Calculating the design depth in accordance with CUR (2013).

The design depth is made up of the nominal depth of the harbour bottom, the maintenance margin down to the planned dredging depth plus dredging tolerances and other allowances, e. g. possible loosening of the bottom while dredging (disturbance zone). The design depth can, therefore, be any value selected by the design team below the dredging depth.

The dredging depth is determined using the following factors:

1. Extent of the silt mass, sand drift, gravel or rubble deposits per dredging period.
2. Depth below the nominal depth of the harbour bottom to which the soil may be removed or disturbed.
3. Costs of every interruption to cargo-handling operations caused by dredging works.
4. Availability of the required dredging device.
5. Costs of dredging work with regard to the depth of the maintenance margin.
6. Extra costs of a quay wall with a deeper harbour bottom.

Additional information can be found in Section 5.2.

Owing to the importance of the above factors, the maintenance margin, representing an addition to the nominal depth, must be stipulated with care. On the one hand, an inadequate margin can lead to high costs for frequent maintenance dredging and the ensuing interruptions to operations. On the other hand, an excessive margin results in higher construction costs and may encourage sedimentation.

It is practical to attain the harbour bottom depth first in at least two dredging operations executed at intervals. A maximum dredging depth of 3.00 m should not be exceeded.

Table 7.1 provides a general guide to the depth of maintenance margins and minimum dredging tolerances to be used for different water depths. More information on dredging tolerances can be found in Section 5.2.2.

The reference values in Table 7.1 already take account of the allowances required by DIN EN 1997-1 (see also Section 5.2.2).

If erosion of the harbour bottom is expected in front of the quay wall, the design depth must be increased, or suitable measures taken to prevent erosion.

Table 7.1 Maintenance margins and minimum dredging tolerances, reference values [m].

Depth of water below lowest water level [m]	Depth of maintenance margin [m]	Minimum dredging tolerance[a] [m]
5	0.5	0.2
10	0.5	0.3
15	0.5	0.4
20	0.5	0.5
25	0.5	0.7

a) Depends on the dredging device.

7.1.7 Strengthening waterfront structures for deepening harbour bottoms in seaports

7.1.7.1 General

Developments in ship dimensions mean that occasionally it is necessary to deepen the harbour bottom in front of an existing quay wall. In these cases, larger crane and imposed loads are often necessary, meaning that the waterfront structures must also be strengthened.

Whether in any individual case it is both possible and economical to deepen the harbour bottom and reinforce the waterfront structure depends on various factors:

First, checks must be carried out to establish whether the required deepening of the harbour bottom is even possible given its current depth. Next, checks must be carried out to establish whether the design and condition of the waterfront structure is suitable for use with a deeper harbour bottom. To this end, design drawings and structural calculations of the wall and the results of soil investigations from the construction phase can be studied, provided they are available. If required, additional soil investigations may reveal more favourable soil properties, resulting from the improvement of the subsoil characteristics due to consolidation since construction. Finally, it is necessary to check whether the required strengthening is economical with respect to the remaining useful life of the waterfront structure compared with the costs of building a new one.

Stability and serviceability must then be verified for the actions due to the new loads and the increased theoretical depth of the harbour bottom. If the stability analyses of the present wall and/or its design drawings are no longer available, one possible approach is to reduce the imposed loads to compensate for the loads on the wall due to the deeper harbour bottom.

7.1.7.2 Design of strengthening measures

There are numerous possibilities to reinforce quay walls to withstand greater loads resulting from deepening of the harbour bottom. It is vital to ensure that the embedment depth of the wall is still sufficient after the deepening. This also applies to the infill piles of combined walls. A few typical solutions are given below.

7.1.7.2.1 Measures to increase the passive earth pressure

The load-carrying capacity of waterfront walls can be improved by increasing the passive earth pressure at the base of the wall. If the in situ soil is soft and cohesive with little

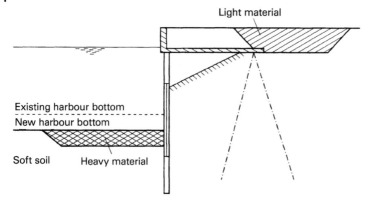

Figure 7.10 Soil replacement in front of and/or behind the structure.

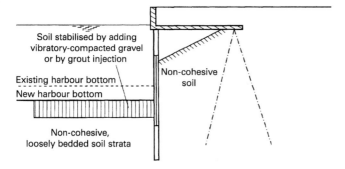

Figure 7.11 Soil stabilisation or soil compaction in front of the structure.

strength, it can be replaced by a non-cohesive material with a high unit weight and high shear strength down to the required depth, as shown in Figure 7.10.

The transition to the in situ soil must ensure a stable filter action. The soil replacement may only be carried out in stages, and it is important to observe deformations of the wall. No cargo-handling operations should be carried out in the area affected during soil replacement activities. If required, the load on the wall can be relieved temporarily by removing the backfill behind the wall. Details of soil replacement can be found in Sections 3.5.12 and 5.9.

If the soil in front of the waterfront wall is non-cohesive and can be compacted, the passive earth pressure can be increased by compaction and, if necessary, by adding gravel or ballast (Figure 7.11). Permeable non-cohesive soils can be stabilised by grout injection.

7.1.7.2.2 Measures to reduce active earth pressure

The active earth pressure on a waterfront wall can be reduced, for example, by building a relieving slab supported on piles (Figure 7.12). Further options include the partial replacement of the backfill with a lighter material (Figure 7.10) or employing grout injection to stabilise the backfill.

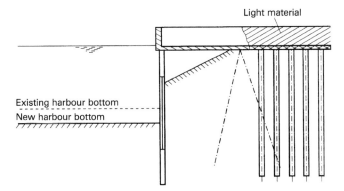

Figure 7.12 Stabilisation with a relieving slab supported on piles.

7.1.7.2.3 Measures involving the quay wall

The waterfront structure itself can be upgraded using additional anchors to carry increased loads (Figure 7.13). If required, it is also possible to drive the existing waterfront structure deeper and extend it (Figure 7.14). This option requires the existing anchors to be temporarily detached and then reattached or replaced by new anchors.

However, in most cases, it will be necessary to build a new wall in front of the old one and to anchor it either with a new superstructure (Figure 7.15) or with raking or horizontal anchors (Figure 7.16).

Provided that there is sufficient space, an entirely new pile-supported slab can be built to provide extra support for the loads from cargo-handling operations (Figure 7.17). This solution also creates a larger cargo-handling area. For more information on the design and calculation of such slabs, see Section 11.2.

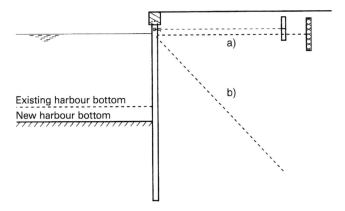

Figure 7.13 Use of additional anchors: (a) horizontal; (b) raking.

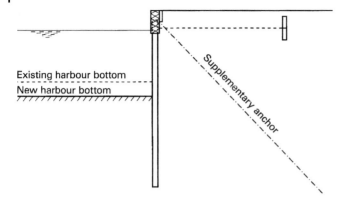

Figure 7.14 Driving the existing waterfront structure deeper and extending it upwards, plus additional anchors.

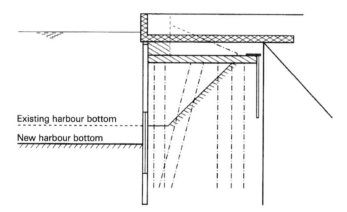

Figure 7.15 New wall in front of an existing one, plus a new superstructure.

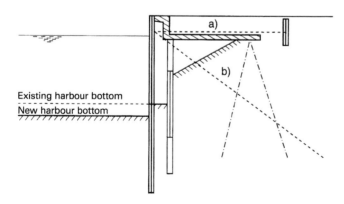

Figure 7.16 New wall in front of an existing one, plus additional anchors (a) or (b).

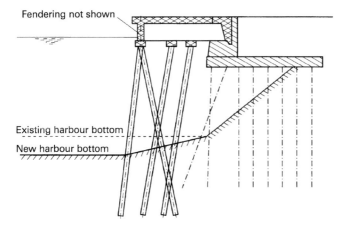

Figure 7.17 Forward extension on piles with an underwater embankment.

7.1.8 Embankments below waterfront wall superstructures behind closed sheet pile walls

7.1.8.1 Embankment loads

Slopes below waterfront wall superstructures can be loaded by currents along the waterfront wall and by flow forces due to groundwater, in addition to active earth pressure. Regarding the stability of slopes, flow forces are disadvantageous when the groundwater level behind the slope is higher than the outer water level. Consequently, the flow forces should be directed away from the slope (Section 3.3.3).

Therefore, the angle of the slope must always be such that the slope is stable for all water levels. In addition, the surface of the slope must be protected against erosion due to currents in the outer water.

7.1.8.2 Risk of silting-up behind the sheet pile wall

In tidal areas there is a risk that silt will accumulate on slopes located beneath superstructures. The ensuing additional loads on the superstructure piles can be substantial.

Silt deposits can only be prevented permanently when no silt-laden water can penetrate the area beneath the superstructure. However, this usually entails considerable extra costs for the waterfront structure.

It is, therefore, generally accepted that outer water will penetrate the area beneath the superstructure. Silt deposits are, therefore, avoided by including outlets at regular intervals along the sheet pile wall just above the base of the slope.

The effect of such measures should be monitored on an individual basis. Should they not achieve the desired results, the silt deposits must be removed regularly.

7.1.9 Re-design of waterfront structures in inland ports

7.1.9.1 General

On canals and rivers with controlled levels, expansion work for a deeper draught may require port facilities to be deepened. In some cases increasing the crane and imposed load capacities may call for the waterfront structure to be re-designed.

In the case of river ports, an increase in water depth is required when the port is located in a side basin and the riverbed is deepened by erosion. In order to ensure access to the port facilities, the harbour bottom must then be deepened.

Where port facilities have sloping banks, deepening the bottom results in a reduction in the width of the basin and water cross-section. If this is unacceptable, a partial slope or vertical quayside are the options. This is then also advantageous for cargo-handling, as the outreach of cargo cranes on partially sloping or vertical expansion projects is shorter than on sloping waterfronts.

Deepening the harbour bottom and/or higher imposed loads results in higher stresses on individual structural components, which might then no longer be adequate under certain circumstances.

7.1.9.2 Re-design options

Given the aforementioned circumstances, it is generally always possible to construct a new waterfront structure in front of or instead of the old one. However, it is often sufficient to renew or reinforce certain parts of the waterfront structure or to implement other upgrading measures as per Section 7.1.7.

Thus, for example, sheet piles can be driven more deeply and a new superstructure built off these. Higher anchor forces can be accommodated with additional anchors. Non-cohesive soils on the harbour bottom can be compacted, which leads to an increase in passive earth pressure. Soil nailing can improve the stability of an embankment.

7.1.9.3 Construction examples

Figures 7.18–7.23 show typical examples of re-designs for waterfront structures for inland ports. All levels are related to MSL.

Figure 7.18 shows an example of increasing the water depth by redesigning a sloping bank as a partially sloped bank with sheet piles.

In the example shown in Figure 7.19, IPB 500 steel beam sections were driven 4.2 m apart behind existing waterfront sheet piles. The sheet pile wall was raised with a reinforced concrete capping beam and re- anchored with driven grouted piles.

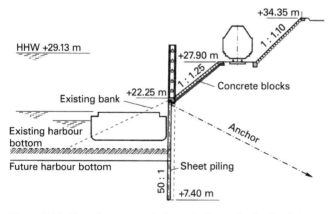

Figure 7.18 Waterfront upgrade by replacing a sloping bank by a partially sloping bank.

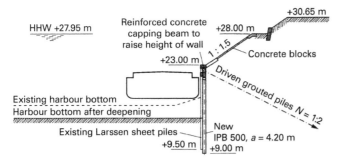

Figure 7.19 Waterfront upgrade achieved by driving anchored IPB 500 beam sections behind the existing sheet piles and raising the height of the sheet pile wall.

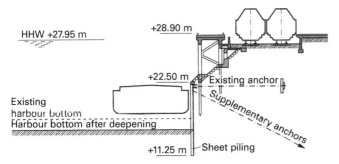

Figure 7.20 Waterfront upgrade by means of additional anchorages to existing sheet piles.

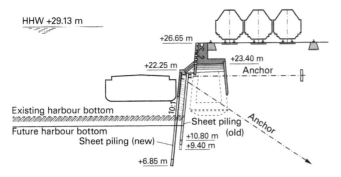

Figure 7.21 Waterfront upgrade by driving new anchored sheet piles in front of the old wall.

In the example shown in Figure 7.20, additional anchorage for the existing sheet pile wall was provided by driven grouted piles, whereas in Figure 7.21 a new anchored sheet pile wall was driven in front of the existing wall.

Figure 7.22 shows a waterfront wall upgraded by increasing the passive earth pressure in front of it – achieved by compacting the in situ soil. In Figure 7.23, soil nailing has been used to stabilise the slope above the waterfront.

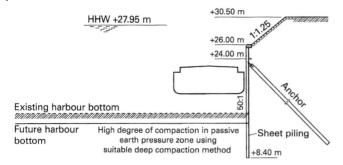

Figure 7.22 Waterfront upgrade by compacting non-cohesive soil in the passive earth pressure area in front of the sheet pile wall.

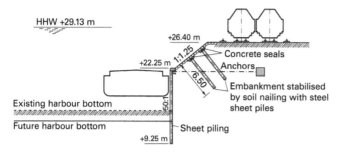

Figure 7.23 Waterfront upgrade with slope secured with soil nailing.

7.1.10 Waterfront structures in regions with mining subsidence

7.1.10.1 General

Waterfront structures in regions liable to mining subsidence must be built to withstand the ground movements expected during their operational life. With regard to mining subsidence, a differentiation is made between vertical ground movements (subsidence) and horizontal ground movements (tensile and compressive ground strains).

As movements in mining subsidence regions generally occur at different times, waterfront structures can be exposed to vertical subsidence and horizontal ground strains in changing sequences.

Local subsidence does not usually influence the groundwater level. Water levels in canals also stay the same and are, therefore, not affected by local subsidence either.

Before building waterfront and other structures in mining subsidence areas, the plans must be submitted to the mining company operating in the area. It is then left to the discretion of the mining company whether to propose safety measures against the effects of mining and to cover the ensuing costs or, alternatively, to cover the costs of rectifying any potential damage resulting from mining subsidence. However, it is actually normally impossible to limit or foresee the extent of any damage in advance.

If the mining company responsible is not willing to undertake measures to prevent damage arising from mining subsidence, or does not consider such measures necessary, and the client is not prepared to provide for such measures, the method of construction selected should be planned and executed so that mining subsidence can be accommodated without severe damage and that any damage can be easily repaired.

Experience has shown that concrete and masonry waterfront structures are frequently seriously damaged by tensile and compressive horizontal ground strains and torsional movements caused by mining subsidence. By contrast, no appreciable damage due to mining subsidence has been seen so far in structures made from steel sheet piles.

For waterfront structures in mining subsidence regions, steel sheet piles can, therefore, be generally classed as suitable, provided several fundamental rules are observed during planning, design and construction.

7.1.10.2 Guidance for planning waterfront structures in mining subsidence regions

The magnitude of the ground movements to be expected must be ascertained from the mining company responsible. Levels and load assumptions are determined on the basis of this data.

Anticipated subsidence, e.g. around canals, can be compensated for by providing a taller waterfront structure. That is generally more economical than raising the height of a wall once subsidence has occurred. If different amounts of subsidence are expected along the length of a waterfront structure, the top of the wall can be raised to match the expected subsidence so that, once subsidence has occurred, the height of the wall is more or less uniform.

In specific cases, it may be appropriate, for reasons of appearance, to increase the heights of quay walls only after subsidence has taken place. However, in these cases, the correspondingly greater loads due to active earth pressure and hydrostatic pressure plus the ensuing increase in anchor loads must be considered during planning in order to avoid subsequent, often very involved, strengthening measures.

Tensile and compressive horizontal ground strains at the level of the waterfront structure do not usually damage U or Z-section sheet piles because the deformation potential of such structures (concertina effect) enables them to withstand ground movements without being overloaded. Compressive horizontal ground strains perpendicular to the waterfront structure displace the wall towards the water. Tensile horizontal ground strains perpendicular to the waterfront structure can increase the loads on anchors when they are very long. However, additional loads on the anchors due to such horizontal strains can usually be accommodated without the anchors failing.

7.1.10.3 Guidance for design and construction

With respect to the loads due to mining subsidence, waterfront structures do not usually have to be designed to withstand higher loads than would be the case outside mining subsidence regions, provided that this is not requested by and paid for by the mining company responsible. This also applies to reinforced concrete capping beams and their reinforcement, provided that the beam remains above the water after the mining subsidence has occurred. Any damage can then be repaired, or a beam completely renewed afterwards.

In order to minimise the susceptibility of quay walls to harmful mining subsidence effects, the portion of the sheet pile wall above the anchors should be as short as possible; anchor bars and walings should be positioned as close as possible to the top of the wall. It is for this reason, and also because of wall deformations caused by tensile horizontal ground strains, that quay walls in mining subsidence areas should be designed to withstand the full active earth pressure and assume that the earth pressure is redistributed.

The steel grades for sheet piles in mining subsidence areas can be chosen according to Section 8.1.2. For capping beams, walings and anchor bars, steel grades S 235 J2 and S 355 J2 to DIN EN 10025 should be used. If walls are anchored with round steel tie rods, upset threaded ends are permissible provided that the requirements of Section 9.1.6.7 are fulfilled. Upset round steel tie rods offer the advantage of greater elongation and greater flexibility than tie rods without upset threaded ends; besides this, they are easier to install and less expensive.

Anchors for waterfront structures in mining subsidence areas should be attached to walings made from pairs of channels, as the waling bolts of such walings can accommodate deformations of the wall more easily. The walings are to be designed so that all effects from mining subsidence can be accommodated without the need for subsequent strengthening. Concerning longitudinal movement of the wall, all walings and steel capping beams should be spliced via elongated holes or holes with sufficient play. If a wall subsequently has to be raised, this must be considered during the design of the capping beams, e. g. by ensuring that the beams can be easily removed. Anchor connections in a capping waling are to be avoided.

Horizontal or slightly inclined anchor rods are advantageous for waterfront structures in mining subsidence areas because only minimal additional stresses are caused in these in the event of differential settlement between anchorages and wall. For the same reason, anchor connections must be hinged. Wherever possible, anchor connections should be installed in the outboard troughs of the sheet piles so that they remain accessible and can be easily inspected.

When accepting delivery of sheet piles according to Section 8.1.2, special attention should be paid to ensuring that the interlock tolerances have not been exceeded.

The deformation potential of sheet piles in the plane of the wall is not seriously impaired by welding the interlocks, e. g. to seal the wall. However, welded interlocks do hinder vertical deformations. Therefore, for sheet piles in mining subsidence regions, do not weld all the interlocks connected on site.

When a sheet pile wall in a mining subsidence area has to be watertight, the interlocks connected in the factory can, therefore, be sealed using an elastic sealing compound that does not hinder vertical movement, and interlocks connected on site should be welded.

Accessible interlocks can also be sealed on site by attaching a plate over an elastic sealing compound in front of the joint. Welding is generally to be avoided as much as possible if it impairs the flexibility of the sheet pile wall.

This requirement applies, in principle, to the interaction of reinforced concrete structural members and sheet piles. It is especially important that the flexibility of the sheet piles is not limited by the presence of heavyweight concrete members. Quay walls and craneways are to be kept separate, with separate foundations, so that they settle independently of each other, and settlement can be directly compensated for. Where a craneway is not laid on sleepers, see Section 7.2.7.2.1, but on reinforced concrete beams instead, the beams should be connected to each other by sturdy ties to maintain the gauge. Electric power should preferably be fed via trailing cables.

7.1.10.4 Monitoring of structures

Waterfront structures in regions at risk of mining subsidence require regular monitoring and reference measurements. Even if the mining company is liable for any damage, the owner of the facility still remains responsible for its safety.

7.2 Equipment

7.2.1 Provision of quick-release hooks at berths for large vessels

Quick-release hooks are provided instead of bollards only in exceptional cases at special berths for large vessels where mooring takes place according to a defined mooring system. The range of movement of the quick-release hook is defined according to the mooring system. Manual and hydraulic release mechanisms with remote control enable simple tying-up and swift casting-off of hawsers, even in the case of heavy hawsers with loads of up to 3000 kN.

Figure 7.24 shows an example of a quick-release hook for a 1250 kN maximum load with a manual release mechanism. It can be used with several hawsers, and it takes little effort to release the hook whether hawsers are carrying full or lower loads.

A quick-release hook is attached to its base via a universal joint. The number of quick-release hooks depends on the line pull to be considered according to Section 4.9 and the directions from which the principal line pulls can occur simultaneously. Several quick-release hooks can be installed on one base. The range of movement must be chosen so that the hook can cover all anticipated operational requirements without jamming. The swivel range is a maximum of 180° horizontally and 45° vertically.

It is easier to attach heavy-duty towing hawsers when the quick-release hook is combined with a capstan.

7.2.2 Layout and design of and loads on access ladders

7.2.2.1 Layout

Vertical ladders are used for vessel embarkation/disembarkation in exceptional circumstances only. They are primarily intended to provide access to mooring equipment and, in emergencies, to enable persons who have fallen into the water to climb onto the quayside. Trained and experienced shipping and operations personnel may also be expected to use the ladders when there are very large water level fluctuations, even in the case of great differences in water levels.

Vertical ladders in reinforced concrete waterfront structures should be placed at approximately 30 m intervals. The position of the ladder depends on the position of the bollard because ladders must not be obstructed by mooring lines. If joints between sections have been included in reinforced concrete quay walls, it is advisable to place the ladders near those joints. In the case of sheet pile waterfront structures, positioning of the vertical ladders in the pile troughs is recommended.

Mooring equipment should be installed on both sides of each ladder (Section 7.2.6).

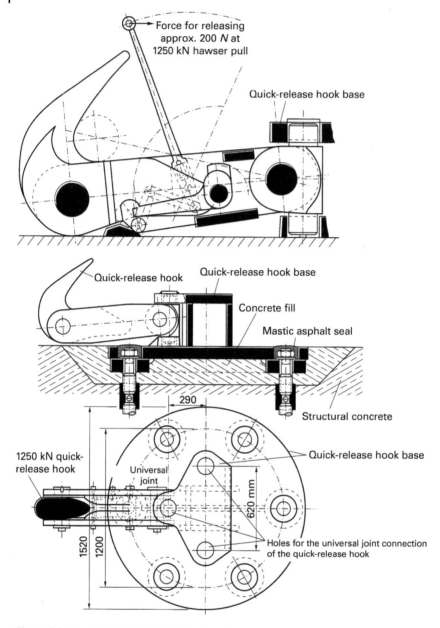

Figure 7.24 Example of a quick-release hook.

7.2.2.2 Design

In order to be accessible from the water at all times, even at low water levels, each ladder must extend down 1.00 m below the lowest low water or lowest astronomical tide. For easy installation and replacement, the lowest ladder mountings are designed as plug-in items into which the stiles can be inserted from above.

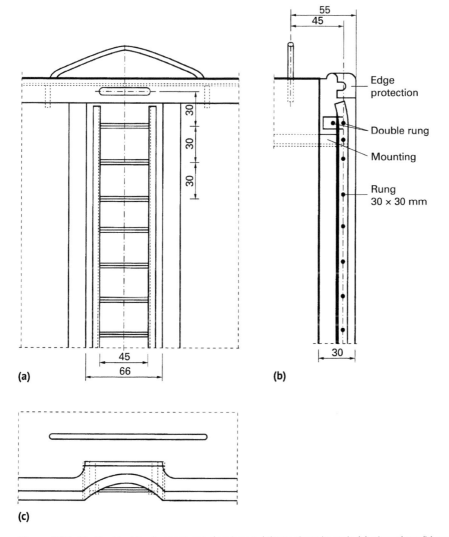

Figure 7.25 Vertical ladder in steel capping beam (dimensions in cm): (a) elevation; (b) section; (c) plan.

Transitions between the top of the ladder and the quayside must be designed to ensure that ascending and descending the ladder can be accomplished safely. At the same time, the ladder must not be a hazard to traffic on the quayside.

Figure 7.25 shows a tried-and-tested design that satisfies these two requirements. Here, the edge protection is dished at each ladder. The detail includes a handrail, made of 40 mm diameter material about 30 cm above the quayside and its longitudinal axis about 55 cm from the face of the wall. If the handrail proves to be an obstacle during cargo-handling, other suitable aids for climbing the ladder must be provided.

Figure 7.26 shows a proven design of this type. The topmost rung of the ladder in this solution is 15 cm below the top of the quay wall.

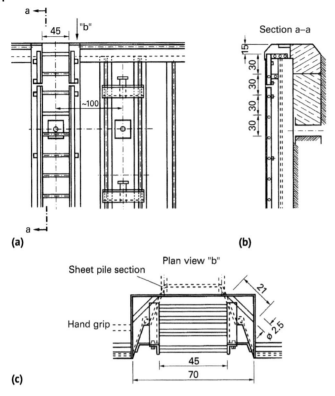

Figure 7.26 Vertical ladder in reinforced concrete capping beam (dims. in cm): (a) elevation; (b) section a–a; (c) plan.

Ladder rungs should be installed with their centre a minimum of 10 cm behind the face of the quay wall and should be made from 30 mm square steel bars that are installed so that one edge points upwards. This reduces the risk of slipping due to ice or dirt. Rungs should be fastened to the stiles at a centre-to-centre distance of 28–30 cm, with a clear width between stiles of 45 cm.

7.2.3 Layout and design of stairs in seaports

7.2.3.1 Layout of stairs

Stairs are used in seaports where persons not acquainted with the conditions in ports require access and where such persons cannot be expected to use vertical ladders to climb from ships to waterfront structures. The upper end of the stair should be placed so that there is little or no interference with foot traffic and cargo-handling. The approach to the stair must be clearly visible and, thus, permit the smooth flow of foot traffic. The lower end of a stair should be positioned so that ships can berth easily and safely with safe passage between the ship and the stair.

7.2.3.2 Practical stair dimensions

Stairs should be a maximum of 1.50 m wide, so that they can be positioned in front of the outboard crane rail on quay walls for seagoing vessels, without projecting into the area of the fixings for the crane rail, which are 1.75–2.50 m from the edge of the quay. The pitch of the stair should be determined using the well-known formula $2s + a = 59-65$ cm (rise s, going a). Concrete steps should have a rough, granolithic concrete finish, and the front edge of each step should be fitted with a steel nosing for protection.

7.2.3.3 Landings
For larger tidal ranges, landings should be positioned at 0.75 m above mean low tide, mean tide level and mean high tide, respectively. Depending on the height of the structure, further landings may be necessary. Intermediate landings are to be positioned after maximum 18 steps; the length of the landing should be 1.50 m or equal to the width of the stair.

7.2.3.4 Railings
Stairs should be fitted with a handrail whose upper edge is 1.10 m above the front edge of the pitch line. Where port operations permit, stairs should be enclosed by a 1.10 m high railing, which can be removable if necessary.

7.2.3.5 Mooring equipment
The quay wall next to the lowest landing should be equipped with mooring hooks. In addition, a recessed bollard or mooring hook should be positioned below each landing. Recessed bollards are used in concrete or masonry quay walls or wall components; mooring hooks are generally used for steel sheet pile structures.

7.2.3.6 Stairs in sheet pile structures
Stairs in sheet pile wall structures are made from steel. The sheet piles at stairs are set back to create a recess large enough to contain the stair. The stair must be protected against underrunning by suitable means (e. g. fender piles).

7.2.4 Armoured steel sheet pile walls

7.2.4.1 The need for armouring, applications
Ever larger vessels, convoy traffic and more powerful engines have resulted in increased operational requirements for waterfronts in inland ports and on waterways. To prevent

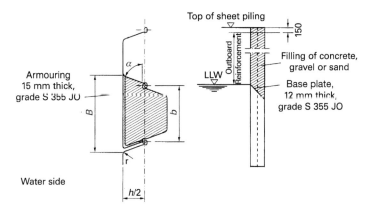

Figure 7.27 Armouring to a U-section sheet pile.

damage, sheet pile waterfront structures must, therefore, have front faces that are as flat as possible. This can be achieved by welding plates in or across the troughs of the sheet piles to strengthen the structure (Figure 7.27). The reinforcement creates a uniformly flat waterfront structure with less elasticity than a wall without such strengthening.

Owing to the technical input and the costs, armouring is only recommended for stretches of waterfront that are particularly exposed to waterborne traffic loads. In inland ports, these are waterfronts with very heavy traffic with push lighters and large motor vessels, waterfronts at changes of direction and the guidance structures at lock entrances.

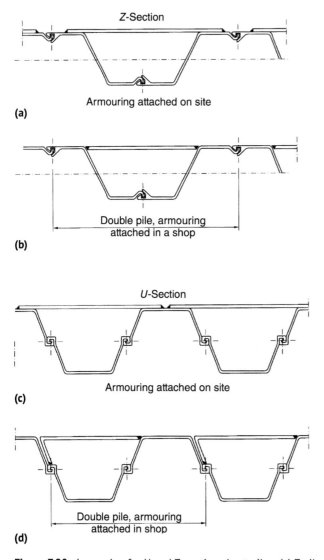

Figure 7.28 Armouring for U and Z-section sheet piles: (a) Z-piles – armouring attached on site; (b) Z-piles – armouring attached at the works; (c) U-piles – armouring attached on site; (d) U-piles – armouring attached at the works.

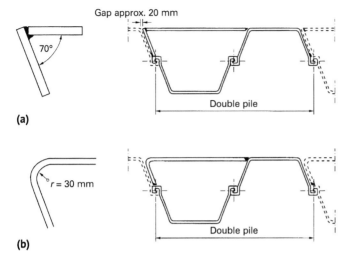

Figure 7.29 Armouring attached at the works: (a) welded; (b) bent.

7.2.4.2 Construction details and the method of armouring

Sheet pile reinforcement is required in the area between the lowest and highest possible water levels where vessel impacts are likely (Figure 7.27). Armouring dimensions depend primarily on the width B of the sheet pile trough. This is determined by the system dimension b of the wall, the angle α of the web of the pile, the section depth h, the radius r between the web and flange of the sheet pile (Figure 7.27) and the section form (U or Z). In addition, it is necessary to distinguish between plating that has been factory-set before driving and armouring attached on site by welding on after driving.

In the case of walls made from Z-sections and armouring attached on site, the plates extend over the full width of two sections right up to the interlocks. This reinforcement projects beyond the section and so also protects the interlocks (Figure 7.28). Welding on the armouring in the workshop cannot be recommended for Z-sections because the armouring stiffens the piles to such an extent that, during driving, they can no longer compensate for unavoidable driving deviations between neighbouring double piles. With walls made from U-sections, the backs of the piles and the free leg of the forward pile can deform elastically.

When installing the armouring on site, pressing and adjustment work will be unavoidable with both types of section. Walls made from U-sections with armouring fitted in the shop are better when it comes to driving and shipping operations. This type of wall armouring creates a completely flat surface (Figure 7.28).

A. Armouring attached at the works for U-piles

When attaching armouring to U-section piles in the fabrication shop, the straight or bent strengthening plates (Figure 7.29) are welded to the rearward pile interlock or on the flange of the forward pile (Figure 7.28). In addition, the interlock of the double pile must be welded to create a rigid connection. Only then can the connecting welds of the armouring survive the driving procedure without being damaged.

Strengthening plates can be straight or bent (Figure 7.29). The gap between the piles is approximately 20 mm wide with straight welded plates so that, related to a system dimen-

7 Configuration of cross-sections and equipment for waterfront structures

sion of 1.0 m, up to about 98% of the wall is closed. In the case of walls with bent plates, the gap is wider due to the minimum radius that must be observed during cold forming.

B. Ladders within the armouring

Sheet pile armouring is generally interrupted around ladders and recessed bollards. If a largely flat surface is also required here, the ladder pile can be provided with recessed

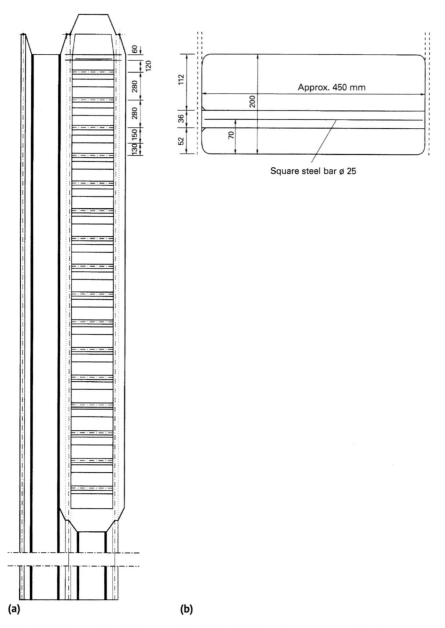

Figure 7.30 Armouring with recessed footholes: (a) elevation on double pile with armouring in the form of a ladder; (b) detail of a foothole box.

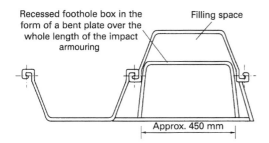

Figure 7.31 Armouring with a recessed foothole box.

footholes (Figure 7.30), or a continuous recessed foothold box can be included as per Figure 7.31.

C. Recessed bollard within the armouring

The positioning of recessed bollards in an armoured wall is shown in Figure 7.32. This solution complies with the requirements of Sections 7.2.6.1 and 7.2.6.2.

D. Filling behind the armouring

When filling the space between armouring and pile, a base plate is generally welded in to retain the filling (Figure 7.27). Sand, gravel or concrete are used as the filling material.

E. Dimensions of the armouring

The plate thickness required for the armouring is derived from the width of the sheet pile trough. As this is always larger than the flange width of the forward pile, the strengthening plates must be thicker than the flange of the pile. To avoid plate thicknesses > 15 mm, the space between armouring and pile can be filled.

Armouring is not taken into account in the structural design of the wall.

7.2.5 Equipment for waterfront structures in seaports with supply and disposal systems

7.2.5.1 General

Supply systems provide public installations and facilities, the businesses in the port and moored ships, etc., with the operational media, power, etc., that they require. Disposal systems serve to drain any wastewater and operational media.

The supply and disposal systems must be located in the immediate vicinity of a waterfront structure, sometimes directly in the structure itself.

Adequate openings for these services must be provided in the structural members of waterfront structures, e. g. in craneway beams. Therefore, consultation among all participants must take place during the planning of supply and disposal systems. Spare openings must be included to allow for any later expansion.

Supply systems include:

- Water supplies
- Electric power
- Communication and remote control systems
- Other systems.

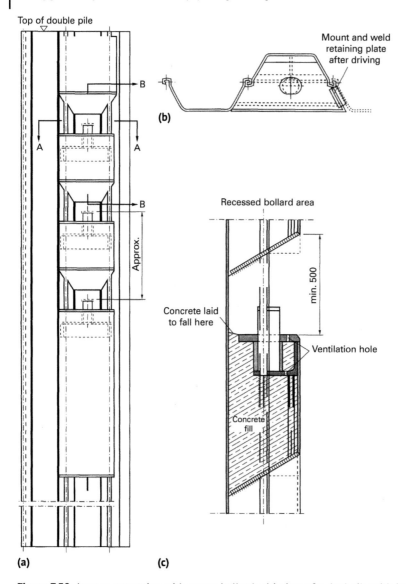

Figure 7.32 Impact armouring with recess bollards: (a) view of a dual pile with impact armouring and bollard recess; (b) Section A–A; (c) Section B–B.

Disposal systems include:

- Rainwater drainage
- Wastewater drainage
- Fuel and oil interceptors.

The respective disposal regulations must be observed.

7.2.5.2 Water supply systems
Water supply systems provide drinking and process water and normally can also be used for extinguishing fires.

7.2.5.2.1 Drinking and process water supplies
In order to safeguard the drinking and process water supply systems in the port, at least two independent supply points are required for each port section, with the lines laid out as ring systems to guarantee a permanent flow.

Hydrants should be installed at approximately 100–200 m intervals; every 60 m along the quayside is a typical spacing for water hydrants for supplying ships. Underground hydrants are placed on quay walls and in paved crane and rail areas so as not to hinder operations. The hydrants must be arranged so that there is no danger of them being crushed by railborne cranes and vehicles, even when standpipes are fitted.

When using underground hydrants, special attention must be paid to protecting the connection coupling against contamination, even in case of any possible flooding of the quay wall. An additional shut-off valve is required to isolate the hydrant from the supply line. Hydrants must be accessible at all times. They must be situated in areas where operations prevent goods being stored.

The pipes are normally laid with an earth cover of 1.50–1.80 m. To protect them against frost, they are also placed at least 1.50 m from the front face of the quay wall. In loaded zones with railway tracks, the lines should be placed in protective ducts.

In quay walls with concrete superstructures, the lines may be placed in the concrete structure. Here, the different deformation behaviour of adjacent structural sections must be taken into account, together with the differing settlement behaviour of structures on deep or shallow foundations. Drinking water is typically supplied via inboard ring mains and branches to the hydrants located at the front of the quay. It must be possible to drain the branches so that, for hygiene reasons, water can be drained from those branches that are not constantly in use.

To reduce expenditure in the case of a burst pipe, pipes used for water supplies should not be laid under areas covered by concrete. Typically, they are located under paved strips reserved for such services.

7.2.5.2.2 Separate fire-fighting water
When there is a high fire risk in a certain section of a port, the recommendation is to supplement the drinking and process water supplies system with an independent fire-fighting system. Water for fighting fires is pumped directly from the port basin. The associated pumping stations can be located within the quay wall below ground so as to not disturb cargo-handling.

It is also possible to feed fire-fighting water into the system from the pumps of the fireboats via special connection points.

In sheet pile quay walls, the suction pipes may be placed in the troughs of the sheet piles, where they are adequately protected against vessel impacts. In concrete superstructures such pipes should be positioned in slots for protection.

The routing of the pipes of the fire-fighting system must satisfy the same requirements as the drinking and process water supplies.

7.2.5.3 Electricity supply systems

Office buildings, port installations, cranes, lighting to railway tracks, roads, operations areas, open areas, quays, berths, dolphins, etc., must all be provided with electric power. Only buried cables may be used for the high and low-voltage supply systems in ports, except during construction. The cables should be laid in the ground in plastic ducts with an earth cover of approximately 0.80–1.00 m in quay walls and operations areas; concrete cable drawpits designed to accept vehicular traffic are also required. The advantage of such a duct system is that the cable installations can be augmented/modified without interrupting port operations.

When there is a risk of frequent flooding of the quayside, power sockets must be mounted on posts raised above the flood level.

Power sockets are generally installed in the top of the quay wall at intervals of approximately 100–200 m. They must be capable of accepting vehicular traffic and be fitted with a drainpipe. These sockets are used to provide the power for welding equipment when carrying out minor repairs on ships and cranes as well as the power for emergency lighting and other purposes.

Ducts for conductor rails, cable troughs and crane power feeding points must be provided in the quayside for power supplies to cranes. The drainage and ventilation of these facilities is particularly important. In quay walls with concrete superstructures, these facilities can be incorporated in the concrete structure.

Special attention is drawn to the fact that electricity supply networks require equipotential bonding facilities. This is to prevent unduly high voltages occurring in crane rails, sheet piles or other conductive components of the quay wall due to a fault in any electrical system (e.g. a crane). Such equipotential bonding systems should be installed about every 60 m.

In the case of craneways integrated into the quay wall superstructure, the equipotential bonding lines are normally concreted in during construction of the superstructure, for reasons of cost. However, they must be laid in protective ducts with sufficient freedom of movement in areas in which differential settlement might be expected.

As a contribution to avoiding air and noise pollution, it may also be necessary to provide ships moored in ports with electricity from the public network. Up until now, connections to an onshore power supply (OPS) were provided in individual instances only (Figures 7.33–7.35). Nevertheless, the recommendation is to consider the possibility of supplying ships with electric power from onshore when building new or converting existing facilities. In addition to taking into account the frequency conversion from the public network (20–100 kV/50 Hz) to the respective onboard system (6–12 kV/60 Hz), it is also necessary to consider the various consumption requirements on board the vessel, which can lie between 5 and 10 MW in the case of very large cruise liners.

Equipment for connecting vessels to onshore power supplies provided at berths within flood zones or tidal areas must be protected against flooding.

7.2.5.4 Other systems

Other systems include all supply systems not mentioned in Sections 7.2.5.2 and 7.2.5.3 but are required, for example, in the quay walls to shipyards. These include gas, oxygen, compressed air, acetylene, steam and condensate lines. The layout and installation of such facilities must comply with relevant regulations, particularly safety regulations. Connections for telephones are usually placed at a spacing of 70–80 m along the front edge of the quay.

Figure 7.33 Example of an onshore power supply arrangement for ferries within the flood zone (Föhr Island, ferry berth).

Figure 7.34 Example of an onshore power supply arrangement for inland waterway vessels with a pontoon within the tidal area (Bremen, Osterdeich berth for inland waterway vessels).

Figure 7.35 Example of an onshore power supply arrangement for cruise liners (Hamburg, Altona cruise liner terminal).

Although the growing availability and use of mobile phones means that such telephone points are being relied on less and less, port authorities nevertheless usually require that a hazardous goods landline telephone be installed.

7.2.5.5 Disposal systems
7.2.5.5.1 Rainwater drainage
The rainwater falling in the quay wall area and also on its land side is drained into the harbour directly via the quay wall. To do this, the quay and operations areas are provided with a drainage system consisting of inlets, transverse and longitudinal channels and a main drain with outlet into the harbour. The sizes of drainage catchment areas depend on local circumstances. As few outlets as possible should be installed in the waterfront structure. They should be designed and installed in such a way that they are not damaged by moored ships or vessel impacts.

To prevent pollution of the harbour water, the outlets are to be provided with gate valves in quay and operational areas where there is a risk of dangerous or toxic substances or contaminated fire-fighting water entering the drainage system.

7.2.5.5.2 Wastewater disposal
It is currently not customary for the port operator to accept the wastewater from seagoing vessels. The wastewater occurring in the port itself is fed through a special waste disposal system into the municipal sewer network. It may not be drained into the port basin. Wastewater drains are, therefore, only found in waterfront structures in exceptional cases.

7.2.5.5.3 Fuel and oil interceptors

Fuel and oil interceptors must be included wherever they are required, in the same way as non-port facilities.

7.2.5.5.4 Disposal regulations for ship waste

Corresponding to the MARPOL convention, ports should provide facilities for the disposal of ship waste, such as liquids containing oils and chemicals, solid ship waste (galley waste and packaging refuse) and sanitary wastewater.

7.2.6 Layout of bollards

See Section 4.9 for loads on bollards.

7.2.6.1 Layout of bollards for sea-going vessels

Bollards for sea-going vessels must be provided to suit requirements. The following recommendations for mooring vessels apply unless port operators specify otherwise:

- Large vessels with a length ≥ 330 m (ultra large container ships, ULCS) require individual bollards at a spacing of approximately 10 m or double bollards every 15–20 m.
- Smaller vessels require individual bollards at a spacing of approximately 15 m or double bollards at a spacing of approximately 30 m.
- Where there are joints between blocks, the bollards should be arranged symmetrically within the blocks.
- In principle, the recommendation for mooring vessels with a length ≥ 330 m or a displacement ≥ 100 000 t is to carry out a dynamic mooring analysis in order to define the bollard loads and spacings. In the analysis, the number of hawsers to be attached to one bollard is a crucial factor. As a rule, the number of hawsers is limited to a maximum of three per bollard, depending on the design loads of the bollard and the hawsers.

Container ships are becoming larger and larger, which is leading to a greater number of hawsers being required for mooring. Currently, large vessels are moored with a 3-3-2 configuration, i.e. three head, three forward breast and two forward spring lines at the bow, and three stern, three aft breast and two aft spring lines at the stern. In future, however, large vessels will require a 4-4-2 mooring configuration at least.

An example of the distance of bollards from the face of the quay wall can be found in Section 7.1. The distance of a bollard from the front edge of the concrete (edge of quay) depends on the type of bollard and the arrangement of the reinforcement in the concrete element. There should be a clear working space at least 1.2 m wide behind every bollard.

7.2.6.2 Layout of bollards for inland waterway vessels

In inland ports, vessels should be moored to the shore by three hawsers (known as lines): a head line, a breast line and a stern line. To this end, an adequate number of bollards must be provided on the bank.

Bollards must be arranged on and above the port operating level, with their top edges extending above the highest navigable water level and, if possible, above the highest high water. The diameter of such bollards must be > 15 cm. If the bollard does not extend sufficiently above highest high water, slippage of the line must be prevented by a cross-rail. In addition to bollards along the top of the bank, river ports must be equipped with other bollards at various elevations corresponding to the fluctuations in the local water level. Only then can the ship's crew moor the ship without any difficulty at every water level and every freeboard height.

In the case of vertical quay walls, the bollards at varying heights should be situated in a vertical line one above the other. The positions of the lines depend on the positions of the vertical access ladders. To avoid mooring lines across ladders, a line of bollards should be located to the left and right of every vertical ladder at a distance of approximately 0.85–1.00 m from the axis of the ladder in the case of concrete or masonry walls and at double the sheet pile spacing in the case of sheet pile walls. The spacing between ladders/lines of bollards should be about 30 m. In the case of steel sheet pile walls, the exact spacing depends on the system dimension of the sheet piles; in the case of concrete or masonry walls the spacing is determined by the block length where there are joints between blocks.

The lowest bollard should be approximately 1.50 m above lowest low water or the mean low water spring level in tidal areas. The vertical distance between the lowest bollard and the upper edge of the quay is divided up by further bollards at a spacing of 1.30–1.50 m (up to 2.00 m in borderline cases).

Bollards on reinforced concrete waterfront structures should be placed in recesses or, in the case of walls set back towards the land, bolted to steel plates cast in flush with the surface of the concrete, with the housings for the plates provided with anchors and cast in. In the case of steel sheet pile walls, the bollards can be bolted or welded in position. The front edge of the bollard post should be 5 cm behind the front edge of the quay. An appropriate gap should be left at the sides, behind and above the bollard post, so that ships' hawsers can be easily looped over and removed again. Edges between bollard recess and front of quay must be rounded to prevent any damage to hawsers or waterfront structure.

In the case of (partially) sloping banks, the bollards should be positioned on both sides next to the steps (Figure 7.36). Steps should form extensions to ladders.

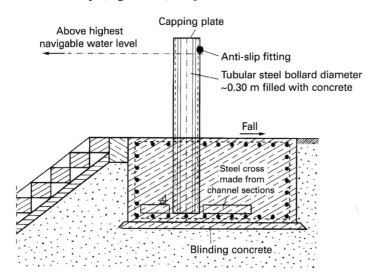

Figure 7.36 Bollard foundation for a partially sloping bank (the drawing shows an example of typical layout; the actual design must be based on structural requirements).

In this arrangement, it is expedient to continue the bollard foundation beneath the steps to create a joint foundation for both bollards.

7.2.7 Foundations to craneways on waterfront structures

7.2.7.1 General

In many cases constructional requirements make it practical to construct a deep foundation for the outboard crane rail as an integral part of the quay wall, whereas the foundation for the land-side rail is generally independent of the waterfront structure. In inland ports the outboard crane rail is also frequently on a foundation separate from the waterfront structure. This facilitates any later modifications that might be required, e. g. different operating conditions due to new cranes or alterations to the waterfront structure. Separate foundations for waterfront structure and craneways may also be necessary in the case of different responsibilities and ownership.

Whether the outboard and/or inboard craneway requires a deep foundation depends on the in situ subsoil and whether the unavoidable settlement associated with shallow foundations can be permitted.

Owing to the fact that the settlement of the ground also affects the decision regarding the type of foundation for the craneway, an appropriate assessment must be carried out (Section 3.2.2).

Please refer to Section 10.3.4 for details of designing long craneway beams without joints. Craneway beams over 2000 m long have been constructed successfully without joints.

7.2.7.2 Design of foundations, tolerances

Craneway foundations may be shallow or deep depending on the local subsoil conditions, the sensitivity of the cranes to settlement and displacements, the crane loads, etc.

The permissible dimensional deviations of the craneway must be taken into account here, distinguishing between tolerances during installation and tolerances during operation.

The installation tolerances mainly relate to the permissible dimensional deviations during the laying and fixing of crane rails and are, therefore, not usually relevant when selecting the type of foundation. Operational tolerances, on the other hand, relate to permissible settlement and differential settlement during operation and are, therefore, critical when deciding on the type of foundation.

Depending on the design of the crane portal, the following values can be taken as a guide for the operational tolerances:

- Level of one rail (longitudinal gradient): 2–4‰.
- Level of rails in relation to each other (cross-level): max. 6‰ of gauge.
- Inclination of rails in relation to each other (offset): 3–6‰

Considerably tighter operational tolerances apply, e. g. 1‰ for the longitudinal gradient, when using special cargo-handling equipment, e. g. container cranes. Operational tolerances should be specified together with the crane manufacturer in every single case.

The reader is referred to HTG (1985) for more information on the relationship between craneways and crane systems.

7.2.7.2.1 Shallow craneway foundations

7.2.7.2.1.1 Strip foundations of reinforced concrete In soils that are not sensitive to settlement, the craneway beams may be constructed as shallow strip foundations in reinforced concrete. The craneway beam is then calculated as an elastic beam on an elastic foundation. An analysis should be carried out to verify the soil pressures beneath the beam. Settlement and differential settlement must also be analysed and compared with the agreed upon operational tolerances.

DIN 1045 applies to the design of the beam cross-section. The action effects due to vertical and horizontal wheel loads – also due to braking along the craneway axis – must be verified.

Craneways with a narrow gauge, e. g. gantry cranes spanning only one track, require the gauge to be maintained with tie beams or tie bars installed at a spacing roughly equal to the gauge. With wide gauges, both crane rails are designed separately on individual foundations. In this case, the cranes must be designed with a pinned leg on one side. Please refer to Section 7.2.8 for the design of the rail fastening.

Settlement of the craneway beam of up to 3 cm can still be accommodated, generally by installing rail bearing plates or by exploiting the adjustment options of rail chairs – work that can even take place during operations without causing serious disruptions. In the case of greater settlement and settlement that abates only slowly during operations, a deep foundation will generally be more economical because it will not be possible to compensate for the settlement merely by inserting bearing plates or adjusting rail chairs. That means costs and longer downtimes.

7.2.7.2.1.2 Sleeper foundations Crane rails on sleepers on a ballast bed are comparatively easy to realign and so are used primarily in mining subsidence regions and where excessive settlement is expected. Even substantial movements can be corrected quickly by realigning level, lateral position and gauge. Sleepers, sleeper spacing and crane rails are calculated according to the theory of the elastic beam on an elastic foundation and in accordance with permanent way standards. Timber, steel, reinforced concrete and prestressed concrete sleepers can be used. Timber sleepers are preferred at facilities for loading/unloading lump ores, scrap and similar cargoes because of the reduced risk of damage from falling pieces.

7.2.7.2.2 Deep craneway foundations

On soil sensitive to settlement or deep fill, craneway beams should be founded on piles. If the piles are installed deep enough, the deep foundation to the craneway beam also relieves the loads on the waterfront structure because the loads from the craneway beam are no longer carried by the structure itself.

Basically, all customary types of pile can be used for deep foundations under craneways. The piles beneath the outboard craneway in particular are loaded in bending, as well due to the deflection of the quay wall. Likewise, larger asymmetric imposed loads can lead to considerable additional, horizontal loads on the piles.

All horizontal forces due to crane operations must either be resisted by the mobilised passive earth pressure in front of the craneway beam, by raking piles or by anchors.

The craneway beam is designed as an elastic beam on an elastic foundation.

Instead of deep foundations on piles, craneway beams can also be mounted on shallow foundations on soil that has been improved or introduced through soil replacement measures.

7.2.8 Fixing crane rails to concrete

Crane rails are to be attached free from stresses but allowing longitudinal movement. Crane rail fixings tested for the respective type of use are available. A number of options that can be used for mounting crane rails on concrete are given below.

7.2.8.1 Supporting the crane rail on a continuous steel plate on a continuous concrete base

When supporting the crane rail on a continuous steel plate, the steel plate is first aligned as flat as possible on the craneway beam and then suitably grouted or bedded on earth-damp, compacted single-sized aggregate concrete. The crane rail is only guided longitudinally on the continuous steel plate but is anchored vertically in such a way so that even uplift forces due to the interaction between bedding and rail can be accommodated. When calculating the maximum moment, anchorage force and maximum concrete compressive stress, the modulus of subgrade reaction method may be used.

Figure 7.37 shows an example for a heavy crane rail. Here, the concrete base was tamped in between the steel angles, levelled and given a levelling coat ≥ 1 mm of synthetic resin or a thin bituminous coating.

If an elastic intermediate layer is placed between the concrete and the continuous steel plate, both rail and anchorage must be calculated for this softer support, which can lead to larger dimensions. The rails should be welded to minimise the number of joints. Short pieces of rail are used to bridge over expansion joints between quay wall sections.

7.2.8.2 Bridge-type arrangement with rails supported centrally on bearing plates

In this arrangement, special bearing plates are used, which assure a concentric transfer of the vertical forces into the craneway beams and also guide the rails, which are able to move longitudinally. The tall rails required for this type of support must be prevented from overturning.

This type of crane rail fixing is used for normal general cargo cranes and, in inland ports, is also preferred for bulk cargo cranes. The heavy-duty design is to be recommended, above all, for the craneways of heavy-duty cranes, very heavy unloaders, unloading gantries, etc. Rail sections S 49 and S 64 are used as running rails in lightweight systems. Heavy-duty installations require PRI 85 or MRS 125 in accordance with parts 1 and 2 of DIN 536, or very heavy special rails made from steel grade St 70 or St 90.

A typical example of a light-duty installation is shown in Figure 7.38. Here, S 49 or S 64 rails are supported on bearing plates according to the K-type permanent way of Deutsche Bahn AG. Rail, bearing plates, anchors and special anchors are fully assembled on the formwork or a special adjustable steel support that can be mounted rigidly. The concrete is then placed with the aid of vibration, so that the bearing plates are supported across their entire area. Occasionally, an intermediate layer of approximately 4 mm thick plastic is placed between bearing plate and underside of rail (Figure 7.38). With cambered bearing plates, the detailing must ensure that the plastic interlayer cannot slide off.

Figure 7.39 shows a heavy-duty craneway in which the support for the rail is cambered upwards in the longitudinal direction, so that the rail is supported on the camber. A non-shrink material is placed or packed beneath the bearing plates. The bearing plates are also provided with elongated holes in the transverse direction, so that changes to the gauge can

246 | 7 Configuration of cross-sections and equipment for waterfront structures

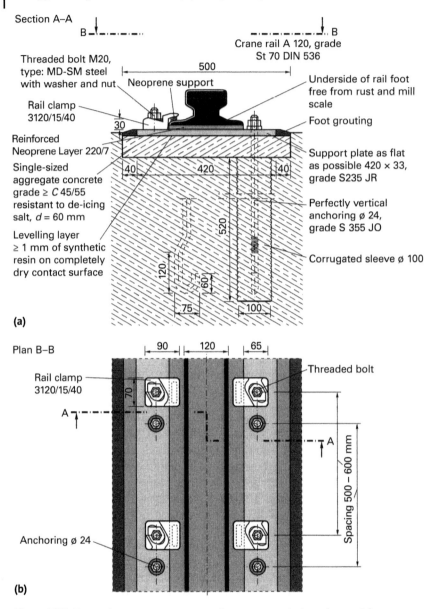

Figure 7.37 Heavy-duty craneway on a continuous concrete base (example): (a) section A–A; (b) plan B–B.

be corrected if need be. This type of support must be provided primarily for rails with a deep web.

However, for crane rails with small section moduli, e. g. A 75–A 120 or S 49, continuous support is recommended for loads exceeding approximately 350 kN because otherwise the spacing between the plates or chairs becomes too small.

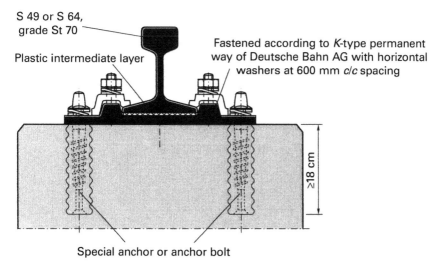

S 49 or S 64, grade St 70

Plastic intermediate layer

Fastened according to K-type permanent way of Deutsche Bahn AG with horizontal washers at 600 mm c/c spacing

≥18 cm

Special anchor or anchor bolt

Figure 7.38 Light-duty craneway on individual supports.

7.2.8.3 Bridge-type arrangement with rails supported on chairs

When rails are supported on chairs, the rail – from a structural point of view – becomes a continuous beam on an infinite number of supports. In order to exploit the elasticity of the rail, an elastic interlayer is inserted between the rail and the chair. This layer is up to 8 mm thick and can be made of, for example, neoprene (for bearing pressures up to $12\,\text{N/mm}^2$) or a textile-reinforced rubber. It also cushions the crane wheels and chassis against impacts and shocks.

The top of the rail chair is cambered, which results in a centralised transfer of the support reaction into the concrete. This cambered bearing lies some distance above the concrete, and the flexibility of the spring washers in the rail fixings allow the rail to expand in the longitudinal direction. Changes in the length of the rails due to temperature changes and rocking movements can, thus, be accommodated (Figure 7.40). Through flexible shaping, the rail chairs can be adapted to any desired requirements. For example, the chairs permit subsequent rail realignment of, for example, $\Delta s = \pm 20$ mm in the transverse direction and $\Delta h = +50$ mm in the vertical direction.

In addition, lateral pockets can be included to accommodate the edge protection angles at crossing points.

The chairs are mounted together with the rails. Following alignment and fixing in position, additional longitudinal reinforcement is inserted through special openings in the chairs and connected to the projecting bars of the substructure (Figure 7.40). The concrete grade depends on structural requirements but should be grade C20/25 at least.

If settlement and/or horizontal displacement of the crane rail necessitate subsequent realignment of the rail, this must be taken into account right at the planning stage by choosing chairs with appropriate adjustment options.

7.2.8.4 Traversable craneways

The demands of port operations frequently require the crane rails to be installed sunk into the quay surface, so that they can be crossed without difficulty by vehicular traffic and

7 Configuration of cross-sections and equipment for waterfront structures

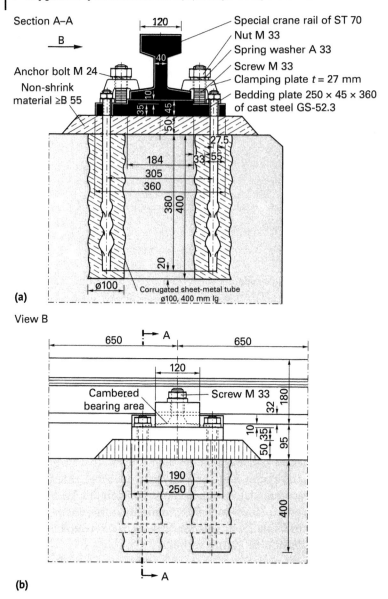

Figure 7.39 Heavy-duty craneway on packed individual supports: (a) section A–A; (b) elevation B.

cargo-handling gear. Crane rails must, therefore, be installed so that they are flush with the quay surface.

1. Traversable heavy crane rails
   Figure 7.41 shows an example of a proven form of construction for a traversable heavy-duty crane rail. The rail is supported on a craneway beam on a bedding of single-sized aggregate concrete, grade C 45/55 at least, levelled off horizontally by means of a flat steel bar. To distribute the loads, the rail, which has a thin coat of bitumen on its underside, is supported on a continuous bearing plate that is bedded on a > 1 mm synthetic

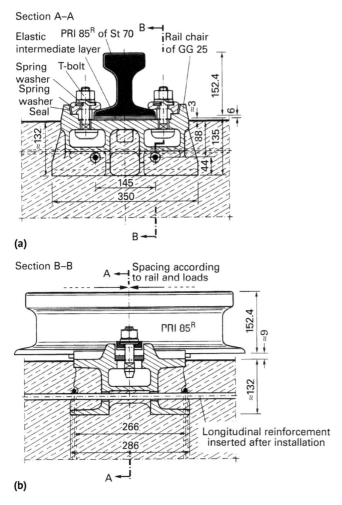

Figure 7.40 Example of a heavy-duty craneway on chairs: (a) section A–A; (b) section B–B.

resin levelling layer. To prevent loads from the longitudinal movement of the rail and bearing plate being transferred to the bolts holding the rail in position, the bearing plate is not connected to the rail fixings. Subsequent installation of the bolts is preferable because this helps to position the bolts exactly. However, this approach must be allowed for when placing the reinforcement in the craneway beam so that enough space is left between the reinforcing bars for the cast-in sheet metal or plastic sleeves for the bolts. If necessary, the holes for the bolts can also be drilled subsequently.

In order to transfer horizontal forces transverse to the rail axis and hold the rail exactly in position, approximately 20 cm wide cleats of synthetic resin mortar are inserted between the foot of the rail and the side of the concrete topping approximately every 1 m.

It is expedient to use a permanently elastic, two-part filler for the top 2 cm of the mastic compound in the reinforced concrete topping joined with stirrups to the rest of the craneway beam.

See Figure 7.41 for further details.

250 | 7 Configuration of cross-sections and equipment for waterfront structures

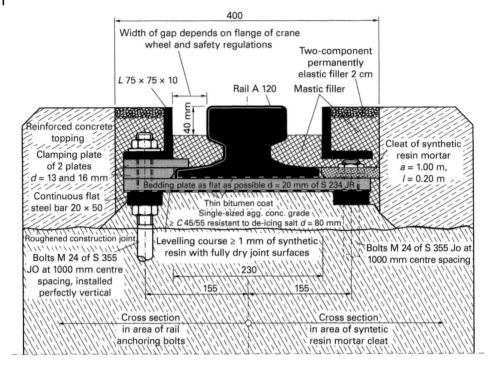

Figure 7.41 Example of traversable heavy-duty craneway (reinforcement omitted for clarity).

2. Traversable light-duty crane rails

   A tried-and-tested example of this is shown in Figure 7.42. Horizontal ribbed plates are fastened to the flat, levelled craneway beam with anchors and screw spikes at a pitch of approximately 60 cm. The crane rail, e. g. S 49, is connected to the ribbed plates with clamping plates and T-bolts in accordance with Deutsche Bahn AG specifications. Levelling plates of steel, plastic, etc., can be installed beneath the foot of the rail to correct minor differences in the level of the concrete surface.

   A continuous steel angle (80 × 65 × 8 mm, grade S 235 JR) is installed to form an abutment for any adjoining reinforced concrete ground slabs. The angle is fitted parallel with the head of the rail, beneath which 80 mm channel Sections 80 mm long are welded at every third ribbed plate. These sections have elongated holes in the (horizontal) web for fastening with T-bolts. At the intermediate ribbed plates, the angle is stiffened by 8 mm thick steel plates. Holes are cut in the horizontal leg of the angle above the fastening nuts, which are covered by 2 mm thick plates after the nuts have been tightened. A bar is welded alongside the rail on the toe of each angle to hold the subsequent mastic filler in place.

   To pave the port area, large reinforced concrete slabs, e. g. Stelcon, are laid loosely up against the steel angle. It is recommended to lay rubber sheets underneath to prevent tilting of the slabs and at the same time to create falls that drain away from the crane rails.

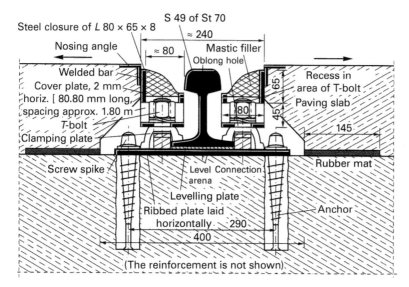

Figure 7.42 Example of a traversable light-duty craneway.

7.2.8.5 A note on rail wear

The wear to be expected for the foreseeable service life of all crane rails must be taken into account in the design. As a rule, a height reduction of 5 mm with good rail support is adequate. Furthermore, more or less frequent maintenance and, depending on the type, checks of the fixings, are recommended during operation to prolong the service life.

7.2.8.6 Local bearing pressure

If local bearing pressure between the rail fastening and the mortar underneath due to travelling loads cannot be prevented by detailing, then this local bearing pressure must be considered when specifying the mortar. If there are sharp parts of screws beneath the bearing plate, then it is essential to verify that the mortar is not damaged by notch effects or elastic deformations.

7.2.9 Connection of expansion joints seal in reinforced concrete bottoms to load-bearing steel sheet pile walls

Expansion joints in reinforced concrete bottoms, e. g. in a dry dock or similar, are protected against large mutual vertical displacements by means of a joggle joint. Only minor mutual vertical displacements are then possible. The transition between the bottom and the vertical load-bearing steel sheet pile wall is formed via a relatively narrow reinforced concrete beam fixed to the sheet piles. The bottom plates separated by the expansion joint are flexibly connected to the beam, also by way of a joggle joint.

The joggle joint also continues in the connection beam.

The expansion joint in the bottom slab is sealed from below with a waterstop with a loop. The waterstop ends at the U-section sheet pile, at the crest of a single sheet pile specially installed for this purpose; see Figure 7.43.

252 | *7 Configuration of cross-sections and equipment for waterfront structures*

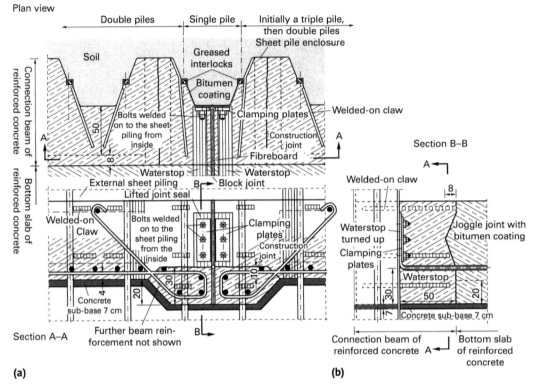

Figure 7.43 Connection of the bottom seal of an expansion joint to U-section sheet piles (example): (a) section A–A; (b) section B–B.

When using Z-section sheet piles, the waterstop ends at a connecting plate welded over the entire trough of the sheet pile as shown in Figure 7.44. The waterstop is turned up and clamped to this.

The interlocks of the connection piles (single piles for U-section, double piles for Z-section) are to be generously greased with a lubricant before installation.

See Figures 7.43 and 7.44 for further details.

7.2.10 Connection of steel sheet piles to a concrete structure

The connection of a steel sheet pile wall to a concrete structure is always a one-off detail and must be designed to suit the actual geometric situation. Good quality work on site is especially important and all work must be carried out properly. The aim should always be to provide the simplest and most robust solution.

The connection between the sheet pile and the concrete structure must allow mutual vertical movements of the overall structure but must also remain permanently watertight in hydraulic engineering applications, for instance.

Figure 7.45 shows examples of the connection between U-section sheet piles and a concrete structure.

If the concrete structure is being newly constructed, a cut individual pile with welded fishtails can be inserted through the formwork and cast into the concrete. The adjoining

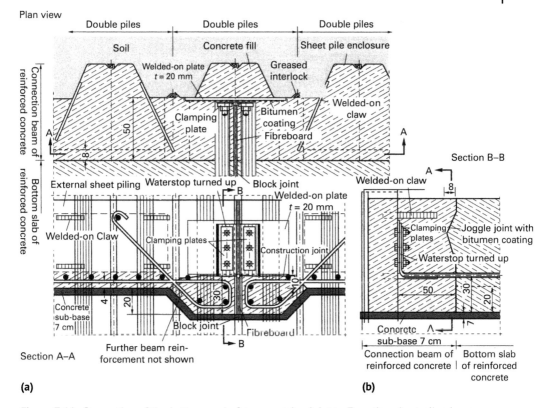

Figure 7.44 Connection of the bottom seal of an expansion joint to Z-section sheet piles (example): (a) section A–A; (b) section B–B.

sheet pile is then threaded into the interlock of the cast-in pile (Figure 7.45a). The connecting interlock must be filled with a plastic compound, so that threading remains possible (see Section 8.1.3.3).

When steel sheet piles are to be connected to an existing concrete structure, a solution such as the one shown in Figure 7.45b is recommended. Here, a U-section sheet pile with an interlock welded to the back is bolted to the concrete structure and the intermediate space filled. Instead of bituminous graded gravel, a backfill in the form of small sacks of wet concrete has proved successful. The next sheet pile is then threaded into the interlock welded to the back of the filled pile. High-pressure grout injection behind the wall may be advisable in order to ensure the permanent watertightness of the connection.

Similar examples for Z-section sheet piles are shown in Figure 7.46.

If the watertightness and/or flexibility of the connection must satisfy a demanding specification, then waterstops must be included in the connection, fixed to the sheet pile with clamping plates and to a steel flat cast into the concrete (Figure 7.47). The embedment depths of the sheet piles must be such that when embedded in a low-permeability soil stratum, groundwater flow is either stopped or the length of the seepage path is long enough to limit the flow to a permissible level and hydraulic heave is ruled out. Movements of the components with respect to each other must be checked carefully when assessing the sealing effect.

Please refer to DIN 18195 parts 1–4, 6 and 8–10.

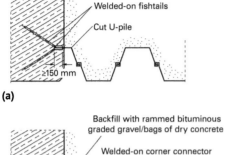

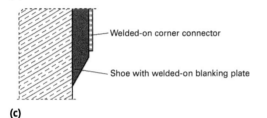

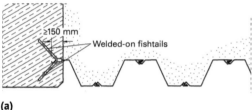

Figure 7.45 Connection of U-section sheet piles to a concrete structure: (a) connection to a concrete structure built later; (b) connection to an existing concrete structure; (c) connection to an existing concrete structure (section).

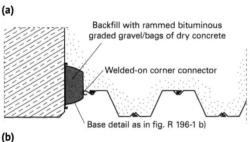

Figure 7.46 Connection of Z-section sheet piles to a concrete structure: (a) connection to a concrete structure built later; (b) connection to an existing concrete structure.

7.2.11 Steel capping beams for sheet pile waterfront structures

7.2.11.1 General

Steel capping beams for sheet pile walls are designed according to structural, operational and constructional requirements. Section 7.2.13 applies for waterfront structures made from reinforced concrete.

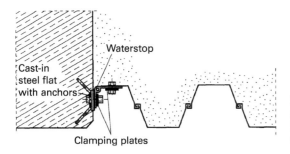

Figure 7.47 Connection of U-section sheet piles to a concrete structure with high demands on the watertightness of the junction.

7.2.11.2 Structural and constructional requirements

A capping beam serves as an upper closure to a sheet pile wall (Figure 7.48). With appropriate flexural stiffness (Figure 7.49), capping beams can also be used to accommodate forces arising during the alignment of the top of the sheet piles and loads from port operations. However, the top of the sheet piles can only be aligned with a capping beam if the length of the unembedded sheet piles is such that it can be deformed sufficiently. With only a short distance between capping beam and waling, the alignment of the sheet piles will be successfully accomplished mostly with the stiffer waling.

During port operations, a capping beam distributes non-uniform loads to the top of the sheet piles and prevents uneven deflections of the head of the wall.

Figure 7.48 shows a standard steel capping beam detail.

The greater the distance between capping beam and waling, the more important the inherent stiffness to align the wall. Berthing loads must also be considered. Figure 7.49 shows a reinforced capping beam waling.

To prevent deflection or buckling, the capping beam shown in Figure 7.48 is strengthened with stiffeners in wide sheet pile troughs, which are welded to beam and pile.

If a capping beam also functions as a waling, it must be designed in accordance with Sections 9.2.1 and 9.2.2.

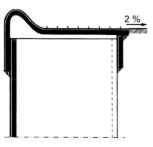

Figure 7.48 Rolled or pressed steel capping beam with a bulb plate, welded to the steel sheet piles.

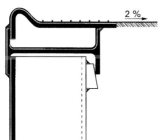

Figure 7.49 Welded capping beam waling with a high section modulus.

7.2.11.3 Operational requirements

The top edge of a capping beam must be designed so that the hawsers trailing over it will not be damaged or damage the top of the quay. Furthermore, it must be ensured that hawsers and lines (e.g. also thin heaving lines) cannot become caught in gaps, joints, etc. For the safety of personnel working on the quay, the front section of the capping beam should be designed as a bulb plate. Horizontal surfaces of steel capping beams should have non-slip finishes (studs, chequered pattern) if possible (Figures 7.48 and 7.49).

Where heavy vehicular traffic operates, welding on a guard rail is recommended to protect the edge (Figure 7.50). If there is an outboard crane rail (Section 7.1.3.4), this can serve as the guard rail.

The outboard side of a capping beam must be smooth. Unavoidable edges are to be chamfered if possible. The design must be such that ships cannot get caught under the capping beam, and there is no danger of beam sections being ripped out by crane hooks (Figure 7.50).

7.2.11.4 Supply, mounting and corrosion protection

All steel capping beam parts are to be supplied without distortion and true to size. During fabrication in the workshop, the tolerances for the width and depth of the sheet pile sections and deviations during driving are to be taken into account. Where necessary, capping beams are to be modified and aligned on site. Capping beam splices are to be designed with the full cross-section.

After mounting the capping beam, it is to be backfilled from behind with compacted sand to avoid corrosion. The backfill is to be replenished when it has settled to such an extent that the beam is exposed.

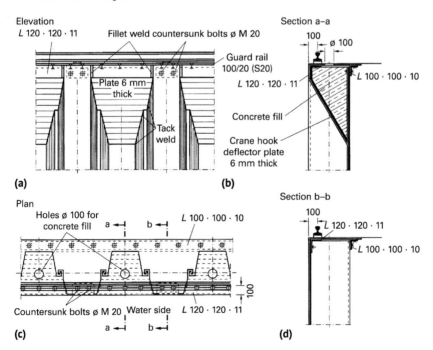

Figure 7.50 Special design of a steel sheet pile capping beam with crane hook deflector plates: (a) elevation, (b) plan; (c) section a–a; (d) section b–b.

If the capping beam is mounted so low that it can be flooded or water can flow over it, lies in the wave action zone or, indeed, is intended to lie below the water level, there is a risk that the backfill designed to prevent corrosion will be washed out. To prevent this, an erosion-proof and watertight connection is required between the steel capping beam and the sheet piles, e. g. a backfilling of concrete behind the beam. This concrete backfill must be secured by means of fishtail anchors or bolts welded to the capping beam. A granular or geotextile filter must be laid beneath the paving of the port operations area behind the capping beam, so that the bedding for the paving cannot be washed away.

7.2.12 Reinforced concrete capping beams for waterfront structures with steel sheet piles

7.2.12.1 General
Structural, constructional and operational considerations govern the design of reinforced concrete capping beams on steel sheet piles.

7.2.12.2 Structural requirements
In many cases, the capping beam not only simply covers the top of the sheet piles, but also stiffens the wall and is, therefore, subjected to horizontal and vertical actions. If, in addition, it functions as a capping beam waling for transferring anchor forces, a sufficiently robust design is required, especially when it also has to support a crane rail directly.

Please refer to Section 9.2.2 for more information concerning horizontal and vertical actions. Additional loads must be taken into account in areas with bollards or other mooring equipment (Sections 4.9 and 4.10), in so far as such loads are not carried by special constructional measures. In addition, where a crane rail is supported directly on a reinforced concrete capping beam (Section 4.11), the beam must be designed for horizontal and vertical crane wheel loads (Figure 7.52). In the structural calculations it is expedient to treat the reinforced concrete capping beam – both horizontally and vertically – as a flexible beam on an elastic foundation (the sheet piles). In the case of heavy-duty capping beams on quay walls for sea-going vessels, the horizontal support can be calculated approximately using a modulus of subgrade reaction $k_{s,bh} = 25 \text{ MN/m}^3$. The vertical support primarily depends on the pile sections, their length and the width of the capping beam. The modulus of subgrade reaction $k_{s,bv}$ for the vertical support must, therefore, be calculated separately for each structure.

As an approximation during the draft design phase, the vertical modulus of subgrade reaction can be taken as $k_{s,bv} = 250 \text{ MN/m}^3$. Limit state considerations are required for the support conditions used in the detailed design of the capping beam. The design should be based on the least favourable case.

Sheet pile structure or bollard foundation anchors connected to the capping beam must be considered in the design.

Special attention should be paid to loads on the capping beam arising from changes in length due to shrinkage and temperature fluctuations. Changes in the length of the capping beam can be severely hampered by the connected sheet piles and the backfilling, resulting in the possibility of corresponding stresses due to shrinkage and temperature changes.

Please refer to Section 10.2 for information on concrete types and reinforcement details for reinforced concrete capping beams.

The vertical loads in the plane of the sheet piles are generally transferred concentrically into the top of the sheet pile. For this reason, the reinforced concrete capping beam should include sufficient tensile splitting reinforcement directly above the sheet piles. On steel sheet piles, the reinforced concrete capping beam can be supported on knife-edge bearings covered by national technical approvals. When transferring large, concentrated loads, e. g. from a crane rail, via the beam into the sheet piles, it should be ensured that the sheet piles can carry the loads via plate action, e. g. by appropriate welding of the interlocks.

The geometric stipulations of port operations may render an eccentric support necessary for the outboard craneway on the capping beam.

It is essential to verify that internal forces are transferred safely from the capping beam to the supporting wall.

7.2.12.3 Buckling and operational requirements

The top of the sheet pile wall must be aligned before the concrete beam is cast. A permanent or temporary steel waling can be used to do this. The reinforced concrete capping beam then forms the visually important alignment of the wall.

If it is necessary to ensure concrete cover of adequate depth over the top of the sheet piles, then this must be taken into account when choosing the width of the capping beam. Contingent on the design, the pile top should have a concrete cover of at least 15 cm on both the water and land sides, and the depth of the capping beam should be at least 50 cm (Figures 7.51 and 7.52). The sheet piles should be embedded approximately 10–15 cm in the concrete capping beam.

The distance of a reinforced concrete capping beam above the water level should be sufficient to allow the sheet piles immediately below the beam to be inspected regularly for corrosion – and renewal of any corrosion protection measures if necessary.

For waterfront structures with an increased risk of corrosion (e. g. in seawater or brackish water), it is advisable to continue the sheet piles on the water side up to the top edge of the quay wall and position the reinforced concrete capping beam behind them. This is an effective way of preventing corrosion at the transition from steel to concrete on the water side.

A capping beam may be designed as shown in Figure 7.52 in order to prevent a ship's hull from catching beneath it. On the water side, the beam is provided with a steel plate bent at

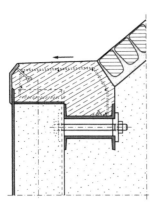

Figure 7.51 Reinforced concrete capping beam for sheet piles without water-side concrete cover on a partially sloped bank.

an angle of 2:1 or steeper. The bottom edge of this plate is welded to the sheet pile or, as shown in Figure 7.50, welded into the outboard troughs of the sheet piles.

The outboard side of a reinforced concrete capping beam can be protected with a steel plate welded to the sheet piles (Figure 7.51). This solution is generally more economical than bolting the steel plate to the sheet piles. Fishtail anchors are then fitted above the sheet pile troughs to anchor the plate in the concrete. The edges of the plate should be chamfered to protect lines trailing over the capping beam. (Figure 7.51). Irregularities in the alignment of the top of the sheet piles of up to about 3 cm can be corrected by inserting small plates to fill the gap. The capping beam is provided with edge protection and rubbing strips as per Section 7.2.12 or DIN 19703. The information given there applies accordingly.

Stirrup reinforcement (shear links) in reinforced concrete capping beams must be designed so that it provides a shear-resistant connection between the parts of the cross-section separated by the sheet piles. To this end, the stirrups should either be welded to the webs of the sheet piles or inserted through holes flame-cut in the piles or fitted into slots in the piles. If the capping beam is reinforced for tensile splitting above the sheet pile wall to carry vertical loads, additional stirrups, e. g. in the troughs on both sides of the sheet piles, must be included so there is shear reinforcement on the underside of the capping beam as well.

Supporting the reinforced concrete capping beam on knife-edge bearings covered by a national technical approval enables a capping beam with a closed cross-section, so that in this case special positioning of the stirrups is unnecessary.

Reinforced concrete capping beams on box sheet piles can also be protected with steel sections (Figure 7.53). In this case, the reinforcement is moved into the sheet pile cells. To this end, the webs and flanges are cut away as necessary and holes flame-cut in them where required. Reinforced concrete capping beams on combined sheet pile walls can be arranged in the same way.

Reinforced concrete capping beams can be strengthened locally so that bollards can be mounted directly on them (Figure 7.54). In such cases, large line pull forces are best carried by heavy-duty round steel tie rods in order to minimise the elongation of the anchor and, hence, the bending moments in the capping beam.

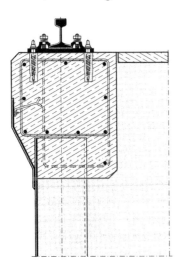

Figure 7.52 Reinforced concrete capping beam for sheet piles with concrete cover on both sides and craneway supported directly on the top surface.

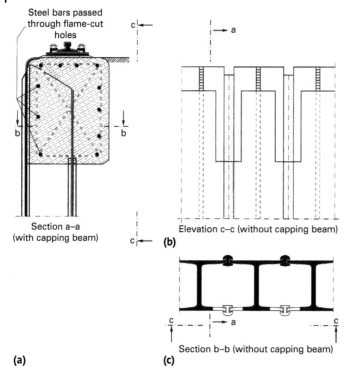

Figure 7.53 Reinforced concrete capping beam on box sheet piles without outboard concrete cover and craneway supported directly on the top surface: (a) section a–a (with capping beam); (b) section c–c (without capping beam); (c) section b–b (without capping beam).

7.2.12.4 Expansion joints

Reinforced concrete capping beams can be constructed without joints when all the actions due to loads and restraint (shrinkage, creep, settlement, temperature) are taken into account, provided that the inevitable cracks are acceptable. The theoretical crack widths must be limited, taking into account the environmental conditions (Section 10.1).

If expansion joints are planned, the lengths of the sections between them should be specified such that no significant restraint forces occur in the longitudinal direction of each section. Otherwise, the restraint forces should be taken into consideration in relation to the substructure or subsoil.

The joints themselves must also be designed so that the changes in length of the reinforced concrete capping beam at these points are not hampered by the sheet piles locally. To this end, it is appropriate, e. g. for sheet pile walls, to position an expansion joint directly above a sheet pile web. The web is then coated with elastic material that accommodates the changes in length of the capping beam without restraint stresses.

Where a capping beam expansion joint is located above a sheet pile trough, the embedment of this pile in the beam should be minimal. To ensure movement is possible, this pile must be covered with a substantial elastic coating that prevents direct force transfer between the pile and the beam and at the same time guarantees a watertight joint between the capping beam and the sheet pile wall. Figure 7.55 shows an example of a capping beam joint above a sheet pile trough.

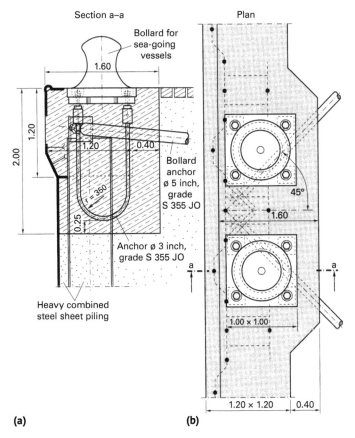

Figure 7.54 Heavy-duty reinforced concrete capping beam on a quay wall for sea-going vessels – details of an anchored bollard foundation: (a) section a–a; (b) plan.

Reinforced concrete capping beams with a sufficiently large rectangular cross-section should include joggle joints to transfer horizontal forces across expansion joints. Steel dowels can be used in capping beams with smaller cross-sectional dimensions.

7.2.13 Steel nosings to protect reinforced concrete walls and capping beams on waterfront structures

7.2.13.1 General

For practical purposes, the edges of reinforced concrete waterfront structures should be provided with carefully designed steel nosings on the water side. This is to protect both the edge and the hawsers running over it from damage caused by port operations. It also serves as a safety measure to prevent line handlers and other personnel working in this area from slipping over the edge. The nosing must be designed so that ships, or crane hooks, are not caught on the underside.

If waterfront structures at inland ports can be flooded, and there is a danger of ships grounding on the structures, the nosing must not have a raised edge.

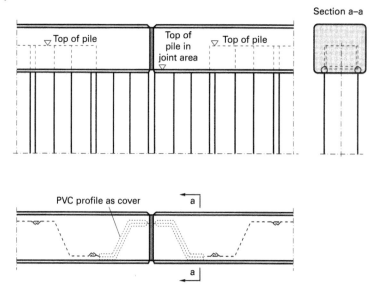

Figure 7.55 Expansion joint in a reinforced concrete capping beam.

7.2.13.2 Examples

Figure 7.56 shows a nosing design that is frequently used for waterfront structures in ports and inland locks. The weephole can be omitted where precipitation is drained towards the land – which is required anyway for waterfront structures handling environmentally hazardous cargos.

The steel nosing shown in Figure 7.56 can also be supplied with angles other than 90° for fitting to waterfront structures with angled front or top surfaces. The separate parts of the nosing are welded together before mounting.

The design in Figure 7.57 depicts a special nosing developed in The Netherlands. It consists of relatively thick plates and strengthened fishtail anchors, so that the raised part does not need to be completely filled with concrete to guarantee the load-bearing capacity. However, the upper ventilation openings, the intention of which is to ensure that the section lies flat on the concrete during concreting, must be closed after concreting to prevent corrosion attacking the inner surface.

The designs shown in Figures 7.58 and 7.59 have proved themselves on numerous German waterfront structures.

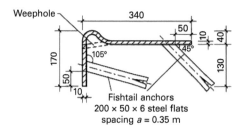

Figure 7.56 Nosing with weephole.

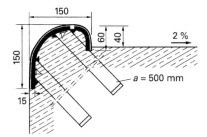

Figure 7.57 Special nosing section frequently used in The Netherlands.

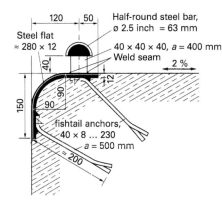

Figure 7.58 Nosing made from a rounded plate, with a foot railing in seaports and without in inland ports.

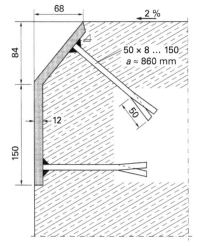

Figure 7.59 Nosing made from bent plate without a foot railing for quaysides liable to flooding in inland ports.

The nosings in Figures 7.56–7.59 must be carefully fixed in the formwork. The nosings in Figures 7.58 and 7.59 must be cast in carefully without any voids. Any rust adhering to the inner surfaces of nosings must be removed with a wire brush before concreting.

7.2.14 Floating berths in seaports

The "Floating Berths" specification of the German Federal Ministry of Transport, Building and Urban Development (BMVI 2012) applies to floating berths on federal waterways. It can be applied correspondingly to seaports as well by taking into account the advice given below.

7.2.14.1 General

In seaports, floating berths are reserved for passenger ferries, port vessels and pleasure craft. They consist of one or more pontoons and are connected to the shore by a bridge or permanent stairs. The pontoons are generally held in place by driven guide piles, and the access bridge/stair has a fixed support at the land end and a movable one at the pontoon. If the floating berth consists of several pontoons, interconnecting walkways ensure that it is possible to move from one pontoon to the next.

7.2.14.2 Design principles

Stipulation of the location of a floating berth must take account of current directions and velocities together with wave influences.

In tidal areas, the highest and lowest astronomical tides should be used as the design water levels. The incline of the access jetty should not be steeper than 1 : 6 at mean tide and not steeper than 1 : 4 at extreme water levels.

Especially when used by the public, the facility must comply with strict requirements, e. g. even under icy conditions. To do this, suitable constructional and organisational measures will be necessary.

The bulkhead divisions of the pontoons must be chosen such that failure of one single cell through an accident or other circumstances will not cause the pontoon to sink. The cells should be vented individually, e. g. with swan-neck pipes. Cells with sounding pipes accessible from the deck are recommended for simplifying the checking of the watertightness. In certain cases, it may be advisable to include an alarm system that warns of an undetected ingress of water. For industrial safety reasons, every cell should be accessible from the deck or through no more than one bulkhead.

Filling the cells with a non-porous foam can also be considered.

A cambered pontoon deck must be provided to ensure surface water run-off.

It is advisable to provide a disconnecting option for the access bridge, e. g. by two piles driven next to the bridge with a suspended cross-member, to guarantee that pontoons can float away rapidly in the event of an accident.

The minimum freeboard required for a pontoon depends on the permissible listing, anticipated wave heights and intended use. For smaller facilities, e. g. for pleasure craft, a minimum freeboard of 0.20 m is adequate for one-sided use, whereas large pontoons require far greater freeboard heights. As a guide, freeboard heights for pontoons up to 30 m long and 3–6 m wide should be about 0.8–1.0 m, whereas pontoons 30–60 m long and up to 12 m wide should have a freeboard height of approximately 1.2–1.5 m.

The freeboard heights must be adjusted to suit the embarkation and disembarkation heights of the vessels, particularly when the facility is used by the public.

7.2.14.3 Loading assumptions and design

As a basic rule, the position of the pontoon should be verified with an even keel, with ballast being provided to compensate where necessary.

An imposed load of 5 kN/m^2 should be assumed when checking the floating stability and listing (one-sided load).

Floating stability verification should also include hydrodynamic loads such as banking-up pressure, flow forces and waves, with calculations being confirmed by tests if necessary. Listing of pontoons and the angles of walkways between pontoons must be checked.

Depending on the pontoon dimensions, listing acceleration and the mutual offset of several pontoons, listing may not exceed 5°; the upper limit is 0.25–0.30 m. Greater listing angles are to be checked on a case-by-case basis.

The ship's berthing force as a load from moored ships is to be taken as 300 kN and 0.30 m/s, or 300 kN and 0.5 m/s for larger facilities (pontoon length > 30 m).

A cushioning effect to reduce the ship's berthing force on pontoons by way of fenders on the outer surface, spring-mounted brackets, rubbing strips and guide dolphins can be considered if these are verified in appropriate investigations. The cushioning effect of guide dolphins can be increased when they are constructed in the form of coupled tubular piles.

7.3 Drainage

7.3.1 Design of weepholes for sheet pile structures

Weepholes in sheet piles are only effective over the long term in silt-free water and when the iron content of the groundwater is so low that it is harmless. If these conditions are not met, there is always the possibility that weepholes become silted up or clogged with iron hydroxide particles and, thus, become ineffective. The iron content in the groundwater of the north German coastal marshlands can be up to 25 mg/l, but in the geest regions it is only 5 mg/l. The risk of iron hydroxide clogging is, therefore, particularly high where post-ice-age cohesive soils are present in the hinterland.

The long-term efficiency of weepholes may, therefore, be assumed for silty water and water with a high iron content when their function can be checked and restored, as necessary. Weepholes can also lose their efficiency in areas with high shell growth.

In the case of very high temporary outer water levels, water infiltrating from outside can cause a higher water level behind the wall. A margin of safety against uplift for all structures behind the waterfront wall must be guaranteed for this higher water level. It should also be noted that in these cases, rising and falling water levels in backfilling can lead to subsidence.

Weepholes must be located below the mean water level, so that they do not become blocked. To guarantee their long-term effectiveness, it is expedient to use gravel filters according to Section 7.3.2.

The weepholes take the form of 1.5 cm wide × approximately 15 cm high slots with rounded ends which are flame-cut in the sheet pile webs (Figure 7.60). In contrast to round holes, these slots cannot be blocked by the grains of gravel used for the filter. The reader is also referred to the last paragraph of Section 3.4.

Weepholes are considerably less expensive than drainage systems with anti-flood fittings. However, experience shows that in tidal areas, they achieve only a minor reduction in the excess water pressure because the water is trapped behind the sheet piles at high tide.

Weepholes are not permitted in quay walls that also function as flood defences.

Weepholes are particularly useful in non-tidal areas, at locations where there is a sudden drop in the surface water level, in intense inflows of groundwater or slope seepage water, or where the structure can become flooded.

It is also generally necessary to investigate the case of ineffective weepholes within the scope of the structural analysis. According to Section 1.1.4, this case may be allocated to design situation DS-A.

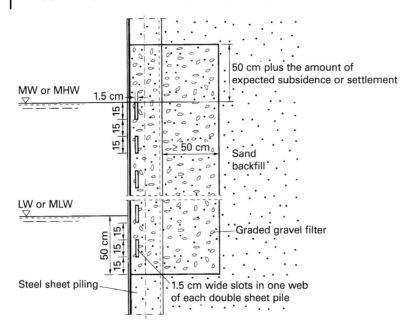

Figure 7.60 Example of weepholes in a steel sheet pile wall.

7.3.2 Design of drainage systems for waterfront structures in tidal areas

7.3.2.1 General

Effective drainage for waterfront structures is only possible where there is non-cohesive soil behind the structure. The risk of clogging with iron hydroxide particles has already been referred to in Section 7.3.1.

If a drainage system is to remain effective over the long-term in harbour water containing suspended matter and silt and also limit the excess water pressure where there is a larger tidal range, it must include branch drains that discharge into main drains fitted with anti-flood valves or flaps that permit the outflow of water from the system into the harbour water but prevent the ingress of silt-laden water. The drainage system must be designed so that it will continue to function reliably even if the backfilling behind the wall settles.

Experience shows that numerous sheet pile drainage systems become less efficient over time due to silting-up as a result of ineffective anti-flood fittings or clogging with iron hydroxide particles. To ensure that this does not occur, the design, installation and maintenance of such drainage systems must be carried out to a high standard.

7.3.2.2 Design, installation and maintenance of drainage systems

Outlets must be fitted with anti-flood valves or flaps to guard against water entering the system from outside and positioned so that they are accessible at mean low tide. Anti-flood fittings must be permanently sealed against wave loads. The spacing between outlets should be about 30 m.

Figures 7.61 and 7.62 show a standard drainage arrangement for the quay walls in Rotterdam. One special technical element is the anti-flood system with a PU sphere to prevent the ingress of seawater.

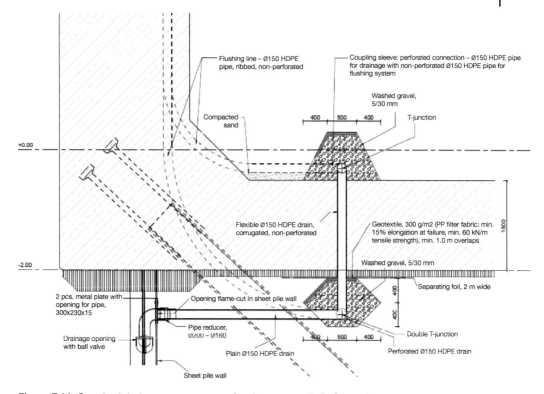

Figure 7.61 Standard drainage arrangement for the quay walls in Rotterdam.

The drains between the outlets should consist of one or more plastic subsoil drainpipes embedded in a gravel filter or ballast sheathed in a geotextile. They must be designed for the surcharge due to the backfilling over the drains.

Connections between pipes and outlets should be designed to resist shearing-through due to settlement of the backfilling behind the sheet piles.

Inspection shafts must be designed and positioned so that the full length of the drainage system can be inspected and, if necessary, cleaned from such shafts.

The drainage system must be inspected and maintained regularly. Inspections and maintenance must be documented.

7.3.2.3 Drainage systems for large waterfront structures

Figure 7.63 shows a groundwater relief system for a larger quay in a tidal area. It consists of four DN 350 subsoil drainpipes made from PE-HD (DIN 19666) that run the entire length of the quay. The pipes are encased in gravel wrapped in a non-woven filter material to protect it against the surrounding soil. The depth was chosen so that the pipes are always within the groundwater, which reduces the risk of clogging due to iron hydroxide particles. The pipes are laid without any falls.

Pumps are used to remove the water from the groundwater relief system of Figure 7.63. This avoids the need for vulnerable anti-flood fittings and the risk of the connection between drains and anti-flood fittings being sheared through.

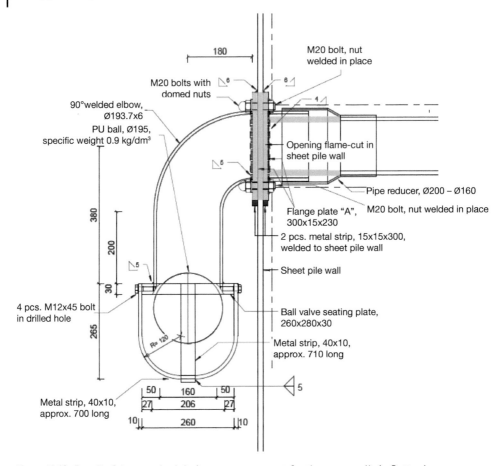

Figure 7.62 Detail of the standard drainage arrangement for the quay walls in Rotterdam.

7.4 Fenders

7.4.1 Fenders for large vessels

7.4.1.1 General

It is customary these days to provide fenders so that vessels can berth safely alongside waterfront structures. Fenders absorb the impact of vessels during berthing and prevent damage to ship and structure while the vessel is moored. For large vessels in particular, fenders are indispensable. Although timber baulks, rubber tyres, etc., are still common, other, modern forms of fender are becoming more and more established. The main reasons for this are:

- The use of fenders increases the service life of the waterfront structure (Section 8.1.10.4).
- The cost of vessels is on the increase, and so ships demand good fenders.
- Ships are growing in size – hence, the surface area exposed to the wind as well.
- The increasing demands placed on moored ships by cargo-handling equipment.
- The strength of the outer hull is being reduced further and further.

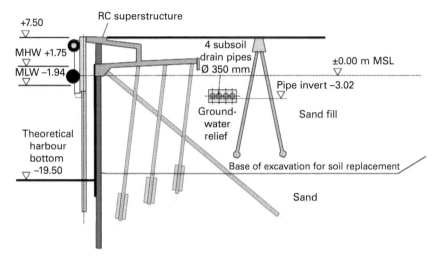

Figure 7.63 Example of a groundwater relief system for a quay in a tidal area.

Fenders are used not only on waterfront structures but also frequently on dolphins and work together with the elastic dolphins to absorb the energy (see also Chapter 12).

7.4.1.2 The fendering principle

A fender is, in principle, an intermediate layer between vessel and waterfront structure which absorbs part of the kinetic energy of a berthing ship; indeed, energy-absorbing fenders absorb most of this energy. In the case of fenders attached to waterfront structures, the energy absorbed by the fender is transferred to the structure. A portion of the berthing energy is absorbed by the ship's hull by means of elastic deformations.

The energy absorption E_f of a fender is shown by its characteristic load–deflection curve (Figure 7.64), which illustrates the relationship between fender deflection s and fender reaction force F_R. The area beneath the curve represents the energy absorption E_f. The energy absorbed at maximum deflection s_{max} is denoted the energy absorption capacity.

All fender designs braced against a rigid waterfront structure are generally characterised by an abrupt increase in the reaction force of the structure once the energy absorption capacity of the fender has been reached. These fender reaction forces must be taken into account when designing the structure.

The dimensions and properties (e.g. characteristic force/load–deflection curve) of the various fenders available can be found in the publications of the fender manufacturers.

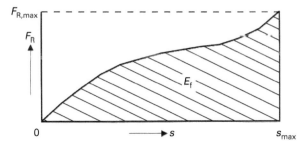

Figure 7.64 Load–deflection curve for a fender.

However, it should be noted that these curves apply only when lateral buckling of the fender is ruled out and when creep deformations under permanent load are not excessive.

Where fenders are used on flexible components and other supporting structures (e.g. dolphins), then it is not only the energy absorption of the fender, but also that of those components that must be taken into account.

7.4.1.3 Design principles for fenders

A fender system must be designed with the same level of care and attention as the entire waterfront structure. The fender system must be considered at the planning stage. Comprehensive guidance on the design and detailing of fenders can be found in PIANC (2020); for berthing velocity, berthing angle and hull pressure, see Roubos et al. (2017, 2018a,b), Hein (2014) and Broos et al. (2018).

A fender system has to satisfy the following main requirements:

- Ships must be able to berth without being damaged.
- Ships must be able to lie at the berth without being damaged.
- Fenders should retain their effectiveness for as long as possible.
- Damage to the waterfront structure must be prevented.
- Damage to dolphins must be prevented.
- Damage to fenders must be prevented.

Therefore, the following steps must be incorporated into the design and detailing of fender systems:

- Compilation of functional requirements.
- Compilation of operational requirements.
- Assessment of the local conditions.
- Assessment of the boundary conditions for the design.
- Calculation of the energy to be absorbed by the fender system.
- Selection of a suitable fender system.
- Calculation of the reaction force and possible friction forces.
- Checking whether the forces transferred to the waterfront structure and the ship's hull can be accommodated.
- Ensuring that all constructional details in the waterfront structure or dolphin can be accommodated, especially fixings, built-in parts, chains, etc., without any damage being caused to the ship or the waterfront structure due to projecting fixings or other parts of the construction

Many different fenders and fender systems are available from numerous manufacturers. The manufacturers frequently offer not only standard products, but also systems tailor-made to suit particular situations.

In order to compare the various fenders available, the quality and system data of the manufacturers should comply with the test methods in PIANC (2020).

Some fenders require considerable maintenance. Therefore, before selecting and installing fenders, the designer is recommended to check carefully whether and to what extent vessels and/or structures are really at risk and which special requirements a fender system will have to fulfil.

When designing a quay wall, pier, dolphin, etc., and the fender support structures, it is not only berthing loads that must be taken into account. The horizontal and vertical movements of the ship during berthing and departure, loading and unloading procedures, swell or fluctuations in the water level, etc., can lead to friction forces in the horizontal and/or vertical direction (provided that these movements are not accommodated by the rotation of suitable cylindrical fenders), which are additional to the berthing forces. If lower values cannot be verified, to be on the safe side a friction coefficient $\mu = 0.9$ should be assumed for dry elastomeric fenders. Polyethylene surfaces result in less friction on the ship's hull; a friction coefficient $\mu = 0.3$ should be assumed in such cases.

A compressive force of a maximum of 200 kN/m² should be considered when designing fenders for large sea-going vessels. All customary fender types can handle such a pressure. According to Broos et al. (2018), the pressure on the hull is not the governing design criterion. Fenders should be designed in relation to the berthing velocity. These developments require further research, and this fact should be taken into account when selecting the type of fender. Certain ships, e. g. naval vessels, require softer fenders.

7.4.1.4 Required energy absorption capacity
7.4.1.4.1 General

The required energy absorption capacity of a fender is the energy that can be absorbed under specified boundary conditions. It must be at least as great as the kinetic energy produced by a berthing ship.

Typically, during berthing, a ship moves transversely and/or longitudinally and at the same time rotates about its centre of mass. At the moment of mooring, contact is generally initially with a single dolphin or fender only (Figure 7.65). The berthing velocity v of the ship at the fender is the critical factor for calculating the energy transferred to the dolphin, the size and direction of which can be calculated from the vectorial addition of the velocity components v and $\omega \cdot r$. In the case of a full frictional connection between ship and fender, during the berthing procedure the berthing velocity of the ship, which is then identical to the deformation rate of the fender, is reduced to $v = 0$. The ship's centre of mass will, however, usually remain in motion, albeit with a different velocity and, potentially, also with a different rotational direction.

The ship, therefore, retains a portion of its original kinetic energy at the moment of maximum fender deformation. Under certain circumstances this can result in the ship turning

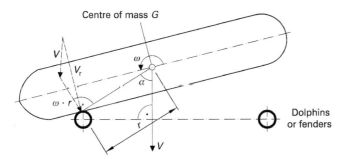

Figure 7.65 Explanatory diagram for calculating the energy to be absorbed due to a berthing manoeuvre.

to make contact with a second fender after striking the first one, which then, at the moment of contact, causes a greater berthing force.

The deterministic method of analysis is normally used for designing fenders. This is based on the energy equation

$$E = \frac{1}{2} \cdot G \cdot v^2$$

where

E kinetic energy of ship [kNm]
G mass of ship, i.e. displacement [t] according to Section 2.1
v berthing velocity of ship [m/s]

The amount of a ship's kinetic energy that a structure (fender and/or dolphin) has to absorb during a berthing manoeuvre represents the energy absorption capacity required to prevent damage to vessel and/or structure. This energy absorption capacity for the example shown in Figure 7.65:

$$E_d = \frac{G \cdot C_m \cdot C_s \cdot C_c}{2 \cdot (k^2 + r^2)} \cdot \left[v^2 \cdot (k^2 + r^2 \cdot \cos^2 \alpha) + 2 \cdot v \cdot \omega \cdot r \cdot k^2 \cdot \sin \alpha + \omega^2 \cdot k^2 \cdot r^2 \right]$$

When $\omega = 0$ (no rotation of ship), the equation simplifies to

$$E_d = \frac{1}{2} \cdot G \cdot v^2 \cdot \frac{k^2 + r^2 \cos^2 \alpha}{k^2 + r^2} \cdot C_m \cdot C_s \cdot C_c = \frac{1}{2} \cdot G \cdot v^2 \cdot C_e \cdot C_m \cdot C_s \cdot C_c$$

In both equations, the following definitions apply:

C_c waterfront structure attenuation factor [–]
C_e eccentricity factor [–]
C_m virtual mass factor [–]
C_s ship flexibility factor [–]
E_d berthing energy to be absorbed [kNm]
G mass of ship, i.e. displacement [t] according to Section 2.1 [t] (The mass of the fully laden ship should always be used – even when operational conditions dictate that only unloaded ships usually berth at the dolphin concerned – in order to cover the case of an unscheduled re-berthing of a ship.)
k radius of gyration of ship [m], generally taken as $0.25l$ for large ships with a high block coefficient
l length of ship between perpendiculars [m]
r distance of ship's centre of mass from point of impact on fender/dolphin [m]
v berthing velocity, i.e. translational movement speed of centre of mass at time of first contact with fender/dolphin [m/s]
α angle between velocity vector v and distance r [°]
ω ship's rotational speed at time of first contact with fender/dolphin [rad/s]

7.4.1.4.2 Information for dolphin berths

If a ship is manoeuvred to a dolphin berth with the help of tugs, it can be assumed that it is hardly moving in the direction of its longitudinal axis and that its side is virtually parallel to the line of the dolphins while it is berthing. Therefore, when designing the inner dolphins within a row of dolphins, the velocity vector v can be assumed to be perpendicular to the distance r, i. e. angle $\alpha = 90°$.

However, this will not be necessary when designing the outer dolphins of a row of dolphins because in this case, the ship's centre of gravity in the direction of the line of dolphins can also approach close to the centre of the dolphins.

The individual factors of the aforementioned equations are defined as follows:

1. Mass of ship/displacement G.
   The mass of the ship, i. e. its displacement, is required for calculating the energy to be absorbed. The recommendation in Section 2.1 includes reference values for the displacement of different types of ship. Provided that no particular values have been specified for the port design, the values in the tables can be used for displacement calculations.
2. Berthing velocity v
   The square of the berthing velocity v is included in the equation for calculating the berthing energy to be absorbed and is, therefore, one of the main parameters to consider when designing fenders and dolphins. The berthing velocity is specified at a right-angle to the waterfront structure or row of dolphins. Measured values for the berthing velocity are not usually available. As a rule, the figure given in Section 4.1 can be assumed.
3. Angle α
   Measurements carried out in Japan resulted in a berthing angle of, generally, $< 5°$ for ships with DWT $> 50\,000$ t (corresponding to $\alpha > 85°$). To remain on the safe side in calculations, it is recommended that the designer assume a berthing angle of $6°$ for such ships (corresponding to $\alpha = 84°$). For smaller vessels, and primarily when berthing without tug assistance, an angle of $10–15°$ should be assumed (corresponding to $75° \leq \alpha \leq 80°$).
4. Eccentricity factor C_e
   The eccentricity factor C_e takes into account the fact that the first contact between ship and fender is not normally in the middle of the ship's side and, therefore, not in line with the vessel's centre of mass either. According to PIANC (2020), the eccentricity factor is calculated as follows (using the factors explained above for the energy equation):

$$C_e = \frac{k^2 + r^2 \cos^2 \alpha}{k^2 + r^2}$$

Assuming a berthing angle of $0°$, i. e. $\alpha = 90°$, is sufficiently accurate for fenders and the inner dolphins in a row of dolphins. The eccentricity factor is, therefore,

$$C_e = \frac{k^2}{k^2 + r^2}$$

The radius of gyration k for large ships with a high block coefficient can usually be taken as $0.25\,l$, where l is the length between perpendiculars.

When designing fenders alongside quay walls, $C_e = 0.5$ can be assumed if more accurate data is not available and for rough calculations, or $C_e = 0.7$ for dolphin fenders. At RoRo berths, $C_e = 1.0$ should be assumed for the end fenders for RoRo ships that dock with bow or stern.

5. Virtual mass factor C_m

    The virtual mass factor takes into account the fact that a considerable quantity of water is moved together with the ship, and this must be included in the mass of the ship in the energy calculation. Various approaches have been used to determine the C_m factor, see PIANC (2020). Assessing this and other approaches in the literature results in C_m values between 1.45 and 2.18.

    PIANC (2020) recommends using the following values:
    - for a large clearance under the keel ($0.5d$): $C_m = 1.5$
    - for a small clearance under the keel ($0.1d$): $C_m = 1.8$,

    where d is the draught of the ship [m].

    Mass factor values for a clearance under the keel between $0.1d$ and $0.5d$ can be found by linear interpolation.

6. Ship flexibility C_s

    The factor for the flexibility of a ship C_s takes into account the ratio of elasticity of the fender system to that of the ship's hull because part of the berthing energy is absorbed by the latter. The following C_s values are normally used:
    - for soft fenders and small vessels: $C_s = 1.0$
    - for hard fenders and larger vessels: $0.9 < C_s < 1.0$.

    Generally, $C_s = 1.0$ can be assumed, which lies on the safe side.

7. Waterfront structure attenuation factor C_c

    The attenuation factor C_c takes into account the type of waterfront structure. With a closed structure (e.g. vertical sheet pile wall), the water between ship and wall already absorbs a considerable portion of the berthing energy as it is accelerated and displaced laterally as the ship approaches. This waterfront structure attenuation factor C_c depends on various influences, e.g.:
    - Arrangement of the waterfront structure
    - Clearance under the keel
    - Berthing velocity
    - Berthing angle
    - Depth of the fender system
    - Vessel cross-section.

    Experience has shown that the following values can be assumed for C_c:
    - Open waterfront structure: $C_c = 1.0$
    - Closed waterfront structure and parallel berthing ($\alpha \approx 90°$): $C_c = 0.9$.

    Values $< C_c = 0.9$ should not be used.

    The attenuation can be considerably less at a berthing angle of just $\alpha = 85°$, i.e. in such cases, assume $C_c = 1.0$.

8. Fender design programs

    Manufacturers can provide computer programs for designing their fenders. Depending on the manufacturer, these programs require influencing parameters to be entered in either metric or non-metric units. Consequently, the results of the calculations may have to be converted for comparisons.

9. Additional factors for exceptional berthing manoeuvres

    Using the detailed information given here allows the designer to calculate the required energy absorption of fenders and dolphins with sufficient accuracy. It is left to the discretion of the design engineer to take account of any potential difficulties caused by exceptional berthing manoeuvres, e.g. by assuming higher berthing velocities or gen-

Table 7.2 Additional factors for exceptional berthing manoeuvres depending on the size and type of vessel.

Type of vessel	Size of vessel	Additional factor
Tanker, bulk cargo	Large	1.25
	Small	1.75
Container	Large	1.5
	Small	2.0
General cargo		1.75
RoRo vessel, ferry		≥ 2.0
Tug, workboat		2.0

eral additional factors when calculating the energy. Exceptional conditions could be, for example, the frequent handling of hazardous goods. The designer should consult PIANC (2002) and Roubos (2018b). Generally, additional factors between about 1.1 and maximum 2.0 are recommended. Table 7.2 provides guidance on additional factors depending on the type and size of vessel in the case of an extraordinary berthing manoeuvre.

10. Selection of fenders

    Once the energy has been calculated, the fenders required can be selected from the relevant manufacturers' publications. However, for detailed planning, the designer is advised to consult the manufacturer because many of the construction details cannot be gleaned from the manufacturers' publications. This concerns, in particular, the construction details regarding the mounting of the fenders.

7.4.1.5 Types of fender system

Diverse fender systems are available on the international market. The various types and models can be seen in the catalogues of the manufacturers.

Cylindrical fenders are the most common type, and these are available in many different sizes. Floating fenders have proved to be worthwhile for quay walls exposed to considerable water level fluctuations as a result of tides. Berths for ferries are frequently custom solutions with polyethylene-coated, low-friction panels on conical fenders or cylindrical fenders loaded along their longitudinal axis.

For a comparison of various types of fender, please refer to PIANC (2020). Further information on the advantages and disadvantages of different fender systems can also be found in the report.

Please note that the designations of the manufacturers can vary for the same type of fender. Test methods for materials and fenders should comply with the data given in PIANC (2020) in order to be able to assess the equivalence or otherwise of the products of different manufacturers. Materials for fenders are these days almost exclusively elastomer or other synthetic products. With the exception of dolphins and rare custom designs, these products guarantee that the energy that occurs during berthing can be transferred to the load-bearing structure in accordance with the calculations and without damage.

For this reason, types of fenders common in the past, e.g. brushwood, vehicle tyres, or timber (rubbing strips, timber fenders, fender piles), cannot be designated as fenders because their insufficiently defined material properties prevent them from being included in the energy absorption calculation. Fenders made from wood or vehicle tyres can, therefore, only be used in construction details, e. g. as nosings or guides.

7.4.1.5.1 Elastomer fenders

7.4.1.5.1.1 General Elastomer fenders are used in many ports for absorbing the impacts of vessels and berthing pressures. These fenders are generally made from a material resistant to seawater, oil and ageing and are not destroyed by occasional overloads, so a long service life can be expected.

Elastomer fender elements are manufactured in various shapes, dimensions and with specific performance characteristics. They meet every requirement, from simple fendering for small vessels to fender structures for large tankers and bulk cargo freighters. Special attention must be given to the particular loads on fenders at ferry terminals, locks, dry docks, etc.

Elastomers are used either alone as a fender material, against which the ships berth directly, or as suitably designed buffers between fender piles or fender panels and the structure. Occasionally, both types of usage are combined. In such cases it is possible to attain the energy absorption capacity and spring constants best suited to any specific requirement using elements made from commercially available elastomers.

7.4.1.5.1.2 Cone fenders with steel plates Most modern fenders are now conical types with a steel plate on which a UHMW-PE plate is mounted (Figure 7.66). The large steel plates are necessary because of the permissible loads on ships' hulls according to PIANC (2020) and should extend from the top edge of the quay down to low water level. Broos et al. (2018) describe a method that can be used instead of the tabulated values for ship hull pressures.

Figure 7.66 Super cone fenders (SCN) with steel plates, Port of Rotterdam.

7.4 Fenders

This type of fender is expensive and vulnerable to damage during maintenance or operations. However, compared with cylindrical fenders, they are less vulnerable to buckling. Their energy and deformation characteristics are similar to those of axially loaded cylindrical fenders.

7.4.1.5.1.3 Cylindrical fenders Thick-walled elastomer cylinders are frequently used. They can have various diameters ranging from 0.125 to > 2.00 m. They have variable spring characteristics depending on the application. Cylinders with smaller diameters are attached with ropes, chains or rods in horizontal, vertical or, where applicable, diagonal positions in front of quay walls.

In the latter case, they are frequently suspended as a protective "garland" in front of a quay wall, mole head, etc.

Cylindrical fenders are usually installed in a horizontal position. To avoid the risk of deflection and tearing at the ends, they must not be hung directly in front of quay walls with ropes or chains, instead they should be threaded over rigid steel tubes or steel trusses made from tubular sections. These are then suspended from the quay wall with chains or wire ropes, or mounted on steel brackets located next to the fenders (Figure 7.67).

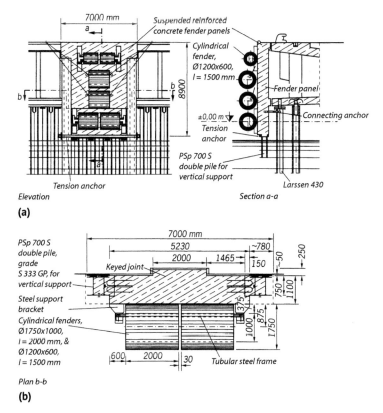

Figure 7.67 Example of a fender arrangement with cylindrical fenders: (a) elevation and section a–a; (b) horizontal section b–b.

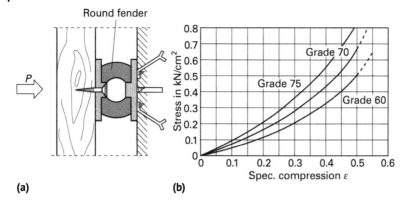

Figure 7.68 General data for round elastomer fenders loaded in the longitudinal direction in Shore A grades 60, 70 and 75 to DIN 53505: (a) example under load; (b) characteristic stress/compression diagram.

7.4.1.5.1.4 Axially loaded cylindrical fenders Cylindrical fenders can also be installed to carry loads in their longitudinal direction. However, owing to the risk of buckling, only shorter cylinders are possible in this case. If the deformation to absorb the berthing energy is not sufficient, several cylindrical fender elements can be arranged adjacent to each other. To prevent buckling of such a row of elements, steel plates in suitable guides can be fitted between the individual elements (Figure 7.68). The reaction force of this type of fender initially climbs rapidly to the buckling load and then drops again as the fender deforms.

The cone fender is a special form that is far less vulnerable to buckling. The energy and deformation characteristics are similar to those of an axially loaded cylindrical fender.

7.4.1.5.1.5 Trapezoidal fenders In order to obtain a more favourable load–deflection curve, special shapes have been developed using special inlays, e. g. textiles, spring steels or steel plates vulcanised into the elements. Metal inlays must be blasted to a bright surface finish and must be completely dry before vulcanising. These elements are frequently made in a trapezoidal form with a height of approximately 0.2–1.3 m. They are attached to the quay wall with dowels and bolts (Figure 7.69).

7.4.1.5.1.6 Floating fenders The great advantage of floating fenders, primarily in tidal waters, is that ships are fendered practically exactly on the waterline and, hence, also roughly in line with their centre of gravity. Floating fenders are available in foam or air-filled versions.

Air-filled fenders are fitted with a blow-off valve, which prevents the fender from bursting if it is overloaded. The valve must be serviced regularly.

Owing to their method of manufacture, foam-filled fenders can be produced in virtually any size and with virtually any properties. They have a core of closed-cell polyethylene foam and a jacket of polyurethane reinforced with a textile. The jacket is easily repaired. Special attention should be paid to the material properties of the jacket because the stresses during deformation are very high.

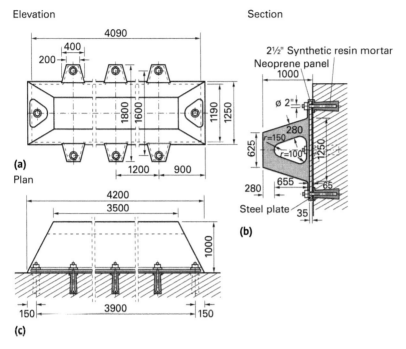

Figure 7.69 Example of a trapezoidal fender: (a) elevation; (b) section; (c) plan.

7.4.1.5.1.7 Fenders made from car tyres and rubber waste Various seaports use old car tyres, mostly filled with rubber waste, as fenders, suspended flat against the face of quays. They have a cushioning effect, but do not possess any appreciable energy absorption capacity. More frequent is the use of several stuffed truck tyres – usually between 5 and 12 – threaded onto a steel shaft with a pipe collar welded on at each end for attaching the suspension and retaining ropes. The ropes hold the fender so that it can rotate on the face of the quay wall. The tyres are filled with elastomer sheets placed crosswise, which brace the tyres against the steel shaft. Any remaining voids are filled with elastomer material (Figure 7.70). Fenders of this type – occasionally in an even simpler design with a wooden shaft – are inexpensive and have generally performed well where the impact energy of berthing ships has not been severe. However, their energy absorption capacity cannot be determined reliably.

Not to be confused with these improvised measures are the accurately designed fenders that are free to rotate on an axle. These fenders are fabricated mostly from very large special tyres that are either stuffed with rubber waste or inflated with compressed air. Fenders of this type are used successfully at exposed positions, such as the entrances to locks or dry docks as well as at narrow harbour or port entrances in tidal areas, where they are suspended horizontally and/or vertically and successfully serve as guides for ships.

Tyres from road vehicles are occasionally used as fenders at ore loading/unloading facilities in the vicinity of open-cast mines. In such cases, the energy absorption capacity should be determined in tests.

7 Configuration of cross-sections and equipment for waterfront structures

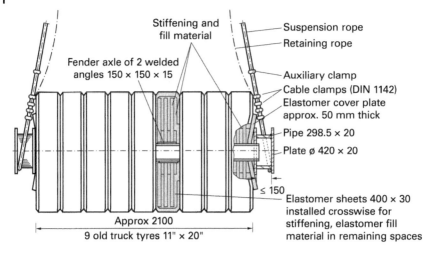

Figure 7.70 Example of a fender made from truck tyres.

7.4.1.5.2 Fenders made from natural materials

Suspended brushwood fenders are still used in countries in which suitable materials are available and/or funding is limited. However, if elastomer fenders are available, then brushwood fenders should be rejected because they involve a higher capital outlay and also higher maintenance costs than elastomer fenders. Brushwood fenders are prone to natural wear and tear due to weather and wave conditions. Their dimensions are adapted to the largest vessels berthed. Unless special circumstances require larger dimensions, fender sizes may be taken from Table 7.3.

7.4.1.6 Construction guidance

Fenders should be located at regular intervals along the quay. The spacing of the fenders depends on the design of the fender system and the vessels anticipated. One important criterion here is the radius of the ship between the bow and the flat side in the middle of the ship. This radius defines – for a given fender spacing – the projecting dimension of completely compressed fenders with the bow making contact with the quay wall between two fenders. To avoid that, the fender spacing must be adjusted to suit.

As a rule, fenders should not be spaced more than 30 m apart.

The fender projection from the structure should not be too large. The maximum load moment of the cranes frequently influences the projection of the fenders.

It is difficult to design a fender system that is equally suitable for both large and small vessels. Whereas a fender designed for a large vessel is sufficiently "soft", this might be too

Table 7.3 Fender dimensions for brushwood fenders.

Size of vessel [DWT]	Fender length [m]	Fender diameter [m]
Up to 10 000	3.0	1.5
Up to 20 000	3.0	2.0
Up to 50 000	4.0	2.5

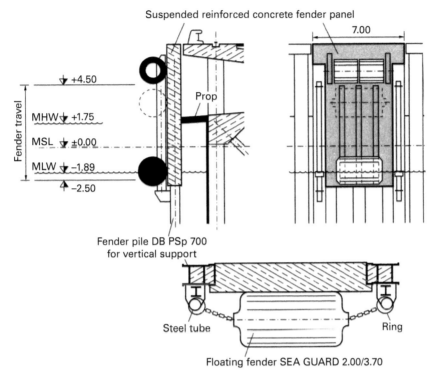

Figure 7.71 Example of a floating fender system at a berth for large vessels with a berthing option for feeder and inland vessels.

inflexible for a small vessel, which can result in damage to the ship. Furthermore, the level of the fenders with respect to the water level is more significant for small vessels than for large ones. In tidal waters, floating fenders can offer considerable advantages.

If container feeder ships or inland vessels are handled at berths for large ships, there is a danger of such vessels becoming caught beneath fixed fenders. In addition, the listing of small ships during loading/unloading procedures at low water can lead to high-level fenders damaging the superstructure and the cargo.

In a new development for the container terminal at Bremerhaven, Germany, floating fenders have been installed in front of fender panels, and these can move up and down with the tide between lateral guide tubes. This solution is shown in Figure 7.71. The fender construction here consists of a fixed upper cylindrical fender (Ø1.75 m) and a moving floating fender (Ø2.00 m) in front of a fixed fender panel. These diameters were chosen to ensure that sufficient listing of smaller vessels is possible at low water levels.

7.4.1.7 Chains

Chains in fender systems should be designed for at least three to five times the theoretical load.

7.4.1.8 Guiding devices and edge protection

7.4.1.8.1 General

Besides the actual fenders, which are specifically designed to absorb energy, there are many elements that are provided merely for constructional reasons, e. g. guiding devices in channels and locks, edge protection, or berthing equipment for smaller vessels not designed for specific situations. Such devices include fender piles, timber fenders, rubbing strips and nosings.

7.4.1.8.2 Timber fenders and fender piles

Timber used in seawater and brackish water can be attacked by the naval shipworm (*Teredo navalis*). Such an attack can lead to total destruction of the timber in a port facility within just a few years. The destruction of the timber is practically invisible from the outside because the worm bores into the wood radially and then extends its path horizontally. The shipworm attacks mainly softwoods, but European hardwoods such as oak, and even tropical woods are also at risk.

The use of timber in load-bearing structures cannot be recommended in seawater and brackish water with a salt content > 5%. When, for example, timber fender piles are used, replacement after infestation by the naval shipworm must be considered.

7.4.1.8.3 Edge protection

Edge protection is made from special sections and fender sections. Owing to their small size and their shape, they possess no significant energy absorption capacity.

7.4.1.8.4 Rubbing strips of polyethylene

In addition to other components such as timber fenders, fender piles, etc., rubbing strips made of plastic, frequently polyethylene (PE), are used in order to reduce the friction forces between waterfront structures and berthing/moored vessels. These components must absorb the loads arising from pressure and friction without fracture and be capable of transmitting them to the waterfront structure via their mountings. To do this they will need to be supported by supplementary load-bearing members in certain cases.

Polyethylene compounds of medium density to DIN EN ISO 1872 (HDPE) and high density to DIN 16972 (UHMW-PE) have proved suitable for use as rubbing strips in hydraulic engineering and seaport construction. Standard forms are rectangular solid profiles with cross-sections between 50 × 100 mm and 200 × 300 mm and lengths up to 6000 mm. Custom sections and lengths can also be supplied. HDPE is cast in moulds and is vulnerable to brittle fractureat low temperatures (< 6 °C). UHMW-PE sections are cut to suit the profile required and, therefore, have smooth edges.

In order to minimise friction forces, rubbing strips should be made of a material that has a very low coefficient of friction together with low abrasion and wear rates, e. g. ultra high molecular weight polyethylene (UHMW-PE).

The shaped parts must always be free from voids and must be produced and processed in such a way that they are free from distortion and inherent stresses. The quality of the processing can be checked by acceptance tests to verify the properties and by additional hot storage tests on samples cut from sections.

Regenerated PE compounds of medium density may not be used because of their reduced material properties.

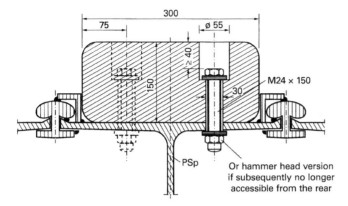

Figure 7.72 Rubbing strip fixed directly to a Peiner sheet pile wall.

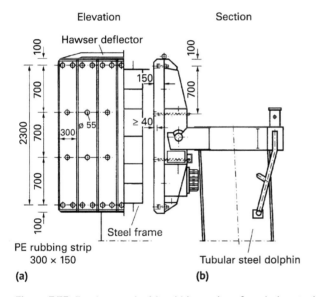

Figure 7.73 Fender panel with rubbing strips of a tubular steel dolphin: (a) elevation; (b) section.

Figures 7.72 and 7.73 show fixing and construction details for typical rubbing strips. The heads of the fixing bolts should be recessed at least 40 mm below the contact surface of the strips. Replaceable bolts should be at least ⌀22 mm, cast-in bolts hot-dip galvanised and at least ⌀24 mm.

7.4.2 Fenders in inland ports

The berthing areas of waterfront structures at inland ports generally consist of concrete, steel sheet piles or faced natural stone. They are constructed either as vertical or with a minimal landward batter (1 : 20 to 1 : 50).

To protect the waterfront structure and the hull of the ship, the ship's crew will normally suspend rubbing strips about 1 m long between quay wall and side of ship.

The designer is recommended to refrain from equipping waterfront structures at inland ports with fender piles or timber fenders.

7.5 Offshore energy support bases

7.5.1 General

Coastal transshipment ports for the individual components of wind turbine installations must be set up when building offshore wind farms. On the water side, such offshore energy support bases consist of berthing facilities for various types of vessel such as supply ships, vessels for transporting the individual components, especially jack-up barges and wind turbine installation vessels (WTIV), and other special vessels. The boundary conditions that apply for the design and construction of the water-side waterfront structures are essentially no different to those for traditional port facilities. Merely the jacking procedures of WTIVs and jack-up barges require the seabed in front of the quay to be built in such a way that jacking can be carried out without problems and without any adverse effects on the waterfront structure. On the land side of the quay, there will need to be appropriate facilities for handling tower parts, nacelles, rotor blades and foundation elements. These separate parts are characterised by heavy loads that need to be transferred to the subsoil by suitable means. The infrastructure should generally be planned depending on the particular boundary conditions of the port that is to serve as an offshore energy support base. The design of the transshipment facilities does not generally affect the design of the quay itself and is, therefore, not covered in this publication. Further requirements regarding the infrastructure, e. g. utility services, transport links, fencing, roads, drainage and lighting, are likewise not dealt with here. The reader should consult suitable specialist publications, so that the appropriate boundary conditions can be included in the respective planning.

7.5.2 Basis for design

The key boundary conditions for waterfront structures at offshore energy support bases are no different from those for waterfront structures at other ports. The other chapters in this publication cover the design principles that continue to apply here:

- Chapter 2 for defining vessel dimensions.
- Chapter 3 for determining active and passive earth pressures and analysing hydraulic heave and ground failure.
- Chapter 4 for the loads on waterfront structures.
- Chapter 5 for earthworks and dredging.
- Chapter 6 for the design of protective and stabilising structures.
- Chapter 8 for the design of the waterfront structure itself.

In addition to the above design principles, it is also necessary to calculate the leg penetration in front of a waterfront structure when analysing jacking procedures. The fundamental boundary conditions for this are described below.

When designing waterfront structures for offshore energy support bases, or verifying existing facilities, it is necessary to realistically ascertain the loads behind the waterfront struc-

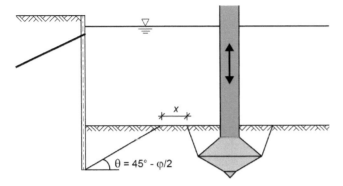

Figure 7.74 Avoiding an overlap between the penetration or extraction mechanism of the jack-up leg and the passive earth pressure failure body.

ture. Besides the vertical loads, there are also horizontal loads due to the movements of lifting and transport equipment. The logistical boundary conditions are specific to every project and can change very quickly, and so the loads must be defined together with the operator.

The loads must be determined for the jacking procedures in front of a waterfront structure. Advice on estimating leg penetration is given below. From this, it is then possible to derive measures and effects, with the following scenarios being possible:

a) The jacking procedure and the leg penetration do not have any effect on the waterfront structure because owing to the clearance between the waterfront structure and the footprint of the leg, there is no risk to the stability of the wall (see Figure 7.74).
b) The leg penetration creates a footprint that requires the waterfront structure to be checked.

In the case of b), suitable soil improvement measures are recommended to reduce the leg penetration. Such measures should be designed together with a soils engineer (see Section 5.9 for further information). Irrespective of this, it is necessary to examine the processes that ensue as a result of multiple jacking operations. In particular, the interaction of successive jacking operations at the same locations should be investigated, although such effects can be avoided by backfilling the footprints with suitable material. Numerical investigations of the penetration of spudcans are presented in Qiu et al. (2013, 2014).

In addition to the above influences, it is also necessary to consider the effects due to scour as a result of vessels standing for longer periods in sandy soils. Special investigations are also necessary if the wind turbine installation vessel employs a jetting system to assist the extraction of the legs from the soil. This procedure should generally be avoided at a port for offshore operations because it disturbs the subsoil in front of the waterfront structure over a wide area and to a great depth.

7.5.3 Nautical requirements

The nautical requirements to be met by offshore energy support bases can be derived from the vessel types associated with offshore operations. Those requirements are determined not only by transport and installation vessels but also by service and feeder vessels.

7.5.3.1 Design vessels

The design vessels for the design of the waterfront structure and the choice of quayside equipment are not fundamentally different from other vessels. When it comes to conventional wind turbine installation vessels, it should be assumed that they always remain at about 10–25 m from the waterfront structure. Mooring lines and fenders are not normally used for WTIVs.

7.5.3.1.1 Wind turbine installation vessels, jack-up rigs

Current WTIVs (Figure 7.75) collect the wind turbine elements to be installed on wind farms from the offshore energy support base. To do this, the vessels raise themselves out of the water on their jack-up legs at the port, so that they can use their very powerful on-board cranes. For example, the crane on board the *Innovation* WTIV has a lifting capacity of 1500 t at a radius of 32.5 m (see Section 2.3).

Irrespective of the particular vessel, the jack-up legs (spuds) are cylindrical or lattice-type steel structures. On the bottom of each leg of most vessels there is a so-called spudcan. A spudcan is an enlargement of the bearing surface, the effect of which is to prevent the jack-up leg penetrating too far into the ground during the jacking operation.

Spudcan sizes vary considerably depending on the maximum load (preload) that the WTIV requires per leg. Table 7.4 lists the maximum preloads, the spudcan sizes and the resultant bearing pressures for a number of known WTIVs.

7.5.3.1.2 Crane vessels

Currently, the two largest crane vessels used for erecting offshore wind farms are the *Stanislav Yudin* and the *Oleg Strashnow*. The latter is equipped with a crane that has a lifting capacity of 5000 t at a radius of 32 m. The ship itself is 183 m long and 47 m wide and has a draught of 8.5–13.5 m (crane operation).

Figure 7.75 Jack-up vessel lifting a monopile.

Table 7.4 Preloads and spudcan dimensions of wind turbine installation vessels.

Ship's name/type	Maximum preload [t]	Spudcan area [m²]	Resultant bearing pressure [t/m²]	Remarks
Seabreeze class (*Victoria Matthias*, *Friedrich Ernestine*)	7 200	11	655	Without spudcans
	7 200	90	80	With spudcans
Innovation	15 000	140	107	
Vidar	13 500	126	107	
Thor	4 677	57	82	
Brave Tern	6 800	106	64	
Aeolus	9 500	50	188	
MPI Discovery	6 475	74	88	6 jack-up legs
SeaJacks Zaratan	2 700	29	93	
SeaJacks Leviathan	5 500	62	89	
A2SEA Sea Installer	9 000	105	86	

7.5.3.1.3 Feeder vessels

Feeder vessels are required when the components of offshore wind turbines have to be delivered (imported) from other locations. For example, the piles for anchoring jacket and tripod structures are meanwhile also supplied from the Far East, e. g. Malaysia. Most feeder vessels are ships of the Panamax class, a category that encompasses all vessels that are just small enough to pass through the locks of the Panama Canal, where the lock chambers are 305 m long × 33.5 m wide × 12.5 m deep.

7.5.3.1.4 Barges

The North Sea barge has become the standard barge in Germany. The dimensions of these barges are standardised at approximately 100 m long × approx. 30 m wide. They have a loading capacity of about 11 000 t. With a minimum freeboard of 2.0 m and a barge side height of 8.0 m, the maximum draught is about 6.0 m. These barges are used relatively frequently when loading and also for the intermediate storage of offshore wind turbine components, which means that sufficient berths for empty and fully laden barges waiting for the arrival of a WTIV must be allowed for in the planning.

7.5.3.1.5 Other vessels

Besides the vessel types mentioned above, floating hotels are also frequently seen at offshore energy support bases. Most of these ships are decommissioned ferries that are used as temporary accommodation for the crew of a WTIV during a change of crew. The *MS Regina Baltica*, with a length of 145 m and a beam of 25 m, is an example of this type of vessel.

7.5.3.2 Overhanging loads and width of navigation channels

Whereas the individual parts of a wind turbine installation (rotor blade, nacelle, tower, foundation) do not normally exceed the width of a WTIV, it can happen that the loading scenarios of certain projects or the need to transport large, prefabricated items results in loads overhanging the ship on one or both sides.

For example, rotor blades or monopiles loaded across a vessel can cantilever by up to 20 m on both sides, and a pre-assembled nacelle plus three rotor blades ("star lift") can reach a total width of about 120 m.

Such widths lead to conflicts in ports and berths when it comes to the clearance between ship and quay (especially in tidal areas) and other structures (especially sea locks).

Various restrictions affecting the journey to the open sea can be expected, depending on the location of the offshore energy support base. The safety clearances to other vessels and manoeuvring room required during the transportation of components pre-assembled for a "star lift" 120 m wide calls for a navigation channel width of 240 m.

Various restrictions and stipulations concerning the transportation of such abnormal cargos can be expected, depending on the degree of development, the width of the navigation channel, the nature and volume of waterborne traffic and the morphological features of the area.

7.5.4 Calculating the leg penetration of WTIVs

The loaded leg penetration behaviour to be expected during preloading[1] can be estimated with numerical methods of calculation or using a simplified method based on the SNAME guideline (2015) and DIN EN ISO 19905-1:2016-11. The calculation of the leg penetration is based on the fact that the penetration of the spudcans (legs) comes to a stop when the resistance to penetration F_V corresponds to the preload (degree of utilisation $\mu = 1$). According to the SNAME guideline (2015) or DIN EN ISO 19905-1:2012, a distinction is made between different modes of failure: a conventional ground failure in the homogeneous monolayer system (non-cohesive soil with effective shear parameters or cohesive soil with undrained shear strengths), punch-through, lateral displacement of the soft strata (squeezing) in a stratified subsoil and a combination of modes.

To calculate the penetration resistance F_V, a circular area equivalent to the area of the spudcans is assumed for simplicity. If the conical tip of each spudcan fully penetrates the subsoil, the maximum area of the spudcans can be assumed when assessing the bearing capacity. For this reason, in the following, the penetration depth z is defined as the distance between the bottom of the watercourse and the tip of the spudcan (the lowest point of the spudcan). However, the distance between the bottom of the watercourse and the maximum area of the spudcan is used to determine the leg penetration. This leg penetration depth is defined as h. The maximum equivalent diameter of the spudcan is D_{eff}.

Determining the leg penetration depth is essentially very dependent on the quality of the soil surveys and the shear parameters assumed. Therefore, a range of soil parameters should always be applied, and minimum and maximum figures calculated for the leg penetration. In some circumstances, it might be necessary to vary the methodology for calculating the ground failure coefficients described below.

---

1) This is understood to be preloading of the legs after the jacking procedure but prior to commencing the actual work operation. The loads applied to the leg should be much higher than the loads applied during normal operation. They are essentially due to the weight of the hull of the ship, including the loads, being distributed over two diagonally opposed legs, while the other legs are relieved of as much load as possible.

7.5.4.1 Penetration resistance in non-cohesive soil

According to the SNAME guideline (2015), the penetration resistance of non-cohesive soils (silica sands) is calculated as a drained process using

$$F_V = (0.5\gamma' D_{\text{eff}} N_\gamma s_\gamma d_\gamma + p'_0 N_q s_q d_q) \cdot A_{\text{eff}}$$

In this equation, γ' is the buoyant unit weight of the soil and N_γ (according to Brinch Hansen based on the SNAME guideline 2015) and N_q are the bearing capacity coefficients. The SNAME guideline (2015) recommendation for spudcan diameters $D_{\text{eff}} > 3$ m is to reduce the angle of internal friction by 5° when calculating the bearing capacity coefficients. In addition, s_γ and s_q are shape factors, d_γ and d_q depth factors (according to the SNAME guideline 2015), p'_0 the effective overburden pressure and A_{eff} the effective area of the spudcans.

Where a soil with an apparently adequate bearing capacity lies above a soil with inadequate bearing capacity, a ground failure (punch-through) can occur in the latter soil.

With the given boundary conditions, this is partly the case at the transition from in situ sands near the surface and clay/silt. Here, the characteristic leg penetration resistance taking into account punch-through with spreading of the load is predicted according to DIN EN ISO 19905-1:2016-11 or the SNAME guideline (2015). The penetration resistance is calculated at the upper edge of the soil with inadequate bearing capacity taking into account a spudcan area fictitiously enlarged by spreading of the load (at an angle of 1 : n) (see Figure 7.76).

For the case without consideration of backflow of the lateral in situ subsoil, the penetration resistance F_V of "non-cohesive soil above cohesive soil" (sand over clay) is calculated as follows:

$$F_V = F_{V,b} - W$$

with

$$F_{V,b} = (c_u N_c s_c d_c + p'_0)\left(1 + \frac{2T}{nD_{\text{eff}}}\right)^2 A_{\text{eff}} \quad \text{and} \quad W = \left(1 + \frac{2T}{nD_{\text{eff}}}\right)^2 A_{\text{eff}} T \gamma'$$

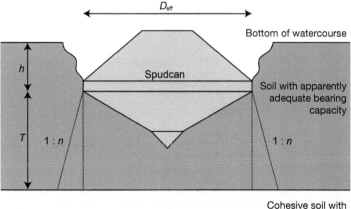

Figure 7.76 Shear mechanism assumed for punch-through.

where T is the distance between the maximum area of the spudcans and the upper edge of the stratum with inadequate bearing capacity and $1:n$ is the load-spreading angle. According to the SNAME guideline (2015) and DIN EN ISO 19905-1:2016-11, values of $3 \leq n \leq 5$ are recommended for sand. According to Dutt and Ingram (1984) and Tomlinson (1986), an angle of spread equal to $\tan^{-1}(1/2)$ can be assumed for coarse-grained materials. In addition, c_u, N_c, s_c and d_c describe the strength behaviour of the underlying cohesive soil.

7.5.4.2 Penetration resistance in cohesive soil

In general, according to the SNAME guideline (2015), the penetration resistance F_V of cohesive soils, without considering backflow of the lateral in situ soil, is calculated as follows as an undrained process:

$$F_V = (c_u N_c s_c d_c + p'_0) \cdot A_{\text{eff}}$$

where c_u is the undrained shear strength of the soil, N_c a ground failure factor, s_c a shape coefficient and d_c a depth factor. Values for N_c, s_c and d_c for conical circular foundations are specified in the SNAME guideline (2015) and DIN EN ISO 19905-1:2016-11.

In cases where there is a soft, cohesive stratum on top of a stratum with a "better bearing capacity" (cohesive or non-cohesive), the possibility of the lateral displacement (squeezing) of the soft stratum should be considered. According to the SNAME guideline (2015) and DIN EN ISO 19905-1:2016-11, squeezing takes place in cohesive soils when

$$D_{\text{eff}} \geq 3.45T \left(1 + 1.1 \frac{h}{D_{\text{eff}}}\right)$$

where T is the distance between the maximum area of the spudcans and the top side of the stratum with a better bearing capacity (see Figure 7.77).

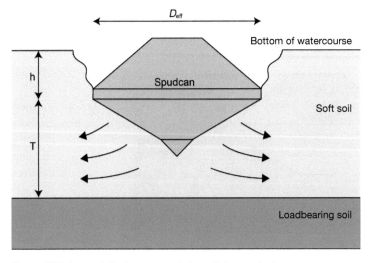

Figure 7.77 Lateral displacement of the soil (squeezing).

If squeezing of the soft stratum is relevant, then the penetration resistance F_V (without backflow) of the soft, cohesive soil should be calculated as follows:

$$F_V = A_{\text{eff}} \left[\left(a + \frac{bD}{T} + \frac{1.2h}{D_{\text{eff}}}\right) c_u + p_0'\right]$$

where a and b are squeezing factors. The squeezing factors $a = 5$ and $b = 0.33$ recommended in the SNAME guideline (2015) should be used in the calculations. However, where squeezing of the soft stratum does take place ($T \ll D_{\text{eff}}$), the bearing capacity of the soft soil cannot be greater than the bearing capacity of the soil below.

7.5.4.3 Recommendations for design

If there is a risk of punch-through in the in situ soils, and, thus, a risk for the WTIVs due to rapid penetration of their legs, then soil replacement measures should be planned in front of the waterfront structure to create a cushioning layer. In the design, it is assumed that punch-through can be ruled out if the predicted characteristic penetration resistance F_V (taking into account the load-spreading method) above the stratum with a lower bearing capacity achieves a value about 1.5 times higher than the specified preload (Lietaert 2011, Dier et al. 2004). The depth of the soil replacement layer d for upgrading the bottom of the watercourse should, therefore, be determined iteratively until the above safety criterion for punch-through is achieved.

The same methods should be used for calculations based on a design vessel without spudcans. To remain on the safe side, skin friction should be neglected when determining the soil replacement measures.

7.5.5 Maintaining and monitoring the jacking surfaces

Jacking surfaces require frequent, regular monitoring and the soil replacement layer requires considerable maintenance. As a rule, soil replacement layers are not compacted during laying for reasons of cost. This means that, depending on the material selected for the soil replacement layer (sand or rip-rap) and the soil mechanics properties of the undisturbed soil below that layer, different penetration values are possible. Therefore, the specific distributed loads and the geometries of the spudcans of the WTIVs have a considerable influence on the penetration dimensions that can be expected.

Penetration dimensions of 0.70–3.50 m have been measured for soil replacement measures that have already been carried out. During a wind farm project, one and the same wind turbine installation vessel will return at intervals to the same jacking position alongside the quay to collect further components. Owing to their good manoeuvrability and dynamic positioning systems, modern WTIVs are able to position their legs in the same place each time with great accuracy.

During the early phases of a new project, the geometries and depths of the footprints should be measured and assessed at regular intervals (see Figure 7.78). The analysis can be simplified by using a multi-beam echo sounder, which can generate a three-dimensional view of the footprint.

In the case of deeper penetration, backfilling the footprints is recommended, even after a short time in order to restore the original level of the soil replacement layer. This precautionary measure is required to prevent the leg of a WTIV from slipping sideways into its

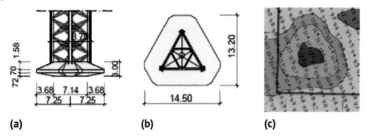

Figure 7.78 (a) and (b) Spudcan geometry of a wind turbine installation vessel; (c) associated footprint on soil replacement layer.

own footprint should the previous jacking position not be met exactly. Such an event could cause stresses in the legs and lead to damage to the hydraulic system for the legs.

At the latest when a different WTIV with legs at different spacings and other spudcan geometries uses the same berth, does such backfilling become absolutely essential. In the offshore industry, shipping companies generally reject the jacking positions that other vessels have already compacted in some areas, with lower levels due to the penetration of the legs of those vessels.

Such areas can be backfilled with the same material used for the soil replacement measures in the first place. Sandy materials can be dumped from barges or trailing suction hopper dredgers. The material can be distributed using water-jetting equipment or a jack-up pontoon.

When using rip-rap, the recommendation is to carry out backfilling by depositing the material carefully from a jack-up pontoon using a suitably equipped hydraulic excavator that can distribute the material underwater to the right level.

7.5.6 Logistical requirements

Owing to their dimensions and the heavy loads involved, the various components for an offshore wind turbine (OWT) place considerable logistical and technical demands on a port. So far, no consistent standards are evident regarding storage, transport and handling of the various OWT components, which means that the most diverse systems are in use, especially for storage. Furthermore, the transport routes, storage areas and pre-parking areas for handling items to be loaded onto WTIVs are characterised by different requirements, which are very much dependent on the respective logistics concept. This section begins by describing all the known dimensions and weights of standard and special components, the different transportation and storage systems and their significance for the distributed loads on the various operating areas.

7.5.6.1 Standard components for offshore wind turbines
7.5.6.1.1 Monopiles/transition pieces

Currently, monopiles reach lengths of about 50–80 m and have diameters of 6–8 m (Table 7.5). The total weight of a monopile is about 400–1300 t. In some projects, large, heavy monopiles are transported by floating them, with their buoyancy improved by including bulkheads, which bring the weight of such monopiles up to about 1300 t.

Table 7.5 Dimensions of monopiles.

Monopiles	Diameter [m]		Length [m]		Weight [t]	
	From	To	From	To	From	To
Current	5	8	50	80	400	1300
Forecast	5	10	50	90	400	2000

Table 7.6 Dimensions of transition pieces.

Transition pieces for monopiles	Diameter [m]		Length [m]		Weight [t]	
	From	To	From	To	From	To
Current	5	8	15	25	200	300
Forecast	5	10	15	30	200	350

In the future, monopiles with diameters of up to 10 m, lengths of about 90 m and weights of up to 2000 t will be needed on projects in water as deep as 40 m. A further increase in dimensions is unlikely in the medium-term, owing to the driving technology available.

Monopiles are stored on concrete blocks, steel trestles or a bed of sand. Monopile foundation structures for offshore wind turbines generally consist of two parts – the actual monopile and a transition piece, which is installed after driving the monopile. The structural connection between the two parts is formed by flanges, grouting or welding. In the future, the monopile structures of some projects will not have a separate transition piece.

The transition pieces have similar dimensions to those of the respective monopile (Table 7.6). They are 15–25 m long and weigh 200–300 t. For the next generation of large monopiles, transition pieces of up to 30 m long and weighing up to 350 t are to be expected.

7.5.6.1.2 Jackets

Jackets have a base size measuring between 25 × 25 m and 35 × 35 m, are 50–70 m high and weigh up to 900 t in total (Table 7.7). In a jacket, the transition piece is an integral part of the foundation structure. The figures for height and weight each include the transition piece.

The foundation piles for anchoring jackets are 2.5–3.6 m in diameter and 40–75 m long. Each pile can weigh up to 250 t (Table 7.8). Similarly to monopiles, these piles are stored

Table 7.7 Dimensions of jackets.

Jackets	Base size [m]		Height [m]		Weight [t]	
	From	To	From	To	From	To
Current	25 × 25	35 × 35	50	70	600	900
Forecast	25 × 25	45 × 45	50	90	600	1400

Table 7.8 Dimensions of foundation piles for jackets.

Foundation piles for jackets	Diameter [m]		Length [m]		Weight per pile [t]	
	From	To	From	To	From	To
Current	2.5	3.6	40	75	90	250
Forecast	2.5	4.5	40	90	90	320

on concrete blocks or steel trestles, which measure about 1.25 × 2.0 m. The weight of a foundation pile is, therefore, transferred to the soil via an area of about 5.0 m².

In the future, jacket foundation structures are expected to be used in water as deep as 65 m. That will require base areas measuring up to 45 × 45 m, lengths of up to 90 m and weights of up to 1400 t.

Besides the construction of symmetrical jackets with four legs, possible alternatives for some circumstances are the three-leg jacket and the twisted jacket.

7.5.6.1.3 Tripods

The tripods currently in use on wind farms have base dimensions of about 30 × 30 m and are about 65 m high. The total weight of these foundation structures is about 900 t. The foundation piles for anchoring the tripods have similar dimensions and weights to those for the jackets.

Tripods are transported on self-propelled modular transporters (SPMT). Concrete blocks and steel trestles are used for storage, which are useful for transferring the tripods to and from the SPMTs. The contact area of the storage trestles/blocks is about 8–10 m² per leg. To pick up a tripod, an SPMT unit with about 12 axles drives beneath each of the three storage trestles/blocks.

7.5.6.1.4 Tripiles

This type of foundation structure (Table 7.9) is unlikely to be used in future projects owing to the costs involved.

7.5.6.1.5 Tower segments

The towers for today's offshore wind turbines with a capacity of 3.6–6.0 MW currently consist of two or three segments. The lower tower segment often houses the electrical equipment and can, therefore, only be transported in the vertical position. The upper segments

Table 7.9 Dimensions of tripiles.

Tripiles	Base size [m]		Height [m]		Weight [t]	
	From	To	From	To	From	To
Current	20 × 20	30 × 30	50	60	750	900
Forecast	It is unlikely that larger structures will be developed.					

can be transported horizontally as well. Tower segments are 4.5–6.0 m in diameter, 12.5–36.0 m high and weigh 50–125 t, depending on the type of installation and number of segments (Table 7.10).

Diameters of up to 8 m, heights of up to 40 m and weights of up to 175 t are expected for future installations with capacities of 8.0–10.0 MW.

The tower segments are stored on supports consisting of two concrete blocks supporting a steel beam. Four supports are required per tower segment. With dimensions of about 1.0 x 1.5 m, the distributed load beneath the supports is about 20 t/m^2.

7.5.6.1.6 Nacelles

The nacelles of modern offshore wind turbines with a capacity of 3.6–6.0 MW are up to 26 m long, 11 m wide and high and can weigh up to 400 t (Table 7.11).

Lengths of up to 34 m, widths and heights of up to 15 m and weights of up to 550 t are expected for future installations with capacities of 8.0–10.0 MW.

Most nacelles include a helicopter platform so that maintenance personnel can reach the nacelle easily and safely.

Owing to the heavy loads involved, nacelles can only be transported with SPMTs. To do this, two SPMT modules, each with eight axles, are used alongside each other. The SPMTs laden with a nacelle results in a total load of about 400–430 t, depending on the manufacturer of the nacelle.

Similarly to the tower segments, nacelles are stored on simple concrete blocks connected via a steel beams. The contact area of one concrete surface is about 1.0 m^2.

7.5.6.1.7 Rotor blades/hubs

The rotor blades for the current generation of offshore wind turbines with a capacity of 3.6–6.0 MW are 55–70 m long, have a maximum profile depth of up to 6.0 m and weigh 15–35 t. Rotor hubs weigh up to 75 t (Table 7.12).

Rotor blades up to 100 m long with profile depths of up to 9 m and a blade weight of up to 60 t plus rotor hubs weighing up to 90 t are expected for future installations with capacities of 8.0–10.0 MW.

A rotor blade can be transported on public roads on a special vehicle designed to carry abnormal loads. Such a vehicle loaded with a rotor blade is 85 m long and about 4 m high, and the total load of this vehicle plus the load is 54 t.

Rotor blades must be stored on specially fabricated steel frames. Such storage allows the rotor blades to be stacked, which can save a considerable amount of space in storage areas.

In the near future, rotor blades will reach lengths of up to 90 m, which means that it will no longer be possible to transport them by road.

7.5.6.2 Transport and handling systems
7.5.6.2.1 Transport systems
7.5.6.2.1.1 Self-propelled modular transporter (SPMT) The self-propelled modular transporter (SPMT) is a means of transport that can be used for the whole range of large components needed for offshore wind turbine installations. With its modular potential, the ability to provide an appropriate number of axles and the great flexibility regarding the size of the platform, it is the only means of transport that can satisfy the specifications for all large items.

Table 7.10 Dimensions of tower segments.

Tower segments	Lower segment						Middle segment						Upper segment					
	Diameter (m)		Height [m]		Weight [t]		Diameter (m)		Height [m]		Weight [t]		Diameter (m)		Height [m]		Weight [t]	
	From	To	From	To	From	To	From	To	From	To	From	To	From	To	From	To	From	To
Current[a]	4.5	6.0	12.5	25.0	50	105	4.0	6.0	17.5	35.0	50	120	3.0	6.0	15.0	36.0	60	90
Current[b]	5.5	6.0	34.0	36.0	110	125	—	—	—	—	—	—	5.5	5.5	34.0	36.0	110	125
Forecast	4.5	8.0	12.5	30.0	50	175	4.0	7.5	17.5	35.0	50	165	3.0	6.5	15.0	40.0	60	165

a) Three segments.
b) Two segments.

Table 7.11 Dimensions of nacelles.

Nacelles	Length [m]		Width [m]		Height [m]		Weight [t]	
	From	To	From	To	From	To	From	To
Current	15	26	6	11	6	11	160	400
Forecast	15	34	6	15	6	15	160	550

Table 7.12 Dimensions of rotor blades and hubs.

Rotor blades/hubs	Rotor blades Length [m]		Profile depth [m]		Weight [t]		Rotor hubs Weight [t]	
	From	To	From	To	From	To	From	To
Current	55	70	4	6	15	35	45	75
Forecast	55	100	4	9	15	60	45	90

These self-propelled vehicles consist of a welded platform frame with four or six axles per module. Each of these modules is provided with a hydraulic coupling, so that any number of modules can be combined and, thus, increase the load-carrying capacity of the configuration. The electronic multi-way steering makes the combination of individual modules extremely flexible and manoeuvrable. In conjunction with a remote control, it is possible to move and position all the units coupled to form one vehicle with millimetre accuracy.

Roll and cargo trailers hauled by tractor units can still be used for smaller items, or rail-borne bogie systems for special goods.

7.5.6.2.2 Handling systems

7.5.6.2.2.1 Fixed slewing cranes (installation vessels) A slewing crane with a very high lifting capacity (up to 15 000 kN) mounted on the installation vessel is ideal for loading/unloading and installing the whole range of large components needed for offshore wind turbines.

7.5.6.2.2.2 Mobile cranes It cannot be assumed that installation vessels will in future always fetch the large components needed for offshore wind turbines from offshore energy support bases. Instead, large components will be transported to the wind farm or installation vessel on feeder vessels, which means that

- land-based slewing cranes with a similar lifting capacity to that of a crane on an installation vessel (primarily for lifting foundation structures and towers), and to a certain extent also
- mobile cranes, primarily for piles and tower segments (tandem lift), but also for nacelles, hubs and rotor blades,

will need to be considered.

7.5.6.2.2.3 Gantry cranes A gantry crane is very limited in its use (e. g. for loading cross-bracing). When loading longer cylindrical components (piles, monopiles, towers, possibly also tower segments), a tandem lift with two gantry cranes will normally be necessary. As a rule, the limited headroom of gantry cranes means that they are unable to load directly onto a wind turbine installation vessel.

7.5.6.3 Imposed loads

The imposed loads necessary for an offshore energy support base depend very much on the various operational areas and the logistical scenarios. It is therefore not possible to specify imposed loads that apply in every case.

Although the use of multi-axle SPMT units results in loads that can be as low as $50 \, kN/m^2$ or even less, even for large components, these days most abnormal load facilities must be able to handle much higher imposed loads. Existing structures are designed for imposed loads of $100–200 \, kN/m^2$ over their whole area.

Some projects currently at the planning stage also envisage special abnormal load areas designed for $400–700 \, kN/m^2$, depending on the type of construction. Crawler-mounted cranes are mostly used in such special abnormal load areas. Such cranes can distribute the load better (e. g. excavator mats), but not reduce it.

Separate load calculations must be carried out in each case to determine the specific loading assumptions for the different handling scenarios.

7.6 RoRo berths

7.6.1 General

Roll-on/roll-off (RoRo) terminals enable RoRo vessels to load and unload wheeled cargos during changing water levels and differing loading conditions. Terminals can have one or more RoRo berths at which RoRo vessels can moor. Such berths include fixed and/or adjustable ramps to/from the quayside, as described below.

This section deals with the hydraulic engineering requirements for RoRo terminals. The requirements on the land side, e. g. barriers, intermediate storage capacities and structures, should be built and maintained to suit the overall terminal functions. However, the reader should consult PIANC (2015/2016) for these requirements.

Generally, a distinction is made between the following types of traffic:

a) Commercial RoRo traffic,
b) Combined RoPax traffic (roll-on/roll-off/passenger traffic),
c) Combined RoRo traffic/railborne traffic.

General data on the dimensions of RoRo vessels can be found in Section 2.1.6. RoRo vessels are distinguished by

a) Their passenger capacities.
b) The different wheeled cargo units, e. g. trailers, railway wagons, all kinds of road vehicles.

Typically, a distinction is made between the following types of vessel:

a) Train and RoRo ferries.
b) Ferries with or without options for carrying passengers.
c) Fast ferries.
d) Car carriers.
e) ConRo ships (combination of container and RoRo ship).

A maximum vessel length of 250 m can be assumed for the Baltic Sea, which leads to a maximum beam of 35 m. The largest ferries currently in operation in the North Sea are 240 m long with a beam of 32 m.

The design of terminals for such traffic calls for special knowledge of the type of traffic envisaged, which in turn determines the loads and kinematics of RoRo berths.

7.6.2 Loading assumptions for RoRo terminals

7.6.2.1 Assumptions for imposed loads

Loading assumptions for the terminal must be specified in the early phases of design depending on the type of vessel and its connection to the quayside. General stipulations regarding the traffic must be the reached in agreement with the statutory regulations, which can be required by standards, directives and the requirements of the local highway authority.

In Germany, loading assumptions are covered by DIN EN 1991-2 (actions on bridges). Load model LM 1 should be used in the structural analysis of linkspans, which are regarded as bridges. Fatigue analyses should be carried out in addition to the typical structural calculations necessary for such structures. Roadway widths and traffic volumes may need to be agreed upon in writing with the client/operator as part of the loading assumptions.

In addition, it might be necessary (depending on the capacity of the vessel) to consider individual heavy-duty vehicles such as forklifts, heavy-duty modular vehicles, RoRo tractor units or similar vehicles. These vehicles represent special loads not covered by load model LM 1. It can be assumed here that movements of abnormal loads on the quayside are monitored and controlled. Accordingly, a simultaneous loading due to special vehicles and LM 1 loads on the individual roadways can usually be ruled out. Typical heavy-duty vehicles are summarised in Table 7.13.

On ships with several decks, it is necessary to consider the imposed loads, depending on the loading assumptions for the individual decks. On the Baltic and North Seas, transporting freight vehicles on the main and upper decks is widely practised these days.

7.6.2.2 Roadway width

When designing a terminal, consideration must be given to providing linkspans with an adequate width, including the zone on the land side. For example, a good recommendation is to increase the roadway width to 4.5 or even 5.0 m to ensure safe movement of traffic to and from the ship.

When considering abnormal loads, a curved approach on the land side or vehicles for special purposes (e.g. offshore rotor transport), it is necessary to investigate the minimum turning circles on the traffic areas available.

Table 7.13 Typical heavy-duty traffic at RoRo terminals.

Type of vehicle	Total load (including payload)	Typical load layouts for individual heavy-duty vehicles
Forklift	634 kN	2-lane traffic
Semi-trailer	890 kN	2-lane traffic
Semi-trailer	1060 kN	2-lane traffic
Trailer	1244 kN	2-lane traffic

Table 7.13 (Continued.) Typical heavy-duty traffic at RoRo terminals.

Heavy-duty modular vehicle	2120 kN	1 lane, special transport, single journey
RoRo tractor unit	1123 kN	1 lane, special transport

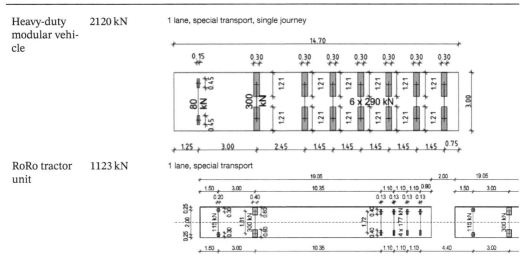

7.6.2.3 Berthing velocity

Generally speaking, the recommendations given in Section 4.1 also apply to ferry terminals. However, higher berthing velocities can occur at ferry berths in frequent use. For example, berthing velocities as fast as 1.00 m/s are not unusual in the design of side fenders. Berthing velocities of up to 0.50 m/s can be necessary when designing stop fenders. The berthing velocities to be assumed for the design of ferry berths must be defined within the scope of a detailed investigation and agreement with the customer and the operator of the ferries.

7.6.3 Kinematics

7.6.3.1 Water level

When designing a RoRo terminal, it is crucial to specify the water levels at the terminal in the following categories:

a) Operational water level,
b) Extreme water levels.

Operational time slots should be quite generous in order to maximise the operational options of the connections to the ferry while maintaining optimum loading and unloading conditions for the RoRo vessels. For example, at a very heavily used RoRo berth in the Port of Rotterdam, an operational time slot is guaranteed for a return period of

- Once in 10 years with an unladen ship at high water.
- Up to once in 10 years with a fully laden ship at low water.

A stable position of the vessel must be ensured outside the envisaged operational time slot.

7.6.3.2 Listing conditions for vessels

To enable a flawless kinematics calculation, listing values of 2° must be used for operational conditions and 5° for extreme operational conditions with restricted loading and unloading.

7.6.3.3 Kinematic design: typical gradients for vehicles on linkspans

The design of fixed and movable linkspans must ensure their safe use by vehicles. Therefore, the majority of design cases to be considered applies to their use by typical road vehicles.

The chosen kinematics for the linkspan depend not only on the geometry of the vehicles themselves, but also on the speed of the vehicles while traversing the linkspan. The faster the vehicles cross over, the more jolt-resistant the transition between ship and quayside has to be. The time available for loading and unloading the RoRo vessel, and hence the options for operational personnel to regulate roll-on/roll-off procedures, also influences the essential kinematics of the linkspan. Changing the ballast condition of the vessel and/or choosing a time with a more favourable water level can improve the loading and unloading of vehicles with high demands regarding ground clearance, maximum ramp gradients and changes to those gradients.

The intended kinematics of the linkspan must be carefully adapted to the intended traffic procedures. The ramp gradients and changes to those gradients given in Table 7.14 have been used for many linkspans around the Baltic Sea. The focus here is on processing the vehicles, trucks and typical abnormal loads given in Table 7.13 in large numbers and within a short time. Limited processing of vessels must still be guaranteed in the case of extreme operational water levels based on extreme values, transition angles and listing conditions.

Patented special solutions, e. g. suspended flaps, can be used in some cases to maintain the roadway gradients given in Table 7.14. Such flaps shorten the linkspan by maintaining the maximum transition angle.

Table 7.14 Recommended gradients/changes to gradients for the Baltic Sea.

Type	RoRo terminal (general) Crest/valley	RoRo terminal (with higher demands regarding crossing speed) Crest/valley	Train ferry terminals
Operational water level	max. 4.5°/4.5° (surging/receding)	max. 1°/2° (surging/receding; recommended for main deck)	2.1°
Extreme operational water level	max. 7.5°/7.5° (surging/receding)	max. 4.5°/4.5° (surging/receding)	2.1°
Fixed ramp, movable passenger bridges (upper deck)	max. 1 : 10 (gradient)	max. 1 : 12.5 (gradient)	—
	max. 3° (change to gradient)	max. 1° (change to gradient)	

7.6.4 Classification of ship-to-shore facilities

7.6.4.1 General

RoRo berths typically consist of fixed and/or adjustable ramps on the land side to which the RoRo vessels can be tied (BS, 2007). BS 6349-8 (2007) defines a fixed shore ramp as the "fixed inclined structure between the normal quay level and a level upon which the shore end of ship ramp can rest". In contrast to this, an adjustable shore ramp, i.e. a linkspan, is an "articulating structure, which can be a floating or a lifted bridge structure, that links a RoRo ship either directly to the shore, to a fixed structure or to a pontoon, or that links a pontoon with the shore".

In principle, many different structures are possible for RoRo ramps and linkspans, which are described below. Fixed land-side ramps are generally used where there are only small fluctuations in the design water level ($\Delta \leq 1.5$ m). The use of movable ramps/linkspans becomes necessary where there are large fluctuations in the design water levels or where it is necessary to keep the changes in gradient small so that vehicles can cross the RoRo ramp.

Furthermore, the following design criteria must be taken into account when choosing a form for a RoRo berth:

a) Type of RoRo traffic.
b) Equipment on board the RoRo vessel.
c) Changes to water levels and design draught.
d) Time spent moored at berth.
e) Availability requirements of the berth (unfavourable weather and tidal conditions, operational readiness).

The various widths of the types of vessel expected to use a berth and the position of the ship's ramp also have a considerable influence on the design of linkspans. The criteria listed in Table 7.15 are among those that have to be considered when designing RoRo ship-to-shore systems.

When it comes to operational and extreme water levels, there are considerable differences between individual ports and harbours, e.g. between those on the North and Baltic Sea coasts, and, therefore, the respective structural solutions also vary considerably. At terminals with passenger traffic, passengers and vehicles should be kept separate for safety reasons, e.g. with separate gangway arrangements or footways cordoned off from roadways.

BS (2007) and PIANC (1995) divide vehicle movements for RoRo ship-to-shore systems into three categories:

a) Direct access with ship's ramps.
b) Fixed shore ramps.
c) Adjustable shore ramps such as systems with pontoons or movable linkspans.

The general advantages and disadvantages of these three different types of RoRo ship-to-shore system are listed in Table 7.16.

Likewise crucial for the design of RoRo ship-to-shore systems is the form of construction of the vessel itself. According to Figure 7.79, we can distinguish between two different types of vessel:

a) Vessels with their own stern flap.
b) Vessels without their own stern flap (e.g. berths for train ferries).

Table 7.15 Design criteria for RoRo ship-to-shore systems.

Conditions at location	• Topography • Bathymetry • Geology/subsoil • Position of berth relative to principal wind direction • Seismicity (if applicable)
Environmental conditions	• Meteorology: normal and exceptional wind, rain, temperatures • Oceanography: normal and exceptional waves, tides, currents, ice, chemical composition of the water, wave states, surging in the port, etc. • Frequency and probability of storm conditions
Operational influences	• Vessels (data, dimensions, types, frequency, approach velocity, time spent at berth, loading and support requirements) • Vehicles (data, dimensions, types, loads, operational dimensions (turning circles etc.)) • Special equipment, equipment for mooring, anchor winches, etc. • Services and equipment, connections to the land (e. g. gangway), lighting and safety, electricity, utility lines • Rails, cranes, loading facilities, permanent way, capacity, weights, clearance, level, speed, reach and lift, etc. • Storage area for cargo
Functional considerations	• Dredging, scouring and silting-up, washout due to ships' propellers • Maintenance methods: cathodic protection, damage repair, etc.
Consideration of navigational aspects	• Width and depth of navigation channel • Approach conditions for vessels (with respect to time)
Restrictions	• Limits to ports, harbours and breakwaters • Imposed by authorities: water quality standard, oil-laden ballast, disposal of dredged material, fill, etc. • Authorisations and licenses (i. e. for radio/Wi-Fi connections between terminal and vessel) • Existing facilities: change of use or extension/limitation

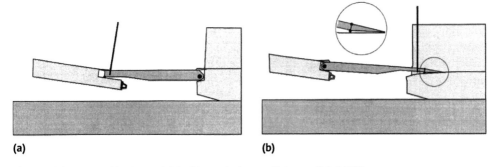

(a)   (b)

Figure 7.79 Vessels (a) with and (b) without their own linkspan (BS, 2007).

7.6.4.2 Ship-to-shore facilities for ships with their own stern flap

Stern flaps vary considerably in terms of their length, width and height. Figure 7.80 shows a ship that can load and unload vehicles at a standardised berth via its own stern flap.

Table 7.16 Comparison of various concepts for RoRo ship-to-shore systems.

General description	Ship's ramp	Fixed shore ramp	Adjustable shore ramp – pontoons	Adjustable shore ramp – movable linkspans
Advantage	Good flexibility	Suitable for vehicles without drivers	Easy to adjust to various water levels in tidal regions	Can be adjusted accurately, accurate adjustment of linkspan reduces waiting time, can serve several decks
Disadvantage		Higher local loads, requires a relatively long ship's ramp	Usage depends on gradient	Usually single-lane traffic	Requires maintenance
Costs	–	±	+	++

Figure 7.80 Ship with its own stern flap alongside a standard berth.

It is normally the case that additional facilities such as pontoons or fixed shore ramps will be required on the land side in order to avoid unfavourable changes in gradient during loading and unloading.

7.6.4.2.1 Adjustable shore ramps – pontoons

Pontoons such as those described in Table 7.17 offer a high degree of flexibility when using a berth for loading/unloading other types of cargo and/or when the method of mooring

Table 7.17 Pontoon ship-to-shore systems.

Types of connection	Advantage	Disadvantage
Floodable pontoon systems	Multiple usage, flexibility (tidal range), can be fitted with additional level control if required	Limited loads, normally single-lane traffic
Linkspan supported on pontoon (BS, 2007)	Inexpensive solution at locations with large tidal range	Limited loads, normally single-lane traffic
Semi-submersible linkspan (BS, 2007)	Inexpensive solution at locations with large tidal range	Limited loads, normally single-lane traffic, unfavourable kinematics

the RoRo vessels changes during the lifetime of the berth. However, the maintenance and operational requirements for pontoons are greater than those for fixed shore ramps.

7.6.4.2.2 Fixed shore ramps

Fixed ramps always require vessels with their own stern flap.

Short, fixed ramps with a limited height can be used as a connection between the quayside and the bridge structure for a movable linkspan.

According to BS 6349-8 (2007), the design of fixed ramps is determined by the tidal range. Where the tidal range exceeds 1.5 m, a greater change in gradient occurs, which normally means it is necessary to install a movable linkspan.

Figure 7.81 shows a fixed ramp at berth 60 in the Port of Rostock. The design of this fixed ramp is based on a comparison of the kinematics of various types of vessel. The ship's stern flap is supported on ramp zones with different gradients depending on the water level and the laden state of the ship.

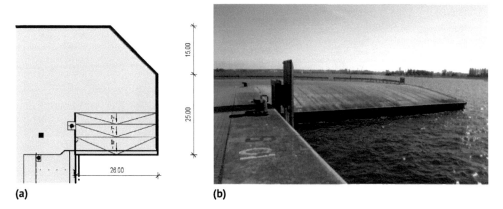

Figure 7.81 Example of a fixed ramp at a port (Rostock, berth 60): (a) part plan view; (b) photo of the fixed ramp.

7.6.4.3 Movable linkspan systems for ships with and without their own stern flap

A distinction is made between two general types of linkspan depending on the type of traffic:

a) Linkspans for road vehicles,
b) Linkspans for rail and road traffic.

Depending on the frequency and importance of the ferry route, these days it is normal to use linkspans with several decks for road vehicles to enable direct loading/unloading of the main and upper decks of the vessel (see Figure 7.82). In a few cases, it is possible to transfer railway wagons to/from the main deck as well.

Appropriate constructional measures, e.g. provision of stop fenders, should be included to prevent vessels colliding with the linkspan.

In addition, movable linkspans at ports covered by EU legislation must comply with the EU Machinery Directive. This means that the design, construction and operation of movable linkspans must comply with certain requirements to ensure safe supervision and precautions during the entire procedure at the terminal.

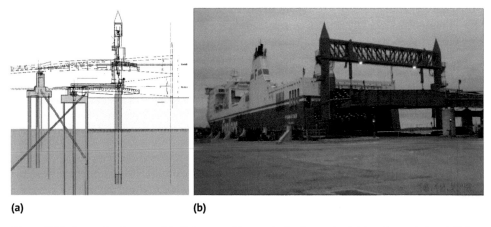

Figure 7.82 Example of a movable linkspan with two decks for road vehicles (Rostock, berth 54): (a) schematic drawing; (b) photo of movable linkspan.

Figure 7.83 Example of a movable linkspan for train ferries (Lübeck, berth 64).

Movable linkspan systems are used for vessels with or without their own stern flap. Figure 7.83 shows a movable linkspan for train ferries at berth 64 in Lübeck. Train ferry berths require the linkspan to be positioned exactly because of the need to ensure continuity of the rails. To do this, a king-pin is normally used to fix the position of the linkspan/vessel.

For reasons of corrosion protection and the need to be able to operate linkspans even in icy conditions, intentional submersion of the linkspan structure under operational conditions should be avoided.

The maximum changes in gradient according to Section 7.6.3.3 must be taken into account when designing a movable linkspan. In the case of a higher design speed for the traffic (e.g. 40 km/h), the maximum changes in gradient under operating conditions should be limited to 2.0° (crest/trough). At train ferry terminals, it is necessary to take into account the maximum possible gradient for railborne traffic (e.g. 2.1° in accordance with permissible local conditions).

Therefore, linkspans for railborne traffic are normally longer, and because of the heavy loads, heavier, which usually results in the need for a through bridge. Such linkspans are always supported directly on the vessel. The loads must be carried on the vessel and so corresponding coordination with the vessel operators is required during the design.

As a result of the much higher loads of train ferry linkspans, it is normal to use a counterweight system to adjust the level. Such systems are moved purely mechanically or in combination with hydraulic drives.

7.6.5 Facilities and equipment on the land side

7.6.5.1 Fender systems at ferry berths
Fender systems must ensure that vessels can berth safely and remain securely in position during loading and unloading. The following fender systems are used for this:

a) Side fenders.
b) Guidance structures for tapered basin solutions.
c) Fenders for berthing the ship's stern (positioning or stop fenders).

As was already described in Section 7.6.2.3, faster berthing velocities must be taken into account for the fender systems. In addition, higher safety factors should be used when designing the fender systems. PIANC (2020) specifies safety factors of "2.0 or higher" for RoRo vessels.

The safety factors and berthing velocities for designing the fender systems must be agreed upon and specified together with the berth operator.

7.6.5.2 Regulations for mooring

The layout of the bollards should be determined based on both operational and safety aspects. Normally, agreements with the user and the regulatory authorities (e.g. harbourmaster) will be necessary within the scope of the design.

In addition to the bollards along the edge of the quay on the water side, ferry berths often require additional bollards on the land side. To tie up the ferries faster, corresponding permanent line extensions are fitted to these additional land-side bollards.

At berths in frequent use with limited waiting times, automated mooring systems may be necessary, possibly in combination with a tapered basin. Such systems require an exact match between the land-side bollards and the facilities on board the vessel (e.g. additional bollards on board or reinforced hull zones). There are two types of automated mooring system on the market:

- Conventional mechanical systems,
- Automatic vacuum systems.

Conventional mooring systems are in widespread use in the German ports bordering the Baltic Sea, e.g. Puttgarden, Lübeck, Rostock. Vacuum mooring systems are in use in a few European ferry terminals, e.g. Dover (UK). They must be kept clear of ice, which can be achieved with heating systems.

When using automated mooring systems, it is necessary to check at the design stage how vessels can be safely moored in the event of failure/overloading of the system. A special operational procedure is required for such a situation.

At berths for general cargo operations that can also be used for loading/unloading RoRo vessels with their own stern flap, the bollards must be demountable, so that there is sufficient clearance for the lowering of a stern flap (see Figure 7.80).

7.6.5.3 Protection against scouring

RoRo terminals are particularly at risk of increased scouring because the vessels are always moored at the same positions due to the fixed and movable ramps on the quayside. Constructional measures at the RoRo berth itself and operational precautions regarding the berthing procedures for RoRo vessels are required to protect the land-side facilities against scouring.

Modern RoRo vessels are generally equipped with powerful drives (e.g. azimuth thrusters) and manoeuvring thrusters, which enable them to berth without tug assistance. Maximum scouring during berthing and casting-off often occurs directly in front of the linkspan or ramp. Damage at various ferry berths, with considerable adverse effects on their stability, has been discovered in recent years. Open foundation solutions instead of sheet pile walls offer the advantage that the energy that builds up is distributed beneath the berth, thus minimising local damage.

Suitable measures for protection against scouring should be determined at the design stage, taking into account the local circumstances in conjunction with the types of vessel expected. Structures can be protected by physical protective measures but also by including a higher scouring allowance in the structural design of the waterfront structure. Regular monitoring of the state of the scouring protection measures is necessary when the berth is in use. Any scouring that exceeds the amounts allowed for in the stability calculations must be backfilled without delay.

7.6.5.4 Gangway facilities for passenger traffic

Safety and/or service aspects require pedestrians to be kept separate from vehicles with the help of gangway facilities and passenger bridges. The design and dimensions of footbridges and passenger bridges should

- Ensure that the different types of vessel expected can be served safely.
- Enable the quick disembarkation of all foot passengers.
- Be suitable for wheelchair users or allow for appropriate special measures (e. g. accompanied by the operator).
- Ensure that luggage and bicycles can also be transported.
- Ensure adequate protection against the weather, which means that heating/cooling and ventilation might be necessary.
- Include moving walkways or luggage conveyor belts in the event of long distances between terminal and ferry.
- Have non-slip floor coverings and gentle gradients.
- Permit easy cleaning and maintenance at frequent intervals.

The footways of gangway facilities are often positioned above the vehicle decks high above the top edge of the quay. If the footway is not fully enclosed, then special measures will be required to exclude any risks to passengers (e. g. anti-climb balustrades). CCTV monitoring for operational safety is especially recommended for long passenger bridges and the final connection to the ship. The direct transition between gangway and ship is usually supervised by instructed personnel. Risk assessments according to the EU Machinery Directive are required for movable gangway facilities.

Fire alarms and escape routes in the event of fire must be included according to the relevant local regulations.

Gangway facilities that lead to an entrance in the side of the ship can make use of one of the forms described in Table 7.18. Passenger bridges that lead to entrances near the bow or stern of the ship are usually combined with the linkspan.

The choice of a suitable passenger access facility depends on

- The type of ship and the dimensions of the ferry
- The tidal range
- The distance between the passenger terminal and the ship,

as well as other factors.

Gangway facility types 1a and 1b (Table 7.18) are suitable for small to moderate tidal ranges, depending on the length and ensuing gradient of the passenger bridge. Type 1a requires the ship's access door to be at a fixed position because the passenger bridge can only move vertically. Besides passenger traffic, type 1b, with its passenger bridge able to move

7.6 RoRo berths

Table 7.18 Various types of gangway system.

Type	Layout		Changes to water levels and draught	Degrees of freedom	Flexibility of system
1a		A: Permanent building B: Passenger bridge C: Elevation frame D: Extending segment	Small to moderate	2	Limited
1b		A: Permanent building B: Passenger bridge hinged vertically and horizontally C: Travelling elevation frame D: Extending segment	Small to moderate	3	Moderate
2		A: Permanent building B: Passenger bridge hinged vertically C: Height-adjustable linkspan D: Movable access platform and passenger transfer bridge	Small to moderate	3	Moderate
3		A: Permanent building B: Passenger bridge hinged vertically C: Travelling (elevation) frame D: Extending segment	Small to moderate	3	High
4		A: Permanent building B: Passenger bridge hinged vertically C: Travelling (elevation) frame D: Extending segment	Moderate to large	3	High

vertically and horizontally, results in increased flexibility because it allows the mooring of various types of ship with access doors at different positions.

Type 2 offers similar flexibility to type 1a, although the form of construction results in less flexibility.

Types 3 and 4 offer maximum flexibility when it comes to the types of ship to be served. The vertically hinged passenger bridge of type 4 also allows suitable gradients even for large changes in water level and draught. However, these two types are very demanding in terms of design, construction and maintenance.

The final connection between the passenger bridge and the ship itself is frequently in the form of an extending bridge segment or a docking cabin that is fitted with a hook or pin that connects to a suitable fitting on the side of the ship. Sturdy balustrades with suitable infill panels are necessary here to guarantee the safety of the passengers on the open transfer section.

Depending on the size of the ships being served and the type of passenger bridge, it might be necessary to design the final connection to the ship such that it can track the motion of the ship actively or passively. Passive movement of the bridge is achieved by the forces that are transferred from the ship via the mechanical connection. The connecting element for such a connection should be designed to fail in a controlled way in the event of an overload. Modern bridge structures are able to measure the motion of the ship and track this actively via automatic drives. Such systems are automatically decoupled in the event of an emergency, so that the bridge can be moved safely away from the side of the ship.

7.7 Jetties

7.7.1 Introduction

In contrast to a quay wall, which provides a vessel with a berth having a direct connection to a terminal and land-side infrastructure, a jetty consists of a loading/unloading facility at a certain distance from the shore.

Jetties are generally in the form of permanent structures, although floating jetties are also used, especially at locations with considerable changes in water level. Figures 7.84 and 7.85 show examples of a permanent and a floating jetty, respectively.

The majority of jetties are used as facilities for handling dry or liquid bulk cargos.

In contrast to this, berths for handling containers or general cargos or intended for other uses, e.g. ferries or cruise liners, are mostly designed and built as permanent quayside facilities in order to take account of the complex logistical requirements of such operations. However, jetties can in some cases provide beneficial solutions for these operations as well.

Jetties generally consist of three basic elements:

- Berthing and mooring facilities that guarantee safe berthing and casting-off manoeuvres plus the provision of suitable berthing conditions.
- Loading/unloading platforms to guarantee trouble-free, efficient transshipment.
- A connection between the transshipment platform and the shore, either in the form of a permanent bridge or an underwater pipeline for liquid bulk cargos.

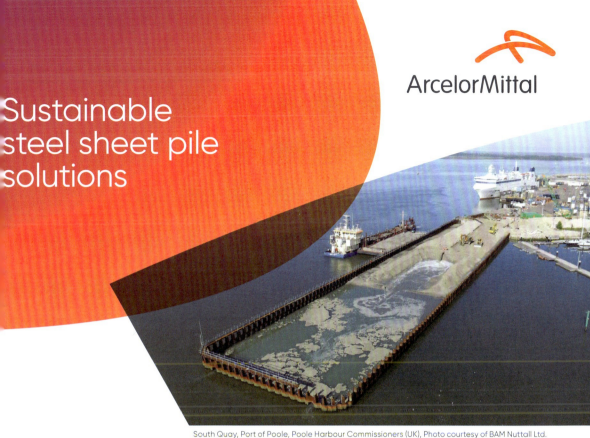

Sustainable steel sheet pile solutions

ArcelorMittal

South Quay, Port of Poole, Poole Harbour Commissioners (UK), Photo courtesy of BAM Nuttall Ltd.

Reduce the environmental impact of your infrastructure projects with **EcoSheetPile™ Plus** steel sheet piles, made from 100% recycled steel and using only certified electricity from renewable sources.

Apply the principles of circular economy:

- **Reduce:** optimise the structure's design with innovative construction methods, higher steel grades and lighter sheet piles profiles;
- **Reuse:** sheet piles can be reused up to ten times in temporary projects;
- **Recycle:** at the end of the structure's service life, the sheet piles can be extracted and recycled without any loss of quality.

Scan here for more information

ArcelorMittal Commercial RPS S.à r.l. | T +352 5313 3105 (Headquarter-Luxembourg)
sheetpiling@arcelormittal.com | sheetpiling.arcelormittal.com

Steel sheet piles 2023

Z Section

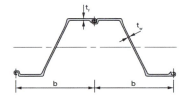

Mass (wall)	from	94	to	249	kg/m²
Wall thickness t_f	from	8.5	to	24.0	mm
Wall thickness t_w	from	8.5	to	17.0	mm
System width b	from	630	to	800	mm
W_{el}	from	1 205	to	5 155	cm³/m

U Section

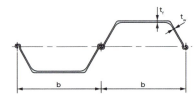

Mass (wall)	from	70	to	197	kg/m²
Wall thickness t_f	from	6.0	to	20.5	mm
Wall thickness t_w	from	6.0	to	11.4	mm
System width b	from	400	to	750	mm
W_{el}	from	625	to	3 340	cm³/m

AS 500® straight web sections

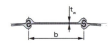

Mass (wall)	from	128	to	158	kg/m²
Wall thickness t_w	from	9.5	to	13.0	mm
System width b				500	mm
$R_{k,s}$	from	3 500	to	6 000	kN/m

HZ®/AZ® combined wall system

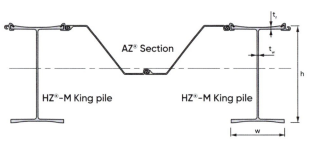

Mass (solution)	from	262.6	to	995.9	kg/m
Wall thickness t_f	from	18.9	to	37.0	mm
Wall thickness t_w	from	13.0	to	22.0	mm
Height h	from	631.4	to	1 087.4	mm
Width w	from	420	to	460	mm
W_{el}	from	4 135	to	46 280	cm³/m

W_{el}: elastic section modulus
$R_{k,s}$: characteristic interlock resistance

All profiles manufactured in Luxembourg are available as EcoSheetPile™ Plus products, made from 100% recycled steel and using 100% renewable electricity.

ArcelorMittal Commercial RPS S.à r.l. | T +352 5313 3105 (Headquarter-Luxembourg)
sheetpiling@arcelormittal.com | sheetpiling.arcelormittal.com

Figure 7.84 Example of a permanent jetty at Lumut, Malaysia (BAM International).

Figure 7.85 Example of a floating jetty at Cologne (Shell Deutschland).

7.7.2 Design of jetties

7.7.2.1 Berth orientation and availability

Jetties are often found at exposed locations and, therefore, cannot provide adequate protection against waves and currents in extreme situations, which means that vessels might have to leave a berth temporarily to avoid dangerous situations. Provided that the frequency and duration of unfavourable boundary conditions lie within acceptable limits, even exposed jetties can generally still fulfil the intended operational requirements regarding the volumes of cargos to be handled.

However, when designing the facility, it is important to make sure that the orientation and configuration of the berths ensure the best possible exploitation of anchor lines and, thus, minimise the loads on the vessels. Optimum utilisation of anchor lines is taken into account in the design by positioning breasting and mooring dolphins at the right places.

To minimise the loads of the vessels on the jetty, the orientation of the vessels at the berths should be coordinated with the prevailing directions of wind, sea state (primarily swell) and currents. It is normally necessary to carry out a dynamic mooring analysis in order to determine the resultant forces in the anchor lines of a vessel that is subjected to the critical combinations of wind, waves and currents acting at the berth. Besides specifying the upper limit to acceptable boundary conditions at a berth, a mooring analysis can use the statistical distribution of the local environmental boundary conditions to estimate the mean probability of occurrence of conditions during which a vessel cannot approach the berth.

Nautical simulations are often used during the front-end engineering design (FEED) in order to define the necessary navigation channels and identify them by way of sea marks and also organise tug assistance for large vessels.

Such analyses can also be employed to investigate and define maximum environmental boundary conditions concerning the safe passages of vessels through access channels plus the berthing and casting-off of vessels. The advantage of real-time simulations is that the human factor can be assessed more realistically in the model studies.

Where several berths are planned on one jetty, it is essential to consider the distances between the berths. The criteria regarding safe minimum distances can be derived from a series of requirements:

- An adequate channel around the vessel to ensure safe navigation with tug assistance corresponding to the results of the nautical study.
- Compliance with safety clearances around the manifolds for transferring combustible gases or liquids, e. g. LPG or LNG.

Where potential berths with a sufficient natural depth of water are only available at a relatively large distance from the shore, that increases the cost of the access bridge, which might render the project uneconomical. To minimise the development costs, in such cases it is necessary to weigh up whether the access bridge can be shortened by dredging an access channel on the sea side. When considering this, it is important that such comparative studies take into account the whole lifecycle costs, e. g. costs for maintenance dredging.

7.7.2.2 Heights of jetties

The heights of jetties and their access bridges must be designed to prevent wave impacts on vulnerable components of the load-bearing structure, also pipelines, conveyor belts, etc. Structural connections and cross-beams between the foundation piles should preferably be positioned above the crests of the waves, if it is not possible to verify that the wave loads can be resisted by the structure without damage.

The maximum water level displacement of the sea state must be determined over the full length of the jetty because it can vary depending on bottom friction and shoaling. As waves enter shallow water, the wave profile deforms such that the maximum wave crest does not necessarily occur close to the sea-side part of the jetty.

7.7.2.3 Influence of environmental conditions on construction work and construction concepts

As mentioned above, jetties are sometimes built in places exposed to wind, waves and currents, and this factor has to be taken into account for the orientation of the berths and the height of the structures. Furthermore, the selection of preferred structural concepts plus suitable device for installing structural components (foundations, pile caps, bridge members, etc.) can be limited by unfavourable environmental boundary conditions.

Floating devices (e.g. crane barges) cannot operate even when wave heights are low, whereas jack-up rigs are affected less by the state of the sea. When it comes to safe construction operations in pronounced sea state conditions, the designer is, therefore, required to carry out a thorough analysis and select structural elements that are reasonable in terms of numbers and weights (optimisation of lifting capacities and lifting cycles).

Therefore, investigating the safe and efficient use of device at the design stage is only one example of the important interaction between structural design and construction on site plus the need to consider aspects relevant to construction in an early phase of the design.

7.7.3 Design of berthing and mooring facilities (ship-to-shore)

The design of breasting and mooring dolphins must be optimised to guarantee safe berthing and casting-off, plus safe conditions for ships lying at berth.

At least one fender (or point of contact) should be available as far apart as possible at the opposite ends of the parallel midship section in order to ensure adequate stability of the moored vessel (see Figure 7.86). It is important to make sure that the points of contact are located in the area of the parallel midship section because otherwise the fenders in contact with the hull of the ship are placed at an angle, which leads to undesirable vertical loads on fender and vessel as the water level or draught of the ship varies (during loading/unloading operations).

Taking the above stipulations, the distance between the points of contact according to Section 12.1.2 is generally 25–40% of the vessel's length.

When planning a berth, compliance with this criterion is required for the whole range of vessel types intended plus all the realistic positions of vessels along the berth (e.g. in order to coordinate the ship's manifold with the manifold on the jetty). This analysis provides the minimum number of points of contact required. In addition, the recommendation is to provide at least two further, redundant, points of contact (fenders) either side of midship,

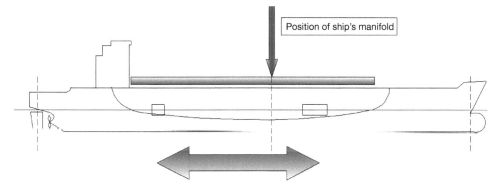

Figure 7.86 Criteria for deriving the recommended dolphin spacing. *Double arrow*: the spacing of dolphins should be chosen to be as large as possible in order to maximise the stability of the moored ship. The spacing is limited by the length of the parallel midship section. The ensuing recommendation for the spacing of dolphins is approximately 0.25–0.40 Loa.

so that it is possible to continue using the berth even in the event of the failure of one single fender.

These same principles also apply at continuous quays handling dry bulk goods. However, it is still important to make sure that the number of fenders is also sufficient for greater berthing angles in order to avoid any contact between the ship's hull and the quayside. The maximum fender spacing should be limited in such situations.

Mooring dolphins should be positioned in such a way that the capacities of the mooring lines on board the ship are used to the full. If forces due to sea state, currents, wind and passing vessels act from changing directions, then the recommendation is to arrange the mooring lines as shown in Figure 7.87.

In some circumstances, such an arrangement of mooring lines is less suitable when a single loading direction prevails. For example, in the presence of a strong current in the longitudinal direction of the vessel, it might be more prudent to limit the vessel's lateral movements.

Furthermore, such an arrangement of mooring lines is unsuitable for bulk goods berths for practical reasons. In this situation, breasting lines cannot be positioned perpendicular to the axis of the vessel because they would restrict the use of cranes and other quayside equipment. Instead, the mooring points must be provided along the berthing line (fender line), although this considerably reduces the ability of the hawsers to resist lateral forces.

The capacity of the mooring system should be assessed according to Section 7.4, also taking into account ice loads according to Sections 4.11 and 4.12.

Mooring equipment for a berth can be determined, for example, on the basis of the maximum forces established by way of a (dynamic) mooring analysis.

On the other hand, in the case of an unprotected berth that cannot be used in all weathers, the capacity of the mooring equipment can be determined by assuming that the winches

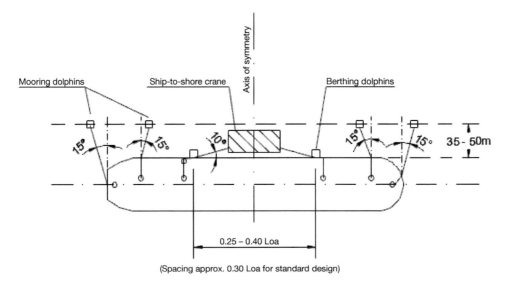

Figure 7.87 Recommended arrangement of mooring lines (OCIMF Guidelines). The horizontal angles should not be exceeded.

give way at 60% of the minimum nominal breaking load of the lines. The corresponding forces should be applied as service loads.

Berthing and mooring equipment for liquid and dry bulk goods are normally based on different concepts. The loading platform at a berth for liquid bulk cargos is generally relatively small. Large loads due to berthing manoeuvres, vessel impact or mooring lines should be ruled out as far as possible because of the risk of damage to the pipelines. Therefore, such berths require mooring and breasting dolphins founded separately from the platform.

On the other hand, equipment at berths for dry bulk goods is less affected by such loads. The platforms at such berths have to be larger in order to handle such goods, and therefore more robust, which means that fenders and bollards can be installed directly on the platform. In regions with strong tides or regions with a significant risk of hurricanes or cyclones plus extreme sea states, the level of the platform will usually have to be such that it is advantageous to separate the supporting structure for the platform and the foundations for the dolphins.

7.7.4 Structural elements of berths

7.7.4.1 Platforms

Jetties are generally designed to handle goods (import/export). A distinction is generally made between terminals for liquid bulk goods (e.g. oil products) and dry bulk goods.

Berths for liquid bulk goods are characterised by the fact that the transshipment takes place at a certain position, i.e. at the ship's manifold. The connection flanges for transferring the product are located here. A system of pumps regulates the flow of the liquid from/to the vessel's tanks. Therefore, compared with berths for dry bulk goods, the loading platform can be relatively small.

Dry bulk goods, on the other hand, can be handled over the entire length of the cargo hatches on the vessel, which must be within reach of the loading/unloading equipment.

A bridge connects the berths with the land. Various platforms, each with specific purposes, complement the main structure.

a) Transshipment platform
   The handling facilities, e.g. screw conveyors for dry bulk goods or loading arms for liquid bulk goods, are mounted on the transshipment platform. Further important installations are also found here, e.g.
   - Technical equipment, e.g. systems for assisting and safeguarding loading/unloading operations (drive units, control desks, fire-fighting installations, etc.).
   - Facilities for delivering goods (supplies to vessels) and transferring crew and personnel to/from vessels (gangway, derrick/crane, access/escape ladders, etc.).
b) Auxiliary platforms
   Auxiliary platforms provide additional facilities for assisting loading/unloading operations. Many different auxiliary platforms are in use; those listed below represent typical uses:
   - Transfer platforms
     Transfer platforms are primarily used to support hoppers that are used to distribute, transfer and redirect bulk goods (e.g. with the help of conveyor belts).
   - Flow control platforms

Equipment for controlling and measuring the volumes transferred at large liquid bulk cargo berths includes flow meter installations for various product lines. In some cases, it is necessary to erect special separate platforms adjacent to the actual transshipment platform.

- Platforms for pipeline expansion loops
  To limit stresses in pipelines caused by temperature fluctuations, it is normally necessary to include several expansion loops – a sort of mechanical spring that minimises the loads in the pipe walls. Expansion loops frequently cantilever beyond the sides of the access bridge and, therefore, require their own support platforms.
- Branching platforms
  Branches are required in some instances when, for example, a single access bridge connects to several berths on both sides of the bridge. Separate platforms are frequently built to simplify the transition and separate the load-bearing systems.
- Control room
  Separate control rooms within view of the loading platforms are frequently built for berths at a long distance from the shore, and these require separate pile foundations.

Travelling ship loaders require a transshipment platform that is roughly the length of the largest vessel expected at the berth, so that all cargo hatches can be reached, and there is also space for positioning equipment adjacent to the hatches.

Berths for liquid bulk cargos can be kept relatively small; a typical platform for a single berth is usually smaller than 40 × 40 m, but the actual size depends on the additional facilities required.

The preferred structural concept for transshipment platforms essentially depends on the location and the prevailing environmental boundary conditions. Most transshipment platforms consist of a concrete deck supported on pile bents or trestles plus raking piles. The method of construction of the deck (precast or in situ concrete etc.) is primarily determined by the accessibility of the facility. Considering the fact that berths are frequently located far from the coast, unprotected on the open sea, safety regulations normally require the use of prefabricated elements as far as possible.

Another trend in the construction industry is the growth in the use of prefabricated modules, right up to the prefabrication of complete platforms on land, which are then set up at the site in one operation (see Figure 7.88). The accessibility of the facility for workers during the construction phase is another important aspect that must be considered.

7.7.4.2 Access bridges

Access bridges provide a connection between the transshipment platform and the shore (see Figure 7.89). Lengths can vary from less than 100 m (where there is deep water directly on the coast) to several kilometres.

The access bridge provides areas for the following installations/facilities:

- Process equipment (oil/gas pipelines, conveyor belts and their supports).
- Technical supply equipment and networks (extinguishing water, drinking water, nitrogen, compressed air, power cables, monitoring/signal/measuring cables, etc.).
- Access facilities (mobile cranes, vehicles, pedestrians).

Figure 7.88 Erection of a prefabricated platform near Darwin, Australia (BAM-Clough).

Figure 7.89 Example of an access bridge, Fujairah, UAE (BAM International).

Besides the operational requirements of an access bridge, which depend on its use for product pipelines, technical equipment and access, it is also necessary to consider the requirements for regular maintenance of and repairs to transshipment facilities when designing the space required and the load-bearing structure. Defining a strategy for the upkeep of the complete facility must play a role when considering the costs and risks involved.

The need for a 50 t crane for maintenance work only, for example, would result in an access bridge that is seriously overdesigned for normal usage. Alternative maintenance concepts based on the provision of a floating device for the relatively rare instances when larger cranes are required enable the access bridge to be designed for the much smaller device needed for regular maintenance.

The operational and other requirements placed on access bridges can vary considerably from project to project and must be taken into account at the draft design stage.

The structure for a very long access bridge is generally broken down into segments about 200 m long. Each segment is regarded as independent and must be stable in itself. Expansion joints are included between the segments in order to accommodate thermal deformations without generating excessive stresses and without inducing loads on neighbouring segments.

The horizontal loads on an access bridge are primarily caused by wind, waves and currents, but can also be due to operational equipment located on the bridge. The starting/braking forces of conveyor belts, the friction forces of pipelines, hydraulic shocks in pipelines due to the sudden closure of valves and the braking forces of vehicles are just a few examples of horizontal loads.

Every pile bent/trestle must include raking piles to resist the horizontal (transverse to the axis) loads acting on an access bridge and limit the ensuing deformations. Loads in the axial direction of an access bridge can be resisted by several raking piles, which are concentrated at the fixed point of each segment.

In regions with a risk of seismic activity, it is also necessary to include earthquake loads when analysing the horizontal stability of the bridge. The reader should consult the specialist literature on seismic design and (local) regulations.

7.7.5 Interaction between load-bearing structure and installations on deck

The plan layout of the transshipment platform is essentially defined by the loading/unloading facilities required, which, in turn, depend on the product being handled and the types of vessel. Specifying the loads on the load-bearing structure is an interactive process between the designers of the installations on the deck (top works) and the jetty structure, and it is necessary to distinguish between facilities for liquid and dry bulk cargos.

7.7.5.1 Loading/unloading of liquid bulk goods

Liquid bulk cargos can be handled with loading arms or hose towers. Examples are shown in Figure 7.90.

(a)

(b)

Figure 7.90 (a) Loading arms and towers for fire-fighting and gangway, Gujarat, India (Shell); (b) gangway, Rotterdam (Cyclomedia Technology B.V.).

The distance between loading arms for liquid bulk varies primarily between 3 m (arms up to 12 in. diameter) and 4 m (arms up to 16 in. diameter). Loading arms must be adequately supported to resist both vertical and horizontal forces and, thus, ensure proper function of the rotary heads, e. g. even during high winds.

Whereas loading arms are generally used at large facilities for correspondingly large vessels with a high loading capacity, hose towers are frequently found at smaller ports. Hose towers represent a less costly investment and can be used for a number of different products, although the loading/unloading rates are much lower. The flanges coupling the pipes on the land site to the ship's manifold are generally connected manually.

7.7.5.2 Loading/unloading of dry bulk goods

Cranes are normally used to load and unload dry bulk goods – normally travelling cranes, so that all the cargo hatches of a vessel can be reached. Loading operations mostly make use of filling nozzles, whereas the unloading of dry bulk products requires other technical equipment – traditionally grabs, which operate reliably, but are relatively slow. Bucket ele-

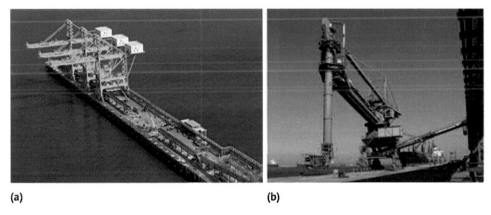

Figure 7.91 (a) Travelling unloading gantries, Lumut, Malaysia (BAM International); (b) bucket elevator (FLSmidth).

Figure 7.92 (a) Unloading facility with screw conveyor (Cyclomedia Technology B.V., Siwertell); (b) radial telescopic loader (Telestack).

vators represent an alternative, but also screw conveyors for bulk goods that are relatively homogeneous, without contamination or clumps. Figures 7.91 and 7.92 show examples of loading and unloading facilities for dry bulk goods.

Radial telescopic or travelling ship loaders are also frequently used. A radial telescopic loader requires a different type of supporting construction, as can be seen in Figure 7.92b. The crane itself stands on a pivot platform, whereas the boom is supported by a crescent-shaped support beam near the line of the berth.

References

BMVI (2012). *Merkblatt Schwimmende Anlegestellen*, 2nd edn. Federal Ministry for Transport, Building & Urban Development, Shipping Dept.

Broos, E.J., Rhijnsburger, M.P.M., Vredeveldt, A.W. and Hoebee, W. (2018). The safe use of cylindrical fenders on LNG, Oil and Container Terminals. *PIANC World Congress*, Panama.

CUR (2013). *Handbook of Quay Walls, Centre for Civil Engineering Research and Codes*, 2nd edn. Leiden: Taylor & Travess.

Dier, A., Carroll, B. and Abolfathi, S. (2004). Guidelines for jack-up rigs with particular reference to foundation integrity. Research Report 289, Prepared by MSL Engineering Limited for the Health & Safety Executive.

Dutt, R.N. and Ingram, W.B. (1984). Jack-up Rig Siting in Calcareous Soils. *Proc. of 16th Annual OTC*, Houston, Paper OTC 4840.

Hein, C. (2014). Berthing Velocity of Large Container Ships. *PIANC World Congress*, San Francisco, 2014.

HTG (1985). Beziehung zwischen Kranbahn und Kransystem. Ausschuss für Hafenumschlagtechnik der Hafenbautechnischen Gesellschaft e. V. *Hansa* 122 (21): 2215.

HTG (1985). Beziehung zwischen Kranbahn und Kransystem. Ausschuss für Hafenumschlagtechnik der Hafenbautechnischen Gesellschaft e. V. *Hansa* 122 (22): 2319.

Lietaert, B. (2011). Design and development of a hazard map for the positioning and siting of large jack-up rigs at the geologically complex areas of the Gulf of Suez. M.Sc. Thesis, Engineering Geology, Delft University of Technology.

OCIMF (2008) *Mooring Equipment Guidelines*, 3rd edn. London: OCIMF MEG 4.

PIANC (1995). Port Facilities for Ferries – Practical Guide, report of PIANC Working Group No. 11.

PIANC (2002). PIANC Report "Guidelines for the Design of Fender Systems: 2002". Report of MarCom, WG 33.

PIANC (2015/2016) PIANC Report "Entwurf von Terminals für RoRo- und RoPax-Schiffe". Report of MarCom, WG 167.

PIANC (2020). PIANC Report "Berthing Velocities and Fender Design". Report of MarCom, WG 145.

Qiu, G., Pucker, T. and Grabe, J. (2013). Penetration of a spudcan near a cavity. *Proc. of Marit. Energy Conf. 2013*, Hamburg, pp. 323–331.

Qiu, G., Drauschke, C., Grabe, J. and Jost, O. (2014). Zur Interaktion von Jack-up Vorgängen vor Spundwänden. *Tagungsband zum Workshop Offshore Basishäfen*, Hamburg, pp. 5–20.

Roubos, A.A., Groenewegen, L. and Peters, D.J. (2017). Berthing velocity of large seagoing vessels in the port of Rotterdam. *Mar. Struct.* 51: 202–219.

Roubos, A.A., Peters, D.J., Groenewegen, L. and Steenbergen, R.D.J.M. (2018a) Partial safety factors for berthing velocity and loads on marine structures. *Mar. Struct.* 58: 73–91.

Roubos, A.A., Groenewegen, L., Ollero, J., Hein, C. and van der Wal, E. (2018b) Design values for berthing velocity of large Seagoing vessels, *PIANC World Congress Panama* 2018.

SNAME (2015). Site specific assessment of mobile jack-up units. Society of Naval Architects and Marine Engineers, Technical Research Bulletin 5-5A, New Jersey, & Corr. 2015.

Tomlinson, M.J. (1986). *Foundation Design and Construction.* London: Longman.

Standards and regulations

BS 6349-1-2000: Part 1: Code of Practice for General Criteria.

BS 6349-8-2007: Part 8: Code of Practice for the Design of Ro-Ro ramps, linkspans and walkways.

DIN 1045: Concrete, reinforced and prestressed concrete structure.

DIN 16972:1995-03: Compression moulded plates made of polyethylene high density (PE-UHMW), (PE-HMW), (PE-HD) – Technical specifications.

DIN 19666:2011-5: Drain pipes and percolation pipes – General requirements.

DIN 19703: Locks for waterways for inland navigation – Principles for dimensioning and equipment.

DIN EN 1991-2:2010-12, Eurocode 1: Actions on structures – Part 2: Traffic loads on bridges.

DIN EN ISO 1872-1:1999-10: Plastics – Polyethylene (PE) moulding and extrusion materials – Part 1: Designation system and basis for specifications (ISO 1872-1:1993).

DIN EN 19905-1:2012-11. Petroleum and natural gas industries – Site-specific assessment of mobile offshore units – Part 1: Jack-ups (ISO 19905-1:2012).

DIN EN ISO 19905-1:2016-11: Petroleum and natural gas industries – Site-specific assessment of mobile offshore units – Part 1: Jack-ups (ISO 19905-1:2016).

8 Sheet pile walls

8.1 Materials and construction

8.1.1 Materials for sheet pile walls

Sheet pile structures consist of separate, interconnected elements driven into the ground, which are subjected to bending and vertical loads. They are used as retaining structures at abrupt changes in ground level (very high in some cases), as quay walls in ports and harbours, and alongside inland waterways. They are subjected to earth and water pressures and transfer vertical loads from superstructures and anchorages into (mostly deeper) soil strata with adequate bearing capacities. Steel has been the dominant material for sheet pile walls for many decades. However, sheet piles made from timber, reinforced concrete, prestressed concrete and polymers can be used in special cases.

8.1.1.1 Timber sheet pile walls

Timber sheet pile walls are advisable only where the existing subsoil favours their installation without damage. The natural boundary conditions limit the thickness of such piles, and therefore the bending stresses should not be excessive. European softwoods and hardwoods exhibit the following properties:

- Compressive strengths $\leq 60 \, \text{N/mm}^2$
- Moduli of elasticity $\leq 14\,000 \, \text{N/mm}^2$

For structures designed for a long working life, the tops of the piles must be at such a level that they remain continuously wet, and steps must be taken to prevent infestation by wood-boring insects. In these conditions, and when other building materials are not available due to local circumstances, timber sheet piles can be considered for protecting the toes of shallow embankments or for landing stages and jetties in marinas for pleasure craft.

DIN EN 1995 should be applied accordingly when planning and designing timber sheet pile walls. Steel connecting elements must be hot-dip galvanised at least or be provided with equivalent corrosion protection. Sections 8.1.1.2–8.1.1.5 of EAU 2012 describe recommendations for driving timber piles, sealing their tongue and groove joints and preservation of the timber.

8.1.1.2 Concrete sheet pile walls

Reinforced or prestressed concrete sheet pile walls are certainly suitable for waterfront structures. A structural connection between the individual elements by way of tongue and groove joints can be designed so that it is possible to fit seals. However, this requires a considerable amount of work. Furthermore, these relatively heavy elements increase the transport and driving requirements over those for steel sheet piles. Their use should, therefore, be restricted to structures where the sealing requirements are low, or where steel sheet piles cannot be used because of a high local corrosion rate or the risk of permanent and strong sand grinding.

Normal weight concretes exhibit the following properties:

- Cylinder compressive strengths of 12–100 N/mm^2
- Moduli of elasticity of 27 000–45 000 N/mm^2

The respective exposure classes corresponding to the local environmental conditions (see DIN EN 1992-1-1) must be taken into account when selecting concretes for concrete sheet piles. Piles should also be designed according to DIN EN 1992-1-1.

Further, more detailed recommendations regarding the design, casting and driving of reinforced concrete sheet piles and sealing their tongue and groove joints can be found in Sections 8.1.2.3–8.1.2.6 of EAU 2012.

8.1.1.3 Steel sheet pile walls

Compared with the above materials, steel has much higher strengths. Steel sheet piles exhibit the following properties:

- Tensile and compressive bending strengths of 240–460 N/mm^2 depending on the grade of steel.
- Modulus of elasticity of 210 000 N/mm^2 regardless of the grade of steel.

A distinction is made between cold-formed and hot-rolled steel sheet piles depending on the method of their production. Cold-formed sections, e.g. lightweight sections and trench sheeting, can support abrupt changes in ground level amounting to a few metres only, whereas hot-rolled sections are suitable for supporting considerable changes in ground level.

8.1.2 Steel sheet pile walls – properties and forms

8.1.2.1 Quality requirements, mechanical and technological properties, dimensional tolerances

DIN EN 10248-1 and DIN EN 10248-2 apply to hot-rolled steel sheet piles, DIN EN 10249-1 and DIN EN 10249-2 to cold-formed steel sheet piles.

If steel sheet piles are stressed in the direction of their thickness (normal to the rolling direction), e.g. junction piles for cofferdams with circular and diaphragm cells, then steel grades with appropriate properties must be ordered from the sheet pile supplier in order to avoid lamellar tearing (see DIN EN 1993-1-10). Hot-rolled steel sheet piles are produced in steel grades with the designations S 240GP to S 460GP.

Steels according to DIN EN 10248 with yield stresses < 355 N/mm^2 require test certificate 2.2, steels with higher yield stresses an inspection certificate 3.1 to DIN EN 10204:2005-01.

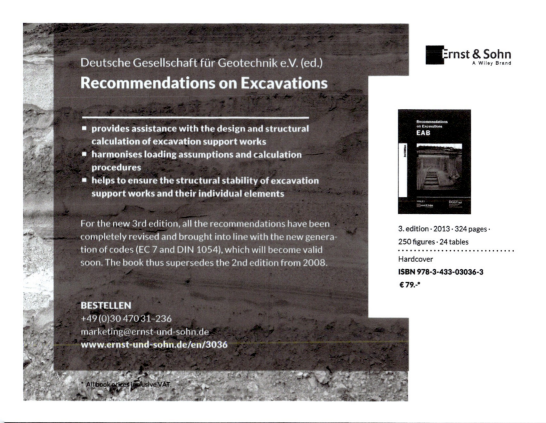

Deutsche Gesellschaft für Geotechnik e.V. (ed.)

Recommendations on Piling (EA Pfähle)

- Recommendations with standard character
- Contains example calculations
- Also applies to the foundations of offshore wind energy structures

This handbook provides a complete overview of pile systems and their application and production. It shows their analysis based on the new safety concept providing numerous examples for single piles, pile grids and groups. These recommendations are considered rules of engineering.

2013 · 496 pages · 260 figures · 91 tables

Hardcover
ISBN 978-3-433-03018-9 € 119*

ORDER
+49 (0)30 470 31-236
marketing@ernst-und-sohn.de
www.ernst-und-sohn.de/en/3018

* All book prices inclusive VAT.

Table 8.1 Requirements for the mechanical properties of steel grades for hot-rolled steel sheet piles.

Steel grade	Minimum tensile strength R_m [N/mm^2]	Minimum yield stress R_{eH} [N/mm^2]	Minimum fracture elongation for measuring length $L_0 = 5.65\sqrt{S_0}$ [%]
S 240 GP	340	240	26
S 270 GP	410	270	24
S 320 GP	440	320	23
S 355 GP	480	355	22
S 390 GP	490	390	20
S 430 GP	510	430	19
S 460 GP[a]	530	460	17

a) As per Table 2 of draft standard DIN EN 10248-1:2006.

In the event of special requirements, it can be agreed that the details of alloying and accompanying elements (C, Si, Mn, P, S, Nb, V, Ti, Cr, Ni, Mo, Cu, N, Al) can be provided.

In special cases, e. g. to accommodate greater bending moments, it is possible to use steel grades with higher minimum yield stresses of up to 500 N/mm^2 in accordance with Section 8.1.1.3. The use of steel grades with minimum yield stresses > 430 N/mm^2 is not covered by the current edition of DIN EN 10248. Therefore, in Germany, a national technical approval will be required when using steel sheet piles made from such grades.

The draft of DIN EN 10248:2006 lists seven grades up to S 460GP for steel sheet piles. The use of steels with higher strengths should be agreed upon separately with the manufacturer of the sections when placing an order.

Steel grades S 235 JRC, S 275 JRC and 355 J0C to DIN EN 10249 can be considered for cold-rolled steel sheet piles.

In special cases, e. g. as stated in Section 8.1.5.7, steels to DIN EN 10025 are used.

Table 8.1 contains information on the requirements regarding the mechanical properties of hot-rolled steel sheet piles. The mechanical properties of cold-rolled steel sheet piles are specified in DIN EN 10025 and DIN EN 10249.

The ladle analysis is binding for the verification of the chemical composition of steel sheet piles (see Table 8.2). The analysis of single bars can be used as an additional test in cases of doubt. If verification of the chemical composition of individual bars is required, a separate agreement must be made for the testing.

8.1.2.2 Interlocking joint forms

When it comes to structural and driving requirements, steel sheet pile walls are suitable for all applications. Such walls consist of individual elements interlocked together to form one structural system.

Examples of proven forms of interlocking joints for steel sheet piling are shown in Figure 8.1. The nominal dimensions a and b, which can be obtained from the suppliers, are measured at right-angles to the least favourable direction of displacement. The minimum interlock hook connection, calculated from $a-b$, must correspond to the values in the figure. Over short sections, the values may not fall below these minimum values by more than

Table 8.2 Chemical composition of ladle/bar analysis for hot-rolled steel sheet piles.

Steel grade	Chemical composition, max. % for ladle/bar				
	C	Mn	Si	P and S	N[a)b)]
S 240 GP	0.20/0.25	—/—	—/—	0.045/0.055	0.009/0.011
S 270 GP	0.24/0.27	—/—	—/—	0.045/0.055	0.009/0.011
S 320 GP	0.24/0.27	1.60/1.70	0.55/0.60	0.045/0.055	0.009/0.011
S 355 GP	0.24/0.27	1.60/1.70	0.55/0.60	0.045/0.055	0.009/0.011
S 390 GP	0.24/0.27	1.60/1.70	0.55/0.60	0.040/0.050	0.009/0.011
S 430 GP	0.24/0.27	1.60/1.70	0.55/0.60	0.040/0.050	0.009/0.011
S 460 GP[c)]	0.24/0.27	1.70/1.80	0.55/0.60	0.035/0.045	0.012/0.014

a) The values stipulated may be exceeded on condition that for every increase by 0.001% N, the max. P content is reduced by 0.005%. However, the N content of the ladle analysis may not exceed 0.012%.
b) The maximum N value does not apply when the chemical composition has a minimum total aluminium content of 0.020%, or when there are sufficient N-binding elements. The N-binding elements are to be specified on the test certificate.
c) As per Table 1 of draft standard DIN EN 10248-1:2006.

Table 8.3 Permissible interlock deviations as per Figure 8.1.

Form	Nominal dimensions (according to section drawings)	Deviations from nominal dimensions		
		Designation	Plus [mm]	Minus [mm]
1	Hook width a	Δa	2.5	2.5
	Interlock opening b	Δb	2	2
2	Button width a	Δa	1	3
	Interlock opening b	Δb	3	1
3	Button width a	Δa	(1.5–2.5)[a)]	0.5
	Interlock opening b	Δb	4	0.5
4	Club height a	Δa	1	3
	Interlock opening b	Δb	2	1
5	Power hook width a	Δa	1.5	4.5
	Interlock opening b	Δb	3	1.5
6	Knuckle width a	Δa	2	3
	Interlock opening b	Δb	3	2

a) Depends on the section.

1 mm. In forms 1, 3, 5 and 6, the required coupling must be present on both sides of the interlock.

Deviations from the nominal dimensions are unavoidable during the rolling of sheet piles and interlocks. The permissible deviations are summarised in Table 8.3.

8.1.2.3 Acceptance conditions for steel sheet piles and steel piles on site; production tolerances required

Although careful and correct construction methods are hugely important for the usability of structures made from steel sheet piles or steel piles, it is also essential that the sections

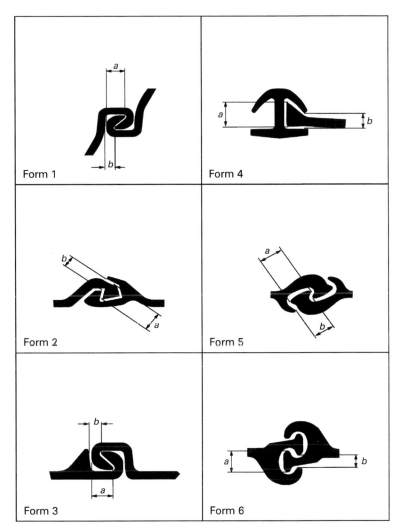

Figure 8.1 Established types of interlock for steel sheet piles. Form 1: a = hook width, b = interlock opening, $a-b \geq 4$ mm; form 2: a = button width, b = interlock opening, $a-b \geq 4$ mm; form 3: a = hook width, b = interlock opening, $a-b \geq 4$ mm; form 4: a = club height, b = interlock opening, $a-b \geq 4$ mm; form 5: a = power hook width, b = interlock opening, $a-b \geq 6$ mm; form 6: a = knuckle width, b = interlock opening, $a-b \geq 6$ mm.

delivered to the building site correspond to specification and comply with certain tolerances regarding dimensions and form. To achieve this, materials are inspected on site to ensure that dimensional and form tolerances have been adhered to. The results must be recorded. As a supplement to the manufacturer's own inspections, a works acceptance can be agreed in each case, which can generally be carried out prior to dispatch when shipping overseas, for example.

The acceptance procedure on site should specify that every unsuitable pile will be rejected until it has been reworked to a suitable standard, unless it is rejected outright. Acceptance of sections on the building site is based on:

Table 8.4 Bending strengths, stiffnesses and dimensions of steel sheet pile walls.

	Corrugated steel sheet pile walls	$W_{y,el,wall}$ cm³/m	G_{wall} kg/m²	$J_{y,wall}$ cm⁴/m	b_{Sys} mm	H_{wall} mm	t_{Fl} mm	t_w mm	Line
Cold-formed DIN EN 10249	U-section	112 – 2032	40 – 120	5100 – 44 700	660 – 1680	90 – 450	$t_{Fl} = t_w$ 4.0 – 10.0		1
	Z-section	600 – 2400	50 – 250	4800 – 129 400	580 – 900	269 – 409	$t_{Fl} = t_w$ 3.0 – 16.0		2
Hot-rolled DIN EN 10248	U-section	500 – 3370	70 – 210	9700 – 75 400	400, 500, 600, 700, 750	150 – 460	6.0 – 21	6.0 – 12	3
	Z-section	1200 – 5200	70 – 253	21 500 – 130 000	580, 630, 670, 700, 750, 770, 800	300 – 510	8.5 – 24	8.4 – 17	4
Jagged walls DIN EN 10248	Jagged pile (at an angle to wall axis)	3800 – 8700	110 – 270	112 000 – 495 000	706 – 1135	750 – 1170	7.5 – 20	6.4 – 12	5

- DIN EN 10248 parts 1 and 2 for hot-rolled sheet piles.
- DIN EN 10249 parts 1 and 2 for cold-formed sheet piles.
- DIN EN 10219 parts 1 and 2 for cold-formed welded hollow sections.

In addition, the values given in Table 8.1 apply to permissible dimensional deviations for interlocks.

DIN EN 10248, Section 6, applies for the limit deviation for straightness of combined sheet pile walls.

The reader is referred to DIN EN 12063, Section 8.3, for more detailed information on the handling and storage of the sections on site.

8.1.2.4 Choice of the section and the form of the wall

Both hot-rolled and cold-formed sheet pile sections are available with different basic shapes. The hot-rolled U and Z-sections differ from the cold-formed sections in terms of their interlocks, the cross-section resistances achievable and their combination options. The hot-rolled sections are further divided into the lighter corrugated and jagged walls and the heavier combined walls.

Corrugated walls include those assembled from

- cold-formed U, Z- and omega-section steel sheet piles,
- hot-rolled U and Z-section steel sheet piles, and
- the jagged walls arranged at an angle to the axis of the wall,

as shown in Table 8.4.

The combined walls include those shown in Table 8.5.

8.1 Materials and construction

Table 8.5 Bending strengths, stiffnesses and dimensions of combined walls and walls with bearing piles.

	Combined steel sheet pile walls	$Wy_{el} = \dfrac{2Iy_{TrPf}}{h \cdot b_{Sys}}$ cm³/m	G_{TrPf} kg/m	Jy_{TrPf} cm⁴	b_{Sys} mm	h mm	a mm	Line
Combined walls	H bearing piles + Z double pile	(single pile) 3480 – 12 240 to 5170 – 20 240 (double pile)	(single pile) 230 to 960 (double pile)	(single pile) 210 000 – 1 390 000 to 420 000 – 2 650 000 (double pile)	(single pile) 1790 to 2540 (double pile)	630 to ≤ 1090	1260 – 1600	6
	Tubular bearing piles + 3 U-piles	2180 – 16 800	210 – 980	245 000 – 5 098 250	2300 – 3600	860 to ≤ 1830	1500 – 1800	7
	Tubular bearing piles + Z double pile	2360 – 18 000			2100 – 3400		1260 – 1600	8
	Box bearing piles + 3 U-piles	3300 – 13 600	265 – 490	380 000 – 2 600 000	1600 – 2550	1000 – 1800	1500 – 1800	9
Bearing pile wall	Bearing pile wall	13 700 – 46 000	(single pile) 230 to 960 (double pile)	410 000 – 2 580 000	(single pile) 470 to 950 (double pile)	630 – 1100	(single pile) 470 to 950 (double pile)	10

8.1.2.5 Cold-formed steel sheet piles

The corrugated sheet pile walls shown in Table 8.4, lines 1 and 2, are unsuitable for building waterfront structures. As a result of the cold-forming production process, they exhibit residual stresses, which would have to be taken into account, for example, when welding on the equipment and anchorage elements needed on port and quay walls. Their interlocks are simple hooks and are not among the tried-and-tested interlock forms shown in Figure 8.1. Compared with those interlocks, they lack the convoluted overlapping interlock forms that reduce seepage through the wall. Their load-bearing capacities are low and lie in the lower range of the hot-rolled sections for corrugated sheet pile walls. Their flange and web slenderness ratios are frequently higher than the limits listed in Table 4.2 of DIN EN 1993-5 for class 3, which means that they can only be taken into account for load-bearing purposes by reducing their resistances, mostly by reducing the yield stress of the steel f_y.

8.1.2.6 Hot-rolled corrugated sheet pile walls

The corrugated sheet pile walls shown in Table 8.4, lines 3 and 4, have proved to be reliable for the construction of waterfront structures over many decades. Corrugated steel sheet pile walls are supplied to site almost exclusively in the form of double piles with un-spliced lengths of up to 31.0 m in order to achieve good driving progress. When necessary, they can be supplied prefabricated with all the necessary fittings for wall corners and junctions. They can also be supplied with a coating to protect against corrosion (see Section 8.1.10.4.1) or sealed interlocks to prevent seepage through the wall (see Section 8.1.3.3) where this is required. Therefore, the sheet pile wall can be supplied to site as a fully prefabricated part ready for driving immediately. This applies to both of the basic elements (U and Z-section single piles) used to construct corrugated steel sheet pile walls.

Generally speaking, the slenderness ratios of the flanges and webs of the majority of hot-rolled sections are such that it is possible to carry out an elastic/plastic analysis according to DIN EN 1993-5. This means that the stresses determined for the elastic system may be applied to the full plastic resistance of the cross-section (class 2).

8.1.2.6.1 Shear-resistant interlocks

The U-section single piles differ from the Z-section piles in that their interlocks lie on the axis of the wall, whereas those of the Z-piles are located at the extreme fibres of the cross-sectional depth of the wall. A Z-pile has section areas that lie either side of the wall axis that are connected by the web, which means they make a contribution when determining their moments of inertia I_y with Steiner components, something that cannot be automatically presumed for the U-section piles owing to their sliding interlock on the axis of the wall. To guarantee that both area's components contribute, both must be linked via a shear-resistant connection in order to achieve the largest possible section resistances at the interlocks.

A shear-resistant connection can be achieved and verified for double piles by way of crimping or welding.

8.1.2.6.2 Verification of interlocking effect by welding

The shear flow in welded interlock connections can be calculated using the following formula:

$$T_d = V_d \cdot \frac{S}{I} \quad \left[\frac{kN}{m}\right]$$

where

V_d design value of shear force [kN]
S static moment of cross-section portion to be connected, related to the centroid axis of the connected jagged wall [m³]
I moment of inertia of jagged wall [m⁴]

For intermittent welds, the shear stress should be set correspondingly higher to DIN EN 1993-1-8, Section 4.9. Verification of the welds is to be carried out according to EN 1993-1-8, Section 4.5, where a plastic analysis – assuming a uniform shear flow – is permitted. For steel grades with yield stresses not covered by DIN EN 1993-1-8, Table 4.1, the correlation coefficient β_w may be obtained through linear interpolation.

Interlock welds should be designed and executed so that, as far as possible, the shear forces are transferred continuously. A continuous seam is best for this. If an intermittent seam is chosen, the length of each seam should be > 200 mm, provided that verification described above does not call for longer welds. In order to keep the secondary stresses within limits, the interruptions in the seam should be ≤ 800 mm.

Continuous weld seams should always be used in areas where the sheet piling is subjected to heavy loads, especially, for example, near anchor connections and at the point where the equivalent force C is introduced at the base of the wall (Figure 8.2).

8.1 Materials and construction

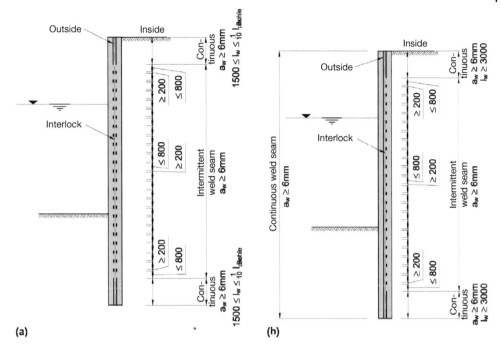

Figure 8.2 Interlock welding principle for walls to resist shear forces when using U-section multiple piles: (a) in easy driving conditions and with only minor corrosion due to harbour water and groundwater; (b) in difficult driving conditions or severe corrosion from outside in harbour water area.

In addition to structural requirements, the loads of driving stresses and corrosion must also be considered. In order to be able to cope with the driving stresses, the following measures are necessary:

- Interlocks are to be welded on both sides at head and toe.
- The lengths of the welds on both sides at head and toe depend on the length of the sheet pile and on any difficulties expected during driving. For easy to moderate driving conditions, they should be at least 1/10 of the length of the pile but not less than 1.50 m.
- For waterfront structures, these weld seam lengths should be 3.0 m for sheet pile walls in difficult driving conditions.
- Moreover, additional seams as per Figure 8.2 are necessary for easy driving conditions; seams as per Figure 8.2b for difficult driving.

In areas where the harbour water is severely corrosive, a continuous weld seam with a thickness $a \geq 6$ mm (Figure 8.2a) is required on the outside down to the sheet pile toe.

If both the harbour water and the groundwater can cause severe corrosion, a continuous seam with a thickness $a \geq 6$ mm is required on the inside of the wall as well.

If, as part of the contract, an additional technical contractual term for sheet piling, piles and anchoring (ZTV-W (LB 214)) is agreed upon, the values given there are to be used for minimum weld seam thicknesses and additions for corrosion.

8.1.2.6.3 Verification of interlocking effect by crimping

Crimping the interlocks achieves only a limited bond because the interlocks can be displaced by a few millimetres at the crimping points when subjected to shear stresses. The number of crimping points per interlock has an influence on the possibility of the relative displacement of the crimped piles and, hence, increases the bond effect. The distribution of the shear forces is determined from the shear force diagram. The shear force P_S at the interlock is calculated as follows:

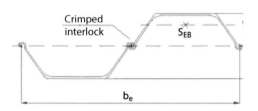

$$P_s = \frac{S_y}{I_y} \cdot b_e \cdot \int_{l_V} V_Z(x) dx \quad [kN]$$

where

V_Z shear force [kN/m]
S_y static moment [cm³/m] related to interlock axis
I_y moment of inertia [cm⁴/m]
b_e loaded width [m]

When checking for failure of the crimping points, the shear force regions with the same sign may be taken as the resistance lengths l_V.

The number n of crimping points required per metre length of interlock over length l_V is

$$n \geq \eta \cdot \frac{P_S}{R_{ser} \cdot l_V} \quad \text{with} \quad \eta = \frac{S_y}{I_y} b_e$$

According to DIN EN 1993-5, 6.4(4), the representative force R_{ser} resisted by one crimping point at the serviceability limit state should be at least 75 kN.

8.1.2.6.4 Reducing the load-bearing capacity and stiffness of U-section double piles

If a shear-resistant connection in a U-section double pile (Figure 8.3) is assumed in the design, then it should be taken into account that the double pile cross-section is asymmetric and, hence, verification of the double pile has to be carried out for its main axis, which results in a lower load-bearing capacity than for verification on the wall axis. To avoid the need for an elaborate analysis, DIN EN 1993-5 specifies the reduction factors β_B and β_D, which depend on the static system, the stiffness/consistency of the soil and the number of single piles connected. These factors are used as an alternative to reduce the section resistances perpendicular to the wall axis (W_y and I_y).

8.1.2.7 Jagged walls made from hot-rolled U-sections

Jagged walls made from U-section double piles with shear-resistant connections and assembled at an angle to the wall axis as shown in Table 8.4, line 5, are used when even the greatest load-bearing capacities of corrugated walls, which are parallel with the wall axis, are no longer adequate. Each one of these double piles includes an omega connector (clutch

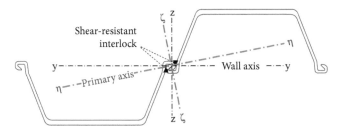

Figure 8.3 Shear-resistant connection in U-section double pile showing main and wall axes.

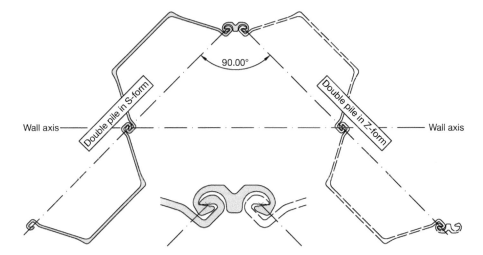

Figure 8.4 Jagged wall with double piles positioned at an angle to the wall axis.

bar) in one of its outer connections, which is normally welded to the pile with the interlock at an angle of 45° to the axis of the double pile and serves for threading the next double pile on site. Consequently, a jagged wall consists of alternating double piles in S form and double piles in Z form, and it closes the load-bearing gap between corrugated walls and combined walls (Figure 8.4).

8.1.2.8 Combined sheet pile walls

Where there is a large abrupt change in ground levels and considerable actions due to earth pressure and excess water pressure plus vertical loads, it might be the case that the cross-section values required (section modulus and second moment of area) can no longer be achieved with corrugated sections. In such situations, different sections must be combined and arranged in order to achieve higher ultimate resistances. Combined steel sheet pile walls, as shown in Table 8.5, are constructed by alternating long bearing piles (primary elements) with shorter and lighter infill piles (secondary elements), as shown in Figure 8.5. The bearing piles are in the form of H-sections, tubes or corrugated sections welded together to form boxes with/without the use of further plates. The secondary elements are generally U or Z-sections in the form of double or triple piles. Triple piles may require adequate stiffeners for structural and constructional reasons. Other suitable sections may also be

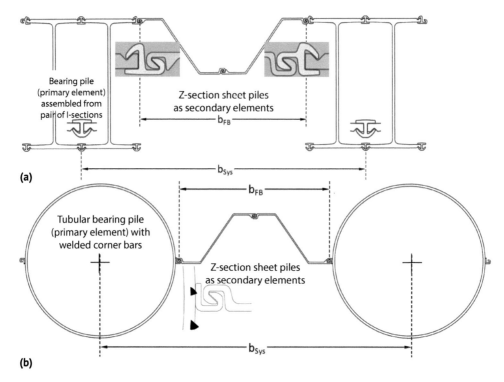

Figure 8.5 Combined sheet pile walls assembled from (a) double H-sections and Z-piles as secondary elements, and (b) tubular piles and Z-piles as secondary elements.

considered for secondary elements if they can properly transmit the loads to the bearing piles and can be installed without damaging interlock connections.

If secondary elements are used with interlock types 1, 2, 3, 5 or 6 as per Section 8.1.2.2 or DIN EN 10248-2, then matching interlocks or sheet pile sections must be attached to the bearing piles by shear-resistant welded joints. The thickness of the inner and outer weld seams of these joints should be $a > 6$ mm. The secondary elements are to be secured against displacement by welding or crimping their interlocks together. To transfer loads from the secondary to the primary elements, it is necessary to check the connecting elements, connectors and load transfer at the connecting point.

The use of secondary elements with interlock type 4 as per Section 8.1.2.2 calls for bearing piles with this type of interlock. Type 4 interlocks are mounted on either the primary or secondary elements prior to driving.

The material properties of the bearing piles should be verified by way of inspection certificate 3.1 to DIN EN 10204:2005-01. The reader is referred to Section 8.1.2.1 for the other components of combined steel sheet pile walls.

If required, forms and sizes deviating from the standard can be agreed separately. Notwithstanding, national technical approvals must still be taken into account with regard to the intended purpose when it comes to tubes made from fine-grained structural steels to DIN EN 10219 and thermomechanically treated steels.

Remarks regarding tubular bearing piles

Tubular bearing piles must have corresponding interlock sections welded on to guarantee adequate connection of the secondary elements. For this purpose, the interlock connection must comply with the tolerances given in Section 8.1.2.2 and be able to transmit reliably the loads from the secondary elements to the tubular sections. At the intersection points between girth and helical welds, interlock sections must fit tightly against the tube. Suitable tube weld seams and interlock section details are required at the intersection point. A welding method test to DIN EN ISO 15614-1 must be performed for interlock welds on tubular sections made from fine-grained steels.

The secondary elements generally consist of Z-section double piles or U-section triple piles. In the case of tubular bearing piles they are usually arranged on the wall axis, which means that such walls do not provide a plain berthing surface.

In combined walls with bearing piles made from box or H-sections, the secondary elements are usually fitted in the water-side interlock connection.

Tubular sections with internal interlocks are particularly useful as bearing piles in combined sheet piling walls if they can be installed using a rotary drilling system, which reduces noise and avoids vibrations. In this case, interlocks are welded flush with the outside face of the tube in the slotted wall of the tube to form an internal interlock.

The steel grade of the tubes should comply with DIN EN 10219 and fulfil all other demands placed on steels for sheet piles.

Tubular bearing piles in combined steel sheet piling are fabricated in the works in full lengths spirally welded, or as individual lengths longitudinally welded and connected by girth welds, using fully or semi-automatic welding equipment. Differing pile thicknesses, stepped on the inside, are possible with longitudinally welded tubes.

Longitudinal and helical line welds must be subjected to ultrasonic testing. Any flaws identified must be colour-coded and repaired manually. The re-testing of repaired weld seams should be recorded within the scope of a new ultrasonic test with the help of a printed log.

Girth welds between the individual tube lengths and transverse welds between the coil ends of spirally welded tubes must be checked by X-ray methods.

The reader is referred to Section 8.2 for the transmission of the axial loads from a tubular bearing pile into the subsoil.

The prerequisite for mobilising the base resistance is the prestressing of the soil within the tube (plug formation) and, hence, the compactability of the soil around the base of the pile. The end bearing pressure that can be mobilised may need to be verified by loading tests.

8.1.2.9 Steel sheet pile sections and straight-web pile sections in tension

Straight web pile sections according to Table 8.6 are not suitable for bending stresses. They have only a very low bending stiffness and can only be loaded in tension. Therefore, they must be stored and handled carefully in order to avoid damage to the sections, their interlocks and any coatings that might have been applied. The manufacturer's instructions for transport, stacking and handling on site must be followed. Straight-web pile sections of > 15 m long should be lifted from their horizontal storage position to their vertical installation position in such a way that in addition to lifting gear attached to the top end of the

Table 8.6 Tensile strengths and dimensions of circular cell walls.

Steel sheet piles used as cell walls	t_w mm	Tensile strength of interlock kN/m	Interlock rotation δ [°]	G_{skin} kg/m²	I_y cm⁴	W_y cm³	B_{sys} mm	h mm
Finger / Thumb (b_{sys})	9.5 to 12.7	3000 to 5500 (6000)	≤ 4.5	128 to 154	168 to 204	46 to 51	500	69 to 75

element, there should be two or three additional support points so that any deformation due to self-weight remains within the elastic range.

The interlocks of straight-web pile sections are able to transfer tensile forces (2000–6000 kN/m) between piles. They are therefore suitable as soil-enclosing membrane walls (cofferdams), which, after filling with a suitable material to form a heavy monolithic block, can withstand large, external horizontal actions generally caused by water pressure, waves and vessel impact, but also active earth pressure, in order to transfer such loads to subsoil with an adequate bearing capacity – in a similar way to gravity walls by the way of resistance to sliding and overturning, as described in Section 8.2.17.1. The hoop tension $F_{t,Ed}$ induced in the straight-web sections by the steady-state earth pressure of the cofferdam filling (possibly also the excess water pressure if the fill material is not drained) is calculated according to Section 8.2.17.1.

Cofferdams can represent economic waterfront structures where deep water (i. e. a considerable abrupt change in ground level) coincides with long berths, and where anchorages are impossible or uneconomic. The larger area of sheet piling required is often compensated for by the weight-savings of the straight-web sections, which are lighter and shorter, and the omission of walings and anchors.

Where the sections shown in Tables 8.4 and 8.5 cannot be driven into a rocky subsoil, cofferdams built from straight-web sections have proved to be an economic solution in many cases. The construction of circular cell cofferdams is described in Section 8.1.5.6.

We distinguish between the following types of cofferdam:

- Cellular cofferdams with circular cells (Figure 8.6a).
- Cellular cofferdams with diaphragm cells (Figure 8.6b).
- Mono-cell cofferdams (Figure 8.6a).

1. Circular cell cofferdams

Circular cells, which are linked by small, connecting arcs, have the advantage that each main cell can be individually constructed and filled, and is therefore stable in itself. The connecting arcs required to seal the structure can be installed later. Junction piles connect them to the stable circular cells. The junction piles generally consist of specially shaped rolled sections or welded or bolted sections in which the angle at the connection can be varied between 30 and 45°. To avoid lamellar tearing, only steel grades with the appropriate properties may be used for welded junction piles (Section 8.1.2). Construction details for welded junction piles can be found in DIN EN 12063.

In order to keep the unavoidable additional stresses in the junction piles low, the clear spacing of the circular cells and the radius of the connecting arcs should be kept to a minimum. If necessary, bent sheet piles may be used for the connecting arcs.

Information on design can be found in Clasmeier (1996).

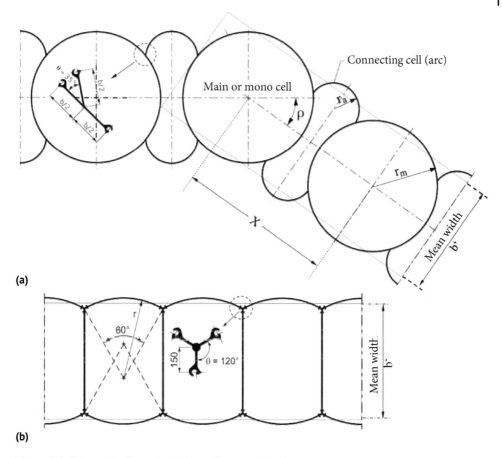

Figure 8.6 Schematic plans of cellular cofferdams: (a) with circular cells; (b) with diaphragm cells.

2. Diaphragm cell cofferdams

Diaphragm cells with straight transverse walls and curved front walls are required when the design value of the hoop tension $F_{t,Ed}$ in a circular cell is greater than the critical design value of the straight-web pile resistances $F_{ts,Rd}$.

Since the individual diaphragm cells are not stable in themselves, a cofferdam of this type must be filled in stages, unless other stabilisation measures are taken. For this reason, the ends of a diaphragm cell cofferdam must be designed as stable structures. The recommendation for long structures is to incorporate intermediate fixed points, especially where there might be a risk of accidents and the consequences of such accidents are significant. All other conditions being equal, diaphragm cell cofferdams require more steel per linear metre than circular cell cofferdams.

3. Mono-cell cofferdams

Mono-cell cofferdams are individual cells that can be used as foundations in open water, e.g. for the ends of breakwaters or guide and fender structures at harbour entrances, or foundations for navigation aids (beacons, etc.).

8.1.3 Watertightness of steel sheet pile walls

8.1.3.1 General

Walls made from steel sheet piles are not completely watertight due to the necessary play in the interlocks. Interlocks threaded in the workshop (W interlocks) are, typically, less watertight than those threaded on site (B interlocks), which are at least partly filled with soil during driving.

Generally, a self-sealing effect due to corrosion with incrustation plus an accumulation of fine particles in the interlocks in sediment-laden water can usually be expected over the course of time. If this is not sufficient, the interlocks can be sealed.

Annex E of DIN EN 12063 can be used to assess the permeability of sheet pile interlocks. The suppliers of steel sheet piles can provide information on the permeability of the various synthetic seals according to Section 8.1.3.3. If very strict requirements have been specified for the watertightness, the interlock seals must be verified in practical tests. Should there be a wide scatter of results, this must be taken into account when determining permeability.

In soil with low permeability, unsealed interlocks function like vertical drains.

If there is fine-grained, non-cohesive soil such as fine sand or coarse silt behind the sheet piling, this can be easily washed out through unsealed interlocks. This is especially the case when there is a high excess water pressure, such as behind enclosures to excavations, and/or the wall is exposed to varying loads due to waves. In these cases, specific measures are usually required for sealing the interlocks.

8.1.3.2 Assisting the self-sealing process

The self-sealing effect of sheet pile interlocks can be encouraged for walls standing in open water when excess water pressure acts intermittently on one side of the wall, e. g. for walls or cofferdams around construction sites. To this end, an environmentally friendly suspended substance, e. g. boiler ash, is poured into the water outside the excavation.

When the maximum possible pumping capacity is used to pump water out of an excavation, there is a difference in water levels between inside and outside. The resulting excess water pressure presses the interlocks together. As the water flows through the interlocks, so the suspended matter is deposited in the interlocks, thus reducing permeability.

If sheet piling can move due to wave action or swell, however, the sealing effect is not permanent because the sealant is crushed between the interlocks and washed out. Similarly, a permanent seal cannot be created using suspended matter when the direction of the excess water pressure alternates.

8.1.3.3 Artificial seals

Sheet pile interlocks can be sealed with artificial seals both before and after installation.

8.1.3.3.1 Methods for sealing sheet piles prior to driving

a) Filling the interlocks with a durable, environmentally compatible, sufficiently plastic compound – for W interlocks in the shop and for B interlocks on site. A noticeable improvement in watertightness is achieved by factory-fitting an extruded polyurethane (PU) polymer seal (commonly called a plastic seal) to B interlocks.

b) The use of this method allows B interlocks that are no longer accessible after driving, e. g. areas below the bottom of an excavation or watercourse, to be sealed as well. Please refer to c) concerning the position of the sealed joint.

c) With both sealing methods, the interlock watertightness achievable depends on the excess water pressure and the method of driving. Impact driving places little stress on the seal, since the movement of the pile in the interlock takes place in one direction only. Vibratory driving places greater stresses on the seal. Therefore, a complete loss of watertightness due to friction and heat cannot be ruled out.
d) Interlock joints of W interlocks are welded watertight, either at the works or on site. In order to avoid cracks in the watertight seam during driving, additional seams are required, e. g. on both sides at the head and base of the pile section, as well as backing seams in the area of the sealing seam. The sealing seam must be placed on the correct side of the sheet piling, e. g. on the air/water side for dry docks and navigation locks.
e) One method known from the rehabilitation of contaminated sites can be used for sealing dykes along waterways with the highest demands on watertightness. In this method, lined holes with a diameter of about 0.1–0.3 m are drilled at a spacing to match the system dimension of the section being used and then filled with a slurry. After the casing has been removed, and before the slurry hardens, the sections welded to the W interlocks are driven.

8.1.3.3.2 Methods for sealing sheet piles after driving

a) Caulking the interlock joints with wooden wedges (swelling effect), rubber or expanding plastic cords (round or profiled) or a caulking compound capable of expanding and curing, e. g. fibres mixed with cement.
The cords are tamped in with lightweight pneumatic hammers and a blunt chisel. The caulking work can also be carried out on water-permeable interlocks. Caulking generally works better on B interlocks rather than crimped W interlocks. Soil particles must be removed from interlock joints prior to caulking.
b) The interlock joints are welded watertight. As a rule, W interlocks are already welded tight at the works, B interlocks on site. Only dry and properly cleaned interlocks can be welded directly. Water-permeable interlocks must, therefore, be covered with steel plates or sections, which are welded to the sheet piling with two fillet welds. Using this method, a fully watertight sheet pile wall can be achieved.
c) On the completed structure, plastic sealing compounds may be placed in accessible joints above the water level at any time, or PU foam injected into the interlocks. Plastic sealing compounds can only be applied to dry surfaces. To achieve this, interlocks must be sealed provisionally beforehand.
In the case of box sheet piles with double interlocks, sealing can also be achieved by filling the emptied cells with a suitable sealing material, e. g. underwater concrete.

8.1.3.3.3 Sealing of penetrations

Apart from taking the watertightness of the interlocks into consideration, special attention must be paid to adequate sealing at the points where anchors, waling bolts, etc., penetrate the sheet piling.

Lead or rubber washers can be placed between sheet pile and plate washer, also between plate washer and nut. To prevent damage to the sealing washers, anchors must be tensioned by means of a turnbuckle and waling bolts with the nuts on the waling side.

Holes in sheet piles for waling bolts and anchors must be properly deburred, so that plate washers fit flush.

8.1.4 Welding steel sheet pile walls

8.1.4.1 Welding suitability, materials

Unlimited suitability of steels for welding cannot be presumed, since the properties of a steel after welding depend not only on the material, but also on the dimensions and the shape plus the fabrication and service conditions of the structural member.

All sheet pile steel grades can be assumed to be suitable for arc welding, provided general welding standards are observed. To ensure weldability, the carbon equivalent (CEV) should not exceed the values given in Table 1 of the draft DIN EN 10248-1:2006 (see Table 8.7).

Sheet piling grades according to Section 8.1.2.1 and steels according to DIN EN 10025 are suitable for welding. Welding suitability must always be verified by inspection certificate 3.1 to DIN EN 10204:2005-01, which indicates both mechanical and technological properties, as well as the chemical composition (Section 8.1.2.1).

Filler metals are to be selected by the welding engineer of the welding contractor appointed to perform the work, taking into account the recommendations of the sheet pile and steel pile supplier. General basic electrodes or filler metals with a high degree of basicity should be used (filler wire, powder).

Steels with an impact strength of 27 J @ −20 °C or 40 J @ −20 °C or 27 J @ −30 °C should be used in cases with unfavourable welding conditions and stresses due to installation (e. g. welding in low temperatures and difficult driving conditions), or three-dimensional stresses and/or predominantly changing loads.

Filler metals are to be selected according to DIN EN ISO 2560, DIN EN 756 and DIN EN ISO 14 341 or according to data provided by the supplier.

8.1.4.2 Design of welded joints in steel piles and steel sheet piles

This recommendation applies to welded joints in steel sheet piles and *all types* of driven steel pile.

The design of the joints is carried out according to DIN EN 1993-1-8. Design and fabrication must comply with the requirements of DIN EN 12063 and DIN EN 12699, or DIN EN ISO 3834 in the case of more demanding requirements. Working areas for welding on site must be protected from wind and weather. Welded joints are to be thoroughly cleaned, kept dry and, if necessary, preheated beforehand.

8.1.4.2.1 *Classification of welded joints*

A butt joint is intended to replace the steel cross-section of a pile or sheet pile as fully as possible. The percentage of effective butt weld coverage is, however, contingent on the type of section, the offset of edges at the joint ends and on the prevailing conditions on site (Table 8.8).

If the butt weld cross-section steel does not match the cross-section of the pile or sheet pile, and if the full cross-section is required for structural reasons, splice plates or additional sections must be used to achieve the full cross-section.

Table 8.7 Carbon equivalent (CEV).

Designation EN 10027		Chemical composition in % by mass Max.[a]												
Steel Name	Steel Number	C		Mn		Si		P		S		N[b]		CEV
		Ladle	Product	Ladle	Product	Ladle	Product	Ladle	Product	Ladle	Product	Ladle	Product	Ladle Product
S 240GP	1.0021	0.2	0.25	—	—	—	—	0.04	0.05	0.04	0.05	0.012	0.014	0.47 0.52
S 270GP	1.0023	0.24	0.27	—	—	—	—	0.04	0.05	0.04	0.05	0.012	0.014	0.47 0.52
S 320GP	1.0046	0.24	0.27	1.6	1.7	0.55	0.6	0.04	0.05	0.04	0.05	0.012	0.014	0.47 0.52
S 355GP	1.0083	0.24	0.27	1.6	1.7	0.55	0.6	0.04	0.05	0.04	0.05	0.012	0.014	0.47 0.52
S 390GP	1.0522	0.24	0.27	1.6	1.7	0.55	0.6	0.035	0.045	0.035	0.045	0.012	0.014	0.47 0.52
S 430GP	1.0523	0.24	0.27	1.6	1.7	0.55	0.6	0.035	0.045	0.035	0.045	0.012	0.014	0.47 0.52
S 460GP		0.24	0.27	1.7	1.8	0.55	0.6	0.035	0.045	0.035	0.045	0.012	0.014	0.47 0.52

a) If necessary to obtain certain properties, some additions of V, Nb, Ti can be made at the discretion of the manufacturer. The maximum value for nitrogen does not apply if the chemical composition shows a minimum total Al content of 0.020% or if sufficient other N binding elements are present.
b) The N binding elements shall be mentioned in the inspection document.

Table 8.8 Effective butt joint coverage expressed as a percentage.

Type of pile of sheet pile		Effective butt joint coverage as % allowance	
		In the fabrication plant	In situ during installation
a) Tubular sections	Calibrated joint ends, root welded through	100	100
b) Piles	Cross-section reduction with material removed from throats		
I-sections		80–90	80–90
Box sheet piles		90–95	
c) Sheet piles	Single sheet piles	100	100
	U and Z-section double piles Interlock With one-sided welding only	80–90	70–80
d) Box piles made from individual sections	Individual sections jointed then assembled	100	50–70
	Box pile to be jointed		70–80

The effective butt coverage is expressed as a percentage and is the ratio between the butt weld cross-section and the steel cross-section of the pile or sheet pile. Possible butt coverage values can be found in Table 8.8.

8.1.4.2.2 Making weld joints

Wherever possible, joints are to be positioned in a lowly stressed part of the cross-section. The joints of adjacent sections must be offset by at least 1 m.

Butt joints under the pile driver are to be avoided as far as possible for economic reasons and because unfavourable weather conditions could have a negative effect on the weld.

If flange splice plates are required to achieve the butt coverage for structural reasons, the following rules must be observed:

- Splice plates should be no more than 20% thicker than the parts of the section being spliced, and should never exceed a thickness of 25 mm.
- The width of the plates should be such that they can be welded to the flanges on all sides without end craters.
- The ends of the plates should be tapered to 1/3 of the plate width at a slope of 1 : 3.
- Before the plate is positioned, the butt weld is to be ground flush.
- Non-destructive tests must be completed before the splice plates are mounted.

If butt joints in service are not subjected to predominantly static loads within the meaning of Section 8.2.7, splice plates over the joints are to be avoided.

8.1.4.2.2.1 Preparation of joint ends

- Each section to be welded should be cut in one plane at right-angles to the axis; an offset in the joint is to be avoided.
- Special attention is to be paid to ensuring a good fit between the cross-sections and, in the case of steel sheet piling, to preserving free movement in the interlocks as well. Differences in width and height between pieces to be welded should not exceed ±2 mm so that offsets in the welded edges will not exceed 4 mm.
- It is recommended that hollow piles assembled from several sections first be fabricated in the full length required and then cut into working lengths after having been suitably coded (e. g. for transportation, driving, etc.).
- The ends intended for each butt joint are to be checked for laminations over a length of about 500 mm.

When preparing I and H-sections for welding, the throat areas of the web are to be drilled out in the shape of a semi-circle facing the flange with a diameter of 35–40 mm (Figure 8.7), so that the flange can be fully welded through with a backing weld. After welding, the edges of these openings must be machined to remove any notches. Run-on and run-off plates must be provided for the flange welds in the vicinity of the butt joint in order to achieve a clean termination at the flange. After removing the plates, the edge of the flange should be ground to remove any notches.

8.1.4.2.2.2 Edge preparation for welding In the fabrication shop, butt welds are generally in the form of V or Y-groove welds. Both edges of the butt joint are to be suitably prepared.

If a butt joint must be welded on site on driven steel sheet piles or steel piles, the top of the driven section must first be trimmed as required by Section 8.1.4.4. The extension piece is to be prepared for a butt weld with or without backing weld.

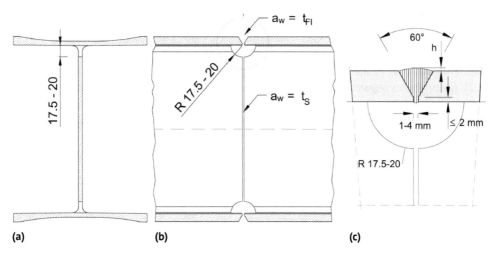

Figure 8.7 Working drawing for a beam splice: (a) section through splice; (b) elevation; (c) detail of weld.

8.1.4.2.2.3 Welding procedure All accessible sides of the butted sections are to be fully connected. Wherever possible, the roots should be gouged and sealed with backing welds.

Root positions that are inaccessible require a high degree of accuracy when fitting together the sections and careful edge preparation.

The proper welding sequence is to be determined so that loads from the welding process do not overlap with those occurring when the structure is in service.

8.1.4.2.2.4 Leaks If sheet piling structures include openings for welding through which the soil could be washed out, such openings must be sealed in a suitable manner (Section 8.1.3.3).

8.1.4.3 Welding on strengthening plates

Strengthening plates are used to increase the flexural strength of the bearing piles and must always be welded to the piles around their full perimeter to prevent corrosion between plate and pile. The ends of the plates should be tapered to reduce the abrupt change in the moment of inertia (Section 8.1.4.2.2). The weld seam thickness a should be at least 5 mm where there is no risk of corrosion, at least 6 mm where corrosion is likely, or should comply with the requirements set out in ZTV-W (LB 214).

Where a strengthening plate spans an interlock located on the flange of a sheet pile, this interlock must be provided with a continuous weld beneath the plate plus an extension of at least 500 mm on either side. This weld must be on the side opposite to that on which the plate is located and the weld thickness should be $a \geq 6$ mm. Under the plate, the weld thickness should be such that the plate makes contact with the pile without the need for any further machining. If this is not done, the welds attaching the plate may be seriously damaged during driving.

If welding of the interlock is not desired, the strengthening plates must be cut in half and welded on either side of the interlock according to the above procedure.

8.1.4.4 Cutting off the tops of driven steel sections for load-bearing welded connections

If load-bearing welded connections (e. g. butt joints, welds for fittings, etc.) are required at the tops of driven steel sections, these may not be located in areas that have been damaged or where the steel has yielded as a result of the driving process, for example. Any brittleness in such regions has a negative effect on the load-bearing capacity of a welded joint. Therefore, the top ends of the piles must be cut off at a point below the extent of any deformation. Alternatively, the welded joints should be positioned outside the deformed area.

8.1.5 Installation of steel sheet pile walls

8.1.5.1 General

When selecting the type of section, its dimensions and the steel grade for steel sheet piles, then the structural requirements, the expectations placed on serviceability, economic considerations and the stresses that will arise while driving the section into the subsoil are all vital considerations. It must be ensured that the section chosen can be installed without damaging the interlocks.

Both an adequate degree of specialist knowledge and careful installation are required, so that an enclosing wall can be constructed with the necessary driving accuracy (see Section 8.1.7.2). The more difficult the soil conditions, the longer the piles, the greater the embedment depth or the deeper any subsequent dredging in front of the sheet piling, the greater will be the need to insist on a high degree of workmanship during construction.

Poor results can be expected if long pile sections are driven one after the other to their final embedment depth because the exposed length of interlock is then too short to ensure adequate initial guidance for the next pile.

A working basis for judging the behaviour of the ground with respect to pile driving is obtained from boreholes and soil mechanics investigations, as well as cone and dynamic penetration tests (Section 3.2.4). In critical cases, driving trials at specific locations are recommended (see Sections 3.2.4.2.5 to 3.2.4.2.7). In this way, it is possible to establish both the drivability of the sections and the potential deviations from their intended positions.

The success and quality of the sheet piling installation depend largely on the driving. This presumes that, in addition to suitable, reliable construction equipment, the contractor has the necessary experience and is, therefore, capable of making the best use of the equipment and the skills of qualified technical and supervisory personnel. It is essential to observe the safety requirements regarding equipment for drilling and foundation work according to DIN EN 16228, parts 1–7. Parts 1, 2 and 4 of this standard apply to equipment for installing sheet pile walls. Furthermore, in accordance with ZTV-W_LB 214:2015, Section 3.1(22), appendix (6a), the contractor must submit details of their method of construction to the client four weeks prior to commencing construction.

The compilation of driving logs in accordance with ZTV-W_LB 214:2015, Section 3.1(22), appendix (6b.1) or (6b.2), should also be agreed between the client and the contractor.

There are currently four methods for the installation of sheet piles in soil:

- Impact driving (caused by the action of a drop weight)
- Vibratory driving
- Pressing
- Threading piles into pre-cut trenches.

Section 3.2.4.2 provides an initial indication for the choice of installation method taking into account the soil strata through which the piles pass.

For structural and constructional reasons, double piles are usually chosen for driving when constructing sheet pile walls with U or Z-sections. Triple piles may also have technical and economic advantages in specific cases. This facilitates pitching and driving of the double piles, so that piles already in place are hardly dragged down.

The individual pile sections assembled to form one double pile element ready for driving should be connected together structurally by crimping or welding their middle interlocks located on the wall axis (Section 8.1.2.6.1).

In difficult conditions, rocky soils and/or with deep embedment depths, technical reasons might demand the selection of sheet piles with thicker walls or a higher grade of steel than required by the structural calculations. The pile toe and, if necessary, also the pile head, may have to be strengthened occasionally, which is particularly recommended for driving in rocky soils or soils with stony inclusions.

ZTV-W_LB 214:2015, Section 3.4(20), calls for the use of a double guide when installing long sheet piles. A leader without a bottom guide is regarded as a single guide. Guides can be in the form of frames, auxiliary trestles, etc. According to Section 8.1.5.2.1, the interlocks of neighbouring piles are not classed as guides.

8.1.5.2 Impact driving

This is probably the oldest method of installing steel sheet piles. It is a widely used, proven method for constructing the steel sheet pile walls shown in Section 8.1.2.4, Tables 8.4 and 8.5.

8.1.5.2.1 Driving equipment

The driving equipment must be designed so that the pile sections can be driven safely and carefully, at the same time being guided adequately. This guiding is particularly important when driving long piles and for deep embedment depths in order to avoid unacceptably large deviations during driving. The size and efficiency of the driving equipment depend on the dimensions and weights of the pile sections, their steel grade, the embedment depth, the subsoil conditions and the driving method selected.

Drop hammers, diesel hammers, hydraulic hammers and rapid-action hammers can all be used for pile driving.

In the case of **drop hammers**, the ratio of hammer weight to weight of pile section with driving helmet should be about 1 : 1 in order to achieve the best degree of efficiency.

Slow-action, heavy hammers can be used for all applications, especially in cohesive soils.

Hydraulic hammers are also suitable for all applications; their energy per blow can be carefully controlled to suit the respective driving resistance and the subsoil (including rock).

Rapid-action hammers tend not to damage the pile section and are especially well suited to driving in non-cohesive soils.

The following factors determine the degree of efficiency and driving performance of pile hammers:

- Total weight of hammer.
- Weight of piston, energy of single blow, type of acceleration.
- Energy transmission, force transmission (driving helmet and guide).
- Pile sections: weight, length, angle, cross-section, form.
- Subsoil (see Section 3.2).

A good degree of efficiency for impact driving is achieved through optimum coordination of these factors.

A driving helmet between pile and hammer is essential for impact driving. Its size and form must be chosen to match the requirements of the equipment and the pile.

As far as possible, the driving blow is to be transferred to the driving element in such a way that, with respect to the resistances, the force is introduced symmetrically and axially. The effect of the interlock friction, acting on one side only, can be countered by adjusting the point of impact.

All driving elements must be guided according to their stiffness and the driving stresses, so that their final position is the intended design position. To guarantee this, the pile driver itself must be adequately stable and set up on firm ground, and the leader must always be parallel to the inclination (rake) of the pile section.

Ensuring the required driving accuracy calls for guiding the driving elements at two points at least, spaced as far apart as possible. A strong lower guide plus spacer blocks for the pile section in this guide are especially important for the accuracy of the driving. The leading interlock of each pile section must also be well guided.

Where pile sections are being driven from floating equipment, the motion of the equipment must be minimised to ensure there is no impact on driving accuracy; a jack-up platform might have to be used.

The first sheet pile section for a quay wall must be positioned with great care, so that good interlock engagement is ensured when driving the following elements. This is especially important for accuracy when driving in deep water.

Choosing hammers and driving methods to match local circumstances and taking greater care when pitching and guiding pile sections reduces the energy required for driving and improves driving progress. The minimum penetration per blow for impact driving should be in accordance with the manufacturer's specification.

8.1.5.2.2 Driving steel sheet piles

Driving methods In the case of difficult subsoil conditions and greater embedment depths, a driving method with two-sided interlock guidance of the pile sections is required to guarantee accuracy. If the driving resistance along the line of the piles increases due to soil compaction, and the sections, therefore, deviate from their intended positions when using the pitch and drive method to drive piles (Figure 8.8), then the piles should be driven in panels (e.g. initial driving with a light hammer and redriving with a heavy one, Figure 8.9), and a staggered method should be used (Figure 8.10), with several pile sections being pitched and then driven in a leap-frog sequence (1-3-5-2-4).

Driving in panels is also recommended for constructing sheet piling enclosures.

The heads of U-section sheet piles tend to lean forwards in the driving direction, those of Z-section piles tend to lean backwards. For U-sections installed by pitch and drive, the leading web (in the driving direction) can be bent outwards by a few millimetres to increase the system dimension somewhat. In the case of Z-sections, the leading web can be slightly pressed in a little towards the trough.

In many cases, this lean can be prevented by using staggered or panel driving. If this does not work, taper piles must be used. These must be designed so that the pile section has the same form at both ends, and the connecting flange with a welded-in taper section is on the driving direction side (Figure 8.11a). This prevents damaging the webs in the soil.

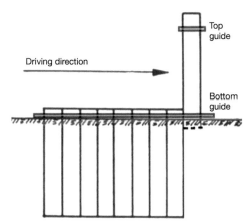

Figure 8.8 Successive driving of individual sheet piles (pitch and drive).

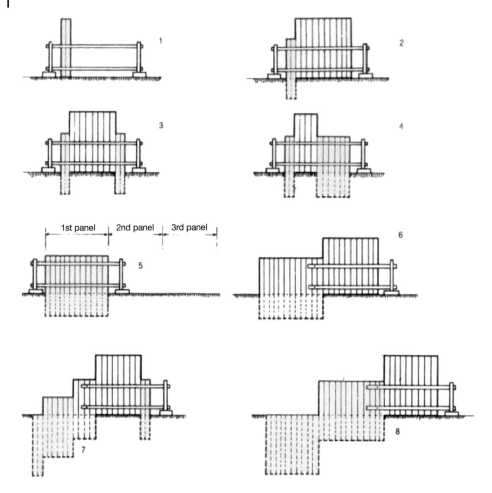

Figure 8.9 Driving sheet piles in panels; 1 – Align the first double pile vertically and horizontally. 2 – Drive the first double pile, pitch and align the other piles. 3 – Drive the double pile at the other end of the panel. 4 – Drive the rest of the panel, working back towards the first double pile. 5 – The first panel partly driven. 6 – The second panel pitched and aligned; the last double pile of first panel becomes the first pile of second panel; the guides are fixed to the double pile driven last. 7 – Drive the first panel to the final depth in stages; drive the last double pile of the second panel. 8 – The first panel is driven to the final depth, the second panel is partly driven, the third panel is pitched and aligned; the last double pile of the second panel becomes the first pile of the third panel.

Chamfering the toes of either U or Z-section sheet piles can lead to damage to the interlocks and is, therefore, not permitted.

Maintaining the width tolerance of the pile sections is essential where the grid-line dimensions of certain stretches of sheet pile walls must be ensured with great accuracy. If necessary, make-up pile sections must be driven (Figure 8.11b).

Driving steel sheet pile walls in rock When bedrock exhibits a fairly thick decomposed transition zone, with strength increasing with depth, or when the rock is soft, experience has shown that steel sheet piles can be driven deep enough into the rock to achieve at least a simple support.

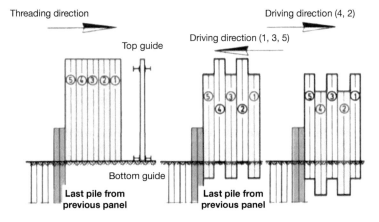

Figure 8.10 Staggered driving of sheet piles.

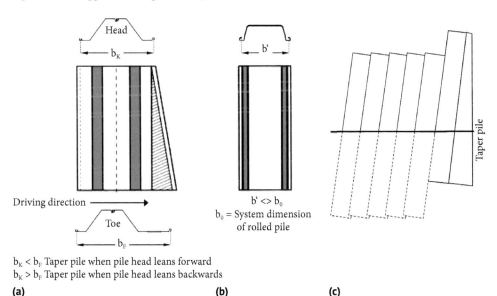

$b_K < b_F$ Taper pile when pile head leans forward
$b_K > b_F$ Taper pile when pile head leans backwards

(a)   (b)   (c)

Figure 8.11 Taper and make-up piles: (a) taper pile; (b) make-up pile; (c) forward lean.

In order to be able to drive sheet piles into bedrock, the piles must be modified and strengthened at the toe, and if need be, also at the head, depending on the pile section and the type of rock. Sheet piling of steel grade S 355 GP (Section 8.1.2.1) is recommended, considering the high driving energy required. Heavy hammers and a correspondingly smaller drop height are very effective for this work. A similar effect can be achieved with hydraulic hammers whose impact energy can be controlled to suit the particular driving energy requirement (see Section 8.1.5.2.1).

Soil investigations and driving trials should certainly be carried out where an unweathered, hard rock extends up to ground level. Special measures may need to be taken to secure the toe of the pile and to guide the piles, e. g. by using rock anchors according to Section 8.12 of DIN EN 12063:1999.

8.1.5.2.3 Driving primary and secondary elements for combined steel sheet pile walls

Impact driving is normally used to install the bearing piles of combined sheet piling (box piles, H-piles or tubular piles) (Section 8.1.2.8). Secondary elements can be installed by means of impact and/or vibratory driving (Section 8.1.5.3). To ensure that the secondary elements can be installed without overloading the interlock connections, the bearing piles must stand parallel to each other within the permissible tolerances at the planned spacing and without any distortion.

When driving tubular piles, any obstacles can be removed by excavation within the tube. However, to do this, the inside diameter of the pile must be at least 1200 mm to allow the operation of suitable excavation equipment, and there can be no construction elements, e.g. internal interlock elements, inside the tube itself (Section 8.1.5).

The risk with impact driving of steel tubes is that the pile heads will buckle, especially in the case of tubes with relatively thin walls. This can mean that the piles cannot achieve their intended depth. In order to avoid buckling in these cases, the pile head must be stiffened. There are various proven methods for this (see Figure 8.12).

Combined steel sheet pile walls are often used to build quay walls at seaports. In view of the (usually) long lengths needed for the bearing piles of such walls, the greatest possible care must be exercised when driving. This is the only way to ensure that the bearing piles achieve their intended positions and the secondary elements installed without damaging the interlocks. This, however, presumes secondary elements with a certain flexibility during installation.

Requirements for wall elements To achieve the flexibility required, the middle interlocks of the secondary elements should not be joined rigidly over their entire length. This allows the interlocks to rotate to some extent. Only the head of each pile should be crimped or welded over a distance of a few decimetres to secure them for transport and lifting purposes.

The use of box piles or double piles made from broad flange or box sheet pile sections, which have a sufficiently high flexural stiffness in both directions and a high torsional stiff-

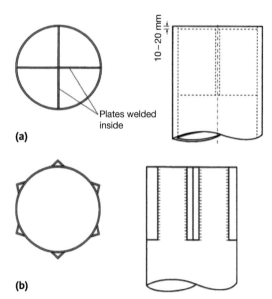

Figure 8.12 Bracing the head of tubular piles (a) with welded plates; (b) with angle sections welded to the outside.

ness, or tubes should be used when bearing piles > 20 m long are required. The increased driving work for such cross-sections must be accepted.

When mounting interlocks on bearing piles, there is a risk that they will fill with soil during driving. Declutching is then more likely when driving the secondary elements. Therefore, in this case, the interlocks are welded at the bottom end before driving and filled, for example, with soft bitumen.

Shear-resistant weld seams ($a \geq 6$ mm) are required if the interlocks are mounted on the bearing piles, which achieves a higher moment of inertia and section modulus at the expense of reduced rotational flexibility of the connection. This causes the centre of gravity of the bearing pile to move away from the centre of the pile in the direction of the flange with the interlocks.

If the interlocks are mounted on the secondary elements, only the upper end is welded in the case of greater embedment depths, so that the rotational flexibility of the connection between primary and secondary elements is maintained during driving and interlock friction is reduced during driving.

The two positions of the interlocks must be coordinated with the length of the weld seam/crimping on the pile, the driving depth, the soil conditions and any driving difficulties that might be expected. Generally, the length of the weld will be 200–500 mm/m.

With especially long piles and/or difficult driving, an additional safety weld at the base of the pile is recommended. With short embedment depths, a shorter transport weld at the top of the piles is generally sufficient.

Care must be taken to ensure that the driving helmet covers the outer connectors when driving the secondary elements deeper than the top edge of the bearing piles while leaving enough clearance to the bearing piles.

If the bearing piles are made from U or Z-section sections connected to each other via web plates, the plates are to be welded continuously to the U or Z-section sections on the outside and at the ends of the bearing piles for at least 1000 mm on the inside. Weld seam thickness must be $a > 8$ mm. Furthermore, the bearing piles must be stiffened at toe and head with wide plates between the web plates, so that the driving energy can be transmitted without damaging the bearing pile.

In addition to the otherwise customary requirements in accordance with Section 8.1.2.3, the bearing piles must be straight with, as a rule, a perpendicular offset no greater than 0.2 % of the pile length as per DIN EN 10248-2. They must be free of warp and, in the case of long piles plus deep embedment, must possess adequate flexural and torsional rigidity.

Permissible twisting tolerances for bearing piles in combined sheet pile walls are not given in DIN EN 10248 or comparable standards. Therefore, these must be agreed upon separately with suppliers or stipulated by the customer upon placing the order.

The top of a bearing pile must be flat and at a right-angle and be shaped such that the hammer blow is transferred over the entire cross-section by means of a well-fitting driving helmet. When driving bearing piles, it is important to ensure that the resultant forces of the driving energy and resistance to penetration will act along the centroid axis of the bearing pile in order to prevent it creeping out of alignment due to an eccentric load.

Elements to strengthen the toe of the sheet pile, e.g. "wings", to increase the axial load-bearing capacity must be positioned such that the resultant force of the resistance to penetration acts along the centroid axis of the pile. Otherwise, the pile could creep out of alignment during driving. Furthermore, wings should continue far enough up the pile to provide guidance during driving.

Secondary elements are to be designed in such a way that they can follow acceptable deviations of the bearing piles from their intended position. Owing to their external interlocks, secondary elements made from Z-sections can adapt to changes in the position of the bearing piles to a limited extent. Secondary elements with interlocks on the axis of the wall (U-sections) can only cope with bearing pile driving deviations by lengthening or shortening the section.

The interlocks between primary and secondary elements must allow easy threading and must have an adequate load-carrying capacity (see Section 8.1.2.2). Interlocks must fit together properly and should not be twisted with respect to each other.

Driving procedure Bearing piles must be installed such that, after driving, they fulfil the following requirements:

- *Parallelism*: bearing piles must be installed essentially parallel, i. e. every pile must stand vertically or adhere to the stipulated inclination.
- *Alignment*: the required driving alignment must be achieved.
- *Distortion/twisting*: distorting and twisting increase the risk of interlock declutching and, therefore, must be prevented as far as possible.
- *Spacing*: the distance between the piles must be equal over their entire length, matching the system's dimension.

These requirements can only be fulfilled by accurately guiding the bearing piles, with double guidance proving expedient. This guidance must be ensured during both the pitching and the driving of the bearing piles. Suitable, heavy, adequately rigid driving equipment, suitable for the length and weight of the piles, set up on a firm base and sufficiently stable in itself, should be used when pitching and driving bearing piles.

The following procedures are recommended for driving combined sheet piling in exposed locations, such as those with extreme weather conditions and high waves, e. g. on sites in river estuaries:

- Jack-up platforms can be used to support driving equipment and driving guides or sufficiently stiff piling frames on supporting piles, which can be extracted afterwards.[1]

Leader-guided driving

- Guide the bearing pile via the piling helmet on the leader, so that it is held in the intended position above the horizontal guide. The play between bearing pile and helmet, as well as between helmet and leader, must be as small as possible during driving.
- The position of jack-up platforms must be checked constantly during driving because driving vibrations can shift the position (e. g. angle).
- Distortion in the bearing piles must be avoided by means of stiff guidance mounted independently of tide, swell, groundswell and waves. Guidance also improves the parallelism and direction of the bearing piles (vertical or raking).

---

1) Note: floating equipment, e. g. working pontoons or half divers, are less suitable; it must be assumed that this type of equipment will move in the groundswell and waves – if only marginally. Floating equipment is only suited to driving combined sheet piling walls in sheltered locations.

- In deep waters, bearing piles can be secured beneath the driving guide by an accompanying underwater parallel guide or cages. Cages are fixed guide frames placed as deeply as possible. It should be noted that extracting cages sometimes requires considerable lifting capacity greatly exceeding the weight of the bearing piles.

Bearing piles are initially positioned using vibratory equipment and then driven to their final depth using heavy driving equipment.

When working in shallow water, guidance can be improved by digging a trench in the base of the watercourse before driving so that the guidance can be used at the greatest depth possible and the driving depth is reduced.

Bearing piles are not driven in a continual order but instead in a "leap-frog" sequence. This ensures that the base of each pile is never driven in soil compacted on one side only. Typically, a driving unit comprising seven bearing piles is driven in the order 1-7-5-3-2-4-6 (large leap-frog sequence). However, the following sequence should be observed at least: 1-3-2-5-4-7-6 (small leap-frog sequence).

Generally, all the bearing piles should be driven to their full depth without interruption

Subsequently, the secondary elements can be pitched and driven in succession. If they are installed (partially) with vibration, it is essential to make sure that the piles are always driven deeper. If progress during vibration suggests that a secondary element is making only marginal progress, driving must be halted immediately in order to avoid any damage to the threading interlocks and the interlock welds. Therefore, as stated in Section 8.1.5.3.4e, driving rates should not be less than 0.5 m/min.

A sufficient wall thickness should be selected depending on the length of the secondary elements. In the case of secondary elements 20 m long, walls should not be less than 12 mm thick.

In soil free from rocks and suitable for jetting, the bearing piles – and possibly the secondary elements, too – can be driven with the aid of water jetting. Jetting equipment is to be installed symmetrically and properly guided in order to counter the lateral deviation of bearing piles from their intended position.

In soil containing boulders and hard soil strata, soil replacement is recommended. For building sites on land, a trench can be excavated along the planned line of the wall, and the sheet piling subsequently placed and driven in this trench.

Special measures, especially in the case of difficult soil conditions, may be required in order to prevent damage to the combined sheet piling during driving. For example, soil replacement, pilot holes, etc., may be appropriate.

The design and construction of combined sheet piling calls for planners, design engineers, fabricators and contractors with considerable experience, with personnel skilled in preparatory works and execution of the works in particular.

Design of piling frames Driving operations in water can be carried out from either a jack-up platform or a pontoon. Both solutions are mobile and can be moved to match construction progress. However, their usage requires a navigable water depth. If the water is too deep, a driving platform can be created with dumped materials. One alternative involves piling frames that provide the pile driver with a stable base and can be moved to match the progress of the works.

356 | 8 Sheet pile walls

Piling frames are working platforms erected on steel, timber or concrete piles (Figure 8.13). They must be designed and built in such a way that they can accommodate the equipment required for driving operations but are still economical. The following boundary conditions should be borne in mind for their design and construction:

- Piling frames can be driven from a floating pontoon or barge although the driving deviations to be expected under such conditions must be taken into account during design.
- The length of the frame is to be selected such that it can be dismantled from an area of the waterfront structure already driven and re-erected ahead of the driving work without interrupting driving operations.

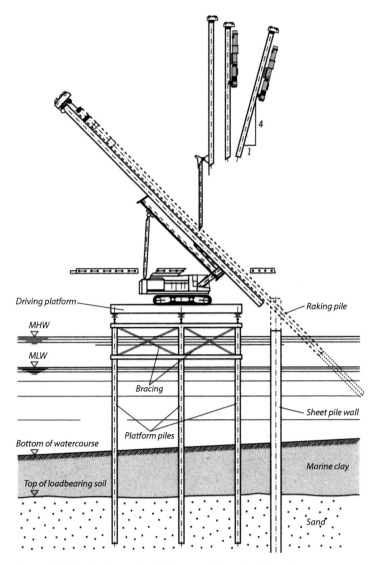

Figure 8.13 Piling frame for driving vertical and raking piles.

- The frame foundation piles are extracted in the dismantling area and re-driven ahead of the current driving operations. If foundation piles cannot be extracted, even with the help of jetting, they must be cut off beneath the planned harbour bottom depth (taking into account dredging tolerances and subsequent deepening work). The frame must be designed to handle the loads arising from the extraction of the frame piles.
- Simple structural systems and configurations should be used for the piling frame, so that the components can be reused many times.
- If the harbour bottom consists of soil at risk of scouring (sand, silt), potential scouring must be taken into account when determining the embedment depth of the frame piles. As the scouring depth around pile groups in tidal currents is particularly difficult to estimate in advance, the scouring must be regularly monitored during construction by means of sounding. If scouring is found to be deeper than that assumed when determining the embedment depth of the frame piles, the scouring must be filled with soil not susceptible to scour (e. g. gravel, stones, cohesive soil).
- The frame piles should be driven well clear of the permanent structure, so that they are not dragged down by the driving of the structural piles. The clearance required depends on the strata of the subsoil and the structural pile types to be used. In cohesive soils, cavities may remain after the frame piles have been extracted. These cavities should be filled if it is thought that they could compromise the stability or serviceability of the final structure.
- Piling platforms built close to the waterfront can consist of a row of piles in the water and supports on the bank for the platform beams. Piling frames in water stand on two or more rows of piles. Piles that form part of the final structure can also be used.

The safety measures and requirements and operating conditions, etc. as per DIN EN 996, Section 4, apply to piling frames.

Together with the loads from the pile driver and the platform, and crane if present, the loads due to currents, waves and ice must be taken into account when designing the supporting piles and frame bracing. If the piling frame is not safeguarded by additional measures (safety dolphin) against, for example, contact with pontoons or other floating construction equipment, additional loads due to vessel impact and, where applicable, line pull forces of 100 kN each applied at the most unfavourable position must also be considered in the design.

8.1.5.3 Vibratory driving of U- and Z-section steel sheet piles

Vibratory hammers (vibrators), which are rigidly clamped to the section being driven, can be used to drive U and Z-section sheet piles into the ground (see Section 8.1.5.3.3). The vibrators generate vertically directed vibrations through eccentrics rotating synchronously in opposite directions, which are then transferred to the pile section and cause the soil to resonate. This reduces the soil's resistance to penetration (skin friction and toe resistance) considerably.

An expert must assess the interaction of vibrator, pile section and soil to determine whether vibratory driving can be successfully used in a given situation. The reader is referred to Section 3.2.4.2.3 for details of how soil and pile section influence vibration.

The effects of vibratory driving on the load-carrying capacity and settlement behaviour of foundation elements and the in situ density of the soil should be evaluated by a geotechnical expert when vibration is being considered. The load-bearing capacity should be assessed in advance by means of loading tests, especially in the case of foundation elements with alternating loads (tension and compression).

8.1.5.3.1 Terms and parameters for vibratory hammers

Important terms and parameters for vibratory hammers are as follows:

1. The type of drive; vibratory hammers can be driven electrically, hydraulically or electro-hydraulically.
2. The driving power P [kW] determines the efficiency of the hammer. At least 2 kW should be available per 10 kN of centrifugal force.
3. Effective moment M [kg m] is the total mass m of the eccentrics multiplied by the spacing r of the centre of gravity of an individual weight from its axis of rotation:

$$M = m \cdot r \quad [\text{kg m}]$$

The effective moment also determines the stroke or amplitude of the vibratory hammer.

4. Revolutions per minute (rpm) n [min^{-1}] of the shafts on which the eccentrics are mounted. The rpm affects the centrifugal forces to the power of two. Electrical vibrators work with a constant rpm, whereas hydraulic vibrators have an infinitely variable rpm.
5. Centrifugal force (exciting force) F [kN]. This is the product of the effective moment and the square of the angular velocity:

$$F = M \cdot 10^{-3} \cdot \omega^2 \quad [\text{kN}] \quad \text{with} \quad \omega = \frac{2 \cdot \pi \cdot n}{60} \quad [\text{s}^{-1}]$$

In practice, the centrifugal force is a benchmark for comparing different vibratory hammers. However, the rpm and effective moment at which the maximum centrifugal force is reached must be taken into account.

Modern vibrators include options for infinite adjustment of speed and eccentric moment during operation. The advantage of such vibrators is that they can be started with an amplitude of zero, free from all resonance. Only upon reaching the preselected speed are the weights extended and adjusted. This avoids undesirable amplitude peaks while starting and stopping.

6. Stroke S [m] and amplitude $\bar{x}$ [m]. The stroke S is the total vertical shift of the vibrating unit in the course of one revolution of the eccentrics. The amplitude $\bar{x}$ is half the stroke. Manufacturers' specifications sometimes list the stroke and, sometimes, the amplitude. The amplitude is the quotient of the effective moment M [kg xm] and the dynamic mass $m_{\text{ham,dyn}}$ [kg] of the hammer:

$$\bar{x} = \frac{M}{m_{\text{ham,dyn}}} \quad [\text{m}]$$

On the other hand, the "working amplitude" $\bar{x}_A$ required in practice is an amplitude that is established during vibration. It is calculated as the quotient of the effective moment M and the total resonating mass m_{dyn}:

$$\bar{x}_A = \frac{M}{m_{\text{dyn}}} \quad [\text{m}]$$

A great tool for anyone designing and building pipe sheet pile walls: The MF-Pipe Selection Lists

MF-Pipe - Vibro-/Impact driving with SteelWall connectors

MF-Pipe - DTH procedure with SteelWall connectors

Find out the most economical combinations for vibration and impact pile driving as well as for the DTH procedure.

Please download MF-Pipe Selection Lists from our website without charge!
www.mf-pipe.com

MF-Pipe is a division of SteelWall ISH GmbH
Tassilostr. 21
82166 Gräfelfing / Germany

Copyright by SteelWall ISH GmbH
www.mf-pipe.com
info@mf-pipe.com

High tensile MF-connectors
for pipe sheet pile walls

MF-Pipe Vibro-/Impact driving with MF63 - MF230

MF63, pipe spacing = 63 mm
Max. tensile strength: 2552 kN/m (FEM)

MF75, pipe spacing = 75 mm
Max. tensile strength: 3419 kN/m

MF100, pipe spacing = 100 mm
Max. tensile strength: 3419 kN/m

MF130, pipe spacing = 130 mm
Max. tensile strength: 3419 kN/m

MF180a, pipe spacing = 180 mm
Max. tensile strength: 3419 kN/m

MF180b, pipe spacing = 180 mm
Max. tensile strength: 2558 kN/m (FEM)

MF230, pipe spacing = 230 mm
Max. tensile strength: 2558 kN/m (FEM)

MF-Pipe - DTH procedure with MF64 / MF64-IC

MF64, pipe spacing = 64 mm
Max. tensile strength: 3419 kN/m

MF64-IC, pipe spacing = 64 mm
Max. tensile strength: 3164 kN/m

For pipe sheet pile walls using DTH driving method we recommend connectors with a minimum tensile strength of 3000 kN/m.

MF-Pipe is a division of SteelWall ISH GmbH
Tassilostr. 21
82166 Gräfelfing / Germany

Copyright by SteelWall ISH GmbH
www.mf-pipe.com
info@mf-pipe.com

The resonating mass m_{dyn} is the sum of the dynamic mass $m_{ham,dyn}$ of the hammer, the mass m_{pile} of the pile section and the mass m_{soil} of the resonating soil volume. The latter is not usually known. Therefore, driving predictions often use $m_{soil} \geq 0.7 \, (m_{ham,dyn} + m_{pile})$. To ensure optimum vibratory driving operations, the calculated working amplitude should be $\overline{x}_A \geq 0.003$ m.

7. Acceleration a [m/s²]. The acceleration of the pile section acts on the granular structure of the surrounding soil. This influences the stresses between the individual grains of non-cohesive soils and, in the ideal case, even cancels them out completely, so that the soil is turned into a "pseudo-liquid" state while the vibratory hammer is operating. This reduces the friction in the soil and soil's resistance to driving. The product of the working amplitude and the square of the angular velocity yields the acceleration "a" of the pile section:

$$a = \overline{x} \cdot \omega^2 \; [m/s^2] \quad \text{mit} \quad \omega = \frac{2 \cdot \pi \cdot n}{60} \; [s^{-1}]$$

Experience shows that $a \geq 100$ m/s² is required for successful vibratory driving.

8.1.5.3.2 Connection between vibratory hammer and pile section

The vibrator must be connected to the pile section via hydraulic clamping jaws to create a connection that is as rigid as possible. This ensures that the energy from the vibratory hammer is ideally transferred to the pile section and from there to the soil and, consequently, that the vibratory driving is a success. As with impact driving, the vibrator should be positioned on the centroid axis of the driving resistance. Therefore, double jaw clamps should be used for the vibratory driving of double (or multiple) piles. The inclusion of a hole for pitching the piles in the direct vicinity of the clamps (which transfer the load) is not recommended because this can lead to cracks. The number and positioning of the clamping jaws should be selected to suit the section.

8.1.5.3.3 Criteria for selecting a vibrator

For the installation of sheet pile sections in uniform, re-arrangeable (non-cohesive) and saturated soils, the vibratory hammer should generate at least 15 kN centrifugal force for each metre of driving depth and 30 kN centrifugal force for each 100 kg mass of pile section. The centrifugal force can, therefore, be calculated from

$$F = 15 \cdot \left(t + \frac{2 \cdot m_{pile}}{100} \right) \; [kN]$$

where

F centrifugal force [kN]
t embedment depth [m]
m_{pile} mass of pile [kg]

The reader is referred to the details of the equipment manufacturer or to computer-assisted prognosis models when selecting a vibrator. Suitability tests are recommended for larger construction projects.

8.1.5.3.4 Experience with the vibratory driving of sheet piles

a) The effects of vibrations on nearby structures and other facilities arising from the vibratory driving of piles and sheet piles cannot be reliably predicted.

b) In principle, structures in the area affected by vibrations are loaded by oscillations, e. g. their foundations and also suspended floors (direct effects). DIN 4150 parts 2 and 3 contain reference values for permissible oscillation velocities. This effect can be minimised by using vibrators with zero resonance on start-up.

In addition, vibrations caused by the operation of a vibrator can cause settlement of the subsoil, which then has indirect effects on nearby structures.

c) Any direct and indirect effects expected to act on structures in the area affected by vibrations should be forecast in advance. Prognosis methods have been compiled by Achmus et al. (2005), for example. Using these forecasts, machinery and operating data can be selected such that the reference values of DIN 4150 parts 2 and 3 are not exceeded and settlements due to indirect effects can be evaluated.

d) If penetration rates $\geq 1\,\text{m/min}$ are achieved, experience suggests that damaging effects on adjacent structures are unlikely. Penetration rates $\leq 0.5\,\text{m/min}$ over longer structures should be accompanied by measurements of the adjacent structures, which are then examined in detail by a geotechnical expert together with a structural engineer to ascertain whether the oscillation velocities generated by the vibratory hammer are acceptable. Reduced penetration rates, e. g. when passing through compacted soil strata, are not usually problematic.

e) In soils that are not very amenable to re-arrangement (weakly cohesive soils, silt) or in dry soils, the effectiveness of vibration is significantly reduced. However, it can be improved by using jetting (Section 8.1.6). Drilling to loosen the soil just in advance of the driving or soil replacement can also be considered as aids to improve driving.

f) Soil compaction through vibration as mentioned in Section 3.2.4.2.3 is more likely to occur at high rotational speeds (Section 8.1.5.3.1). In order to avoid compaction, it can be expedient to carry out the work with a vibrator that generates the same centrifugal force but operates with a lower rpm and therefore with a lower acceleration.

g) More information on the watertightness of interlocks pre-sealed with synthetic materials can be found in Section 8.1.3.3.

h) Section 8.1.5.2.3 applies for vibratory driving records. The driving logs should contain the time taken for each 0.5 m penetration in addition to the operational data of the vibrator. The vibration work should preferably be recorded by continuous logging of operational data, vibration times and penetration.

i) Vibration is generally a low-noise driving method. Higher noise levels can occur with defective vibratory action as a result of the sheet pile wall vibrating as well and the clamps hitting each other. Vibration of the wall can be intensive in the case of long piles and staggered or panel driving. The use of driving assistance in accordance with Section 8.1.6 or padded clamping jaws can provide a remedy.

j) The risk of settlement in the vicinity of existing structures must be taken into account even when using modern high-frequency vibrators with variable eccentrics.

k) It should also be noted that low penetration rates and prolonged vibrating with powerful vibrators can lead to the interlocks being heated to such an extent that they fuse together. Where the penetration rate is reduced temporarily, water cooling of the pile, especially around the interlocks, can prevent this welding due to overheating.

l) During vibratory driving, obstacles in the soil are, in most cases, identified directly through a significant decrease in the penetration rate.

8.1.5.4 Pressing of steel sheet piles
8.1.5.4.1 General
Pressing meets the increasing demands for low noise levels and no vibrations during the driving of steel sheet piles.

Limits for using the pressing method to drive steel sheet piles in different soils are given in Section 3.2.4.2.4.

The cost of pressing is higher than that of methods employing hammers or vibrators. However, in many cases this is at least partly offset by the fact that, for example, sound-proofing measures (Section 8.1.9) are then unnecessary.

8.1.5.4.2 Pressing equipment
A distinction is made between presses suspended from cranes, presses with leader guidance and presses supported on sheet piles already driven.

Presses suspended from cranes require a frame that guides the piles at two levels. In the case of presses with leader guidance, the piles are guided at the top by the leader and at the bottom by a frame. All presses for the above two methods are equipped with several adjacent rams on the axis of the wall which jack the piles into the ground in a given sequence. The reaction forces for the pressing forces are provided by the equipment, the weight of the sheet piles and their skin friction in the soil.

Presses supported on sheet piles already driven are popular because of the small working space that they require. They require no further guidance frames or means of support. The sheet pile to be pressed in is both aligned and inserted by the ram in the head of the jack. This type of press activates the reaction forces via the adjacent piles already driven. However, the installation methods described in Section 8.1.5.2.2 cannot be used in this case.

Drilling or water jetting can be used to loosen the soil and thus assist pressing. The fittings required for this are already integrated into some presses.

Pressing in sheet piling is only successful when the penetration resistance of the piles is suitably estimated and the pressing forces and stiffness of the sections are matched to this.

Current sheet pile presses on the market are capable of generating a maximum pressing force of about 1500 kN. In the case of pile-supported presses, the reaction forces are provided solely by adjacent piles. Therefore, pressing forces of approximately 800 kN should not be exceeded with such presses, as otherwise this could result in piles already driven being extracted again.

8.1.5.4.3 Sheet piles for pressing
The majority of the machines available on the market today can only install single U or Z-section sheet piles. Presses suspended from a crane and those with leader guidance can operate with loosely assembled double, triple or quadruple piles, which are then driven as single piles.

Where such equipment is used to drive U-section piles, then compliance with DIN EN 1993-5/NA, NDP 6.4(3) is required.

Sheet piling presses for installing double piles can be assisted during the driving procedure by drilling pilot holes in the trough of the pile.

In order to minimise interlock friction in non-cohesive, fine-grained soils, the open interlocks of the sheet piles should be filled with a material to displace the soil, e. g. hot bitumen or similar material.

Sections to be pressed should be selected for both the structural requirements in the finished structure as well as the stresses during pressing. Experience has shown that sections being pressed in should not be too soft.

8.1.5.5 Installing steel sheet piles in pre-cut trenches supported by a suspension

Installing steel sheet pile sections in a pre-cut trench filled with bentonite slurry is a good solution in the case of difficult geological conditions, deep piles or in the proximity of existing structures vulnerable to damage. If the piles have to pass through legacy contamination, or if such contamination has to be sealed off, then this method creates a virtually impermeable wall, with the bentonite-cement grout protecting the load-bearing sheet pile wall against corrosion. To ease the installation of sheet piles, prefabricated multiple piles are worthwhile for corrugated walls or prefabricated subassemblies for combined walls. If necessary, an additional vertical load can be attached to assist pitching and installation.

8.1.5.6 Pitching straight-web sections and the principle behind the circular cell cofferdam

Cellular cofferdams may be constructed on load-bearing subsoil only. Soft strata, especially if they occur near the bottom of the cofferdam, reduce the stability significantly due to the formation of fixed failure planes. Such soils should be replaced with sand inside the cofferdam or drained with vertical drains. If none of these measures are taken, the circumferential tensile force (hoop tension) increases after filling as a result of the excess pore water pressure that occurs, which has a detrimental effect when analysing the failure of the sheet pile section (STR).

Fine-grained soil to DIN 18196 or DIN EN ISO 14688-2 may not be used for filling the cells.

The filling should be particularly water-permeable in the case of enclosures to excavations in order to guarantee the drawdown level for the dewatering.

Therefore, in order to minimise the dimensions of the cofferdam and to achieve adequate stability, a soil with a high unit weight γ or γ' and high internal angle of friction φ_k should be used. Both these soil parameters can be improved by vibrating the filling.

8.1.5.6.1 *Cellular cofferdams as excavation enclosures*

In excavation enclosures founded on rock it must be possible to lower the water in the cofferdam at any time with a drainage system (monitored by observation wells) to such an extent that it satisfies the stability analysis. Drainage openings at the bottom of the exposed wall, filters at the level of the base of the excavation and good permeability of the entire fill are essential.

Experience has shown that the permeability of sheet piling interlocks subjected to tensile stresses is low, so no particular measures need to be taken here.

The part of the excavation enclosure subjected to the external hydrostatic pressure must be adequately watertight. It can be practical, in some cases, to provide additional sealing measures on the outboard side of the cofferdam, e. g. underwater concrete.

8.1.5.6.2 Cellular cofferdams as waterfront structures

In waterfront structures much of the cell filling is submerged. Therefore, deep drainage systems cannot be used. When there are large and rapid fluctuations in the level of the water table, however, providing drainage systems in the cell filling and the backfill to the structure can be advantageous because this prevents a larger resultant water pressure (Figure 8.14). In these situations the planned efficiency of the drainage measures is critical for the stability and useful life of the waterfront structure.

The superstructure should be designed and constructed so that

- The risk of local damage to the cofferdam cells due to ship impacts is prevented. This can be achieved, for example, through components designed to spread the load.
- The inclusion of appropriate measures, e. g. fenders with a high energy absorption capacity, reduces the magnitude of the global effect of a ship impact to such an extent that the stability of the cofferdam cells is not at risk.
- Vertical loads from superstructures or other head details are founded on the fill, so that they generate only hoop tension in the straight-web pile sections.

Figure 8.15 illustrates a corresponding wall head detail.

Components subjected to significant vertical actions, e. g. due to cranes, should be built on separate foundations, e. g. an additional pile foundation, which can be positioned either adjacently to or even within the cofferdam. This avoids an increase in the hoop tension

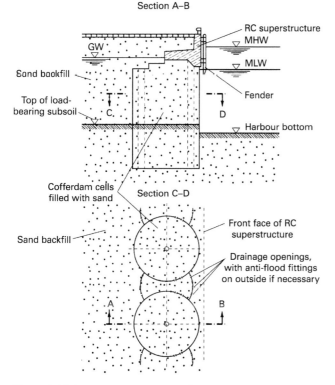

Figure 8.14 Schematic illustration of a waterfront structure constructed using circular cells, with drainage.

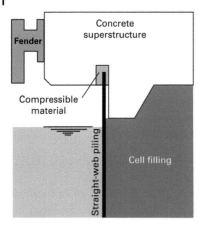

Figure 8.15 Detail of head of wall for circular cell cofferdam with superstructure.

and larger eccentricities for the action resultants. Load-bearing piles should be driven into the load-bearing strata beneath the base of the cofferdam so that they do not need to be considered when analysing the stability of the cellular cofferdam.

8.1.5.6.3 Construction

Cofferdams are recommended for the conditions stated in Section 8.1.2.9. They can be constructed both on land and in the water regardless of their particular purpose. Generally, a cofferdam built to provide protection against flooding or to enclose an excavation is built in the dry.

By contrast, a cofferdam designed exclusively as a waterfront structure is usually built from a floating or jack-up platform over the water.

8.1.5.6.3.1 Construction on land The traditional case for constructing a cofferdam on land is for a weir or power station site on a river that is to be dammed. The cofferdam is built on dry ground in the floodplain during periods of low water in order to carry out excavations and building work within the cofferdam. The cofferdam elements (double-wall or cellular cofferdam, U/Z-section or straight-web sheet piles) are set up on guide frames and, depending on the subsoil, driven as much as several metres into the ground. The inside of the cofferdam is filled with non-cohesive soil, which is subsequently compacted to a high degree and, if possible, drained. If the cofferdam is to be used to enclose an excavation, it must be driven into the ground to the appropriate depth. Filling of the cells is not, or not necessarily, required in this case, as the surrounding soil represents the filling. However, it may be necessary to improve the soil in the cofferdam through compacting it by means of vibroflotation, or by installing vertical drains to relieve the pore water pressure and, hence, increase stability.

8.1.5.6.3.2 Construction in water In essence, the difference between construction in water and construction on land is that when working in water, all the operations have to be carried out from a floating or jack-up platform. When building a double-wall cofferdam, pile-driving frames have to be set up first, or otherwise jack-up platforms or pontoons will be required. Cellular cofferdams require working platforms to be built in the water. It may also be possible to preassemble a complete cell around a working platform erected on land.

A floating crane is then used to lift the cell with the working platform and place it in position. This avoids the need to assemble the individual pile sections on the open sea, which is often a laborious process.

Generally, superstructures in the form of cantilever retaining walls or wave chambers are built on cofferdams to incorporate all the necessary facilities for the berthing of sea-going vessels.

The success or failure of the construction of waterfront structures comprising cellular cofferdams is very much influenced by the preparatory work. Individual cells without any filling are highly sensitive, non-load-bearing, easily damaged elements, and, therefore, a number of remarks concerning straight-web sections, the driving platform and driving methods are given below.

8.1.5.6.3.3 Diameter of circular cells Irrespective of the structural calculations, the actual diameter of a circular cell depends on the number of individual piles plus their rolling tolerances and the play at the individual interlocks. Therefore, the diameter can vary between two values, which are important for the size of the guide ring. A minimum diameter must be specified for fabricating guide rings.

8.1.5.6.4 Installation of straight-web sections

The proper installation of the piles requires at least two, on high cofferdams three, guide rings. These are placed around the driving platform, which is generally constructed as a space frame and suspended from several driven piles or pile bents/trestles supported on a load-bearing seabed.

All the piles are positioned around the driving platform and the rings tensioned accordingly. The piles are subsequently driven step by step, each about 50 cm per circuit of the vibrator or rapid-action hammer. It is also possible to lower the entire cell by using a multitude of vibrators, each one acting on several piles simultaneously.

Prior to installation, the connections of all interlocks should be checked, with the help of divers if necessary.

On very tall cells it may be necessary to divide the piles into several lengths for easier handling. Basically, as only tensile forces from the internal pressure or downward wall friction forces due to the earth pressure have to be accommodated, a corresponding stagger can be provided without leading to any drop in the load-bearing capacity.

After all the straight-web sections of the cell have been pitched, the individual sections should be driven step by step, e. g. using a driving procedure that makes numerous circuits of the cell, with the increment in the penetration being reduced each time.

One critical situation for the stability of a circular cell can occur after it has been set up and the driving platform has been dismantled following installation of the piles. External actions, e. g. due to wave pressure, can cause a circular cell to collapse. Therefore, the interlocks at the top of the cell are welded and the inside filled to about two-thirds of the height as quickly as possible. The recommendation is to fill the entire circular cell cofferdam at the same time as dismantling the driving platform.

8.1.5.7 Driving of steel sheet piles and steel piles at low temperatures

Steel sheet piles of all steel grades are suitable for driving at temperatures above 0 °C. If driving has to be carried out at lower temperatures, then special care is required when handling and driving the pile sections.

In easy driving conditions, driving is still possible down to temperatures of about −10 °C, particularly when using S 355 GP and higher steel grades. However, fully killed steels to DIN EN 10025 should be used when difficult driving with a high energy input is expected and when working with thick-walled sections or welded pile sections.

At temperatures below −10 °C, steel grades with enhanced cold workability must be used.

8.1.6 Driving assistance

- If the results of the soil investigations (see Chapter 3) indicate that difficult or even extremely difficult driving conditions can be expected, or where certain penetration rates must be maintained, or the steel sheet piles must be installed especially carefully, then this is often only possible with the help of driving assistance. Driving assistance methods prepare the subsoil in such a way that economic penetration rates can be achieved. At the same time, they avoid overloading the driving equipment and the piles themselves, and guarantee that the required driving depths will be reached. Driving assistance reduces the energy requirements for driving the sections. The desired outcome is less noise pollution and fewer vibrations.
- Suitable preliminary investigations (e.g. numerical prognosis models) are definitely recommended for estimating the effects of driving vibrations and noise expected in each specific case. If necessary, driving trials might prove useful for calibration. DIN 4150 parts 2 and 3 contain guidance on the values for permissible oscillation velocities when driving.
- The reader is referred to Section 8.1.9 with regard to noise control.

Apart from the pitching of piles in suspension-filled trenches (Section 8.1.5.5), other driving assistance methods used are water jetting, pre-drilling and blasting.

8.1.6.1 Water jetting to assist the driving of steel sheet piles

Sections 3.2.4.2.5, 8.1.5.2.3, 8.1.5.3 and 8.1.9.4 refer to water jetting to assist the driving of piles. Jetting with water can be used with impact driving, vibrating and pressing in order to:

- Generally facilitate installation.
- Prevent overloading the equipment and overstressing the pile sections.
- Achieve the necessary embedment depth.
- Reduce vibrations in the ground.
- Reduce costs through shortening installation times, reducing power requirements and/or enabling lighter equipment to be used.

The jetting pressure can be varied to suit the soil structure and strength. However, water jetting is not permitted below the design depth.

8.1 Materials and construction

8.1.6.1.1 Low-pressure jetting

- Low-pressure jetting involves directing a jet of water at the base of the pile section. The subsoil is loosened by the water injected under pressure and the loosened material is carried away in the flow. Essentially, this reduces the toe resistance of the piles to be inserted. Depending on the soil structure, the skin friction and the interlock friction can also be reduced by the rising water.
- The low-pressure method is limited by the strength of the subsoil, the number of jetting lances and the pressure and volume of water required. In order to establish the necessary operating data for low-pressure jetting, trial driving operations are recommended.
- Low-pressure jetting can be used in non-cohesive, densely bedded soils, and especially in dry, uniform non-cohesive soils, sands and gravels.
- Jetting lances have a diameter of 45–60 mm, and the pressure at the pump is between 5 and 20 bar. Constricting the nozzle or using special nozzles can create a jet action with a correspondingly greater flushing effect. Water for jetting is usually supplied by centrifugal pumps, and water consumption can reach approximately 1000 l/min.
- Depending on how difficult driving is, jetting lances are inserted next to the pile section or fixed to the pile section and carried down with this into the ground. Reductions in the strength of the soil and settlement may occur as a result of introducing relatively large volumes of water.
- Low-pressure jetting processes in which the jetting process is combined with vibration and jetting pressure is restricted to 15–40 bar have proven effective for driving sheet piling in very compact soils that would normally present extremely difficult driving conditions.
- The low-pressure jetting method may be considered as environmentally compatible and, thus, suitable for use in residential and inner-city areas.
- The success of this method essentially depends on the optimum match between the jetting, the vibrator and the in situ subsoil. It is, therefore, important to assess local experience or, where this is unavailable, to carry out trials beforehand. Vibrators with variable effective moments and rpm control are ideal because their operating data can be adjusted to the specific penetration resistance.
- Usually, between two and four lances are fixed to the pile section (double pile), with the tip of the lance flush with the toe of the pile. The optimum arrangement is one pump per lance. Jetting begins simultaneously with vibration in order to prevent the nozzle from becoming clogged by the ingress of soil material.
- At penetration rates ≥ 1 m/min, jetting can continue until the embedment depth required by the calculations is reached. The soil properties previously determined for the sheet piling calculation then generally apply. However, the angles of inclination for active and passive earth pressures should be limited to $\delta_a = +1/2\varphi$ and $\delta_P = -1/2\varphi$, respectively. If high vertical loads are to be borne by a section inserted with the help of jetting, the vertical load-bearing capacity must be assessed in loading tests.

8.1.6.1.2 High-pressure jetting

The use of high-pressure jetting eases the driving of sheet piles in soils of varying compactness; indeed, in some cases, without jetting, driving would not be possible at all. Precision pipes 30 mm in diameter are used as jetting lances. The pressure at the (reciprocating) pump is between 250 and 500 bar. Jetting lances are fitted with screw-on nozzles, the cross-section of which can be adjusted to suit the surrounding soil. Water consumption lies between 30 and 150 l/min per nozzle.

High-pressure jetting is primarily suited to firm, overconsolidated cohesive soils such as silt and clay rocks and friable sandstone.

However, high-pressure jetting is only economical when the lances can be extracted and reused. To this end, the jetting lances cannot be permanently attached to the sections being driven. Instead, they are guided through clips welded to the section. The jetting nozzles must be about 5–10 mm above the bottom end of the pile and should be chosen to maximise their working life.

When using high-pressure jetting, the pressure, number of lances and type of nozzle must be matched to the specific in situ soil. Where soil conditions vary, these adjustments must also be carried out during driving operations.

The angles of inclination for active and passive earth pressures should be limited to $\delta_a = +1/2\varphi$ and $\delta_P = -1/2\varphi$, respectively.

If high vertical loads are to be borne by a section inserted with the help of high-pressure jetting, the vertical load-bearing capacity must be assessed in loading tests.

8.1.6.2 Shock blasting

Blasting is frequently used to loosen up rocky soils in advance. The blasting fragments the rock along the planned line of the sheet pile wall in such a way that a vertical, ballast-filled trench is created, into which the sheet piles can be installed – preferably by vibratory methods. The loosening should extend as deeply as the intended base of the sheet pile wall and should be wide enough to accommodate the sheet pile wall section. On either side of the trench, the rock remains stable.

In principle, every type of rock can be blasted. The choice of a suitable blasting method, the detonation sequence, the layout of the charges, the type of explosive (waterproof, highly explosive) and, in particular, accurate positioning, spacing and inclination of the blast holes are all critical for the achievement of the aims of the blasting.

8.1.6.2.1 Blasting method

Trench blasting with short-period detonators and inclined blast holes, as shown in Figure 8.16, has proved to be a successful way of achieving the aims of blasting while causing minimum vibration. Trenches up to 1 m wide can be produced by this method (Dynamit Nobel 1993).

The blasting sequence begins by drilling stab holes in a V-arrangement (shown on the right of the longitudinal section in Figure 8.16) to relieve the stress in the rock by creating a second free face in addition to the surface of the trench, against which the rock can be thrown. This improves the effect of the blasting and leads to lower vibration than is the case with vertical blast holes.

Following the V-cut, the other charges detonate in succession at intervals of, usually, 25 ms. The first round of detonations creates space for the rock to be loosened up by the subsequent detonations. In addition, the detonations bounce off each other so that the rock along the line of the blast holes in the trench is thrown backwards and forwards several times, thus reducing the size of the fragments.

The explosive acts in a V-form (included angle 90°) in the direction of the free face (see the cross-section in Figure 8.16). In order for the sheet pile wall to be driven to the intended

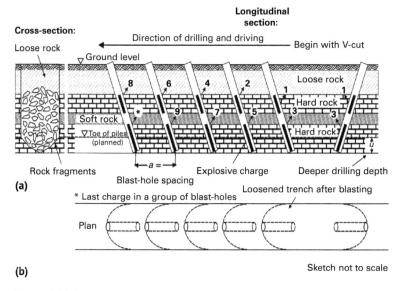

Figure 8.16 Principle of trench blasting with inclined blast holes (sketch not to scale): (a) section; (b) plan.

depth despite the narrow tip of the cone of debris, the blast holes for the explosives must extend below the planned toe of the sheet pile wall. Above the cone of debris, a ballast-filled trench ensues, into which the sheet piles can subsequently be vibrated without damage.

The spacing of the blast holes a in Figure 8.16 roughly corresponds to the average width of the trench. Common blast hole spacings for inclined drilling are 0.5–1.0 m

The trench should be only a few decimetres wider than the sheet pile sections being used. Trenches should not be too wide because the base of the sheet piling is assumed to be fixed in the structural analyses.

In rocks with changing strengths, the individual charges should be placed in the hard rock segments in order to achieve an optimum blasting effect.

8.1.6.2.2 Blasting advice

The following basic advice should be taken into account when planning and carrying out blasting work:

1. Before beginning work it is necessary to investigate the subsoil by means of trial blast holes and trial blasting. The aim of these trials is to determine the optimum blast-hole spacing and the explosives required. Ultrasound measurements taken prior to and after blasting can be used, for example, to estimate the volume loosened.
2. As vibratory driving of sheet piles causes less damage than impact driving, the sections should be vibrated into the trenches of loosened material. Compaction of the loosened rock during the vibration process is not generally a risk. Impact driving is only permissible in exceptional cases.
3. Sheet pile sections should be vibratory driven immediately after blasting because the geostatic surcharge and, possibly, groundwater flow forces (hydraulic compaction) can partially reverse the loosening effect in the trench.

4. The driveability of the blasted rock and the possibility of installing sections with vibration can be assessed with the results of dynamic penetration tests (DPH). A high number of blows ($N_{10} > 100$) indicates that difficulties will be experienced when driving the sheet piles.
5. If difficulties occur while driving the sections, additional blasting will be necessary. The sheet pile sections must be extracted prior to this.
6. Simply splitting the rock permits impact driving only.
7. Evidence of the vibrations due to blasting occurring at the nearest structures should be comprehensively collected with vibration measuring instruments to DIN 45669. These measurements can be carried out within the scope of the contractor's own monitoring procedures.
8. Data on the drilled blast holes must be recorded in blast hole logbooks. These should state the boundaries of the strata, contact pressure during drilling, flushing losses and water-bearing strata. This information provides details of clefts, voids and narrowing of the blast hole. The angle and depth of every blast hole must be recorded. The logbooks should be made available to the chief blasting engineer prior to placing the charges in the holes to ensure optimum placement of the explosive charges in each blast hole.
9. Spot checks should be carried out to assess the accuracy of the blast holes, especially at the start of the drilling. Precise blast hole measurement systems are available (e.g. BoreTrack). A drilling accuracy of 2% is feasible. The axes of the blast holes and the sheet pile wall must lie in one plane throughout the intended depth, so that the sheet piles are always driven into the rock loosened by the blasting.
10. Only water should be used for flushing because the changing colour of the water and the drilling debris allows changes of strata to be identified, whereas flushing with air produces only a uniform dirty cloud of dust. Another problem is that the air introduced under high pressure is forced into softer strata and existing clefts, which may create undesirable paths. This can, in turn, lead to blowouts during blasting that are remote from the desired blasting point and, thus, diminish the success of the blasting.
11. Casings should be used when drilling in soft cohesive soils and non-cohesive soils because otherwise there is a risk of material collapsing into the inclined blast holes.
12. The blasting parameters should be systematically optimised on the basis of the records of the first blasts. Separate blasting trials are then unnecessary. Strict coordination of the drilling and driving works and a constant exchange of information on the work carried out so far are necessary and improve the success of the operations.
13. The maximum amount of explosive per detonation stage should be established on the basis of the vibration calculations prior to any blasting. On no account should this amount be exceeded.

8.1.6.3 Pre-drilling

Uncased holes 105–300 mm in diameter can be drilled at a spacing corresponding to the width of the sheet piles to perforate and relieve the subsoil and, thus, facilitate the driving of sheet piles.

The same effect can be achieved by using high-pressure jetting in rock with changing strength and similar subsoils (Section 8.1.6.1.2).

If sheet piles must be driven deep into bedrock, accurate blasting can be used to loosen the rock along the line of the sheet piling and thus facilitate driving. When selecting the section and the grade of steel, possible irregularities in the subsoil and the ensuing driving stresses must be taken into account. The reader is referred to Section 8.1.6.2 for details of blasting.

8.1.7 Monitoring pile driving operations

During the installation of steel sheet piling, the positions and condition of the pile sections must be constantly checked and suitable measurements carried out to ascertain when the intended embedment depth has been reached. Together with the correct starting position, adherence to tolerances must also be checked in intermediate phases, especially after the first few metres of penetration. This should make it possible to detect even small deviations from the design position (angle, out-of-plumb, distortion) or deformations of the pile head so that early corrections can be made and, if necessary, suitable countermeasures initiated.

In difficult driving conditions with obstacles, the penetration, line and position of the pile sections are to be observed frequently and with particular care. If a pile section no longer moves, i.e. unusually slow penetration, the driving/vibration is to be stopped immediately. Subsequent piles can then be inserted first. Later, a second attempt can be made to drive the protruding pile more deeply.

Observations when driving primary and secondary elements are to be carried out for each pile section (see Section 8.1.7.4). All records should be made available immediately so that decisions regarding further work or other measures can be made promptly.

Individual pile sections that become very difficult to drive just before reaching their design depth should not be forcibly driven further, as there is a risk of damaging the section, interlocks and weld seams. In individual cases, a shorter embedment depth can be accepted if this avoids damage to the section. However, it is important to ensure that, due to the shallower depth of an individual pile section, neither the stability (passive earth pressure, hydraulic heave failure), nor serviceability (e.g. water penetration) of the structure as a whole are compromised.

In the case of unusually large embedment depths for the bearing piles of combined sheet piling, and when the bearing piles are not driven to a final set, it may be necessary to remove the piles, compact the soil and redrive them so that they can transmit the vertical loads assigned to them. The piles may need to be lengthened in some cases.

If observations during driving, e.g. distortion or out-of-plumb sections, indicate that sections have been damaged, an attempt should be made to inspect the piles by digging them partially free or extracting them to investigate the cause of the distortion or misalignment, e.g. by examining the subsoil for obstacles.

8.1.7.1 Declutching, signal transmitters

Damage caused by accidents can be repaired in most cases. Constructional measures can be employed to seal the interlocks of sheet pile walls (Section 8.1.8). Declutching on sheet piling arises when the interlock of a section being driven becomes detached from that of a section already driven. Possible causes of declutching include obstacles in the soil and, particularly in the case of combined sheet piling, deviation of driven bearing piles from the intended position both on plan and in the plane of the wall. Therefore, adhering to driving tolerances is the most important precautionary measure for preventing declutching. However, declutching, particularly in the case of combined sheet piling walls, cannot be

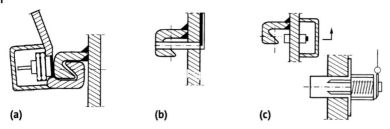

Figure 8.17 Signal transmitters: (a) proximity switch;, (b) electric pin contact; (c) mechanical spring pin.

completely ruled out even when driving is performed carefully and driving tolerances are observed.

Only in exceptional cases is it possible to identify a case of declutching from the driving data (driving energy, driving progress). It is, therefore, necessary, after the driven pile has been revealed through excavation, to inspect it for declutching, possibly with the help of divers, and to carry out any necessary repairs.

Where the stability or serviceability of a structure could be compromised by declutching (e. g. excavations in open water), declutching detectors (signal transmitters) can be used, which make it possible to identify declutching during driving (Figure 8.17).

A proximity switch such as the one shown in Figure 8.17a can continually measure whether the interlock connection is still intact during driving. If the interlock of the threading pile makes contact with the electric contacting pin shown in Figure 8.17b, it is sheared off, which then indicates that the threading interlock is still in the interlock of the driven pile. Similarly, the mechanical spring pin shown in Figure 8.17c indicates that the interlock connection is still intact.

In addition, there is also the robust option – tried and tested in practice – of driving a short piece of interlock with a welded-on cable ahead of the interlock to the required depth, which, thus, indicates that the threading interlock is still running in the driven interlock.

8.1.7.2 Driving deviations and tolerances

The following tolerances for sheet piling should be included in the calculations at the planning stage for deviations of the piles from the design position in accordance with DIN EN 12063:

- ±1.0% of the embedment depth for normal soil conditions and driving on land.
- ±1.5% of the embedment depth for driving in water.
- ±2.0% of the embedment depth for difficult subsoil.

The deviation is to be measured in the top metre of the pile section.

The deviation of the top of the sheet piling perpendicular to the axis of the wall must not exceed 75 mm for driving on land and 100 mm for driving in water.

The tolerances given in DIN EN 12063 for sheet piling are not suitable as a measure of the required accuracy of bearing piles in combined sheet piling walls. The driving of bearing piles in a combined sheet piling wall must be significantly more accurate than the tolerances for sheet piling given in DIN EN 12063. Consequently, the tolerances given above for driving deviations do not apply to bearing piles in combined sheet piling walls. The tolerances for such piles must be agreed upon on an individual basis. The reader is referred to

Figure 6 in DIN EN 12063 for details of establishing tolerances for bearing piles in combined sheet piling walls.

Owing to the risk of declutching, bearing piles in combined sheet pile walls must be straight, vertical (or at the stipulated inclination), parallel to each other, undistorted and at the planned spacings (Section 8.1.5.2.3).

8.1.7.3 Measuring driving deviations

The correct starting position, and also intermediate positions of the pile sections, can be checked by using two measuring devices: one checking the position in the y direction, the other in the z direction. These measurements should generally be prescribed for driving bearing piles in combined sheet piling walls. The tolerances to be maintained when driving combined sheet piling must always be determined and agreed upon between designer, client and contractor as per DIN EN 12063, taking into account the rolling tolerances stated in DIN EN 10248 and, if required, the additional section tolerances stated by the product supplier, e.g. increase/decrease in width, limits for interlock twist, etc. When constructing quay walls in exposed locations, the agreed upon tolerances for each bearing pile must be checked after driving and removing guides, not only at the pile head and directly above the lowest possible waterline, but also by divers at the depth of the watercourse bottom.

If the permissible tolerances have been exceeded, the bearing piles must be extracted and re-driven. Alternatively, correction piles must be installed as secondary elements. The make-up piles are fabricated either in accordance with the measurements or as particularly flexible elements (spring piles). This flexibility can be achieved by removing the middle interlock of a secondary element and welding on a half shell.

If the vertical positioning of pile sections is checked with spirit levels, these must be long enough (at least 2.0 m) and used with a straightedge, if necessary. The checks are to be repeated at various points to compensate for local irregularities.

8.1.7.4 Records

Records of the driving observations are to be kept according to DIN EN 12699, Section 10. The list in this standard corresponds to the pre-printed forms of ZTV-WLB 214:2015, Section 3.1(22), appendix (6b.1) and (6b.2), which has been withdrawn. Under difficult driving conditions, the driving energy for the whole driving procedure should be recorded for the first three pile sections and thereafter for every 20th section.

Modern pile drivers record the driving data fully on data media so that the information can be quickly evaluated using special software. When working in difficult driving conditions in changing soils, complete records of all the driving data are recommended.

When installing pile sections using vibratory driving, the penetration achieved should be continually recorded and documented in order to identify any irregularities.

8.1.8 Repairing interlock declutching on driven steel sheet piling

8.1.8.1 General

Interlock declutching can occur during driving of steel sheet piles or might be due to other external actions. However, the better the recommendations for the design and construction of sheet piling structures are observed with care and diligence, the more the risk is reduced. The reader is referred to the recommendations in Sections 8.1.2 and 8.1.5.

However, interlock declutching, especially on combined sheet piling, is still possible even when strictly observing these recommendations. Interlock declutching in quay walls in tidal areas is particularly critical, as these walls are regularly exposed to excess water pressures during low water, which can wash out the backfilling via the gaps. Cavities then form in the backfill, which, due to the arching effect, can remain stable for long periods of time but then collapse unexpectedly, causing subsidence in the port operations area.

Therefore, sheet piling in water must be exposed by excavating and checked by divers for interlock declutching. Any damage found must be repaired. Repairs with steel offer many options in line with the respective boundary conditions.

8.1.8.2 Examples of repairs

If observations during driving suggest interlock declutching over an extended area of the wall and the sheet piling cannot be extracted due to lack of time, repairs in the form of grouting a large area of the soil behind the wall is a popular option. This prevents the soil from escaping through the wall and, thus, causing subsidence behind the wall. High-pressure injection has proved particularly effective for grouting soil (see Figure 8.22). Afterwards, the damaged areas can be permanently sealed with steel sections or plates attached to the front of the wall.

Individual cases of interlock declutching are usually only discovered during assessments by divers after the soil in front of the wall has been excavated. Declutching must be repaired in such a way that not only is the damaged area sealed, but that the wall can continue serving its purpose without any restrictions. This is especially important for sheet pile walls consisting of only U- or Z-piles.

The sealing of interlock declutching depends, above all, on the size of the opening and on the sheet pile section.

Repair work is generally carried out on the water side. Smaller interlock openings can be closed with wooden wedges. Large openings can be temporarily sealed, e.g. with a rapid-setting material, such as rapid-hardening cement or a two-part mortar, both placed in sacks. However, in order to seal the damaged areas permanently, it is necessary to close off the damaged area down to at least 0.5 m, but preferably 1.0 m, below the bottom of the harbour floor by attaching steel sections or plates to the water side of the sheet piling (Figure 8.22). In addition, the damaged area must be concreted in order to prevent sand from being washed out at a later date, resulting in subsidence behind the wall. The concrete used may need to be reinforced to protect it against vessel impacts. Installing an additional protective layer, e.g. of ballast, on the bottom of the watercourse is recommended in front of the damaged section when the soil in this area is at risk of scouring. Otherwise, the repairs should extend below the calculated scouring depth.

Most of this work has to be carried out underwater and, therefore, always requires diver assistance. Very high demands are, therefore, placed on the technical abilities and reliability of the divers. In difficult conditions, the plates can be fitted in the dry using underwater enclosures (limpet dams), which are held against the wall by water pressure.

The smoothest possible sheet piling surface on the water side is to be strived for. For this reason, protruding bolts, for example, must be burned off after the damaged area has been concreted, and they have fulfilled their function as a formwork element. The steel plates and the sheet piling are to be joined to the concrete with anchors in the form of splayed fishtails.

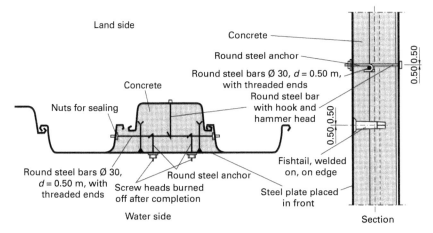

Figure 8.18 Example of repairing damage at a small opening in a wall.

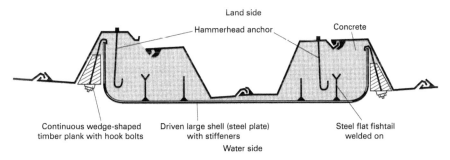

Figure 8.19 Example of repairing damage at a large opening in a wall.

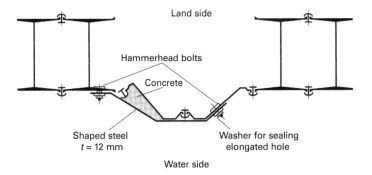

Figure 8.20 Example of repairing damage at an opening in combined sheet piling.

The solutions shown in Figures 8.18–8.22 for repairing underwater declutching have proved successful in practice, but other methods are also possible.

Interlock declutching on combined sheet piling is repaired, if possible, by driving further secondary elements behind the wall as per Figure 8.21 or on the water side as per Figure 8.20. In the former, i. e. in accordance with Figure 8.21, the excess pressure behind the wall presses the pile section against the intact parts of the sheet piling during dredging on the water side.

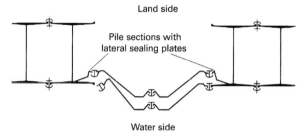

Figure 8.21 Example of repairing damage by driving a sheet pile section behind the wall to close off an opening in combined piling.

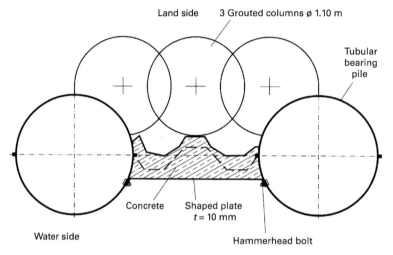

Figure 8.22 Example of repairing declutching in combined sheet piling with high-pressure injection and a plate attached in front of the wall.

The repairs shown in Figures 8.18–8.20 presume that no soil is washed out through the damaged area, e. g. because there is no excess water pressure behind the wall or the damaged area has already been successfully sealed temporarily. However, this is normally not the case, meaning that the repair shown in Figure 8.22 must take place in two steps. The soil behind the wall must first be grouted using high-pressure injection, with the damaged area plugged using a generously dimensioned frontal embankment of dumped material so that the high-pressure injection suspension does not escape through the damaged area. Afterwards, the dumped material can be dredged away and the declutching permanently sealed with concrete and specially fabricated steel parts reaching below the design harbour bottom.

When driving in front of the wall, suitable measures must be taken to ensure that the base of the pile section is always pressed against the bearing pile; in other words, it has to be located as close to the bearing pile as possible. Before dredging takes place, the head of the pile section must be fixed to the bearing pile, e. g. with hammerhead bolts. The elevation of the first dredging cut must be coordinated with the bearing capacity of the pile section between this upper fixed support and the lower, more flexible earth support. After driving, the pile section must be fixed to the bearing pile in the exposed area, etc.

Constant investigations of the repaired areas by divers are necessary during dredging operations. Any local leaks can be sealed by additional injections.

8.1.9 Noise control – low-noise driving

8.1.9.1 General remarks on sound level and sound propagation

The aim – right from the planning phase of the driving works – must be to keep the anticipated environmental nuisance of the planned construction site to a minimum. This includes shifting the times of construction work involving high levels of noise to times of the day when it will be less of a nuisance and strictly observing driving breaks during the early morning, midday rest periods and the evenings. If this means accepting a reduction in daily output, then this must be taken into consideration in the tender.

The sound emissions of a sound source (e.g. a machine) are characterised by the sound power or sound pressure level measured at a defined distance. The sound power is the sound energy emitted per unit of time. It does not depend on the ambient conditions and hence is a parameter of the sound source, which is assessed according to subjective criteria (e.g. human hearing).

The principal method of assessment is the A-weighted sound power level, which is 10 times the logarithm of the ratio of the sound power to the reference power ($P_0 = 1\,\text{pW} = 1 \times 10^{-12}\,\text{W}$). The A-weighting reflects a filter for the frequency response of the human ear. The type of assessment is identified either by using the index LWA or by adding a suffix to the unit of measurement dB(A).

The sound power level, being 10 times the logarithm of the ratio of the sound pressure of the sound source to a reference pressure (in air: $p_0 = 20\,\mu\text{Pa}$), is a variable that depends on the measuring distance and the ambient acoustic conditions.

Other important parameters that can be used for estimating the sound emissions of a sound source are the sound distribution over various frequency bands (third-octave, octave, narrow) and possible chronological fluctuations and directional characteristics.

Owing to the logarithmic calculation in acoustics, it is easy to appreciate that doubling the number of sound sources raises the sound pressure level by 3 db(A). However, studies have revealed that the sound pressure level must rise by 10 dB(A) in order for the noise to be perceived by the human ear as "twice as loud". Figure 8.23a shows the increase in the total sound pressure level upon superimposing a number of equally loud sources.

Sources with different sound levels have a totally different influence on the overall sound level. If the difference between the sound pressure levels of two sources is greater than 10 dB(A), the quieter source actually has no influence on the overall sound level (see Figure 8.23b).

Consequently, measures to control noise can only be effective when, as a first measure, the loudest individual noise levels are reduced. Eliminating the weaker individual noise levels only makes a minor contribution to reducing noise.

Given ideal free-field propagation in an infinite semi-spherical space, and on account of the geometric propagation of sound energy, the sound pressure of a point-like sound source diminishes with the distance from the source by

$$\Delta L_P = -20 \cdot \lg \frac{S}{S_0}$$

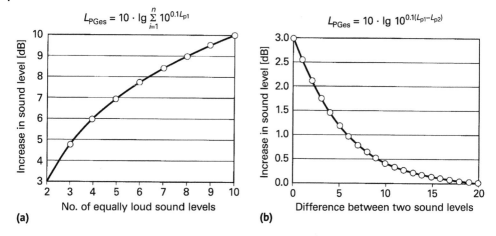

Figure 8.23 (a) Increase in sound level when several equally loud sound levels are heard simultaneously; (b) increase in sound level for two levels of different loudness.

where

ΔL_p change in sound pressure [dB]
S distance 1 to sound source [m]
S_0 distance 2 to sound source [m]

Therefore, doubling the distance reduces the sound pressure by 6 dB(A). In addition, the sound is attenuated by up to 5 dB(A) over larger distances over natural, uneven terrain on account of air and ground absorption, vegetation and buildings.

On the other hand, it must be taken into account that a single sound reflection on a structure in the vicinity of the sound source or concrete or asphalt surfaces can increase the sound level by up to 3 dB(A) depending on the extent to which surfaces absorb or scatter sound. When there are several reflecting surfaces, each can be substituted by a theoretical mirror sound source with the same loudness level as the original sound source and the resulting increase in level calculated taking account of the calculation rules for the interaction of several sound sources (see Figure 8.23a).

When sound propagates over larger distances, it is also necessary to take into account the fact that the decrease in sound level due to meteorological influences, e. g. wind currents and temperature stratification, can have both a positive effect, i. e. in the sense of a greater decrease in sound level, and a negative effect. For example, a positive temperature gradient (increase in air temperature with altitude = inversion) amplifies the sound level because the sound waves are deflected back to the ground at locations more than approximately 200 m away from the sound source. This effect is particularly noticeable over areas of water, which are generally colder than the ambient air (which heats up more quickly) and, thus, give rise to a positive temperature gradient. A rapid cooling of the land after sunset also results in a positive temperature gradient.

In interaction with ground reflections, the curvature of the sound waves can also have the effect that the propagation of the sound remains limited to a corridor between the ground and the inversion layer, so that the geometric propagation attenuation is reduced by 50%.

The influence of the wind is comparable with that of temperature. Here again, the smaller reduction in sound level in the wind direction is caused by a change in the horizontal wind speed with increasing altitude and the resulting downward deflection of the sound waves. This effect is particularly noticeable on cloudy or foggy days when the wind with a speed of up to 5 m/s exhibits an essentially laminar flow. On the other hand, turbulence and vertical air circulation, caused mainly during the day by solar radiation, can result in a greater reduction in sound level due to the scatter and refraction of the sound waves.

8.1.9.2 Regulations and directives for noise control
The following regulations and directives apply to noise control:

- Allgemeine Verwaltungsvorschrift zum Schutz gegen Baulärm – Geräuschimmissionen (regulation for protection against construction noise – noise immissions). Carl Heymanns Verlag KG, Cologne, 1971.
- Allgemeine Verwaltungsvorschrift zum Schutz gegen Baulärm – Emissionsmessverfahren (regulation for protection against construction noise – emission measuring methods). Carl Heymanns Verlag KG, Cologne, 1971.
- Council Directive 79/113/EEC of 19 December 1978 on the approximation of the laws of the Member States relating to the determination of the noise emission of construction equipment and equipment (OJ L 33, 8 Feb 1979, pp. 15–30).
- 15th Regulation for Enforcement of the Bundesimmissionsschutzgesetz (federal imission control act) of 10 November 1986 (Baumaschinen-LärmVO, construction equipment noise act).
- Directive 2000/14/EC of the European Parliament and of the Council of 8 May 2000 on the approximation of the laws of the Member States relating to the noise emission in the environment by equipment for use outdoors (OJ L 162, 3 Jul 2000 pp. 1–78; corr.: OJ L 311, 12 Dec 2000, p. 50).
- ISO 9613-1 (1993): Acoustics – attenuation of sound during propagation outdoors – Part 1: Calculation of the absorption of sound by the atmosphere.
- DIN ISO 9613-2 (1999): Acoustics – attenuation of sound during propagation outdoors – Part 2: General method of calculation (ISO 9613-2:1996).
- Third regulation for Gerätesicherheitsgesetz (equipment safety act) – Maschinenlärminformationsverordnung (machine noise information act) of 18 Jan 1991 (Federal Gazette 15. 146; 1992, p. 1564; 1993, p. 704).
- VDI Directive 2714 (Jan 1988): Outdoor sound propagation.

Furthermore, the regulations of federal state legislation regarding noise levels to be observed at night and during public holidays must be complied with, together with the laws and regulations regarding occupational safety (safety regulations UVV, GDG in conjunction with equipment safety act).

The permissible noise exposure in the area of influence of a sound source is stipulated graduated according to the need to protect the surrounding areas from construction site noise. The need for protection results from the actual use of the areas as stipulated in the local development plan. If no local development plan exists or the actual use deviates considerably from the use intended in the plan, the need for protection results from the actual use of the areas.

According to the regulation covering construction noise (AVV Baulärm), the effective level generated by the construction equipment at the place of exposure may be reduced by 5 or 10 dB(A) if the average daily operating period is less than 8 or 2.5 h, respectively. On the other hand, a nuisance surcharge of 5 dB(A) is to be added when the noise includes clearly audible sounds such as whistling, singing, whining or screeching. If the rating level of the noise caused by construction equipment determined in this way exceeds the permissible recommended exposure value by more than 5 dB(A), measures must be initiated to reduce the noise. However, this is not necessary when the operation of the construction equipment does not cause any additional dangers, disadvantages or nuisances as a result of not merely occasional effective extraneous noises.

The emissions measuring procedure serves to ascertain and compare the noise levels of construction equipment. For this purpose, each item of construction equipment undergoes a minutely prescribed measuring procedure during various operating procedures under defined boundary conditions. As part of the standardisation of EU regulations, the noise emissions of construction equipment are now stated as sound power levels L_{WA} related to a semi-spherical surface of 1 m². The sound pressure level L_{PA} is still frequently used and relates to a radius of 10 m around the centre of the sound source or – in combination with occupational safety regulations – at the position of the equipment operator.

Recommended emissions levels have been defined for the certification or use of various items of construction equipment. Using the latest equipment prevents these levels from being exceeded. Up until now, no mandatory values have been stipulated for pile drivers.

8.1.9.3 Passive noise control measures

Screens represent a method of passive noise control which prevents the propagation of sound waves in certain directions. The screen is lined with sound-absorbent material on the side facing the sound source in order to avoid reflections and so-called standing waves. The effectiveness of a screen depends on its effective height and width and the distance from the source that needs to be screened off. Basically, the screen should be erected as close as possible to the sound source. To remain effective, there should not be any gaps (e.g. open joints) in the screen.

So-called encapsulated solutions employ enclosures or baffles to surround the sound source completely with soundproofing material.

A soundproof enclosure around pile driver and pile can reduce the noise level during driving. However, such enclosures complicate working procedures considerably. As encapsulation makes it impossible to monitor the driving procedure and thus increases the risk of accidents, the use of this type of passive noise control is very limited. In addition, encapsulation is expensive, adds considerable extra weight to the equipment and is vulnerable to damage.

If some form of a screen cannot be avoided, a U-shaped mat of textured sheeting suspended over the hammer and piling element is preferable and still enables the pile driver operator to watch the driving procedure. A reduction of up to 8 dB(A) can be achieved in the screened direction. Up until now, so-called sound chimneys have been used for smaller pile drivers only.

8.1.9.4 Active noise control measures

The most effective and, as a rule, also least expensive way of reducing noise levels both on and around pile-driving sites is to use low-noise construction equipment. For example, compared with hammers, vibratory driving can reduce the sound level considerably. Hydraulic pressing of sheet piles and the installation of bearing piles in pre-drilled holes can certainly be classed as low-noise methods. But the scope for using these low-noise driving methods essentially depends on the properties of the in situ subsoil.

Active noise control measures also include construction methods that ease the driving of piles or sheet piles into the subsoil, thus reducing the energy required for the driving procedure. Besides water jetting or drilling or blasting to loosen the soil, these methods also include limited soil replacement in the area of the pile sections plus threading sheet piles into pre-cut, bentonite-filled trenches. However, such noise control measures can be used only if they are suitable for the in situ subsoil and construction conditions.

8.1.10 Corrosion of steel sheet piling, and countermeasures

8.1.10.1 General

Steel in contact with water undergoes a natural process of corrosion, which is influenced by numerous chemical, physical and, occasionally, biological parameters. There are different corrosion zones over the exposed height of a sheet pile wall, which are characterised by the type of corrosion (surface, pitting or tuberculation) and its intensity.

To measure the degree of corrosion, we can assess the decrease in wall thickness [mm] or – related to the time in use – the rate of corrosion [mm/a].

Typical mean and maximum values for wall thickness reduction are shown in Figures 8.25 and 8.26. These diagrams are based on numerous wall thickness measurements on sheet pile walls, piles and dolphins in the North Sea and the Baltic Sea as well as inland waters and can be assigned to the following corrosion zones: LWZ (low water zone), PIZ (permanent immersion zone) and SZ (splash zone). Owing to the multitude of factors that influence corrosion, the measured values are subjected to a very wide scatter.

8.1.10.2 How corrosion influences the load-bearing capacity, serviceability and durability of steel sheet piling

Corrosion influences the stability, serviceability and durability of unprotected steel sheet piling, and so the following aspects must be considered:

1. Corrosion decreases the design value of the component resistance corresponding to the different reductions in wall thickness in the various corrosion zones. Depending on the local bending stresses, this can reduce the load-bearing capacity and serviceability of the structure (DIN EN 1993-5, Sections 4–6).
2. In analyses of the load-bearing capacity and serviceability according to EN 1993-5, the section modulus and the cross-sectional area of the sheet piles are to be reduced in proportion to the mean values of the wall thickness losses according to Figure 8.25a (freshwater) or Figure 8.26a (seawater).

   Unprotected sheet piling should be designed in such a way that the maximum bending moment lies outside the zone of maximum corrosion.

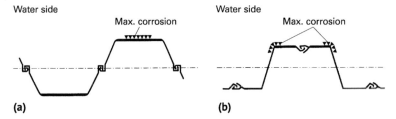

Figure 8.24 Zones where (a) U- and (b) Z-section sheet piles can rust through in the low water zone (seawater).

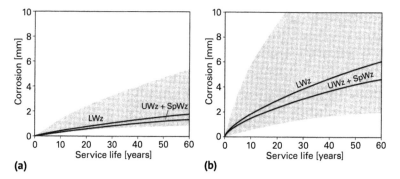

Figure 8.25 Decrease in thickness as a result of corrosion in freshwater: (a) mean values; (b) maximum values.

3. In the light of experience gained over the past few decades, it would seem that corrosion can limit the durability of sheet pile walls (DIN EN 1993-5, Section 4) to a useful life of 20–30 years, especially in the seawater of the North Sea and the Baltic Sea (Alberts 2001). Once the steel is rusted through, the soil behind the wall can be washed out, thus leaving voids that can suddenly collapse, causing subsidence at the surface. This is linked with considerable safety risks and restrictions on port operations. U-sections frequently rust through in the middle of the flange of the front pile, Z-sections frequently at the junction between flange and front web (see Figure 8.24).

The basis for assessing the durability of sheet pile structures (estimating the useful life until the first occurrence of rusting) are the maximum values for wall thickness losses in accordance with Figure 8.25b for freshwater and Figure 8.26b for seawater.

Unprotected sheet pile walls should be planned and designed taking into account wall thickness losses as per the regression curves shown in Figures 8.25b and 8.26b unless other values are available for the location. A decision must be made as to whether the design is to be based on the mean or maximum regression curve values. To avoid uneconomic designs, the recommendation is to use the corrosion values above the regression curve only if local experience suggests this is necessary.

For older, unprotected sheet pile walls, stability assessments should always be based on ultrasound measurements of the wall thickness in order to assess the local corrosion influences. Information on performing, analysing and troubleshooting such ultrasound measurements can be found in Alberts and Schuppener (1991) and Alberts and Heeling (1996).

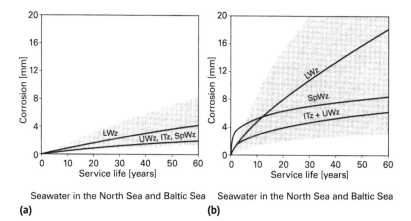

Figure 8.26 Decrease in thickness as a result of corrosion in seawater:
(a) mean values; (b) maximum values.

8.1.10.3 Design values for loss of wall thickness in various media

The following empirical figures for wall thickness loss through corrosion in different media can be used as design values for new sheet pile structures and for checking existing sheet pile walls unless other values are available for the location.

1. Freshwater
   The design values for the reduction in wall thickness for sheet pile walls in freshwater can be taken from the regression curves of Figure 8.25, depending on the age of the structure. The area shaded grey represents the scatter of those structures investigated.
2. Seawater of the North and Baltic Seas
   The design values for the reduction in wall thickness for sheet pile walls in the seawater of the North and Baltic Seas can be taken from Figure 8.26, depending on the age of the structure. The area shaded grey represents the scatter. The measured values in Figures 8.25 and 8.26 are comparable with those from international literature on the subject (Hein 1990). They are, however, on the whole somewhat higher.
3. Brackish water
   Brackish water zones are a mixture of freshwater from inland areas and salty seawater. The design values for the reduction in wall thickness can be estimated on the basis of the figures for seawater (Figure 8.26) and freshwater (Figure 8.25), depending on the location of the structure within the briny water zone.
4. Corrosion above the splash zone (atmospheric corrosion) The rate of corrosion is low above the splash zone on waterfront structures, amounting to a corrosion rate of about 0.01 mm/a (C1 to C5 as per DIN EN ISO 12 944). Higher values can be expected where de-icing salts are in use or in areas involving the storage and handling of substances known to attack steel.
5. Corrosion in the soil, microbiologically induced corrosion (MIC)
   The aggressiveness of soils and groundwater can be roughly assessed using DIN 50929. In soils the corrosion can be further aggravated by the activity of bacteria that attack steel (microbiologically induced corrosion, MIC) (Binder and Graff 1995, Graff et al. 2000). Microbiologically induced corrosion can be expected where organic substances reach

the rear face of the sheet pile wall, either through circulating water (e. g. in the vicinity of landfill sites for domestic waste or in wastewater percolation areas) or through soils with a high organic content. In such cases, high corrosion rates and, typically, non-uniform degradation can be expected. Aggressive soils such as humus, carbonaceous soils, waste washings and slag must, as a matter of principle, be avoided whenever possible when backfilling sheet pile structures. This also applies to aggressive water.
6. If steel sheet piling is embedded in non-aggressive, natural soil, then the expected corrosion rate on both sides of the piling is very low (0.01 mm/a). Corrosion rates in the same order of magnitude are to be expected when the sand filling behind a sheet pile wall is such that the troughs of the piling are also completely embedded.

8.1.10.4 Corrosion protection

The corrosion protection required for a specific sheet pile structure is ascertained by assessing the following boundary conditions and usage requirements:

- Intended purpose and design life of the structure.
- General and specific corrosion loads at the location of the structure.
- Experience with corrosion phenomena at adjacent structures.
- Options for configuring and designing the structure to inhibit corrosion.
- Costs of premature repairs to unprotected sheet pile walls, e. g. adding plates (Binder 2001).

Subsequent protection measures or complete renewal are extremely difficult, so particular care must be taken when planning and applying protective systems. Additional, specific protection measures may be required depending on the nature and intensity of the corrosion and also on the demands placed on the corrosion protection system.

In essence, we distinguish between the following methods of corrosion protection.

8.1.10.4.1 Corrosion protection with coatings

Previous experience shows that a coating prolongs the design life of a sheet pile structure by up to 25 years (DIN EN ISO 12 944).

The prerequisite is that the surfaces to be protected are blast-cleaned to standard Sa 2 1/2 before applying the coating and that the coating system selected is suitable for the specific application.

Guidance on selecting coating systems depending on the local conditions is published in the BAW's list of recommended systems. BAW-tested coating materials can be found in the list of tested systems for corrosivity categories Im1 or Im2/Im3 (BAW).

The list also specifies the compatibility of the coating system with cathodic corrosion protection (CCP) according to Section 8.1.10.4.2. The laboratory tests required for inclusion in this list are specified in the BAW guideline on testing coating systems for corrosion protection in steel hydraulic engineering (BAW).

Regulations for preparing the surfaces, carrying out and monitoring the coating work and other contractual provisions can be found in ZTV-W, LB 218 (additional technical contractual conditions for corrosion protection in steel hydraulic engineering [service phase 218], 2009 ed., published by the Federal Ministry for Transport, Building & Urban Development).

Information on repairing coating systems can be found in the current research reports published by the BAW.

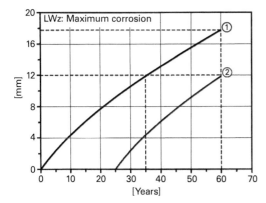

Figure 8.27 Corrosion in seawater in the low water zone, maximum corrosion rate; curve 1: uncoated steel, curve 2: coated steel.

When it comes to the *serviceability* of sheet pile structures, it is the rate of corrosion in the low water zone (LWZ) that governs. Figure 8.26 (maximum values) shows that a 12 mm thick sheet pile in seawater without a coating is completely rusted through after 35 years (Figure 8.27, curve 1). If a coating that is assumed to last 25 years is applied, the first cases of rusting will appear after a working life of 60 years (Figure 8.27, curve 2).

When it comes to the *load-bearing capacity*, the mean corrosion rate should be selected. Maximum bending stresses in sheet pile structures in ports and harbours are mostly found in the permanent immersion zone (PIZ). Therefore, when assessing the effect of corrosion on the stability of sheet piling structures, the (uniform) mean corrosion rate according to Figure 8.26a should be used. Accordingly, a structure suffers 2.0 mm of corrosion during a 60-year working life. Coating reduces the rusting to 1.4 mm over the same period.

In the low water zone (LWZ), according to Figure 8.28b, an uncoated structure suffers 4.0 mm of corrosion, a coated structure 2.6 mm. Coatings should be applied to steel sheet piles completely at the works so that only transport and installation damage needs to be repaired on site.

The choice of coating system should make allowance for the possibility that cathodic corrosion protection (CCP) might be installed subsequently. Attention must be paid to the compatibility of the coating materials with the CCP. More information on BAW-approved systems can be found at (www.baw.de). This list is based on laboratory tests, so local experience should be taken into account. Where coatings have to protect steel sheet piling against sand abrasion, the abrasion value A_w of the coating material governs (RPB 2011, Guidelines for the Testing of Coating Systems for the Corrosion Protection of Hydraulic Steel Structures). Fenders must be installed to protect the coating from vessel impacts.

8.1.10.4.2 Cathodic corrosion protection (CCP)

The corrosion of steel sheet piling below the waterline can be substantially eliminated by cathodic corrosion protection (CCP) with stray current or sacrificial anodes. Additional coating or partial coating is an economic measure and usually indispensable in the interests of good current distribution and lower power requirements.

CCP systems are ideal for protecting those sections of sheet piling where renewing any protective coatings or repairing corrosion damage on unprotected sheet piling is impossible or at best complicated and costly, e. g. a tidal low water zone.

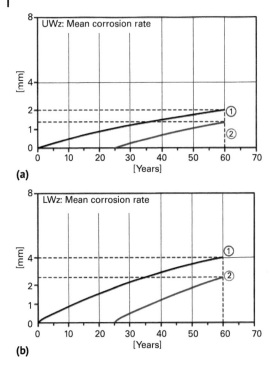

Figure 8.28 Reduction in wall thickness in seawater: (a) permanent immersion zone (PIZ); (b) low water zone (LWZ); curve 1: uncoated steel, curve 2: coated steel.

Sheet pile structures with CCP require special constructional measures (Wirsbitzki 1981). Therefore, any potential subsequent expansion of the CCP system must be taken into account during the planning phase of the sheet pile structure. Combined CCP and coating also guarantees permanent protection against the soil behind sheet pile structures.

When operating a CCP system with a parasitic current, it is important to make sure that the potential permitted for any additional coating is maintained.

8.1.10.4.3 Alloying additives for steel sheet pile steels

Differences in corrosion behaviour have not been observed for sheet pile steels according to DIN EN 10248 and the steels given in DIN EN 10025 (structural steels), DIN EN 10028 and DIN EN 10113 (higher-strength fine-grained structural steels). Furthermore, the much publicised positive effect of adding small amounts of copper together with nickel and chromium have not been confirmed to date.

8.1.10.4.4 Corrosion protection by oversizing

The useful life of a steel sheet pile structure can be extended by oversizing the thickness of the pile sections. In these cases, when analysing the ultimate limit state (ULS) and serviceability limit state (SLS), the mean wall thickness loss values as per Figures 8.25 and 8.26 are to be assumed unless experience of local corrosion rates is available. Unless the client specifies otherwise, the ULS analyses for wall thicknesses expected at the end of the wall's service life which take into account corrosion can be assigned to design situation DS-A. When checking corrosion, the maximum wall thickness loss values as per Figures 8.25 and 8.26 can be assumed unless values based on local experience are available.

8.1.10.4.5 Further information on corrosion protection

As far as corrosion attack is concerned, structures in which the back of the sheet piling is only partially backfilled, or not at all, are at a disadvantage.

Surface water should be collected and drained clear of the rear face of a sheet pile wall. This particularly applies to quays handling aggressive substances (fertilisers, cereals, salts, etc.).

Free-standing, open piles are exposed to corrosion over their whole periphery, whereas essentially only the outer surfaces of closed piles, e. g. box piles, are exposed. Experience has shown that the inner surfaces of closed piles are protected against corrosion when the pile is filled with sand.

In the case of free-standing sheet piling, e. g. flood defence walls, the coating of the sheet piling must be extended deep enough into the ground. Subsequent settlement of the soil must be taken into account.

A sand bed is recommended around round steel tie rods; inhomogeneous fill material must be avoided. A coating on or other protection for the tie rod is essentially superfluous as the design according to Section 9.1.6.7 includes a reserve for corrosion. Once a coating has been applied, it must not be damaged, because such damage encourages pitting. The connections of round steel tie rods must be carefully sealed.

When backfilling coated sheet piling structures, damage to the coating cannot be completely avoided. However, this risk can be reduced to a negligible level in most cases by using backfill material free from stones.

Ship berths should always be equipped with timber rubbing strips, fender piles and permanent fender systems to protect sheet piling against constant chafing and scuffing caused by pontoons, ships or their fenders. This protection should ensure that there is never any direct contact between the surface of the sheet pile wall and a ship or pontoon. Otherwise, greater decreases in thickness and further corrosion must be expected – clearly exceeding the values given in Figures 8.25 and 8.26.

Steel in concrete is a particularly active cathode and can therefore increase corrosion. Increased corrosion should, therefore, always be expected at the transition from steel to reinforced concrete (e. g. at concrete capping beams); see Heiß et al. (1992) for details. Section 7.2.11 is to be taken into account for steel capping beams.

8.1.11 Risk of sand grinding on sheet piling

When steel sheet piling is used, it must be coated with a system that can permanently withstand sand grinding at the location. Assessment of the necessary abrasion resistance of the coating is performed in accordance with RPB (2011) (Guidelines for the Testing of Coating Systems for the Corrosion Protection of Hydraulic Steel Structures).

8.2 Design of sheet pile walls

8.2.1 General

DIN EN 1997, part 1, Section 2.4.7.3.4, prescribes three possible design approaches for analysing the load-bearing capacity. This takes the form of combining the actions A, resistances R and materials M (geotechnical parameters) and using the associated partial safety factors given in annex A of DIN EN 1997. With one exception, applications in Germany are based on design approach 2.

To distinguish this from the variants also permissible under design approach 2, in which the partial safety factors are applied to the actions, the German design approach is denoted 2*.

The load-bearing capacity analyses for retaining structures include the following limit states:

- STR (structure failure, limit state for failure of structure or structural component),
- GEO-2 (geotechnical failure, limit state for failure of the subsoil), and
- GEO-3 (analysis of overall stability).

The analysis of the sheet pile component under bending moments and normal forces forms part of the STR limit state. Load-bearing capacity analyses for sheet pile made from

- Reinforced concrete are carried out to DIN EN 1992.
- Timber are carried out to DIN EN 1995.
- Steel are carried out to DIN EN 1993-5, taking into account Section 8.2.7.

DIN EN 1993-5 also permits plastic-plastic design and other methods for steel sheet pile structures. Such a design approach, which exploits both the plastic cross-section design and the plastic system load-bearing capacity (in the case of statically indeterminate systems) at the ultimate limit state, can also be appropriate for waterfront structures in certain cases. However, the elastic-elastic and elastic-plastic design approaches explained below are normally used. With regard to the plastic-plastic design of waterfront structures, there is a lack of sound experience relating to the active earth pressure distribution due to the redistribution of internal forces when utilising the system reserves to the full. The classic active earth pressure distribution and the earth pressure diagrams given in this book cannot be used as a basis for design using a plastic-plastic approach.

The following analyses are included in the GEO-2 limit state for retaining walls:

- Ground failure in the passive earth pressure zone due to horizontal action effects from soil support $B_{h,d}$ (analysis format $B_{h,d} \leq E_{ph,d}$, see DIN 1054, "note to 9.7.4").
- Axial sinking of the sheet pile wall in the subsoil due to vertical action effects $\Sigma V_{i,d}$ (analysis format $\Sigma V_{i,d} \leq \Sigma R_{i,d}$, see Section 8.2.5.6).
- Ground bearing capacity.
- Stability of the lower failure plane for anchored retaining walls, see Section 9.3.

The analysis of safety against slope failure forms part of limit state GEO-3. DIN 1054, "note to 9.7.2", describes the minimum conditions for which the analysis of slope failure is to be carried out for retaining walls. The analyses of the horizontal base support for a retaining wall in the soil and the ability to carry the vertical components of mobilised passive earth pressure are used to calculate the required embedment depth of the retaining wall in the subsoil. As part of both of these analyses, redistribution of the earth pressure on the load side of the retaining wall can be considered as per Section 8.2.3.1.

In the analysis of a sufficient horizontal base support for the retaining wall, the partial safety factor $\gamma_{R,e}$ may not be reduced for passive earth pressure. Reduced partial safety factors are discussed further in Sections 8.2.1.2 and 8.2.1.3.

The serviceability limit state (SLS) embraces conditions that will cause the structure to become unusable but without it losing its load-bearing capacity. In the case of waterfront structures, this analysis must be carried out such that no wall deformations damage the structure or the surroundings. Further information on SLS analyses can be found in DIN 1054, "note to 9.8".

8.2.1.1 Partial safety factors for loads and resistances

When designing sheet pile walls, as well as anchor walls and anchor plates for round steel tie rods, the following partial safety factors govern for STR limit state analyses:

- γ_G and γ_Q for actions according to Table 1.1.
- $\gamma_{R,e}$ and $\gamma_{R,h}$ for resistances according to Table 1.2.
- $\gamma_{R,e,red}$ for passive earth pressure as per Section 8.2.1.2.
- $\gamma_{G,red}$ for resultant water pressure according to Section 8.2.1.3.

For more information on applying the reduced partial safety factors $\gamma_{R,e,red}$ and $\gamma_{G,red}$, please see Sections 8.2.1.2 and 8.2.1.3.

8.2.1.2 Determining the design values for bending moments

As a rule, considerable displacement of a structure is required to mobilise the full passive earth pressure in front of a waterfront structure. The amount of displacement is chiefly dependent on the embedment depth, the in situ density of the soil and the type of movement. DIN 4085 provides advice on the horizontal active earth pressure E'_{pgh} achieved depending on the amount of displacement s.

Redistribution of the active earth pressure according to Section 8.2.3.1 is carried out down to the design bottom.

When calculating bending moments, a reduced partial safety factor $\gamma_{R,e,red}$ may be used for reducing earth pressure as per Table 8.9, if there is non-cohesive soil beneath the design bottom with at least medium strength ($q_c \geq 7.5\,\text{MN/m}^2$) or cohesive soil with at least stiff consistency.

If there is initially soil with a soft consistency or low strength beneath the design depth (intended depth plus dredging tolerance, precautionary prior dredging, scour allowance, etc.), this cannot be used as horizontal support for the wall. Such strata can only be applied as a surcharge p_0 on the new design bottom, which is situated at the level of the beginning of the load-bearing, i.e. at least very stiff or firm, soil.

The reduced partial safety factor $\gamma_{R,e,red}$ for calculating bending moments may only be applied in the load-bearing strata (Figure 8.29).

If the soil in the passive earth pressure zone is of low strength or soft consistency, then the action effects must be calculated using partial safety factors $\gamma_{R,e}$ that have not been reduced.

Table 8.9 Reduced partial safety factor $\gamma_{R,e,red}$ for passive earth pressure for determining bending moments.

GEO-2	DS-P	DS-T	DS-A
$\gamma_{R,e,red}$	1.20	1.15	1.10

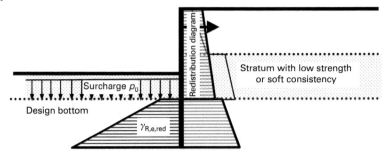

Figure 8.29 Loading diagram for determining bending moments with reduced partial safety factors in soils with low strength or soft consistency.

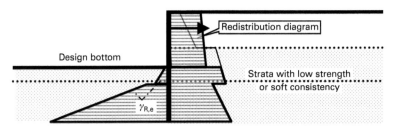

Figure 8.30 Loading diagram for determining bending moments without reduced partial safety factors in soils with low strength or soft consistency below the design bottom.

If the reduced partial safety factor $\gamma_{R,e,red}$ is not used, the active earth pressure redistribution diagram corresponding to Figure 8.30 need only be continued down to the bottom of the watercourse/excavation.

Where the stipulated boundary conditions for using the reduced partial safety factor $\gamma_{R,e,red}$ are met, the degree of fixity for the full utilisation of the embedment depth/pile length calculated with non-reduced partial safety factors can be used to calculate the bending moments. The internal forces (moments and shear forces) calculated with $\gamma_{R,e,red}$ play an important role in the sheet pile analysis.

For structures that fall within the remit of this book, anchor forces and the required embedment depth of the wall may only be determined using the full partial safety factor $\gamma_{R,e}$.

The redistribution of the active earth pressure according to Section 8.2.3 is always carried out down as far as the design bottom defined beforehand.

8.2.1.3 Partial safety factor for water pressure

Actions due to water pressure are to be calculated according to Sections 3.3.2, 3.3.4 and 3.5.7.

When the boundary conditions given below apply, it is possible to reduce the partial safety factor γ_G used for calculating the design values of loads due to water pressure. The partial safety factors $\gamma_{G,red}$ are given in Table 8.10.

Reducing the partial safety factors for actions due to water pressure is only permitted when at least one of the following three conditions is satisfied:

- Verified measurements are available regarding the positional and chronological relationships between groundwater and outer water levels to guarantee the water pressure used

Table 8.10 Reduced partial safety factor $\gamma_{G,red}$ for calculating the design value of water pressure actions.

STR	DS-P	DS-T	DS-A
$\gamma_{G,red}$	1.20	1.10	1.00

in the calculations and also to serve as a basis for assigning to design situations DS-P, DS-T and DS-A.
- Numerical models of bandwidth and frequency of occurrence of the true water levels, and hence water pressures, are analysed and lie on the safe side. These forecasts are to be checked through observations, beginning with the construction of the sheet pile wall. Where the measurements are larger than those predicted, the values on which the design was based must be guaranteed by appropriate measures such as drainage, pumping systems, etc. These must be monitored permanently.
- There are geometric boundary conditions present that limit the water level to a maximum value – as is the case of where the top edges of sheet pile walls designed as flood defence walls that limit the depth of the floodwaters, for example. Drainage systems installed behind the sheet pile wall do not represent a clear geometrical limit to the water level in the meaning of this stipulation.

8.2.2 Free-standing/cantilever sheet pile walls

8.2.2.1 General
Contingent on the bending resistance of the wall, sheet pile walls fully fixed in the ground and without an anchorage can be economical if there is comparatively little difference in ground levels. Such an arrangement can also be used for larger differences in level if the installation of an anchor or other pile head support would be very involved and if relatively large displacements of the pile head can be regarded as harmless in terms of serviceability.

8.2.2.2 Design, calculations and construction
In order to attain the necessary stability of free-standing sheet pile walls (i.e. without anchors), their design, calculations and construction must satisfy the following requirements:

- Ascertain all actions as accurately as possible, e.g. including the compacted earth pressure in backfill according to DIN 4085. This applies in particular to those actions applied to the upper section of the sheet pile because these can substantially affect the design bending moment and the embedment depth required.
- It must be possible to establish an exact assignment to design situations DS-P, DS-T and DS-A taking into account, for example, unusually deep scouring and unusual excess water pressures.
- In the calculations, the design bottom level in the passive earth pressure zone must include the additional depths required to allow for any scouring and dredging work.
- The structural calculations may be carried out in accordance with Blum (1931), with the active earth pressure applied in a classic distribution.
- The theoretical embedment depth required, taking into account Sections 8.2.9 and 8.2.10, must be reached during construction.

- In the serviceability state (i. e. in an analysis with characteristic actions), the deformation of the wall is to be included in addition to the internal forces. The deformation values that occur must be investigated to ensure their compatibility with the structure and the subsoil, e. g. with respect to the formation of gaps on the active earth pressure side in cohesive soils, which could fill up with water. The compatibility of the deformations must also be checked with all other aspects of the project. This approach is particularly important for larger differences in ground levels.
- Influences from wall deformations can be compensated for by driving at a suitable angle in order to avoid an unattractive overhang at the top of the wall.
- The top of a free-standing sheet pile wall, at least in a permanent structure, should be provided with a capping beam or waling of steel or reinforced concrete to distribute the actions and prevent non-uniform deformations as far as possible.

8.2.3 Design of sheet pile walls with fixity in the ground and a single row of anchors

8.2.3.1 Active earth pressure

Calculations for anchored retaining structures in ports and harbours can be carried out using the active earth pressure. Under certain conditions the minimum earth pressure as per DIN 4085 is to be used in cohesive strata.

The resulting earth pressure force calculated with the classic distribution over the height H_E – which, if necessary, is increased when considering the minimum earth pressure in cohesive strata – may be redistributed over the height H_E for the ULS and SLS analyses. It must be redistributed for determining the anchoring force.

The ratio of the position of the anchor head a to the redistribution depth H_E serves as a criterion for deciding the case when selecting the redistribution diagrams (Section 8.2.3.2). The distances H_E and a and the classic earth pressure distribution $e_{agh,k}$ due to soil dead load or minimum earth pressure from a variable ground surcharge of up to $10\,kN/m^2$ over a large area are defined in Figures 8.31 and 8.32. Other actions (block loads or further ground surcharges over a large area) are to be redistributed while taking into account the actual load-bearing behaviour of the wall. In particular, it should be noted that stiffer building components attract loads.

Figure 8.32 shows an example of a superstructure with depth $H_Ü$ and a reinforced concrete slab to shield the earth pressure.

The following definitions apply:

H_G total difference in ground levels.
$H_Ü$ depth of superstructure from ground level to underside of shielding slab.
H_E depth of earth pressure redistribution zone above design bottom (for a superstructure with a shielding slab, depth H_E begins at the underside of the slab).
a distance of anchor head A from top edge of redistribution depth H_E.

Beneath the design bottom, the non-redistributed active earth pressure is applied to the actions side.

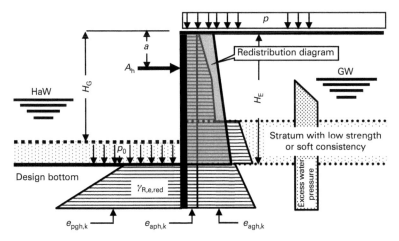

Figure 8.31 Example 1: redistribution depth H_E and anchor head position a for calculating bending moments with $\gamma_{R,e,red}$.

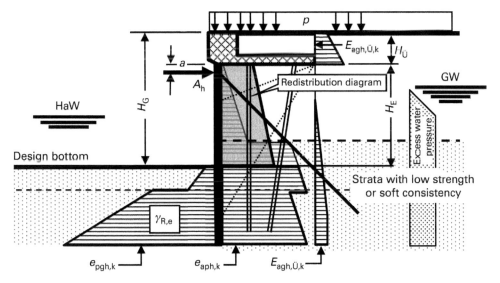

Figure 8.32 Example 2: Redistribution depth H_E and anchor head position a for calculating bending moments with $\gamma_{R,e}$.

8.2.3.2 Earth pressure redistribution

The earth pressure redistribution is selected depending on the method of construction:

- Trenching in front of wall (cases 1–3, Figure 8.33).
- Backfilling behind wall (cases 4–6, Figure 8.34).

We distinguish here between three ranges for anchor head distance a:

- $0 \leq a \leq 0.1 \cdot H_E$
- $0.1 \cdot H_E < a \leq 0.2 \cdot H_E$
- $0.2 \cdot H_E < a \leq 0.3 \cdot H_E$

8 Sheet pile walls

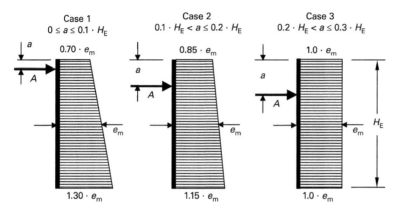

Figure 8.33 Earth pressure redistribution for the "trenching in front of the wall" method of construction.

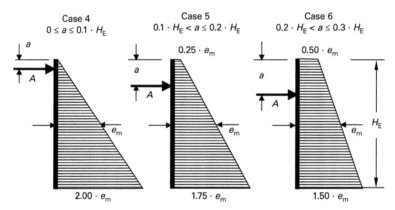

Figure 8.34 Earth pressure redistribution for the "backfilling behind the wall" method of construction.

Besides the designations shown in Figures 8.31 and 8.32, in Figures 8.33 and 8.34 the magnitude of the mean value e_m of the earth pressure distribution over the redistribution depth H_E is as follows:

$$e_m = \frac{E_{ah,k}}{H_E}$$

The loading diagrams of Figures 8.33 and 8.34 include all the anchor head positions a in the range $a \leq 0.30 \cdot H_E$. These redistribution diagrams do not apply to anchors at lower levels, and in such cases, the appropriate earth pressure diagrams must be determined separately.

If the ground surface is a short distance below the anchor, the earth pressure may be redistributed according to the value $a = 0$.

The loading diagrams for cases 1–3 are only valid on the condition that the earth pressure can redistribute to the stiffer support areas as a result of adequate wall deformation. A "ver-

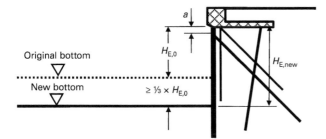

Figure 8.35 Additional excavation depth required for earth pressure redistribution according to the "trenching in front of wall" method of construction.

tical earth pressure vault" thus forms between anchor and soil support. Consequently, cases 1–3 may not be used when:

- The sheet pile wall is backfilled to a large extent between the bottom of the watercourse and the anchorage, and subsequent excavations in front of the wall are not deep enough for adequate additional deflection (guide value for adequate deflection is an excavation depth of approximately 1/3× the redistribution depth $H_{E,0}$ of the original system corresponding to Figure 8.35).
- There is cohesive soil behind the sheet pile wall that is not yet sufficiently consolidated.
- A retaining wall with increasing flexural stiffness does not exhibit the wall deflections necessary for the formation of a horizontal arch, as is the case, for example, with reinforced concrete diaphragm walls. (In this situation, the designer must check whether the displacement of the support at the base as a result of mobilisation of the passive earth pressure is sufficient for an earth pressure redistribution according to the "trenching" method cases 1–3.)

If loading diagrams cases 1–3 are not permitted for the above reasons, then it is possible to use cases 4–6 belonging to the a/H_E value for the "backfilling behind wall" method of construction.

8.2.3.3 Passive earth pressure

When designing a sheet pile wall using the Blum method (Blum 1931), the anticipated soil reaction is entered into the calculation with a linear increase opposite to the true progression. At the same time, and in the case of retaining walls with fixity, an equivalent force C is required to maintain equilibrium.

Here, the characteristic soil support $B_{h,k}$ required for determining the embedment length is formed by the mobilised passive earth pressure $E_{ph,mob}$, which must exhibit a progression not unlike that of the characteristic passive earth pressure $E_{ph,k}$ and may not be redistributed.

Further information can be found in Section 8.2.5.5.

8.2.3.4 Bedding

The soil support to a sheet pile wall can also be analysed using a horizontal bedding; see Laumans (1977). It should be noted that the soil reaction stress $\sigma_{h,k}$ at the design bottom as a result of characteristic actions may not be greater than the characteristic – i. e. maximum possible – passive earth pressure stress $e_{ph,k}$ according to DIN 1054, Eqs. (A9.3) and (A9.4). More information can be found in the 5th edition of the EAB (2012).

8.2.4 Design of sheet pile walls with a double row of anchors

In contrast to Section 9.2.4, which deals with issues regarding auxiliary anchors, this section covers sheet pile structures with a double row of anchors, i.e. anchors at two different levels.

The total actions acting on the sheet pile wall due to earth and water pressures and variable loads are assigned to the two anchor positions A_1 and A_2 as well as soil support B. Owing to the distribution of the actions over the given structural system, the majority of the total anchorage force is carried by the lower anchor A_2.

Where round steel tie bars are used as anchors, it is advisable to connect both anchor positions to a common anchor wall at the same level and apply the direction of the resultant of anchor forces A_1 and A_2 as the anchor direction when checking stability at the lower failure plane according to Section 9.3.

Anchors not connected to a common anchor wall (e.g. grouted anchors to DIN EN 1537) require the two rows of anchors to be analysed independently when checking stability.

8.2.4.1 Active and passive earth pressures

Take into account the active and passive earth pressures in the same way as for a wall with a single row of anchors.

8.2.4.2 Loading diagrams

The loading diagrams shown in Section 8.2.3 for a wall with a single row of anchors are valid for determining the internal forces, support reactions and embedment length of a sheet pile wall with double anchors. Here, the level of anchor head a required for determining the a/H_E value and for specifying the loading diagram is taken to be the average level between the two anchor positions A_1 and A_2. The earth pressure is then redistributed over depth H_E down to the design or model bottom in a similar fashion to a sheet pile wall with one row of anchor.

8.2.4.3 Considering deformations due to previous excavations

As earlier deflections of sheet pile walls due to slippage of the soil on the earth pressure failure plane can be only partly reversed, the effects of temporary construction conditions on the stresses in the final condition must then be taken into account when they are critical for verifying serviceability. This might be the case, for example, when considering the wall deflection at the level of anchor A_2, which for a sheet pile wall temporarily anchored only at point A_1 should be considered as a yielding support for the structural system in the final condition.

8.2.4.4 Bedding

As for a sheet pile wall with a single row of anchors, a wall with two rows can be designed using horizontal bedding forces as a soil support (Section 8.2.3.4).

8.2.4.5 Comparative calculations

As a comparison, a calculation according to Section 9.2.4 must be carried out when designing the top of the wall and positioning upper anchor A_1. If this calculation results in higher stresses, they govern the design.

8.2.5 Applying the angle of earth pressure and the analysis in the vertical direction

The magnitude of the earth pressure angle selected, or permissible, depends on the largest angle of internal friction between building material and subsoil (angle of wall friction) that is physically possible, the equilibrium conditions and the relative displacements of the sheet pile wall with respect to the soil.

The earth pressure angle has an influence on the analysis of the wall in the vertical direction. The following equilibrium and limit state conditions must be satisfied:

- Analysis of vertical component of mobilised passive earth pressure as per Section 8.2.5.5.
- Failure due to vertical movement as per Section 8.2.5.6.

In the case of untreated wall surfaces, the earth pressure angle to be applied depends on the properties of the wall. The following distinctions must be made:

- The back of a wall is classed as "joggled" when its form gives it such a large surface area that it is not the wall friction acting directly between soil and wall material that is critical, but rather the friction on a straight rupture surface in the soil that is only partly in contact with the wall. This is usually the case, for example, with walls constructed from bored cast-in-place piles. Diaphragm walls made from sheet piles or soldier piles lowered into a hardened cement-bentonite suspension can be classed as joggled. This also applies to driven, vibrated or pressed sheet pile walls.
- Untreated steel, concrete and timber surfaces can generally be regarded as "coarse", especially the surfaces of soldier piles and infill panels.
- The surface of a diaphragm wall can be classed as "less coarse" when there is only marginal filter cake formation, e. g. diaphragm walls in cohesive soil. Experience has shown that this also applies to diaphragm walls in non-cohesive soils if the length of time in which the trenches supported by the suspension is kept short (assuming excavation in accordance with the general rules for such work).
- The back of a wall is always classed as "smooth" when the clay content and consistency of the in situ soil are such that activation of a significant amount of friction cannot be expected.

The earth pressure angles can be applied depending on the wall properties as per Sections 8.2.5.1 and 8.2.5.2.

With treated surfaces, soft subsoil that can form a lubricating layer on the surfaces of sections being driven and sections installed with the help of water-jetting, the friction can be reduced to such an extent that the angle of wall friction as per Sections 8.2.5.1 and 8.2.5.2 cannot be established. In these cases, the angle of wall friction must be restricted to a maximum of half the angle of internal friction, $|\delta_k| \leq 0{,}5 \cdot |\varphi'_k|$. Alternatively, a higher wall friction angle will have to be verified by a geotechnical expert.

The influence of reinforcement at the toe of the pile on the angle of wall friction must be assessed by a geotechnical expert.

8.2.5.1 Angle of inclination $\delta_{a,k}$ of active earth pressure

Active earth pressure is typically calculated assuming straight failure planes. The angle of inclination $\delta_{a,k}$ of the earth pressure can be used depending on the surface properties of the wall up to the following limits:

Wall surface property:

Joggled wall	$\|\delta_{a,k}\| \leq (2/3) \cdot \varphi'_k$
Coarse wall	$\|\delta_{a,k}\| \leq (2/3) \cdot \varphi'_k$
Less coarse wall	$\|\delta_{a,k}\| \leq (1/2) \cdot \varphi'_k$
Smooth wall	$\delta_{a,k} = 0.$

8.2.5.2 Angle of inclination $\delta_{p,k}$ of passive earth pressure

Passive earth pressure is usually calculated for curved failure planes. In doing so, the angle of inclination $\delta_{p,k}$ of the passive earth pressure can be applied within the following limits:

$$-\varphi'_k \leq \delta_{p,k} \leq +\varphi'_k$$

To simplify the calculation, the approach using straight failure planes is permissible between the following limits:

$$-\frac{2}{3} \cdot \varphi'_k \leq \delta_{p,k} \leq +\frac{2}{3} \cdot \varphi'_k$$

provided that the angle of internal friction φ_k, angle of wall inclination α_k, angle of ground inclination β_k and inclination angle $\delta_{p,k}$ lie within the following limits:

$\varphi'_k \leq 35°$
$\alpha_k \leq 0°$ (sign definitions as per Figure 8.36)
$\beta_k \geq 0°$ for $\delta_{p,k} \geq 0°$ or $\beta_k \leq 0°$ for $\delta_{p,k} \leq 0°$

For these conditions, there is no significant difference between the K_{ph} values for the respective limit min $\delta_{p,k}$ from the methods with straight (min $\delta_{p;k} = -2/3\varphi'_k$) and curved (min $\delta_{p;k} = -\varphi'_k$) failure planes.

Depending on the surface properties, the following additional limits apply to the passive earth pressure angle of inclination:

Wall surface property:

Joggled wall	$\|\delta_{p,k}\| \leq \varphi'_k$
Coarse wall	$\|\delta_{p,k}\| \leq \varphi'_k - 2.5° \leq 30°$
Less coarse wall	$\|\delta_{p,k}\| \leq (1/2) \cdot \varphi'_k$
Smooth wall	$\delta_{p,k} = 0°.$

8.2.5.3 Angle of inclination $\delta_{C,k}$ of equivalent force C_k

When designing walls with full or partial fixity in the ground according to the Blum approach (Blum 1931), the soil reaction below the theoretical base TF on the actions side is used to resist the equivalent force C. The extra depth required to resist this reaction force is calculated according to Section 8.2.9 as a surcharge Δt_1 on the embedment depth t_1. As part of this approach, the direction of action of equivalent force C is inclined at an angle $\delta_{C,k}$ to the horizontal.

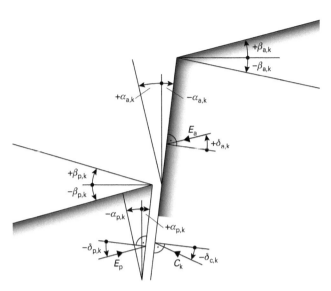

Figure 8.36 Sign definitions for angle of wall inclination α_k, angle of ground inclination β_k, angles of inclination of active $\delta_{a,k}$ and passive $\delta_{p,k}$ earth pressures and angle of inclination $\delta_{C,k}$ of equivalent force C.

The angle of inclination $\delta_{C,k}$ can be used within the following limits:

$$-\varphi'_k \leq \delta_{C,k} \leq +\frac{1}{3}\cdot\varphi'_k$$

However, depending on the surface properties of the wall, the angle cannot be larger than the limits for $\delta_{p,k}$ stipulated in Section 8.2.5.2.

8.2.5.4 Magnitude of equivalent force C

The magnitude of equivalent force C_k for fixed walls, calculated using Blum's approach for passive earth pressure, is determined using the equilibrium condition $H_k = 0$ for all characteristic actions and support reactions. For all analyses in the vertical direction, it must be remembered that this method results in an equivalent force C that is too large, because the full passive earth pressure mobilised for soil support B_k is assumed to act down to the theoretical base of the sheet pile wall (TF). When the true progression of the soil reaction B_k is taken into account, the equivalent force C is only about half its theoretical magnitude. At the same time, the associated soil support B_k is reduced by exactly this value (see Figures 8.37 and 8.40).

In order to compensate for this error, the horizontal components of the equivalent force and the soil support according to Blum ($C_{h,k}$ and $B_{h,k}$) are each reduced by $1/2C_{h,k}$ when calculating the associated vertical components.

8.2.5.5 Analysis of vertical component of mobilised passive earth pressure
1. Analysis format
   This analysis, which must be carried out for each characteristic combination of actions, ensures that the angle of inclination $\delta_{p,k}$ selected for calculating the passive earth pressure will actually be established in practice. The angle of inclination $\delta_{p,k}$ can only be applied to that negative value for which it has been proved that the downward char-

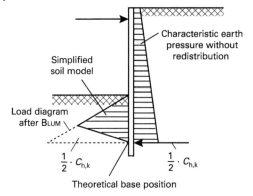

Figure 8.37 Effective portion of soil reaction when fixed in the soil according to Blum (1931).

acteristic actions $\Sigma V_{i,k}$ are greater than or equal to the upward characteristic actions $B_{v,k}$ (DIN 1054, section A9.7.8). The analysis format required for this is as follows for a simple support:

$$\sum V_{i,k} \geq |B_{v,k}| \quad \text{mit} \quad B_{v,k} = \sum B_{hi,k} \cdot \tan \delta_{pi,k}$$

and for a fixed wall:

$$\sum V_{i,k} \geq |B^*_{v,k}| \quad \text{where} \quad B^*_{v,k} = \sum B_{hi,k} \cdot \tan \delta_{pi,k} - \frac{1}{2} C_{h,k} \cdot \tan \delta_{pr,k}$$

The analysis is to be carried out with the same angles of inclination used to calculate the active and passive earth pressures beforehand.

In order to carry out this analysis, the angle of inclination $\delta_{p,k}$ must be modified, if required, until reaching the positive limit value for $\delta_{p,k}$ as per Section 8.2.5.2. This results in a significant reduction in passive earth pressure $E_{ph,k}$ and, hence, also in a greater required embedment depth t_1 for the wall.

The ground failure analysis in the passive earth pressure zone due to load $B_{h,d}$ from the soil support (see Section 8.2.1) must be carried out for the modified angle of inclination $\delta_{p,k}$.

In the case of fixed walls, for the calculation is the soil support B_k with $i = 1$ to r strata down to the depth of the theoretical base TF is to be applied in the analysis according to Figure 8.40 taken into account.

2. Vertical component $V_{Q,k}$ from variable actions

The vertical component $V_{Q,k}$ due to variable actions Q may not be used for this analysis if it does not make a significant contribution to the action effect of soil support B_k. This is the case, for instance, for actions that act directly at the top of the wall, e. g. the support reactions $F_{Qv,k}$ of the superstructure due to cranes and stacked loads and the downward vertical components $\Delta A_{Qvi,k}$ of the anchor force due to horizontal, variable actions in the area around the top of the wall or above the anchor position, e. g.

- Crane lateral impact and storm locking device.
- Line pull forces.
- Earth pressure from variable actions on the wall area above the uppermost anchor position.

3. **Vertical force components $V_{i,k}$**
   The vertical force components $V_{i,k}$ are to be used with angle of inclination δ (positive upwards, negative downwards):

$V_{G,k}$	$\Sigma F_{G,k}$ due to *constant* axial effects F
$V_{Av,k}$	$P_{v,k,min}$ due to the anchor force for anchors inclined downwards
$P_{v,k,min}$	$P_{v,k}\Delta P_{Qv,k}$ ($P_{v,k,max}$ for anchors inclined upwards or raking piles in compression)
$V_{Eav,k}$	$\Sigma(E_{ah,i,k} \tan \delta_{ai,k})$ due to earth pressure E_{ah} with $i = 1$ to r strata down to the depth of the theoretical base TF
$V_{Cv,k}$	$1/2 C_{h,k} \tan \delta_{C,k}$ due to equivalent force $C_{h,k}$ (see also Section 8.2.9)

8.2.5.6 Failure due to vertical movement

In addition to the analysis of the horizontal load-bearing capacity of the soil support and the vertical component of the mobilised passive earth pressure according to the model concept of active and passive sliding wedges of soil, an analysis must also be carried out with regard to the failure of soil-supported walls due to vertical movement as per Section 8.2.1.

This analysis ensures that the wall has adequate stability to resist sinking as a result of downward vertical actions.

In walls made from steel beam sections (I-sections) and walls made from a combination of steel sections and sheet piles (combined walls), the analysis of the axial load-bearing capacity of the load-bearing elements (open steel tubular sections, box sections, single and double steel beam sections) is carried out according to *EA-Pfähle* (2012) in conjunction with the *Annual Technical Report EAP, TJB* (2014). See Becker and Lüking (2015) for the derivation of the analysis concept. In this concept, the essential features are the geometry-based idealisation of prestressing effects (plug formation) and their occurrence depending on the method of installation. The empirical values for external skin friction and end bearing pressure on plugs can be found in Becker and Lüking (2015).

For simplicity, when assessing the vertical load-bearing capacity of load-bearing elements in waterfront structures with boundary conditions according to Section 8.2.5.6 (6), it is possible to use model 2, i.e. substituting internal skin friction for a possible plug formation.

1. **Loading diagrams** One of the loading diagrams shown in Figure 8.38 (system sketch 1 or 2) may be chosen for the analysis. Taking into account Section 8.2.5.6 (4), either the vertical component of the soil support $B^*_{v,k}$ (Figure 8.38, system sketch 1) or the skin friction $q_{s,k}$ (Figure 8.38, system sketch 2) may be assumed to act on the passive earth pressure side.

The recommendations given in this section apply below the theoretical base (TF) for system sketch 1. However, the recommendations apply to the entire embedment depth for system sketch 2.

The effective skin surfaces should be assumed according to Figures 8.38 and 8.39.

Special note for system sketch 2: Above the theoretical base, the skin friction $q_{s,k}$ may not be assumed to act as a resisting force on the surface loaded by active earth pressure. In the case of walls constructed according to Figures 8.33 and 8.34, the deformation of the sheet pile wall leads to a higher horizontal stress state. Therefore, the skin friction according to Tables 8.11–8.14 should be doubled for such walls. As a simplification, the full periphery of the loadbearing section may be assumed from the design bottom down to the theoretical

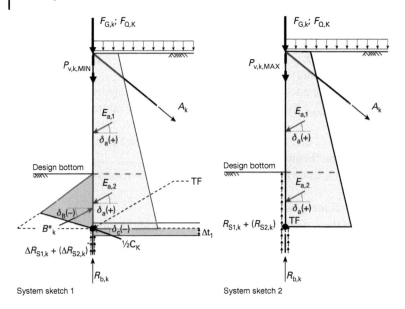

Figure 8.38 Approach for actions and resistances using the example of a wall fixed at the base.

base. However, in this case, the increased skin friction may not be assumed on the passive side. For more information, see Becker (2017).

Below the theoretical base, the peripherally developed surface may be assumed for both system sketches.

2. Analysis format When analysing the safety of soil-supported walls against failure due to vertical movements in the subsoil (DIN EN 1997-1, "note to 9.7.5"), all downward axial actions ΣV_i and axial resistances ΣR_i must be taken into account with their design values. The total load V_d may not exceed the axial resistance $\Sigma R_{i,d}$. The analysis for the limit state condition is

$$V_d = \sum V_{i,d} \leq \sum R_{i,d}$$

3. Vertical load V_d In this case, V_d is the design value of all downward axial actions on a wall or soldier pile base according to DIN 1054:2010, Section 9.7.5.

To calculate the load all downward characteristic axial part-actions are multiplied by the partial safety factors for limit state GEO-2 as per Table 1.1 for permanent (G) and variable (Q) actions for the respective design situation, separated within the combinations of actions according to cause,

$$V_{F,d} = \sum (V_{F,G,k} \cdot \gamma_G + V_{F,Q,k} \cdot \gamma_Q)$$

due to axial downward actions F

$$V_{Pv,d} = \sum (V_{Pv,G,k} \cdot \gamma_G + V_{Pv,Q,k} \cdot \gamma_Q)$$

due to anchor force component P_v

$$V_{Eav,d} = \sum (V_{Eav,G,k} \cdot \gamma_G + V_{Eav,Q,k} \cdot \gamma_Q)$$

due to the sum of stratum-by-stratum resultants E_{av} arising from the earth pressure distribution of all strata down to the depth of the theoretical base TF.

4. Design values of axial resistances $R_{i,d}$ The design values $R_{i,d}$ of upward axial resistances are calculated by dividing the characteristic value $R_{i,k}$ of the individual resistance by the partial safety factors for limit state GEO-2 valid for the respective design situation.

In a similar way to the pile design, the partial safety factors for piles are used for skin friction and end bearing pressure.

The partial safety factor for passive earth pressure $\gamma_{R,e}$ is used for friction resistances $R_{Bv,k}$ or $R^*_{Bv,k}$ and $R_{Cv,K}$ due to the characteristic horizontal components of the soil support force $B_{h,k}$ or $B^*_{h,k}$ and half the equivalent force $1/2 C_{h,k}$.

There are two options for the load-bearing capacity analysis:

a) Consideration of soil support (see Figure 8.38, system sketch 1)
   The following resistances are to be used:

   $R_{Bv,d} = (B_{h,k} - 1/2 C_{h,k}) \cdot \tan \delta_B / \gamma_{R,e}$ Wall friction resistance calculated from mobilised soil support $B_{h,k}$

   $R_{Cv,d} = 1/2 C_{h,k} \cdot \tan \delta_C / \gamma_{R,e}$ Wall friction resistance calculated from half the equivalent force $C_{h,k}$

   $R_{b,d} = R_{b,k}/\gamma_b$ Design value of pile base resistance calculated from characteristic value of pile base resistance $R_{b,k}$ for open steel tubular sections, box sections, single and double steel beam sections: according to EAP, TJB (2014) for sheet pile walls:

   $$R_{b,k} = A_W \cdot q_{b,k}$$

   $\Delta R_{s,d} = \Delta R_{s,k}/\gamma_s$ Additional skin friction calculated from wall friction for open steel tubular sections, box sections, single and double steel beam sections: according to EAP, TJB (2014):

   $$\Delta R_{s,k} = A_S \cdot q_{s,k}$$

b) Consideration of skin friction and end bearing pressure (see Figure 8.38, system sketch 2)
   The following resistances are to be used:

   $R_{b,d} = R_{b,k}/\gamma_b$ Design value of pile base resistance calculated from characteristic value of pile base resistance $R_{b,k}$ for open steel tubular sections, box sections, single and double steel beam sections, according to EAP, TJB (2014) for sheet pile walls:

   $$R_{b,k} = A_W \cdot q_{b,k}$$

   $R_{b,d} = Q_{b,k}/\gamma_b$ Design value of pile base resistance calculated from base resistance $Q_{b,k}$ when applying $Q_{b,k}$ obtained from pile loading tests.

$R_{s,d} = R_{s,k}/\gamma_s$ Skin friction calculated from wall friction for open steel tubular sections, box sections, single and double steel beam sections: according to EAP, TJB (2014) for sheet pile walls:

$$R_{s,k} = A_S \cdot q_{s,k}$$

$R_{s,d} = Q_{s,k}/\gamma_s$ Skin friction as a result of wall friction $Q_{s,k}$ when applying $Q_{s,k}$ obtained from pile loading tests.

The magnitude of the negative angle of earth pressure δ_B in the analysis of failure due to vertical movement may be taken to be $|\delta_B| \leq \varphi'_k$ depending on the roughness of the wall and is not dependent on the analysis according to Section 8.2.5.5.

The characteristic value of pile base resistance $R_{b,k}$ is calculated by multiplying the cross-sectional area of the wall section A_W by the characteristic value of resistance at the base of the wall $q_{b,k}$. The values given in Table 8.11, which are dependent on q_c, may be taken as empirical values for the characteristic base resistance.

The magnitudes of the partial safety factors γ_b and γ_s, which are independent of the design situation, depend on the calculation of the characteristic base resistance $q_{b,k}$. If $q_{b,k}$ and the wall friction $q_{s,k}$

- Are obtained from empirical values given in EA-Pfähle (2012), then $\gamma_b = \gamma_s = 1.40$.
- Are described in the geotechnical report and have been confirmed by way of pile loading tests, then $\gamma_b = \gamma_s = 1.10$.

The skin friction $R_{s,k}$ is calculated by multiplying the developed surface of the wall section A_S by the wall friction. The developed surface is assumed depending on the model chosen (see Figure 8.38). To mobilise any additional wall friction resistance required ΔR_s, it is necessary to extend the wall beyond TF.

5. Corrugated sheet pile walls – resistances to be assumed   Skin friction and base resistance should be determined on the basis of static and dynamic pile loading tests. Where results from pile loading tests are unavailable, for driven sheet pile walls it is possible to use the empirical values for skin friction and base resistance given in Tables 8.11 and 8.12 in the design. Intermediate values may be obtained by linear interpolation.

The empirical values given in Tables 8.11 and 8.12 depend on the resistance q_c determined in cone penetration tests and averaged over the depth, or the shear strength of the undrained soil. When specifying the numerical value

Table 8.11 Empirical values for characteristic pile base resistance $q_{b,k}$ and characteristic skin friction $q_{s,k}$ of sheet pile walls driven into non-cohesive soils.

Mean resistance q_c in cone penetration test [MN/m²]	Base resistance $q_{b,k}$ at ultimate limit state [MN/m²]	Skin friction $q_{s,k}$ at ultimate limit state [kN/m²]
7.5	7.5	20
15	15	40
≥ 25	20	50

Table 8.12 Empirical values for characteristic pile base resistance $q_{b,k}$ and characteristic skin friction $q_{s,k}$ of sheet pile walls driven into cohesive soils.

Shear strength $c_{u,k}$ of undrained soil [kN/m²]	Base resistance $q_{b,k}$ at ultimate limit state [MN/m²]	Skin friction $q_{s,k}$ at ultimate limit state [kN/m²]
60	—	15
100	1.00	20
150	1.75	25
≥ 250	2.50	35

- The critical zone for base resistance extends from $1 \cdot D_{eq}$ above the base of the wall to $4 \cdot D_{eq}$ below it.
- The critical zone for skin friction extends along the embedment depth of the wall (see Figure 8.39).

If the stratification of the subsoil exerts a significant influence on the CPT resistance or the undrained shear strength, then several zones should be specified for the skin friction.

The empirical values given in Tables 8.11 and 8.12 were derived from the results of pile loading tests carried out on driven sheet pile walls prior to backfilling or trenching operations.

6. Combined walls – resistances to be assumed Model 2 (see Figure 8.39) may be used for single and double steel beam sections plus tubular sections with a diameter ≥ 1400 mm. This model takes into account both the internal and external skin friction, although the external skin friction $q_{s,k}$ should be assumed to act no higher than at the design bottom.

Owing to settlement caused during installation of the piles, the internal skin friction $q_{is,k}$ should be assumed to act over max. 80% of the embedment length. The embedment length here should be restricted to the area below the design bottom even if – due to the type of installation (e.g. walls with trenching on one side) – the sections can be backfilled with soil above the design bottom. This ensures that it is possible to use the empirical values given in Tables 8.11 and 8.12. EAP, TJB (2014) should be consulted when determining the resistances for smaller diameters.

Skin friction and characteristic pile base resistance should be determined on the basis of static and dynamic pile loading tests. Where results from pile loading tests are unavailable, for open tubular steel piles, box section piles and single and double steel beam sections it is possible to use the empirical values for the characteristic pile base resistance $q_{b,k}$ and (external) skin friction $q_{s,k}$ given in Tables 8.13 and 8.14.

The empirical values given in Table 8.13 have been derived from the results of pile loading tests performed on open tubular steel piles, box section piles and single and double steel beam sections driven into non-cohesive soils.

In the case of combined walls, the volumetric earth pressure is to be applied when calculating the soil support, e.g. in accordance with Weißenbach (1985), when this is smaller than the continuous passive earth pressure for a selected embedment depth. As mentioned in Section 8.1.4.2, for simplicity, the full passive earth pressure can be assumed when using a clear bearing pile spacing of a maximum of 1.80 m and a minimum embedment depth of

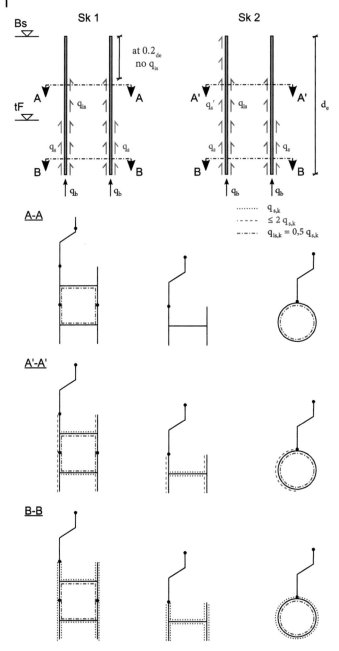

Figure 8.39 Assumptions for effective skin surface and characteristic pile base resistance according to model 2 (system sketches Sk1 and Sk2); EAP, TJB (2014).

5.00 m in the passive earth pressure zone, even when the embedment depth of the intermediate piles is less than that of the bearing piles.

Table 8.13 Empirical values for characteristic pile base resistance $q_{b,k}$ and characteristic external skin friction $q_{s,k}$ for open steel tubular sections, box sections and single and double steel beam sections in non-cohesive soils.

Mean resistance q_c in cone penetration test [MN/m²]	Base resistance $q_{b,k}$ at ultimate limit state [MN/m²]	External skin friction $q_{s,k}$ at ultimate limit state [kN/m²]
7.5	9	30
15	18	60
≥ 25	25	80

Table 8.14 Empirical values for characteristic pile base resistance $q_{b,k}$ and characteristic external skin friction $q_{s,k}$ for open steel tubular sections, box sections and single and double steel beam sections in cohesive soils.

Shear strength $c_{u,k}$ of undrained soil [kN/m²]	Base resistance $q_{b,k}$ at ultimate limit state [MN/m²]	External skin friction of single bearing element $q_{s,k}$ at ultimate limit state [kN/m²]
60	—	20
100	1.0	27
150	1.75	35
≥ 250	2.50	50

8.2.6 Taking account of unfavourable groundwater flows in the passive earth pressure zone

Designs must allow for the influence of a flow around the sheet pile wall as a result of different water levels in front of and behind the wall (Section 3.3.4).

Irrespective of this, it is necessary to verify the stability of the bottom against a hydraulic heave failure for limit state HYD.

Any risk to the stability caused by subsurface erosion of the bottom as a result of groundwater flows is to be investigated and, if necessary, ruled out by taking suitable measures.

8.2.7 Verifying the load-bearing capacity of a quay wall

1. Predominantly constant loads
   Verification of the load-bearing capacity for all types of sheet pile wall is to be carried out according to DIN EN 1993-5. According to this standard, the analysis format for safety against loss of the load-bearing capacity of the sheet pile wall section with the design value E_d for internal forces and design value R_d for the resistance of the section is

   $$E_d \leq R_d$$

   When it comes to methods of designing bearing piles in combined walls, DIN EN 1993-5 refers to DIN EN 1993-1. In the design, flexural-torsional buckling (which is only necessary for I-shaped bearing piles in combined walls) need not be checked when the following boundary conditions are met:

- It is a combined wall fully backfilled with non-cohesive soil having a medium or higher density or cohesive soil having a stiff or better consistency.
- It is a combined wall made from double I-section bearing piles and said bearing piles are embedded on at least three sides in load-bearing subsoil and the maximum free-standing length is 7.5 m.
- Double I-section bearing piles are welded together adequately to form a box section with a high resistance to flexural-torsional buckling.

2. Predominantly variable loads
   Non-backfilled sheet pile walls free-standing in water are loaded primarily cyclically by the impact of waves. In this situation, a large number of load cycles take place over the lifetime of the wall, so that verification of fatigue strength to DIN EN 1993-1-9 is required.
   To prevent adverse effects caused by notch effects, such as from structural weld seams, tack welds, unavoidable irregularities on the surface due to the rolling process, pitting corrosion and the like, killed steel to DIN EN 10025 should be used in such cases.

3. Simplified design method for combined walls with a single row of anchors
   The bearing piles in a combined wall that is anchored near the top and embedded in load-bearing subsoil (see Section 8.1.2.8) may be designed using a simplified method, provided that
   - The boundary conditions are met for the omission of the flexural-torsional buckling check described in 1) above.
   - Uniaxial bending plus a normal force (compression) is present.
   - Single, double or multiple I-section or tubular bearing piles are being used.
   - Verification for cross-section class 3 (elastic-elastic) is carried out.
   - The compression complies with $N_{Ed} \leq 0.25 N_{pl}$ (essentially bending action effects).

   If the foregoing boundary conditions are met, verification of the cross-sections and stability of the bearing piles may be carried out together in an extended normal stresses verification:

$$\sigma_{x,Ed} = \frac{N_{Ed}}{A} + \frac{M_{y,Ed}}{W_{el,y}} + \frac{N_{Ed} \cdot w_{z,d}}{W_{el,y}} \leq \frac{f_y}{\gamma_{M1}}$$

where

$\sigma_{x,Ed}$	design value of normal stress acting in longitudinal direction
$M_{y,Ed}$	design value of moment acting about the y axis
N_{Ed}	associated design value of normal force acting (compression)
$w_{z,d}$	largest deflection due to the design actions of the actions multiplied by γ
A	cross-sectional area of bearing pile
$W_{el,y}$	elastic section modulus of bearing pile about the y axis
f_y	critical yield stress
γ_{M1}	partial safety factor for load-bearing capacity of components in the case of stability failure $\gamma_{M1} = 1.1$

The local effects of water pressures are to be taken into account according to DIN EN 1993-5, Section 5.5.4, by assuming a reduced yield stress. Additional shear and comparative stress analyses should be carried out if required. If the loads are primarily variable, consider 2) above.

Where driving imperfections have been included as specified in DIN EN 1993-5, Section 5.5.1, these must be considered in a suitable way. This can be done, for example, by increasing $w_{z,d}$.

It is possible to apply the method to class 4 cross-sections, provided the yield stress used in the calculations is reduced to such an extent that the section can be classified as a class 3 cross-section.

Bearing pile cross-sections are classified according to Table 5.2 of DIN EN 1993-1-1.

If the resultant bearing pile length based on designing with characteristic actions (DIN EN 1997, design approach 2*) is taken as the member length when carrying out the design with the actions multiplied by γ, then the support in the ground is assumed to be at the end of the member.

4. Checking for safety against buckling is unnecessary for tubular bearing piles when these are filled with non-cohesive soil over their full length.
5. In combined walls, the horizontal and vertical loads are transferred into the subsoil solely by the bearing piles. The infill piles (secondary elements) close off the wall and transfer the earth pressure (partly) and the excess water pressure directly to the bearing piles.

Verification that the loads are transferred from the secondary elements to the bearing piles (primary elements) by way of local plate bending of the flange may be carried out in accordance with DIN EN 1993-5, section D.1.2.

According to DIN EN 1993-5, Sections 5.5.1 (2) and 5.5.4 (1)P, the bearing piles must be designed for the horizontal and vertical loads acting over the system dimension (= width of secondary element + width of bearing pile). The bearing piles are restrained by an anchor at the top and by the support in the subsoil at the bottom.

Where material with a medium or higher density is used as the backfilling to the wall, the secondary elements are loaded primarily by excess water pressure only because the largest part of the earth pressure is carried directly by the bearing piles through the formation of a horizontal arch. If this condition is met, then experience shows that non-welded secondary elements at least 10 mm thick and maximum 1.60 m wide for Z sections and maximum 1.80 m wide for U sections can transfer excess water pressures of up to 40 kN/m² to the bearing piles of combined walls without any further verification. With bearing piles at greater spacings and/or higher loads due to excess water pressure or in cases where the formation of a horizontal arch behind the bearing piles cannot be assumed, it will be necessary to verify that the loads are transferred from the secondary elements to the bearing piles.

When using wider secondary elements, intermediate horizontal walings can be used as additional supports to accommodate the excess water pressure.

For secondary elements up to 1.80 m wide and embedment depths of at least 5.00 m, it is permissible to assume the full passive earth pressure in front of the bearing piles for simplicity, even if the secondary elements do not extend as far as the underside of the bearing piles.

Where secondary elements are > 1.80 m wide and/or the embedment depths of the bearing piles are < 5.00 m, it will be necessary to check whether instead of the full passive earth pressure in front of the continuous wall, the volumetric passive earth pressure in front of the narrow compressive surfaces of the bearing piles is critical according to DIN 4085:2007-10, Section 6.5.2.

In combined walls with an eccentric arrangement of the secondary elements, considering them as part of a composite cross-section according to Section 8.1.2.6.1 is only worthwhile when the displacement of the centroid axis is compensated for by strengthening the bearing piles on the opposite side.

8.2.8 Selection of embedment depth for sheet piles

Structural, constructional, operational and economic matters may be relevant to the embedment depth of sheet pile walls in addition to the corresponding load-bearing analyses and the supplement required by Section 8.2.9. Any foreseeable future deepening of the harbour bottom and any possible danger of scour below the design bottom should be considered to the same extent as the required margin of safety against slope, foundation, heave and erosion failures.

These latter requirements usually result in such a large minimum embedment depth that partial fixity is available at least – apart from the special case of foundations in rock. Even if a simple support would be adequate theoretically, it is often advisable to increase the embedment depth because this can have economic advantages as well. The resistance of the section is utilised more uniformly over the length of the sheet pile wall and, therefore, at least partial fixity of the sheet pile wall is advisable when using the Blum method (Blum 1931).

If the sheet pile wall also has to transfer vertical loads into the subsoil, not all the piles have to continue as far as the load-bearing stratum. Instead, it can be sufficient to make the embedment length of only some of the piles long enough so that they are effective as vertical load-bearing piles, provided that it can be shown that this number of piles can carry the loads without sinking into the subsoil.

8.2.9 Determining the embedment depth for sheet pile walls with full or partial fixity in the soil

If a sheet pile wall is designed according to Blum (1931), then assuming full fixity in the soil (degree of fixity $\tau_1 = 100\,\%$), the entire embedment length below the design bottom consists of the embedment depth t_1 down to the theoretical base plus the extra depth Δt_1 (additional driving depth). The extra length Δt_1 is necessary in order to accommodate the design value of the (mobilised) equivalent force R_C (corresponding to the equivalent force C of Blum (1931)) actually acting on the theoretical base TF as a soil reaction force distributed over depth Δt_1 (Figures 8.40 and 8.41).

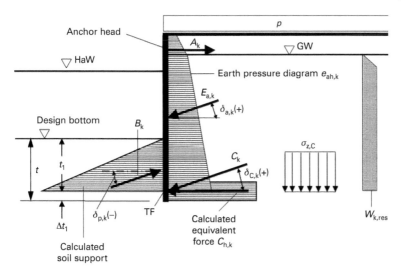

Figure 8.40 Actions, calculated support and soil reactions of a sheet pile wall with full fixity in the soil.

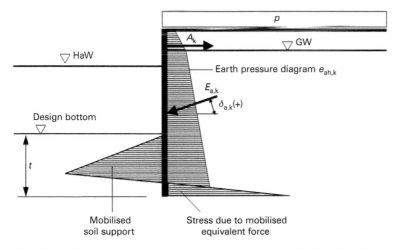

Figure 8.41 Mobilised support and soil reactions of a sheet pile wall fixed in the soil.

Notation for Figures 8.40 and 8.41:

t	required total embedment depth $t = t_1 + \Delta t_1$ for the sheet pile wall fixed in the soil [m]
TF	theoretical base of sheet pile wall (load application point for equivalent force C)
t_1	distance between TF and design bottom [m]
Δt_1	additional depth for accommodating equivalent force $\frac{1}{2} C_{h,d}$ via a soil reaction force below TF [m]
$\sigma_{z,C}$	vertical soil stress at TF on equivalent force side [kN/m²]
$\delta_{p,k}$	angle of inclination of passive earth pressure [°]
$K_{pgh,C}$	passive earth pressure factor at TF on equivalent force side for angle of inclination $\delta_{C,k}$
$\delta_{C,k}$	angle of inclination of equivalent force C [°]

If the more exact analysis given below for calculating Δt_1 is not carried out, the additional depth for sheet pile walls fully fixed in the ground can be simplified to

$$\Delta t_1 = t_1/5$$

However, this is only possible when the actions do not include any significant water pressure component.

The design value $C_{h,d}$ is

$$C_{h,d} = \sum(C_{Gh,k} \cdot \gamma_G + C_{Qh,k} \cdot \gamma_Q)$$

Or, in the case of a reduced partial safety factor for the hydrostatic pressure component and separation according to equivalent force portions,

$$C_{h,d} = \sum(C_{Gh,k} \cdot \gamma_G + C_{Gh,W,k} \cdot \gamma_{G,red} + C_{Qh,k} \cdot \gamma_Q)$$

The calculated equivalent force portions are:

$C_{Gh,k}$ due to permanent actions G
$C_{Gh,W,k}$ due to permanent water pressure actions
$C_{Qh,k}$ due to variable actions Q

The associated partial safety factors are:

γ_G for permanent actions
$\gamma_{G,red}$ for water pressure with a permissible reduction
γ_Q for variable actions

At failure, the characteristic value $E_{phC,k}$ of the soil support for accommodating the actual equivalent force $\frac{1}{2}C_{h,d}$ is the magnitude of the passive earth pressure on the equivalent force side beneath the theoretical base TF:

$$E_{phC,k} = \Delta t_1 \cdot e_{phC,k}$$

The characteristic value of the passive earth pressure stress $e_{phC,k}$ on the equivalent force side at the level of TF is

$$e_{phC,k} = \sigma_{z,C} \cdot K_{pgh,C} \quad \text{in non-cohesive soils, and}$$
$$e_{phC,k} = \sigma_{z,C} \cdot K_{pgh,C} + c'_k \cdot K_{pch,C} \quad \text{in cohesive soils}$$

(taking into account the respective consolidation state as a result of the shear parameters $c_{u,k}$ or φ'_k and c'_k).

The vertical soil stress $\sigma_{z,C}$ is to be calculated at the level of the base TF on the equivalent force side.

The design value $E_{phC,d}$ of the soil support for accommodating the equivalent force $1/2 C_{h,d}$ is calculated with the partial safety factor $\gamma_{R,e}$ for passive earth pressure:

$$E_{phC,d} = \frac{E_{phC,k}}{\gamma_{R,e}}$$

The analysis format for complying with the limit state condition for accommodating the equivalent force $C_{h,d}$ as a soil reaction is

$$\frac{1}{2}C_{h,d} \leq E_{phC,d}$$

From this limit state condition, we get the magnitude of the required additional depth Δt_1 beneath the theoretical base TF for walls fully fixed in the soil:

$$\Delta t_1 \geq \frac{1}{2} C_{h,d} \cdot \frac{\gamma_{R,e}}{e_{phC,k}}$$

The above equation for the extra depth in the case of full fixity in the ground (degree of fixity $\tau_1 = 100\,\%$) is also used to determine the extra depth for sheet pile walls with only partial fixity in the ground, i.e. for any degree of fixity over a possible range $\tau_1 = 100\,\%$ to $\tau_0 = 0\,\%$ (simply supported in the soil).

The degree of fixity designated here with τ_{1-0} for a partially fixed sheet pile wall becomes $\tau_{1-0} = 100 \cdot (1 - \varepsilon(E)/\max\varepsilon(E))\,[\%]$ with the end tangent angle $\varepsilon(E)$ of the deflection curve for the theoretical base TF selected and the end tangent angle $\max\varepsilon(E)$ for a simple support in the soil. The embedment depth associated with the degree of fixity τ_{1-0} is designated t_{1-0} and the extra depth Δt_{1-0}.

Partial fixity in the soil is associated with – compared with full fixity – smaller values for the equivalent force component $C_{h,d}$ and, hence, also additional depth $\Delta t_{1-0} < \Delta t_1$. In the case of a simple support for the sheet pile wall in the soil ($\tau_0 = 0\,\%$), $C_{h,d} = 0$ and $\Delta t_0 = 0$ apply.

A required minimum value Δt_{min} must be maintained for the additional embedment depth, which is defined depending on the degree of fixity present ($100\,\% \geq \tau_{1-0} \geq 0\,\%$):

$$\Delta t_{min} = \frac{\tau_{1-0}}{100} \cdot \frac{t_{1-0}}{10}$$

8.2.10 Steel sheet pile walls with staggered embedment depths

8.2.10.1 Applications

Sheet piles (generally double piles) are frequently driven to different depths for technical and, in the case of fully fixed walls, for economic reasons, too. The permissible extent of these alternating embedment depths, known as a staggered arrangement, depends on the stresses in the bases of the longer piles and on construction issues. For driving reasons, staggering within a single pile unit is not recommended for sheet piles.

At failure, a uniform, continuous passive earth pressure zone (which depends on the geometrical boundary conditions according to Section 8.1.2.8) forms in the region of the soil support of a staggered sheet pile wall (similarly to the wedge of soil in front of closely spaced anchor plates). The full soil reaction down to the bottom of the deeper sheet piles, therefore, applies when determining the loads without taking into account the staggering. The bending moment at the bottom edge of the shorter piles must be resisted by the longer piles alone. Therefore, in order to limit the stresses in the deeper piles, only adjacent pile units (double piles at least) in sheet pile walls are staggered (Figures 8.42 and 8.43). A length of 1.0 m is usual for staggered arrangements. In practice, it has been found that a structural check of the longer piles is unnecessary. For a greater staggered length, it will be necessary to verify the load-bearing capacity of the deeper piles with respect to multiple stresses due to bending moments combined with axial and shear forces.

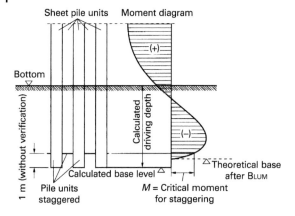

Figure 8.42 Staggered sheet piles for a sheet pile wall fully fixed in the soil.

8.2.10.2 Sheet pile walls fixed in the soil

According to the Blum method (Blum 1931), sheet pile walls fully fixed in the soil ($\tau_1 = 100\,\%$) may exploit the entire stagger dimension s to save steel; the longer sheet piles are driven to the depth determined according to Section 8.2.9 (Figure 8.42), the shorter piles terminate at a level higher by the stagger dimension s.

In the case of walls partially fixed in the soil ($100\,\% > \tau_{1-0} > 0\,\%$), the steel saving depends on the degree of fixity present. A corresponding saving in steel is achieved by driving the longer sheet piles below the level of the theoretical base of the wall by a certain fraction of the stagger dimension s_U, with the shorter piles again terminating at a level higher by the stagger dimension s.

Dimension s_U depends on the degree of fixity τ_{1-0} [%]:

$$s_U = (100 - \tau_{1-0}) \cdot \frac{s}{2 \cdot 100}$$

8.2.10.3 Sheet pile walls simply supported in the soil

When the sheet pile is simply supported in the ground, the stagger dimension s no longer leads to a saving in steel (owing to the equation for dimension s_U from Section 8.2.10.2, which also applies for degree of fixity $\tau_0 = 0\,\%$). Instead, it leads to an enlargement of the soil zone that can be mobilised for support B, but this cannot be used in the calculations.

In this case, the longer sheet piles (see Figure 8.43) must be driven below the theoretical underside of the wall by a distance $s_U = s/2$.

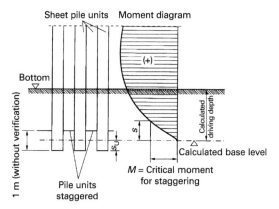

Figure 8.43 Staggered sheet piles for a sheet pile wall simply supported in the soil.

With a stagger dimension $s > 1.0$ m, the load-bearing capacity of the longer piles must be verified according to Figure 8.43.

The same applies to reinforced concrete or timber sheet piles, provided that the joints between the piles have sufficient strength to guarantee that the longer and shorter sheet piles work together.

8.2.10.4 Combined walls

Walls made up of bearing and intermediate piles (primary and secondary elements, see Section 8.1.2.8) must take into account the excess water pressure due to flows around the wall, such that the required margin of safety against hydraulic heave failure (Section 3.4) is guaranteed in front of the shorter, intermediate piles. Longer piles should be used when there is a risk of scouring.

Where the subsoil at the bottom of the watercourse includes soft or very soft strata, the embedment depth of the shorter, intermediate piles should be determined by special investigations.

8.2.11 Horizontal actions on steel sheet pile walls in the longitudinal direction of the quay

8.2.11.1 General

Combined and corrugated steel sheet pile walls react comparatively flexibly to horizontal actions in the longitudinal direction of the quay. If such actions occur, a check must be carried out to verify that the resulting horizontal action effects parallel to the quay can be accommodated by the sheet pile, or whether additional measures are required. In many cases, in-plane stresses in sheet pile structures due to earth and water pressures can be avoided by choosing an appropriate design. Such designs can involve, for example, crossing anchors at quay wall corners according to Section 9.2.6. Another example is radial arrangements of round steel tie rods, with an anchor plate at the centre of the curve, for curving sections of quay walls or pier heads; further tie rods then continue back from this central plate to an anchor wall of sheet pile sections. The anchor force resultants of the anchors connected here must act in the same direction as the anchor force resultant of the radial anchors so that the central anchor plate remains in equilibrium.

8.2.11.2 Transferring horizontal forces into the plane of the sheet pile wall

Available construction components such as capping beams and walings can be used to transfer horizontal forces provided that they have been designed to do so. Otherwise, additional measures are required, e.g. the installation of diagonal bracing behind the wall. Welding the interlocks in the upper section of the wall will suffice in the case of smaller longitudinal forces.

The action components parallel to the quay due to line pull forces act at the mooring points, the maximum actions due to wind at crane wheel locking points and those due to ship friction at the fenders. The load application points of these friction forces can occur at any place along the wall. This also applies to horizontal loads as a result of crane braking, which must be transferred from the superstructure into the top of the wall. Longitudinal forces can be carried over a longer length of the wall provided that the distributing construction components have been designed accordingly.

For this reason, the flanges of steel walings should be bolted or welded to the land-side sheet pile flanges (Figure 8.44).

Longitudinal forces can also be transferred via cleats that are welded to the waling and braced against the sheet pile webs (Figure 8.45).

When a waling consists of two channel sections, the waling bolts can only be used to transfer longitudinal forces when the two channels are joined by a vertical drilled plate welded in place on the land-side sheet pile flange. The force from the waling bolts is then accommodated by the plate via bearing stresses, whereas the bolts are subjected to shearing stresses (Figure 8.46).

When loads act parallel to the plane of the sheet pile wall, the capping beams and walings, including their splices, should be designed for bending combined with axial and shear forces.

In order to transfer horizontal actions in the direction of the quay from a reinforced concrete capping beam to the top of a sheet pile wall, the latter must be adequately embedded in the beam. The design of the reinforced concrete cross-section must take account of all the global and local loads that occur in this area.

8.2.11.3 Transferring horizontal forces acting parallel to the line of the sheet pile wall from the plane of the wall into the subsoil

Horizontal longitudinal forces in the plane of the sheet pile wall are transferred into the ground by friction on the land-side sheet pile flanges and by resistance in front of the sheet pile webs. The latter, however, cannot be greater than the friction in the ground over the length of the sheet pile trough.

In non-cohesive soils the force can, therefore, be accommodated entirely by friction, which requires the use of a reasonable mean value of the friction coefficient between soil and steel, as well as between soil and soil. The effectiveness of this force transfer in non-cohesive soils increases with the angle of internal friction and the strength of the backfill; in cohesive soils the force transfer improves with the shear strength and the consistency of the soil.

When transferring horizontal forces acting parallel to the plane of the sheet pile wall from a capping beam or waling into the subsoil, additional bending moments occur transverse to the principal load-bearing direction of the sheet pile wall. These bending moments can be calculated with the numerical model for a fixed or simply supported anchor wall. Instead of the resisting soil reaction as a result of the mobilised passive earth pressure, however, the aforementioned mean wall friction force, or a corresponding shear resistance, is assumed in this case.

As a rule, only double piles joined with shear-resistant welds should be considered as load-bearing elements carrying these additional loads.

Where welding is not used, the piles should be considered as single piles only.

When accommodating horizontal forces in the longitudinal direction of the quay, the sheet piles are stressed by bending in two planes. When superimposing the resulting stresses, DIN EN 1993-1-1 permits the use of the design value for the equivalent tensile stress $\sigma_{v,d}$, which – for the boundary conditions for individual corner stresses given in the standard – may exceed the maximum permissible normal stress $\sigma_{R,d}$ by 10%.

By taking into account friction resistances in the horizontal direction, only a reduced angle of earth pressure inclination $\delta_{a,k,red}$ may be used in the sheet pile calculations. In

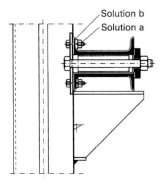

Figure 8.44 Transferring longitudinal forces by means of close-tolerance bolts in the waling flanges (solution a) or by welding (solution b).

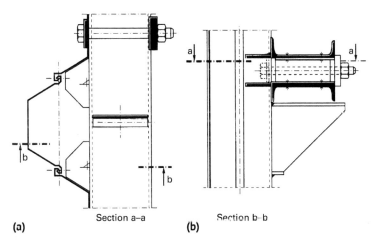

Figure 8.45 Transferring longitudinal forces by means of steel cleats welded to the waling: (a) section a–a; (b) section b–b.

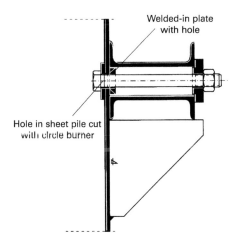

Figure 8.46 Transferring longitudinal forces by means of waling bolts and welded plate with hole drilled prior to welding.

doing so, the magnitude of the resultant of the vector addition of the two friction resistances orthogonal to each other may not exceed the maximum possible wall friction resistance value of the sheet pile wall with respect to the soil.

8.2.12 Design of anchor walls fixed in the ground

Obstacles in the ground, such as ducts, pipes, cables, etc., sometimes mean that it is not possible to connect round steel tie rods to the centre of the anchor wall. In such cases, the tie rods must then be positioned at a higher level and connected to the upper area of the anchor wall.

The calculations for the anchor wall in such cases are to be carried out as for a non-anchored sheet pile wall for limit state STR or GEO-2 (Figure 8.47). The anchor force A_k of the sheet pile wall to be anchored is entered into the calculation as the characteristic value F_k of the tensile force acting at the top of the anchor wall.

The partial safety factors for earth pressure and, if applicable, water pressure actions, are applied according to Section 8.2.1. The factor for the tensile force F_k is the quotient of the design value A_d and the characteristic value A_k of the anchor force, both of which are taken from the calculations for the sheet pile wall.

The extra embedment depth Δt_1 required is calculated according to Section 8.2.9.

In this form of construction, the sheet piles of the anchor wall are considerably longer than an anchor plate loaded centrally and also have a larger cross-section.

Staggering the anchor wall according to Section 8.2.13 is only permissible at the base of the wall, but a stagger of up to 1.0 m for deep anchor walls is possible without any special verification.

In the case of predominantly horizontal groundwater flows, a number of weepholes must be provided in the anchor wall if the water pressure acting on the wall needs to be reduced. The resultant water pressure must be taken into account in the design of the sheet piles for the anchor wall.

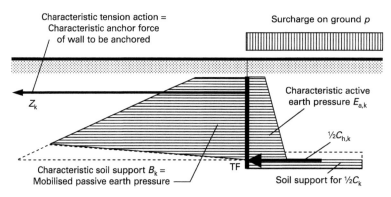

Figure 8.47 Actions, soil reaction and equivalent force for an anchor wall fully fixed in the ground as required for stability verification for limit state GEO-2.

8.2.13 Staggered arrangement of anchor walls

In order to save materials, anchor walls may be staggered in the same manner as the waterfront sheet pile wall. Both ends may be staggered in the same wall. However, the stagger should not exceed 0.5 m top and bottom. When the wall is staggered top and bottom, all the double piles can be of the same length, i. e. 0.5 m shorter than the overall height of the anchor wall. The double piles are driven so that one double pile is driven to the bottom of the anchor wall, the next double pile driven so that its top matches the top of the anchor wall and so on, alternately. A stagger > 0.5 m is permissible for deep anchor walls only, provided that the bearing capacity of the soil support and the load-bearing capacity of the section in terms of bending moment, shear force and axial force have been verified. Such verification is also required for a stagger of 0.5 m, however, if the overall height of the anchor wall is < 2.5 m. In such a case, it must be verified that the bending moments can be transferred from the deeper ends to the neighbouring piles.

The same applies to reinforced concrete or timber sheet piles, provided that the joints between the piles have sufficient strength to guarantee that the longer and shorter sheet piles work together.

8.2.14 Waterfront sheet pile walls in unconsolidated, soft cohesive soils, especially in connection with non-sway structures

For various reasons, ports, harbours and industrial facilities with associated waterfront structures must sometimes be built in areas with poor subsoil. Alluvial cohesive soils, possibly with peat inclusions, are, thus, subjected to higher loads due to increases in the level of the terrain and, hence, placed in an unconsolidated state. The resulting settlement and horizontal displacements call for special construction features and a structural design treatment that is tailored to the particular site.

In unconsolidated, soft cohesive soils, sheet pile structures may only be designed as "floating" structures when neither the serviceability nor the stability of the entire structure and its parts will be endangered by the resulting (differential) settlement and horizontal displacements. To assess this and initiate the necessary measures, the expected settlement and displacements must be calculated.

If a quay wall is constructed in unconsolidated, soft cohesive soil in connection with a structure on a practically immovable foundation, e. g. on a pile trestle with vertical piles, the following solutions are possible: The sheet pile may be anchored or supported so that it is free to move in the vertical direction but such that the connection to the structure remains fully effective and load-bearing even in the case of the maximum theoretical displacements.

This solution is quite straightforward apart from the settlement and displacement calculations. For operational reasons, however, it can generally only be used for a rearward sheet pile wall for designs involving pile trestles. The vertical friction force at the support must be taken into account in the design of the trestle. Slotted holes are not sufficient at the anchor connections of sheet piles in front of the structure. In fact, a sliding anchorage is then required.

The sheet pile wall is supported against vertical displacements by driving a sufficient number of pile sections deeply enough to reach the load-bearing subsoil deep in the ground.

In this situation, the load-bearing capacity of the sheet pile wall with respect to the following vertical loads must be guaranteed by the deeper piles alone:

- The self-weight of the wall.
- Soil clinging to the sheet pile as a result of negative skin friction and adhesion.
- Axial actions on the wall.

This solution is practicable in technical and operational terms when the sheet piles are located in front of the structure. Since the cohesive soil clings to the sheet pile wall during the settlement process, the active earth pressure decreases. If the supporting soil in front of the base of the sheet pile wall also settles, however, the characteristic passive earth pressure, and hence the potential soil support, also decreases as a result of negative skin friction. This must be taken into account in the sheet pile wall calculations. When calculating the vertical load on the sheet pile wall arising from soil settlement, the negative skin friction and adhesion for the initial and final states are taken into consideration.

Apart from the anchoring or support against horizontal forces, the sheet pile wall should be so suspended from the structure such that the aforementioned actions are transferred to the structure and from there to the load-bearing soil.

In this solution, the sheet pile wall and its upper suspension are calculated using the aforementioned information.

If the load-bearing soil is located at a reasonable depth in construction terms, the entire wall can be driven down into the load-bearing stratum.

The passive earth pressure in this stratum is calculated with the usual earth pressure angles of inclination and the partial safety factors as per Table 1.2. When calculating the soil reaction of the overlying soil with a lower strength or consistency, only a reduced characteristic passive earth pressure may be assumed. Serviceability must be verified.

8.2.15 Design of single-anchor sheet pile walls in seismic zones

8.2.15.1 General

The findings of soil investigations and soil mechanics tests must be carefully checked to establish what effects the vibrations caused by a critical earthquakes may have on the shear strength of the subsoil. The results of these checks can be critical for the design of the structure. For example, when soil conditions are such that liquefaction is to be expected as per Section 3.5.14.2, anchorages with high-level anchor walls or anchor plates are not permitted unless the mass of earth supporting the anchoring is adequately compacted in the course of the construction work and the danger of liquefaction, thus, eliminated. Please refer to Section 3.5.14.2 for the magnitude of the seismic coefficient k_h and other actions, as well as the design values for loads and resistances, plus the safety factors required.

Taking into account the sheet pile walls loads and support reactions determined according to Section 3.5.14.2, the calculation can be carried out as per Section 8.2.3, but without redistribution of the active earth pressure.

The characteristic active and passive earth pressures determined with the fictitious angles of inclination for the reference and ground surfaces are used as a basis for all calculations and analyses. However, tests have revealed that the rise in the active earth pressure due to an earthquake does not increase linearly with the depth; instead, it is higher near the surface. Anchorages must, therefore, be generously dimensioned.

Verification of stability for anchorages at the lower failure plane is to be carried out according to Section 9.3. This must take into account the additional horizontal forces occurring at reduced variable loads due to acceleration of the mass of earth to be anchored and the pore water contained therein.

8.2.16 Sheet pile waterfronts on inland waterways

8.2.16.1 General

Where canals are to be constructed or extended in areas where only limited space is available, waterfront structures of anchored steel sheet piles are frequently the best engineering solution and – after considering the land acquisition and maintenance costs – also the most economical. This is especially true for stretches requiring an impervious bottom. If necessary, sheet pile interlocks can be sealed, see Section 8.1.3. An example of an application is shown in Figure 3.4 in Section 3.3.2.

Where shipping conditions allow, the top edge of the sheet pile should remain below the water level for reasons of corrosion protection and landscaping. Please refer to the Federal Ministry of Transport, Planning, and Housing for more information on standard cross-sections.

8.2.16.2 Stability analysis

To verify stability, the waterfront structure and its components are analysed and designed according to the pertinent recommendations. Please refer to Sections 3.3.2 and 1.1.4 for water pressures. In the case of vertical imposed loads, in contrast to Section 4.2, a uniformly distributed surcharge of $10\,kN/m^2$ (characteristic value) should be assumed (Figure 3.4).

In addition, please also refer to Section 8.2.10 for guidance on staggered embedment depths and Section 8.2.8 for guidance on selecting embedment depths.

8.2.16.3 Design situations

The actions and action effects assigned to the design situations are characteristic values. In design situation DS-P, the resultant water pressure to be expected is due to the unfavourable canal and groundwater levels that occur frequently. This also includes a lowering of the canal water level by 0.80 m in front of the sheet pile due to passing vessels.

In design situation DS-A and in the case of a rapid and severe drop in the canal water level, it will be necessary to investigate the two load cases given below (Figure 3.4):

a) Canal water level 2.00 m below groundwater level.
b) Canal water level at canal invert with groundwater level 3.00 m higher.

Where the failure of a waterfront structure would cause bridges and loading installations, etc., to collapse, then a sheet pile wall must be designed for the "canal empty" load case or must be adequately secured by means of special structural measures.

In the structural investigations, the design canal invert level or base of excavation (e. g. the underside of the bottom protection) may be assumed to be the theoretical level for calculation purposes. Excavation down to 0.30 m below the design bottom level is generally permissible without special calculations when the wall is fully fixed in the ground (see Section 7.1.6). However, this does not apply to walls with and without anchors, which have a simply supported base. If in exceptional cases greater deviations are to be expected and

8 Sheet pile walls

severe scour damage due to ship propeller action is likely, the calculations should assume a bottom depth that is at least 0.50 m below the design bottom level.

8.2.16.4 Embedment depth

If soil with a low permeability is encountered at an attainable depth in dam or dyke stretches that are to be made watertight, the sheet piles should be driven to such a depth that they are embedded in this stratum. A lining to the canal bottom is then unnecessary. However, this course of action must not have a negative impact on the wider flow of groundwater.

8.2.17 Calculation and design of cofferdams

8.2.17.1 Verification of ultimate limit states STR, GEO-2 and GEO-3

Failure of the cofferdam in the subsoil must be checked for the GEO-2 and GEO-3 ultimate limit states using the following four analyses:

1. Verification against failure due to overturning and sliding (GEO-2)
   Verification of stability against failure at the ultimate limit state GEO-2 is to be carried out for a cofferdam using the actions $W_ü$, E_a, any variable external actions, resistances G, E_p and, if required, the cohesion at the failure plane, all of which are shown in Figures 8.48–8.50. In the case of circular and diaphragm cell cofferdams, the mean width b' as shown in Figure 8.6 is to be used as the theoretical width of the cellular cofferdam. This width results from converting the actual plan form into an equivalent rectangular area.

   Where a cofferdam rests directly on rock (Figure 8.48), upon failure, a convex failure plane forms between the bases of the cofferdam walls. As a first approximation, the curve for this failure plane can be generated by a logarithmic spiral for the characteristic value of the friction angle φ_k.

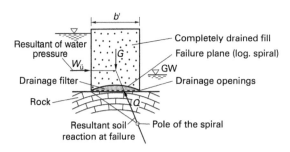

Figure 8.48 Cofferdam simply supported on rock, with drainage.

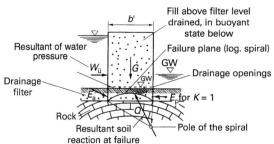

Figure 8.49 Cofferdam on rock overlain with other soil strata, with drainage.

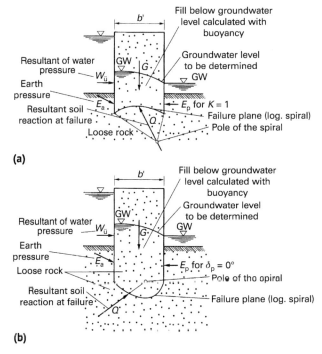

Figure 8.50 Cofferdam embedded in load-bearing soil, with drainage:
(a) shallow embedment; (b) additional investigation for deep embedment.

If the cofferdam is supported on rock that is overlain with other soil strata (Figure 8.49) or is embedded in load-bearing loose rock (Figure 8.50), the actions are increased by the additional active earth pressure of these soil strata and the resistance by the additional passive earth pressure. A reduced passive earth pressure is applied to take account of the minor shape changes – usually with $K_p = 1.0$ – and in the case of deeper embedment in loose rock, then with K_p for $\delta_p = 0$.

The resulting moments about the pole of the spiral are designated as a characteristic load due to actions and as characteristic resistances due to the resistant variables. To determine the design value of the moment M_{Ed} as a result of horizontal actions $W_{ü}$, E_a and variable external actions, e.g. line pull, the characteristic values of the individual moments are multiplied by the partial safety factors γ_G and γ_Q and added together.

The design value of the resisting moment M_{Rd} as a result of the vertical action G (the dead load of the cofferdam fill), passive earth pressure E_p (for embedment in loose rock) and possible cohesion at the failure plane is determined by dividing the characteristic individual moments by the partial safety factors $\gamma_{R,h}$, $\gamma_{K,e}$ and γ_c and adding them together,

$$M_{Ed} = M_{kG} \cdot \gamma_G + M_W \cdot \gamma_G + M_{kQ} \cdot \gamma_Q \leq \frac{M_{kG}^R}{\gamma_{R,h}} + \frac{M_{Ep}^R}{\gamma_{R,e}} + \frac{M_c^R}{\gamma_{R,h}} = M_{Rd}$$

where

M_{kG} characteristic value of effective single moment due to earth pressure E_a
M_W characteristic value of effective single moment due to resultant water pressure $W_ü$
M_{kQ} characteristic value of effective moment due to variable external force
M_{kG}^R characteristic value of resisting single moment due to dead load of fill G
M_{Ep}^R characteristic value of resisting single moment due to passive earth pressure E_p
M_c^R characteristic value of resisting single moment due to cohesion at the logarithmic failure plane

Use the partial safety factors for the ultimate limit state GEO-2 according to Section 1.2.4 for actions and resistances depending on the respective design situation.
According to Table 1.2, $\gamma_c = 1.0$ is the partial safety factor for limit state GEO-2. In order to ensure additional safety in the cohesion term, then according to DIN EN 1997-1, a partial safety factor $\gamma_{R,h} = 1.1$ should be used here in a similar way to verification for sliding.
Verification of stability against failure is satisfied when

$$M_{Ed} \leq M_{Rd}$$

The most unfavourable failure plane for the analysis is the logarithmic spiral that results in the smallest M_{Rd}/M_{Ed} ratio.
The main action affecting cofferdams is generally the resultant water pressure $W_ü$. This results from the difference in water pressures acting on the outer and inner walls of the cofferdam and is applied down to the bottom of the outer, i.e. loaded, wall. The water level within the cofferdam enclosure need not always correspond with the level of the bottom.
The resistance of the cofferdam against failure at the ultimate limit state GEO-2 "overturning and sliding" can be increased by
- Widening the cofferdam.
- Choosing a filling material with a higher unit weight and greater angle of internal friction.
- Draining the cells.
- In the case of founding on loose rock, by embedding the cofferdam piles deeper in the subsoil.

Where a cofferdam with a deeper embedment is selected, verification against failure is to be carried out with both a convex failure plane (Figure 8.50a) and a concave failure plane (Figure 8.50b).
In the latter case, the position of the spiral must be chosen such that its pole lies below the line of action of E_p for $\delta_p = 0°$ (Figure 8.50), which is also a condition for deep embedment.
The above verification confirms both safety against failure due to overturning and also sliding.

2. Verification of safety against base failure (GEO-2)
For cofferdams that are not founded on rock, verification of safety against heave failure is to be carried out to DIN 4017 on the basis of DIN 1054, using the mean width b' as the cofferdam width.

3. **Verification of safety against failure due to loss of overall stability (GEO-3)**
   According to DIN EN 1997-1, Chapter 11, verification of failure due to loss of overall stability (ground failure) is to be carried out to DIN 4084 in the case of backfilled cofferdams that are part of a waterfront structure. To carry out the analysis, position the failure plane on the load-side, theoretical limit to the cofferdam width, which coincides with the above value for the mean width b'.
4. **Additional analyses in the case of water flows:**
   - Any flow force present is to be considered in the analyses called for in 1.–3. above.
   - Carry out an analysis to verify safety against failure of the subsoil as a result of a hydraulic heave failure.
   - Carry out an analysis to verify safety against failure of the subsoil as a result of piping.
   - In order to rule out the aforementioned failure modes, special sealing measures are required at the base of the sheet pile wall where cofferdams are founded on jointed rock or rock with varying strength.

Afterwards, it will be necessary to check for failure of the straight-web section as a result of hoop tension forces at the ultimate limit state (STR).

When designing a cellular cofferdam it can be assumed that the stresses caused by external actions such as water pressure and, if applicable, earth pressure can be accommodated by the monolithic block action of the cofferdam filling. To verify safety against failure of the straight-web section, it is sufficient in this case to determine the hoop tension at the level of the base of the excavation or watercourse because this is generally where the critical internal pressures occur.

However, in some circumstances, it may be necessary to check the resistance to the hoop tension at several levels when the cofferdam is surrounded by or embedded in cohesive strata. Hoop tension values greater than those at the base can occur in these areas due to the abrupt rise in internal pressure or the smaller passive earth pressure, as well as the possibility of pore water pressure.

Design values for hoop tension are calculated using the formula $F_{t,Ed} = \sum p_{i,d} \cdot r$. They are calculated by multiplying the characteristic actions inside the cell due to excess water pressure, earth pressure at rest (where $K_0 = 1 \sin \varphi_k$) and variable loads ($\sum p_{a,K}, \sum p_{m;k}$) by the appropriate partial safety factors (see Section 1.2.4).

The design values for hoop tension ($F_{tc,Ed}, F_{tm,Ed}, F_{ta,Ed}$) in the individual wall elements (Figure 8.51) may be determined using a simplified method according to DIN EN 1993-5, Section 5.2.5(9), as follows:

In the common wall: $F_{tc,Ed} = \sum p_{a,d} \cdot r_a \cdot \sin \varphi_a + \sum p_{m,d} \cdot r_m \cdot \sin \varphi_m$
In the main cell wall: $F_{tm,Ed} = \sum p_{m,d} \cdot r_m$
In the connecting arc: $F_{tc,Ed} = \sum p_{a,d} \cdot r_a$

The analysis of the sheet pile sections and the welded junction piles is carried out according to DIN EN 1993-5, Section 5.2.5.

Verification of the wall sections is given when the resisting tensile strength $F_{ts,Rd}$ of the web and interlock is equal to or greater than the design values of the hoop tension ($F_{tc,Ed}, F_{tm,Ed}, F_{ta,Ed}$):

$$F_{ts,Rd} \geq F_{tc,Ed} \quad \text{or} \quad F_{tm,Ed} \quad \text{or} \quad F_{ta,Ed}$$

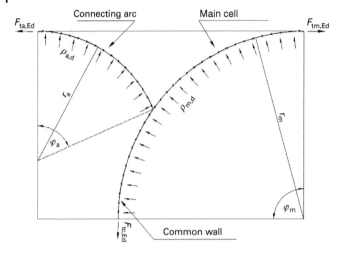

Figure 8.51 Hoop tension $F_{t,Ed}$ in the individual wall elements of a cellular cofferdam.

where

$$F_{ts,Rd} = \beta_R \cdot \frac{R_{k,S}}{\gamma_{M0}} \quad \text{(interlock)} \qquad F_{ts,Rd} = t_w \cdot \frac{f_{yk}}{\gamma_{M0}} \quad \text{(web)}$$

and

$R_{k,S}$ characteristic value of interlock tensile strength
f_{yk} minimum yield stress of steel to Section 8.1.2.1, Table 8.1
t_w web thickness of the straight-web piles Table 8.6
γ_{M0} partial safety factor for sheet pile material
β_R reduction factor for interlock tensile strength (DIN EN 1993-5: $\beta_R = 0.8$)

Assuming that the junction pile is welded based on DIN EN 12063, safety is deemed to be adequate when the resisting tensile strength of the junction pile ($\beta_T \cdot F_{ts,Rd}$ for interlock and web) is greater than or equal to the hoop tension design value $F_{tm,Ed}$:

$$\beta_T \cdot F_{ts,Rd} \geq F_{tm,Ed} = \sum P_{m,d} \cdot r_m$$

The reduction factor β_T according to DIN EN 1993-5, Section 5.2.5(14), can be taken as

$$\beta_T \cdot F_{ts,Rd} \geq F_{tm,Ed} = \sum P_{m,d} \cdot r_m$$

8.2.17.2 Double-walled cofferdams as excavation enclosures and waterfront structures

In double-walled cofferdams the parallel Z- or U-section steel sheet pile walls are driven into or otherwise placed on the bottom to suit the subsoil conditions and the hydraulic and structural requirements. They are then tied together. If the double-walled cofferdam is supported on rock or does not achieve a sufficient embedment, at least two rows of anchors are required.

The inclusion of cross-walls to form anchor cells as shown in Figure 8.52 can be expedient for the work on site. In long permanent structures they also limit any damage caused by

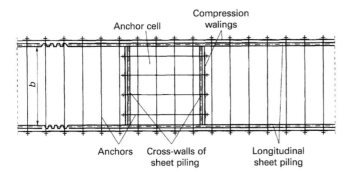

Figure 8.52 Plan of a double-walled cofferdam with anchor cells anchored in themselves.

vessel impact. The spacing of the cross-walls (or anchor cells) determines the length of the individual sections for the construction of the cofferdam, including tie rods and fill.

The remarks regarding the fill in Section 8.1.5.6 apply here. When verifying the stability of a cofferdam with large external actions due to water pressure, e.g. excavation enclosures in open water, the permanently effective drainage of its soil filling is crucial for minimising its dimensions. The fill is drained towards the excavation, for which weepholes according to Section 7.3.1 are sufficient.

Drainage can also be useful when building a cofferdam to serve as a waterfront structure. Such cofferdams are drained towards the water side. However, if there is a danger of pollution, drainage systems with anti-flood fittings according to Section 7.3.1 must always be used.

In the following, the external sheet pile wall to a cofferdam that is subjected to the actions due to earth and water pressures plus variable actions is designated as the load-side wall, whereas the opposite sheet pile wall that is not subjected to such loads is designated as the air, excavation or water-side wall, depending on the particular use of the cofferdam.

The following checks should be carried out for the sheet pile wall component at the ultimate limit state (STR).

In a filled cofferdam, the load transfer into the subsoil relies on the cofferdam functioning as a compact block of soil. The moment as a result of horizontal actions (water and earth pressures) with respect to the point of rotation is transferred into the load-bearing subsoil by means of vertical stresses in the soil through the cofferdam filling acting as a monolithic block. These vertical stresses in the soil change linearly over the width of the cofferdam and reach a maximum at the side opposite to the forces, i.e. the air or water-side sheet pile wall, which is, therefore, subjected to a pressure higher than the active earth pressure. Experience has shown that this increase in active earth pressure can generally be taken into account with adequate accuracy by increasing the active earth pressure – calculated with $\delta_a = 2/3 \varphi'_k$ – by 25%. Another action affecting the air side of the sheet pile wall is the resultant water pressure, which results from a possible water level difference between the drawdown level within the fill and the air or water-side water level.

If the cofferdam fill is installed by using hydraulic filling methods and compacted, the active earth pressure can rise to match the hydrostatic pressure as a result of the filling effect. When using in situ soil as the cofferdam filling, however, auxiliary failure planes can develop within the fill following excavation work. Redistribution of the active earth pressure according to Section 8.2.3 is then permissible.

The air-side wall is designed as an anchored sheet pile wall taking into account all actions. If the sheet pile wall is embedded in load-bearing loose rock, the supporting passive earth pressure can be calculated with an angle of inclination as per Section 8.2.5. The determination of the design values for action effects can be carried out according to Section 8.2.3 for a sheet pile wall with one row of anchors and Section 8.2.4 for two rows.

The load-side wall can be built using a different section and can be shorter than the air or water-side wall, provided that the latter is checked for the individual construction phases, or if the requirements pertaining to watertightness and limiting flow around the wall are satisfied.

Various actions and resistances must be considered when calculating the design values for the stresses in the load-side wall:

- Transferring the anchor force of the air-side sheet pile wall (or anchor forces with two rows of anchors).
- External water pressure.
- External active and, where applicable, also passive earth pressure.
- Ship impacts, line pulls and other horizontal actions.
- Support provided by the soil filling.

The distribution and magnitude of the earth support over the height of the wall must be applied such that the equilibrium condition $\sum H = 0$ is satisfied. If the load-side wall is fixed in the ground, the equivalent force C must be considered when checking the equilibrium.

Furthermore, the following analyses are necessary for failure of double-walled cofferdams at the ultimate limit state (GEO-2).

1. Verification of stability at the lower failure plane (internal stability)
   The stability at the lower failure plane is to be verified according to Section 9.3. Here, the course of the lower failure plane can be approximated as follows for a single row of anchors:
   - Air- or water-side wall
     From the theoretical base for a simply supported water-side wall.
     From the point of zero shear in the zone of fixity for a fixed wall.
   - Load-side wall
     - For simplicity, at the base of an equivalent anchor wall with height $2 \cdot a$, which is assumed to be simply supported (Figure 8.53a).
     - At the point of zero shear of a fixed equivalent anchor wall (Figure 8.53).
       The point of zero shear for a fixed equivalent anchor wall is given by the anchor force and the course of the partial passive earth pressure $\overline{E_p}$.
       As a simplification, a triangular distribution may be selected for the partial passive earth pressure $\overline{E_p}$ with its maximum ordinate $\overline{e_p}$ at the intersection of the failure plane starting from the top of the air-side wall at an angle ϑ_p for $\delta = 0°$ and the axis of the load-side wall. This triangular partial passive earth pressure is applied from the top edge of the cofferdam to the level of the point of zero shear on the air-side wall. The maximum ordinate of the partial passive earth pressure results from the moment equilibrium about the lower peak of the partial passive earth pressure that has been determined beforehand.

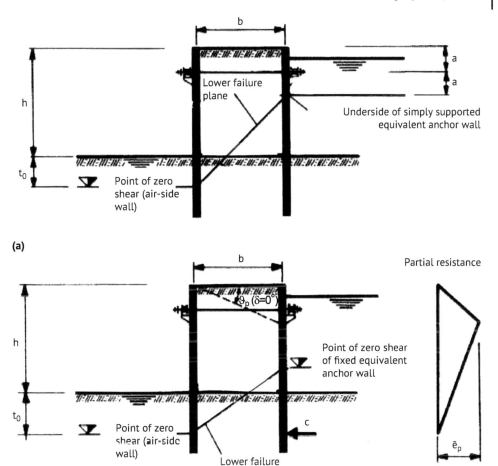

Figure 8.53 Course of lower failure plane within a double-walled cofferdam: (a) assumption for a simply supported equivalent anchor wall; (b) assumption for a fixed equivalent anchor wall.

In the case of a fixed equivalent anchor wall, in addition to verifying stability at the lower failure plane, it will be necessary to carry out the following checks:
- Verification of equivalent force C.
- Verification of loads and stresses acting on the wall.

The latter verification is unnecessary if at least the same section is used for the load-side wall as for the air-side wall.

An equivalent anchor wall may be assumed in the calculations where there are multiple rows of anchors. The course of the lower failure plane may be specified similarly to that for a single row of anchors, although the point of zero shear must be considered to be below the lowest row of anchors.

2. Verification against failure due to overturning and sliding (external stability)

The theoretical width of a double-walled cofferdam is taken as the centre-to-centre distance b between the two sheet pile walls. For verification of the stability of double-walled cofferdams, the same principles essentially apply as for the stability of cellular cofferdams.

In contrast to Figure 8.50a,b, the passive earth pressure E_p in front of the air-side sheet pile wall – owing to its greater deflection potential – may be assumed to correspond to that of a customary anchored sheet wall according to Section 8.2.5, with an angle of inclination $\delta_p < 0°$. The following applies to the position of the failure planes to be investigated:

- Air-side sheet pile wall
  For a sheet pile wall simply supported in the ground, the logarithmic spiral of the failure plane intersects the base of this wall.
  For a sheet pile fixed in the ground, the logarithmic spiral of the failure plane intersects the point of zero shear force.
- Load-side sheet pile wall
  The starting point of the logarithmic spiral is in this case generally at the same level as that of the air-side wall. If the load-side wall is shorter than the air-side wall, the failure plane on the load-side wall must continue to the existing base.

A concave failure plane usually governs in a double-walled cofferdam because of the deeper embedment of the sheet pile walls in the subsoil. The resistance to failure of a double-walled cofferdam can be improved by employing one or more of the following measures:

- Widening the cofferdam.
- Selecting a fill material with better γ_k, γ'_k and φ'_k values.
- Compacting the cofferdam fill, possibly the subsoil as well.
- Driving the cofferdam sheet piles deeper if this generates a concave failure plane that satisfies the limit state conditions against failure of the soil.
- Providing further anchor levels (but the advisability of the installation of another anchor level with its associated difficulties, e. g. underwater with the help of divers, should be checked).

3. Base failure
   See Section 8.2.17.1 (2).
4. Ground failure
   See Section 8.2.17.1 (3).
5. Additional analyses in the case of water flow
   See Section 8.2.17.1 (4).

The following constructional measures should be considered:

1. Excavation enclosures
   The information given in Section 8.1.5.6.1 should be taken into account here, but excluding the details regarding interlocks in tension.
   The weepholes near the base of the air- or water-side sheet pile wall should be located in the webs of the sheet pile sections.
   The walings for transferring the anchor forces are mounted on the outer side of the sheet pile as compression walings, provided that shipping operations do not preclude this. Waling bolts are not needed with this solution, and there are also advantages for the installation of the anchors. The anchor penetration on the load side, i. e. water side, must be made watertight.

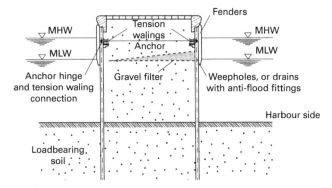

Figure 8.54 Schematic illustration of a mole structure in the form of a double-walled cofferdam.

2. Waterfront structures, breakwaters and moles
   The designs in Section 8.1.5.6.2 apply here, accordingly (Figure 8.54).
3. Special aspects concerning the construction of double-walled cofferdams
   Special attention must be given to the installation of any lower anchors. As these are usually underwater and can only be installed with the help of divers, simple but effective, connections to the sheet pile wall are vital.
   Prior to filling the cofferdam, the bottom surface inside the cofferdam should be cleaned in order to avoid increased earth pressure and, hence, higher loads on the low-level anchors.

8.2.17.3 Narrow moles built from sheet piles

Narrow moles built with sheet pile walls are double-walled cofferdams in which the spacing of the sheet piles is only a few metres and, thus, considerably less than a customary double-walled cofferdam (Section 8.2.17.1). The main loads on these moles are excess water pressure, vessel and ice impacts, line pull, etc.

The sheet pile walls are tensioned against each other at or near the top and are also braced to resist compression and, thus, ensure joint transfer of the external actions.

The space between the sheet piles is filled with sand or gravelly sand with at least a medium density.

To accommodate the external forces acting perpendicular to the axis of the mole and transfer them into the subsoil the mole should be considered as a free-standing structure fully fixed in the ground and consisting of two parallel, interconnected sheet pile walls. The influence of the soil fill between the two sheet pile walls as a result of silo action is neglected when determining the stiffness of the system. The two walls are interconnected by an essentially hinged connection at or near the top. A rigid connection is also possible but leads to large bending moments at the top of the sheet piles and, thus, the need for elaborate connections. In addition, it leads to axial forces in the sheet pile walls, which in some circumstances reduce the passive earth pressure that can be mobilised for a sheet pile wall in tension.

The full magnitude of the mobilisable passive earth pressure cannot be used because this is partly required to resist the active earth pressure and, possibly, excess water pressure due to the filling of the mole. This portion must be determined in advance and deducted from the total mobilisable passive earth pressure.

As both sheet pile walls exhibit an essentially parallel deflection curve due to the external actions, the bending moment in the total system can be distributed over both walls in the ratio of their flexural stiffnesses. The ultimate limit state conditions according to DIN EN 1993-5 are to be checked separately for each of the two sheet pile walls using the applicable design values of the loads (M_{Ed}, V_{Ed}, N_{Ed}) and the section resistances.

The individual sheet pile walls are loaded by active earth pressure from the fill and the surcharge on the fill, as well as by external actions. In addition, resultant water pressures have to be considered if the water level inside the filling can be higher than in front of the sheet pile walls. Generally, a receding water level action should also be considered for the resultant water pressure, which takes into account flooding of the mole with a subsequent, brief lowering of the outer water level. In some circumstances, an internal water level that drops at a much slower rate will cause a higher resultant water pressure. The anchors tying together the sheet pile walls of the mole are to be designed for the tension loads resulting from the aforementioned actions.

Walings, anchors and bracing must be designed, detailed and installed in accordance with the relevant recommendations.

Special attention should be paid to the load transfer points for accommodating the external actions on the mole. An analysis of the forces acting parallel with the centreline of the mole is to be carried out according to Section 8.2.11.

Cross-walls or anchor cells are to be provided as per Section 8.2.17.1.

References

Achmus, M., Kaiser, J. and Wörden, F.T. (2005). Bauwerkserschütterungen durch Tiefbauarbeiten, Grundlagen, Messergebnisse, Prognosen. *Mitteilungen Institut für Grundbau, Bodenmechanik und Energiewasserbau (IGBE), Universität Hannover Heft 61*.

Alberts, D. (2001). Korrosionsschäden und Nutzungsdauerabschätzung an Stahlspundwänden und -pfählen im Wasserbau. *1st conf. „Korrosionsschutz in der maritimen Technik"*, Germanischer Lloyd, Hamburg.

Alberts, D. and Heeling, A. (1996). Wanddickenmessungen an korrodierten Stahlspundwänden; statistische Datenauswertung. *Mitteilungsblatt BAW Nr. 75*, Karlsruhe.

Alberts, D. and Schuppener, B. (1991). Comparison of ultrasonic probes for the measurement of the thickness of sheet-pile walls. In: *Field Measurements in Geotechnics*, (ed. by Sørum). Rotterdam: Balkema.

Baumaschinen-LärmVO. 15. Verordnung zur Durchführung des BImSchG vom 10.11.1986 (Baumaschinen-LärmVO).

Binder, G. (2001). Probleme der Bauwerkserhaltung – eine Wirtschaftlichkeitsberechnung. BAW-Brief No. 1, Karlsruhe.

BAW. www.baw.de, accessed: 21 Jun 2020.

BAW. https://www.baw.de/DE/service_wissen/publikationen/qualitaetsbewertung/qualitaetsbewertung.html, accessed: 21 Jun 2020.

BAW (2011). Richtlinie Prüfung von Beschichtungssystemen für den Korrosionsschutz im Stahlwasserbau (RPB), Federal Waterways Engineering & Research Institute.

Becker, P. (2017). Zum Nachweis der Abtragung von Vertikalkräften bei Verbauwänden. *Bautechnik* 94: 190–199.

Becker, P. and Lüking, J. (2015). Harmonisierung der Berechnungsverfahren der axialen Tragfähigkeit für offene Profile nach EA-Pfähle und EAU. *Bautechnik* 92 (2): 161–176.

Binder, G. and Graff, M. (1995). Mikrobiell verursachte Korrosion an Stahlbauteilen. *Materials and Corrosion* 46 (11): 639–648.

Blum, H. (1931). *Einspannverhältnisse bei Bohlwerken*. Berlin: Ernst & Sohn.

Clasmeier, H.-D. (1996). Ein Beitrag zur erdstatischen Berechnung von Kreiszellenfangedämmen. *Mitteilung des Instituts für Grundbau und Bodenmechanik, Universität Hannover Heft 44*.

Dynamit Nobel (1993). *Sprengtechnisches Handbuch*. Troisdorf: Dynamit Nobel Aktiengesellschaft (ed.).

EA-Pfähle (2012). *Recommendations on Piling* (ed. by Deutsche Gesellschaft für Geotechnik e. V.). Berlin: Ernst & Sohn.

EAB (2012). *Empfehlungen des Arbeitsausschusses "Baugruben"* (ed. by Deutsche Gesellschaft für Geotechnik e. V.). Berlin: Ernst & Sohn.

EA-Pfähle, TJB (2014). Jahresbericht 2014 des Arbeitskreises „Pfähle" der DGGT. *Bautechnik* 91 (12).

EAU (2015). *Recommendations of the Committee for Waterfront Structures, Harbours and Waterways EAU 2012* (ed. by Deutsche Gesellschaft für Geotechnik e. V. and Hafentechnische Gesellschaft e. V.). Berlin: Ernst & Sohn.

Grabe, J. and Heins, E. (2016). Diskussionsbeitrag zur axialen Traglast von Wänden im Grenzzustand des Versinkens. *Bautechnik*, 93: 304–311, https://doi.org/10.1002/bate.201500060.

Graff, M., Klages, D. and Binder, G. (2000). Mikrobiell induzierte Korrosion (MIC) in marinem Milieu. *Materials and Corrosion* 51 (4): 247–254.

Heiß, P., Möhlmann, F. and Röder, H. (1992). Korrosionsprobleme im Hafenbau am Übergang Spundwandkopf zum Betonüberbau. *HTG-Jahrbuch* 47: 170–174.

Hein, W. (1990). Zur Korrosion von Stahlspundwänden in Wasser. *Mitteilungsblatt BAW* 67: 1–40, Karlsruhe.

Hettler, A., Becker, P. and Kinzler, S. (2018). Bericht des Arbeitskreises Baugruben: Entwurf EB 85 und Anhang A 10. *Bautechnik* 95: 684–692, https://doi.org/10.1002/bate.201800049.

Laumans, Q. (1977). Verhalten einer ebenen, in Sand eingespannten Wand bei nichtlinearen Stoffeigenschaften des Bodens. Baugrundinstitut Stuttgart, Mitteilung 7.

Lüking, J. and Becker, P. (2015). Harmonisierung der Berechnungsverfahren der axialen Tragfähigkeit für offene Profile nach EA-Pfähle und EAU. *Bautechnik* 92: 161–176, https://doi.org/10.1002/bate.201400062.

Weißenbach, A. (1985). *Baugruben, Teil II, Berechnungsgrundlagen*, 1st reprint. Berlin: Ernst & Sohn.

Wirsbitzki, B. (1981). *Kathodischer Korrosionsschutz im Wasserbau*. Hamburg: Hafenbautechnische Gesellschaft e. V.

ZTV-W (LB 214) (2015). *Zusätzliche Technische Vertragsbedingungen – Wasserbau für Spundwände, Pfähle, Verankerungen*. Karlsruhe: Bundesanstalt für Wasserbau.

ZTV-W (LB 218) (2009). *Zusätzliche Technische Vertragsbedingungen – Korrosionsschutz im Stahlwasserbau*. Karlsruhe: Bundesanstalt für Wasserbau.

Standards and regulations

DIN 1054: Subsoil – Verification of the safety of earthworks and foundations – Supplementary rules to DIN EN 1997-1.

DIN 4017: Soil – Calculation of design bearing capacity of soil beneath shallow foundations.

DIN 4084: Soil – Calculation of embankment failure and overall stability of retaining structures.

DIN 4085: Subsoil – Calculation of earth pressure.

DIN 4150: Vibrations in buildings.

DIN 18196: Earthworks and foundations – Soil classification for civil engineering purposes.

DIN 45669: Measurement of vibration immissions.

DIN 50929: Corrosion of metals – Corrosion likelihood of metallic materials when subject to corrosion from the outside.

DIN EN 1537: Execution of special geotechnical works – Ground anchors.

DIN EN 1992 Eurocode 2: Design of concrete structures.

DIN EN 1993 Eurocode 3: Design of steel structures.

DIN EN 1995 Eurocode 5: Design of timber structures.

DIN EN 1997 Eurocode 7: Geotechnical design.

DIN EN 10025: Hot rolled products of structural steels.

DIN EN 10028: Flat products made of steels for pressure purposes.

DIN EN 10204:2005-01: Metallic products – Types of inspection documents; German version EN 10204:2004

DIN EN 10219: Cold formed welded structural hollow sections of non-alloy and fine grain steels.

DIN EN 10248: Hot-rolled steel sheet pile.

DIN EN 10249: Cold-formed steel sheet pile.

DIN EN 12063: Execution of special geotechnical work – Sheet pile walls

DIN EN 12699: Execution of special geotechnical works – Displacement piles.

DIN EN 16228: Part 1–7 Drilling and foundation equipment – Safety.

DIN EN ISO 2560: Welding consumables – Covered electrodes for manual metal arc welding of non-alloy and fine-grain steels.

DIN EN ISO 3834: Quality requirements for fusion welding of metallic materials.

DIN EN ISO 12944: Paints and varnishes – Corrosion protection of steel structures by protective paint systems

DIN EN ISO 14341: Welding consumables – Wire electrodes and weld deposits for gas shielded metal arc welding of non-alloy and fine grain steels – Classification

DIN EN ISO 14688: Geotechnical investigation and testing – Identification and classification of soil.

DIN EN ISO 15614: Specification and qualification of welding procedures for metallic materials – Welding procedure test.

ns and sliding, to carry horizontal loads from earth and water pressure as well as loads from
9
Anchorages

9.1 Piles and anchors

9.1.1 General

As a rule, waterfront structures need appropriate anchoring elements to prevent overturning and sliding, to carry horizontal loads from earth and water pressure as well as loads from the superstructure such as line pull and vessel impact. For smaller changes in ground level, these loads can be carried by appropriately designed structures supported on pile bents or trestles if necessary. Larger changes in ground level, such as those in modern seaports and inland ports, require special anchoring solutions. The main components used for anchoring waterfront structures are as follows:

- Piles, especially displacement piles, micropiles, special piles (load transferred to subsoil over full length, no "unbonded length").
- Anchors (load transferred to subsoil locally via anchorage elements, e. g. anchor plates, local body of grout of limited length, "unbonded anchor length").

9.1.2 Displacement piles

9.1.2.1 Installation

The installation of displacement piles is regulated by DIN EN 12699 "Displacement piles" in conjunction with DIN SPEC 18 539. No soil is removed during installation; instead, the soil immediately around the pile is forced aside into the surrounding soil and possibly compacted. Displacement piles are installed using impact-driving, vibratory, pressing or drilling methods. Three typical types are described below; more detailed information can be found in the *Recommendations on Piling* (2012). Displacement piles can be installed as raking piles (generally up to 45°).

9.1.2.2 Steel piles (without grouting)

Steel piles are rolled products that can be supplied as I-sections in long lengths. With appropriate planning, their geometrical shape and grade of steel can be adapted to suit the particular structural, geotechnical and installation circumstances of the project. Steel piles can be driven or vibrated as "plain" piles and are relatively unaffected by obstructions in the ground and difficult driving conditions. Where required by the local soil conditions, they can be lengthened at the pile head. Steel piles are easy to weld to other structures made from steel or reinforced concrete.

The steel grades used should comply with DIN EN 1993-1-1 and DIN EN 10025.

Under special circumstances, steel wings can be welded to steel piles to improve their load-bearing capacity. However, such piles should only be used in soils – preferably non-cohesive – free from all obstacles, and must be embedded sufficiently deeply in load-bearing subsoil. Where there are cohesive strata, the wings should be located below these, and any open driving channels should be closed off, e. g. by grouting. Wings should be designed and positioned such that they do not render driving too difficult and can themselves resist the driving process intact. The shape of the wings and level at which they are attached should, therefore, be adjusted to suit the particular soil conditions. It should be noted that saturated cohesive soils are, indeed, displaced during driving, but not compacted. In non-cohesive soil, a highly compacted solid plug can form, primarily around the wings, as a result of the vibrations of driving, which then makes further driving more difficult.

Wings should be positioned symmetrically about the pile axis just above the toe of the pile so that there is enough space to fit a min. 8 mm thick weld seam between wing and end of pile. The upper end of the wing must also have a correspondingly strong transverse weld. These weld seams link up with approximately 500 mm long joints on both sides of the wing in the longitudinal direction of the pile. Between these, intermittent weld seams suffice.

Considering the restraint forces, the wing attachment area should be sufficiently wide (generally at least 100 mm). Depending on the soil stratification, wings can also be positioned higher up the pile shaft.

9.1.2.3 Grouted piles

This term also includes driven piles with a grouted skin and grouted displacement piles. Grouted piles consist of steel sections as described in Section 9.1.2.2. However, they have a special toe form for introducing the grout at the base of the pile. A box-shaped cutting shoe made from welded sheet steel is attached to the toe of the pile.

The grout is fed through a line fitted to the pile and extending down to the toe of the pile, where it is ejected and, thus, forced into the voids left by the cutting shoe.

The hardened cement mortar or concrete with a high fines content in the cavity creates the bond between the steel pile and the subsoil. Depending on the subsoil, the skin friction that can be activated can be three to five times greater than that of a non-grouted steel pile.

A further increase in the load-bearing capacity of the grouted pile results from the displacement of the soil during driving, which creates three-dimensional tension in the subsoil. Grouted piles are generally installed at rakes between 2 : 1 and 1 : 1.

Grouted piles are particularly suited to non-cohesive soils with a relatively high pore volume.

9.1.2.4 Vibratory driven grouted piles

Vibratory driven grouted piles have certain similarities to grouted piles. They consist of steel sections as described in Section 9.1.2.2. At the base, to widen the I-shaped pile cross-section, approximately 20 mm thick steel plates are welded to the web and flanges on all sides. These create a cavity along the pile shaft equal to the plate thickness, which is filled

Pile instead of anchor?
Without prestressing?
TITAN Micropile.

- Comparable factor of safety – even without prestressing
- One installation method – for all ground conditions
- Flexible system – also ideal where access is difficult

Further details: www.ischebeck.com

FRIEDR. ISCHEBECK GMBH
Loher Str. 31-79 | DE-58256 Ennepetal | Germany

Bill Addis (Ed.)

Physical Models

Their historical and current use in civil and building engineering design

- the book summarizes the history of model testing by design and construction engineers in a single volume for the first time
- model testing is alongside knowledge of materials and structural behaviour a major driver in progress in civil and building engineering

The book traces the use of physical models by engineering designers from the eighteenth century, through their heyday in the 1950s-70s, to their current use alongside computer models. It argues that their use has been at least as important in the development of engineering as scientific theory has.

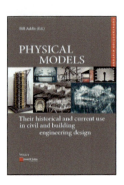

2020 · 1114 pages · 896 figures · 14 tables

Hardcover
ISBN 978-3-433-03257-2 € 149*

eBundle (Print + PDF)
ISBN 978-3-433-03305-0 € 249*

ORDER
+49 (0)30 470 31-236
marketing@ernst-und-sohn.de
www.ernst-und-sohn.de/en/3257

* All book prices inclusive VAT.

with cement mortar during vibratory driving in a similar way to the grouted pile to increase the shaft resistance.

Vibratory driven grouted piles can be installed vertically and at rakes of up to 1 : 1. However, it is difficult to achieve satisfactory penetration (pile penetration per unit of time related to the vibration energy applied) with vibratory driven raking piles. It has proved beneficial to apply a "prestress" of approximately 100 kN axial compression to raking vibratory driven grouted piles. This enables the pile to maintain continuous frictional contact with the subsoil, and the energy input can then be transferred more effectively to improve the penetration rate.

9.1.3 Load-bearing capacity of displacement piles

9.1.3.1 Internal load-bearing capacity

The internal load-bearing capacity of a pile should be verified depending on the materials used. *Recommendations on Piling* (EA-Pfähle, 2012), Section 5.10, contains a detailed description of the analyses required. In grouted displacement piles the force transfer is achieved by way of the bond stresses between the steel pile and the grout. These bond stresses have been proved to be higher than the skin friction values of the grout and the subsoil and, therefore, verification of this interface is unnecessary. The reader is expressly referred to the corresponding literature (the paper by Kempfer, Luking, Mardtfeld). In addition, these results are also included in appendix A5 of the *Recommendations on Piling* (EA-Pfähle, 2012).

9.1.3.2 External load-bearing capacity

The resistance of a displacement pile in tension is calculated according to DIN EN 1997-1. According to DIN EN 1997-1, the external load-bearing capacity of a tension pile should always be verified by loading tests.

The procedure for checking the external load-bearing capacity according to DIN EN 1997-1 is dealt with in detail in EA-Pfähle (2012), Section 6.3. Pile tests and any necessary pile loading tests are also covered in EA-Pfähle (2012).

For preliminary design, the external load-bearing capacity of tension piles can also be derived approximately from the resistance measured in cone penetration tests. To ensure that the in situ soils are correctly identified, the local skin friction should always be measured as well during penetrometer tests. The unequivocal allocation of the values obtained from penetrometer tests to the in situ soil types requires the help of at least one borehole to calibrate the results of the penetrometer tests. Reference values for characteristic pile resistances depending on the types of soil and their strengths can be found in EA-Pfähle (2012).

9.1.3.3 Additional stresses

A wide variety of additional stresses may be encountered when anchoring waterfront structures with raking piles, e. g. due to consolidation settlement, compaction settlement, block load-bearing behaviour, ground deformation due to the deflection of a sheet pile wall, etc. With ductile piles, no verification of the internal stability in the span is required for the additional stresses referred to above and when the wall connection detail is sufficiently flexible or ductile. Instructions on how to take account of these additional stresses can be found in EA-Pfähle (2012), Section 4.6.2.

9.1.4 Micropiles

9.1.4.1 Installation

The installation of micropiles is regulated by DIN EN 14199 in conjunction with DIN SPEC 18539. Micropiles can be installed in the form of bored piles (with a maximum shaft diameter of 300 mm) or as displacement piles (with a maximum cross-section dimension of 150 mm). The tension member (tendon) is either inserted into a pre-drilled hole or, in the case of self-drilling micropiles, itself functions as a sacrificial drilling rod. Micropiles are generally not prestressed. Unlike grouted anchors, micropiles transfer the load into the subsoil over their entire length, provided that there is no unbonded segment.

9.1.4.2 Composite piles

Micropiles in the form of composite piles have a continuous, prefabricated steel load-bearing tendon filled and surrounded by pressure-injected cement mortar. Grouting takes place under increased hydrostatic pressure (min. 5 bar) on the grouting material. With appropriate equipment it is possible to post-grout the micropiles using sleeve pipes or tubes à manchette. Post-grouting is not possible with self-drilling micropile systems. Composite piles require an approval. The tendons can be made from ribbed steel reinforcing bars, solid steel bars or steel pipes with cut or rolled threads or welded-on wires. This makes it easy to create connections to reinforced concrete or steel wall structures. Approvals are also required for these connection details.

9.1.4.3 Cast-in-place concrete piles

Micropiles in the form of cast-in-place concrete piles have a shaft diameter of 150–300 mm. They are inserted into pre-drilled holes, include longitudinal reinforcing bars over their full length and are filled with concrete with a high level of fine aggregate or cement mortar. Grouting is generally carried out with compressed air on the exposed cast concrete surface. Post-grouting is possible by inserting sleeve pipes with each pile.

9.1.4.4 Load-bearing capacity of micropiles

9.1.4.4.1 Internal load-bearing capacity of micropiles

The internal load-bearing capacity of micropiles in the form of composite piles in accordance with Section 9.1.4.2 is determined by the material and cross-section parameters of the steel used and regulated by the approval required.

The internal load-bearing capacity of micropiles in the form of cast-in-place piles in accordance with Section 9.1.4.3 should be verified in line with the design rules for reinforced concrete.

9.1.4.4.2 External load-bearing capacity of micropiles

The resistance of micropiles in tension is calculated according to DIN EN 1997-1. In order to verify the external load-bearing capacity, loading tests should always be carried out afterwards.

The external load-bearing capacity of micropiles for preliminary designs may be calculated using the empirical values given in EA-Pfähle (2012). However, these values, too, should be verified in loading tests. According to DIN EN 1997-1, loading tests for micropiles in tension should be carried out on a minimum of 3% of the piles (but no fewer than two piles).

9.1.4.4.3 Bond stresses in grouted micropiles
Verifying the bond stresses at the pile shaft/grout interface of a grouted micropile is always carried out for each particular type during investigations for the approval.

9.1.4.4.4 Additional stresses
Basically, the information given in Section 9.1.4 also applies to micropiles. According to *Recommendations on Piling* (2012), Section 4.6.2(4), the corrosion protection to the load-bearing tendon in a micropile is not impaired by additional stresses applied transversely to the pile axis when the tendon is housed in a continuous plastic sleeve filled with cement mortar.

9.1.5 Special piles

Special solutions for anchoring quay walls are always appearing on the market. These primarily involve a combination of different, well-known building processes and methods that have been developed for a specific method of construction, e. g. jet grouting combined with driven or bored steel sections. These piles have proved worthwhile in individual cases; however, there is insufficient experience that can be generalised to discuss them separately in the context of this recommendation. Where such special solutions are to be used, their feasibility should always be verified on a case by case basis.

9.1.6 Anchors

9.1.6.1 Construction
Unlike piles, anchors always have a well-defined anchorage length or anchoring point via which the load is transferred into the ground as well as an equally clearly defined unbonded length over which no load is intended to be transferred into the ground. Anchors are installed by inserting them into cased or uncased holes (grouted anchors). Alternatively, they can be positioned close to the surface, fixed to anchor plates and then covered (e. g. anchors made from round steel rods or rolled sections). The head details of grouted anchors are regulated by approvals. See Section 9.2 for anchor connection details.

9.1.6.2 Grouted anchors
The construction of grouted anchors is regulated by DIN EN 1537 in conjunction with DIN SPEC 18537. Grouted anchors consist of an anchor head and a tension member (tendon) with a defined unbonded anchor length and an anchorage length. Structural steel sections, reinforcing bars and prestressing steels (bars or strands) can be used for the tendons. Grouted anchors in accordance with DIN EN 1537 can be prestressed. All grouted anchors for long-term use (> 2 years) require an approval. The approval covers all aspects from the factory production of the tendon, the corrosion protection, the installation of the tendon through to testing and prestressing the anchors.

9.1.6.3 Anchoring with anchor plate
As with the grouted anchors referred to in Section 9.1.6.2, this type of anchor consists of an anchor head and a defined unbonded anchor length. The load is not transmitted into the subsoil via the bond between anchor and soil but via a large anchor plate (anchor wall).

Figure 9.1 Example of sheet pile wall anchor plates.

Upset T-heads and threaded ends, anchor chairs, coupling sleeves and turnbuckles, as well as anchor plates made from sheet pile sections or precast concrete elements (see Figure 9.1) allow for various structural solutions to address a wide variety of situations. The angle of installation for these anchors is normally limited to a maximum of 8–10° because, otherwise, the earthworks required to excavate the anchor trench and then backfill it once again are uneconomic.

Horizontal anchors in the form of "dead-man" anchors are suitable for special structures such as pier heads, quay walls and headlands provided that the anchor can be tied to the opposite quay wall. In such cases, it is not usually possible to install anchors in the form of raking piles.

9.1.6.4 Prefabricated inclined anchors

A prefabricated inclined anchor is a factory-fabricated steel tension element consisting of a steel pile section plus an anchor plate, which is rotated into position on site. The angle of installation is between 0° and 45° from the horizontal (see Figure 9.2).

During prefabrication, a hinged steel fitting is mounted on the anchor head, so that the anchor can later be attached to the quay wall but still rotated into position (see Figures 9.2 and 9.3). To accommodate this anchor head detail, each bearing pile is fitted with a compatible steel mounting that allows the anchor to be rotated into position on site. The anchor plate is welded to the buried end of the anchor at 90° to the anchor axis (see Figure 9.3).

Lifting gear is required to install such prefabricated inclined anchors. While suspended from a crane, the anchor head is attached to the sheet pile wall so that it can rotate but still

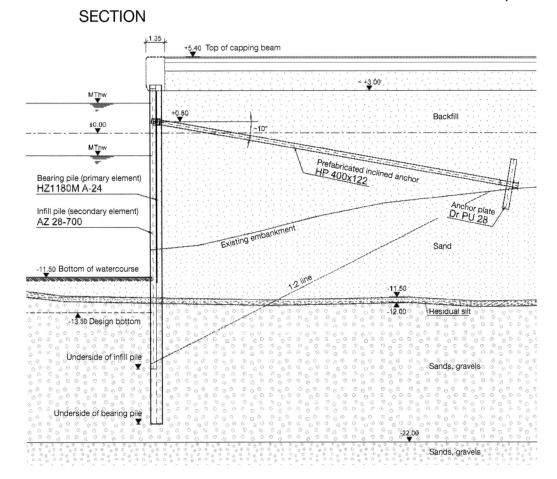

Figure 9.2 Example of a prefabricated inclined anchor (section).

constitute a structural connection. The anchor shaft rotates about the hinge at the wall, thus allowing the anchor with its anchor plate, still supported by the crane, to be lowered to the bed of the watercourse. This is then the final, "rotated" position of the inclined anchor. Backfilling of the area around the anchor can then commence. It is essential that the first loads of fill material are placed directly in front of and on the anchor plate so that the anchor resistance can be mobilised right from this stage. Vibrating each anchor plate into the bed of the watercourse for a distance equal to half its height with the help of an underwater vibrator improves the load-bearing capacity of such prefabricated inclined anchors.

9.1.6.5 Load-bearing capacity of anchors
9.1.6.5.1 Internal load-bearing capacity

The internal load-bearing capacity of anchors is determined by the material and cross-sectional parameters of the steel used.

9 Anchorages

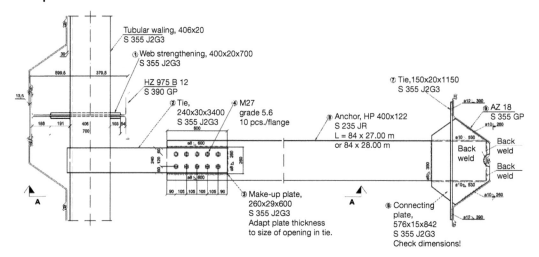

Figure 9.3 Example of a prefabricated inclined anchor (plan).

9.1.6.5.2 External load-bearing capacity

Suitability tests should always be carried out to determine the external load-bearing capacity of grouted anchors. The characteristic pull-out resistance is determined from the results of tests according to DIN EN 1997-1.

In addition, all grouted anchors must pass an acceptance test in accordance with the German Eurocode 7 manual prior to being used. Random tests, as used for piles and micropiles, are not enough.

9.1.6.6 Load-bearing capacity of anchor walls

Verification of the stability of anchorages with anchor plates was dealt with in Section 8.2.7 (1).

Predominantly variable loads

Section 8.2.7 (2) applies when verifying the load-bearing capacity. Close-tolerance bolts of grade 4.6 or higher are to be used for bolted connections to walings and capping beams.

9.1.6.7 Round steel tie rods and waling bolts

The design of round steel tie rods and waling bolts is carried out in accordance with DIN EN 1993-5, Section 7.2, using the reduction factor κ and the core cross-sectional area A_{core}.

Predominantly static loads

Grades of steel for round steel anchors and waling bolts are the same as those given in Section 8.1.2.1 for sheet piles.

The analysis format for the ultimate limit state to DIN EN 1993-5 is

$$Z_d \leq R_d$$

The design values are calculated using the following variables:

Z_d design value for anchor force, $Z_d = Z_{G,k} \cdot \gamma_G + Z_{Q,k} \cdot \gamma_Q$
R_d design resistance of anchor, $R_d = \min[F_{tt,Rd}; F_{tg,Rd}]$
$F_{tg,Rd}$ $A_g \cdot f_y / \gamma_{M0}$
$F_{tt,Rd}$ $A_g \cdot f_y / \gamma_{M0}$
A_g cross-sectional area of shaft
A_{core} core cross-sectional area of thread
$f_{y,k}$ yield stress
$f_{ua,k}$ tensile strength
γ_{M0} partial safety factor to DIN EN 1993-5 for anchor shaft
γ_{M2} partial safety factor to DIN EN 1993-5 for threaded segment
κ reduction factor ($\kappa = 0.6$)

The design resistance of the anchor R_d is generally taken to be the resistance of the threaded part multiplied by the reduction factor κ. The reduction factor $\kappa = 0.6$ also takes into account any potential additional stresses due to anchor installation under the less-than-ideal conditions of building sites and the ensuing unavoidable bending stresses in the threaded part. Notwithstanding, it is still necessary to provide constructional measures to ensure that the anchor head can rotate sufficiently.

The additional serviceability analyses called for in DIN EN 1993-5 are already implied in the limit state condition $Z_d \leq R_d$ owing to the κ value selected and the customary upsetting relationships between shaft and thread diameters; the additional analyses are, therefore, unnecessary. Round steel tie rods can have cut, rolled or hot-rolled threads to Section 9.1.6.8.

A prerequisite for a proper design is a suitable detail for the anchor connection, which means anchors must be connected via a form of hinge. Anchors must be installed at a higher level so that any settlement or subsidence does not cause any additional stresses.

Upsetting of the ends of the anchor bars for the threaded segments and T-heads, as well as tie rods with eyes is permissible when

- Using steel grade groups J2 and K2, if necessary, in the normalised/normalising rolled condition (+N) – however, not thermomechanically rolled J2 and K2 group steels (see Section 8.1.2.1).
- Other steel grades, e. g. S 355 J0, are used and accompanying tests ensure that the strength values do not fall below those in DIN EN 10025 after the normalising procedure of the forging process.
- The upsetting and the provision of T-heads and eyes are carried out by specialist fabricators, and it is ensured that the mechanical and technological values in all parts of the round steel tie rod are in accordance with the steel grade selected, the alignment of the fibres is not impaired during the machining process, and detrimental microstructure disruptions are reliably avoided.

The "failure of an anchor" analysis does not need to be performed for round steel tie rods and anchors.

As mentioned in Section 8.1.10.4.5, round steel tie rods should be installed without an anti-corrosion coating. The close-tolerance, continuous threads standardised in ISO 13 (see

Figures 9.2 and 9.3, Section 9.1.6.4) offer relatively good protection against subsurface corrosion.

After installation in fill, round steel tie rods must be surrounded by a sufficiently thick layer of sand over their full length.

If it is necessary to coat round steel tie rods to prevent corrosion, measures will be necessary on site to ensure that the coating is not damaged. If, despite this, damage does occur, the coating must be repaired, so that the original quality is restored.

9.1.6.8 Screw threads for sheet piling anchors
The following thread types are used for sheet piling anchors:

9.1.6.8.1 Cut thread (machined thread) to Figure 9.4
The outside thread diameter is equal to the diameter of the round steel bar or the upsetting.

9.1.6.8.2 Rolled thread (non-cut thread produced in cold state) to Figure 9.5
After the thread has been rolled, the outer diameter of the thread is marginally greater than the diameter of the anchor rod. The diameter or the upsetting diameter of tie rods with rolled threads can, therefore, be smaller than anchors with cut threads for the same load-bearing capacity.

When using steel grades S 235 JR and S 355 J2, any upsetting must be turned down to the nominal thread diameter before rolling the thread in order to achieve a thread conforming to the standard. Drawn steels (up to Ø36 mm) do not need to be pre-machined.

9.1.6.8.3 Hot-rolled thread (non-cut thread produced in hot state) to Figure 9.6
Hot-rolling produces two rows of thread flanks on the thread shaft, which are opposite to each other but complement each other to form a continuous thread. The nominal diameter

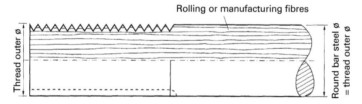

Figure 9.4 Cut thread.

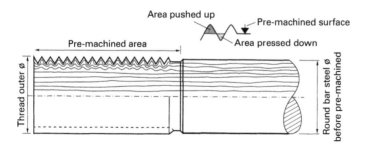

Figure 9.5 Rolled thread.

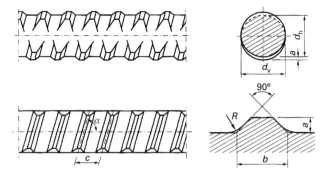

Figure 9.6 Hot-rolled thread.

governs the load-bearing capacity, although the actual diameter can easily deviate from this. Components with the same thread type must be used for end anchorages and butt joints.

9.1.6.9 Required safety margins
For more information on verifying the load-bearing capacity of round steel tie rods with threads of various types plus the structural design of anchorages, see Section 9.1.6.7.

9.1.6.10 Further information on rolled threads
- Rolled threads have a high profile accuracy.
- Rolling a thread is a type of cold-forming. Cold-forming increases the strength and yield stress of the thread root and flanks, which has a favourable effect on transferring anchor forces concentrically via the thread.
- The thread root and flanks are particularly smooth on rolled threads and, therefore, have a higher fatigue strength under dynamic loads.
- In contrast to cut threads, the fibre orientation in the steel is not disrupted in rolled or hot-rolled threads.
- Rolled threads with larger diameters are primarily suited to concentrically loaded anchors with dynamic loads.
- In the case of round steel tie rods with rolled threads, it is important that also the nuts, couplers and turnbuckles do not have rolled threads, because internal threads are always stressed less than external ones. When the internal thread is loaded, hoop tension is generated, which supports the thread. Therefore, anchors with rolled external threads can be combined with nuts and turnbuckles having cut internal threads without hesitation.

The aforementioned measures reduce the risk of anodic areas on the round steel tie rods and any ensuing pitting.

The design and installation of sheet pile wall anchorages with grouted round steel tie rods is covered by DIN 1054 and DIN EN 1537.

9.1.6.11 Predominantly fluctuating loads
Anchors are generally mainly subjected to static loads. Primarily fluctuating loads only occur in anchors in rare, special cases, but more frequently in waling bolts.

Only fully killed steels to DIN EN 10025 may be used where fluctuating loads are likely. Verification of fatigue strength is to be carried out according to DIN EN 1993-1-9.

If the basic static load is less than or equal to the reversed load amplitude, the recommendation is to apply a permanent, controlled prestress that exceeds the stress amplitude to the anchors or waling bolts. This ensures that the anchors or waling bolts remain under stress and do not fail abruptly when the stress increases again.

A prestress, which is not defined exactly, is applied to anchors and waling bolts, in many cases during the installation procedure. In cases without controlled prestressing, a stress of only $\sigma_{R,d} = 80\,N/mm^2$ may be assumed for the threads of anchors or waling bolts, regardless of the design situation and steel grade, and neglecting the prestress.

Always ensure that the nuts of waling bolts cannot loosen due to repeated changes in stress.

9.2 Walings and pile and anchor connections

Walings are normally installed as tension members on the inside of sheet pile walls, so that the outer face of the wall remains smooth. On anchor walls they generally act as compression members fitted on the back of the wall.

Walings and pile and anchor connections must be robust because they have to resist additional actions due to, for example, wall alignment, berthing pressures or the varying stiffnesses of anchors in addition to the loads for which they are actually designed. Therefore, the forces calculated for the anchors to sheet pile walls should be increased by 15%.

9.2.1 Design of steel walings for sheet piling

9.2.1.1 Design

It is expedient to provide walings in the form of two channels.

Splices between waling sections should be positioned at points of minimum stress. A full cross-section splice is not required but it must be capable of carrying the calculated internal forces.

9.2.1.2 Fixings

Walings are either supported on welded brackets (Section 8.2.11, Figures 8.44–8.46) or – especially with limited working space beneath the walings – suspended from the sheet piling. The fixings should be designed so that vertical loads on the walings can be transferred to the sheet piling. Brackets simplify the mounting of the walings. Suspension details must not weaken the walings and should, therefore, be welded to the walings or attached to the plate washers of the waling bolts.

The anchor force is transmitted through the waling bolts into the walings. Bolts are placed in the centre between the two waling channels and transfer their load through plate washers welded to the walings (Section 8.2.11, Figures 8.44–8.46). Overlong waling bolts are used to help align the sheet piling against the walings.

Owing to the risk of corrosion and the stresses caused by wall alignment and berthing pressures, waling bolts must be at least 38 mm in diameter.

9.2.1.3 Extra waling

Sheet piling that has become severely misaligned during driving can be locally realigned with an extra waling, which remains as part of the structure.

9.2.2 Verification of steel walings

Walings, waling bolts (see also Section 9.1.6.7) and plate washers are designed to DIN EN 1993-1 or DIN EN 1993-5. Heavier walings made from steel grade S 235 JR are preferable to lighter ones made from S 235 J2, as they are more robust and can, therefore, be used to align the wall. Splices, stiffeners, bolts and connections must be designed in accordance with structural steelwork standards, and the components should permit easy welding. To allow for possible corrosion, load-bearing weld seams must be at least 2 mm thicker than required by the structural calculations. Additionally, the welds must be designed to resist all anticipated horizontal and vertical actions and transfer them to the anchors or the sheet pile wall (anchor wall). The following actions are to be considered for the design check.

9.2.2.1 Horizontal actions
1. The design value of the horizontal component of the anchor tension from the sheet pile wall calculations plus 15%.
2. The design values of line pull forces applied directly to the waling.
3. The design value of the berthing load depending on the size of the vessel, the berthing manoeuvres and current and wind conditions. Ice impact may be neglected.

9.2.2.2 Vertical actions
1. The dead load of the waling members including stiffeners, bolts and plate washers.
2. The portion of the soil surcharge calculated from the rear surface of the sheet piling to a vertical plane through the rear edge of the waling.
3. The portion of the imposed load on the quay wall between the rear edge of the sheet piling capping beam and a vertical line through the rear edge of the waling.
4. The vertical component of the active earth pressure from underside of waling to ground surface and acting on the vertical plane passing through the rear edge of the waling.
5. The vertical component of the inclined anchor tension plus 15%.

The design values of the loads given in points 1–5 above are to be considered for limit state GEO-2.

With several walings one above the other, the vertical loads are divided among the walings. In order to ensure the secure mounting of the waling brackets, the actions are applied at the rear edge of the waling.

9.2.2.3 Method of calculation

The actions are split into component forces vertical and parallel to the surface of the sheet pile wall (principal waling axes). In the calculations, it should be assumed that the walings are supported by the anchors to carry forces perpendicular to the plane of the wall and by the brackets or suspension details for actions parallel to the wall. The support and span moments resulting from the design value of the sheet piling support reaction are – considering the end bays – generally calculated using the equation $q \cdot l^2/10$.

9 Anchorages

9.2.3 Reinforced concrete walings to sheet pile walls with driven steel piles

9.2.3.1 General

Anchors in the form of steel anchor piles driven with a rake of 1:1 have frequently proved practicable and quite economical for quay walls.

This is particularly the case when the upper strata behind a wall consist of soil types that make anchoring with round steel tie rods difficult or even impossible, but also when the wall must be anchored before backfilling.

If the anchor piles are driven before the sheet pile wall and the sheet piles suffer forward or backward lean during driving, then the anchor piles already installed are not always in the right position relative to the sheet piling.

However, inaccuracies of this kind can be compensated for with reinforced concrete walings in which the reinforcement allows for the positions of the anchors (Figure 9.7).

It is advisable to mount a temporary steel waling on the sheet piling if the reinforced concrete waling is a considerable distance above the existing terrain. It is removed after the piles have been connected and the reinforced concrete waling has reached its full strength.

9.2.3.2 Construction of reinforced concrete walings

To ease construction, the dimensions of a reinforced concrete waling should not be smaller than those shown in Figure 9.7.

Reinforced concrete walings are connected to the sheet piling by round or square steel bars that are welded to the sheet pile webs (Figure 9.7, Nos. 4 and 5). Extra reinforcement is generally only used at expansion joints. The pile forces are also transferred via round or square steel bars (Figure 9.7, Nos. 1–3).

The steel connecting bars welded to the sheet piles and the anchor piles are generally grade S 235 JR. Round steel bars of grade BSt 500 are also used. Square steel bars can be welded directly to the wall and the anchor.

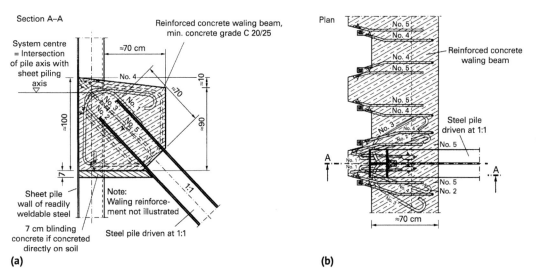

Figure 9.7 Reinforced concrete waling for steel sheet piling: (a) section A–A; (b) plan view.

Welding may only be carried out by qualified welders under the supervision of a welding engineer. Only materials whose suitability for welding is known and which are of uniformly good quality and are compatible with each other are to be used (see Section 8.1.4).

9.2.3.3 Connection between pile and waling

If no large-scale settlement or subsidence are to be expected in the backfill behind the wall, the anchor piles can be directly fixed at the reinforced concrete waling.

This economical solution can also be used when the restraint stresses resulting from minor settlement (e. g. thin soil strata liable to settle, or well-compacted backfill with non-cohesive soil) are included in the design of the waling.

Figure 9.8 shows a favourable connection solution with flat-head bars – as already used for many years for bollard anchor details. Here, one end of the round steel bar is upset to form a circular disc at the head, which is up to three times the diameter of the bar. The end of the round steel bar to be welded to the tension pile is flattened to ensure a good weld.

End anchoring in the concrete can also be achieved, however, by welding cross-bars or plates of a suitable size to the round and square anchor bars.

The pile should be embedded in the reinforced concrete by an amount equal to roughly twice its width/diameter.

It is necessary to check the transfer of the internal forces in the steel pile at the connection to the reinforced concrete waling. In doing so, the combined stresses in the top of the pile due to normal force, shear force and bending moment should be taken into account. If necessary, strengthening plates may be welded to the sides of the steel pile to improve the transfer of the tensile forces. The anchoring bars, in the form of loops, are then connected to these plates. The voids that tend to form alongside the webs of the anchor piles as a consequence of this detail must be carefully filled with concrete to avoid corrosion.

Only specially killed steels (FF) resistant to brittle fracture, e. g. grade S 235 JG or S 355 JR, may be used for piles and their connections in quay walls anchored with piles subjected to larger bending stresses.

If anchors are placed in areas of soil with thicker strata of soil that are highly sensitive to settlement or if the backfill cannot be compacted, then it is better to include a hinge in the pile connection detail.

9.2.3.4 Design of pile connections

The calculated pile force plus 15% is assumed to act at the intersection between the sheet pile wall axis and the pile axis. The waling, including its connections to the sheet piling, is taken to be uniformly supported. The dead loads, vertical surcharges, pile forces, bending moments and shear forces from the anchor piles are actions and are included in the calculations as design values.

The forces acting at the pile connection, which result from the soil surcharges on the steel pile due to backfill or settlement, are calculated for an equivalent beam assumed to be fixed at the waling and in the load-bearing soil. The fixity moment acting at the pile connection plus the shear force acting at that position must be verified for design situation DS-A when analysing the connection between the waling and the sheet pile wall. With respect to the analyses of the wall itself, these loads need only be taken into account when the anchor piles are considered to spread the load on the sheet piling.

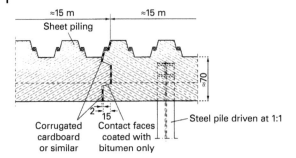

Figure 9.8 Joggle joint in a reforced concrete waling.

A weakening of the pile cross-section at the point of fixity at the waling in order to reduce the connection moment and the associated shear force is not permissible because such a reduction can easily lead to pile failure, especially in the case of poor execution of the work.

If steel piles are connected via hinged connections, then the hinge must also be verified for additional loads due to settlement and/or subsidence.

9.2.3.5 Expansion joints

Reinforced concrete walings can be cast with or without expansion joints. Their design is based on Sections 9.2.5.4 and 10.3.2. See Section 10.3.1 for details of construction joints.

Where expansion joints are specified, they should be arranged so that they do not hinder the changes in length of the sections.

To provide mutual support for the separate sections of the structure in the horizontal direction, expansion joints should have a joggled form, with dowels, if necessary, as per Figure 9.8. The horizontal joggle is placed in the slab in the case of quay walls on pile trestles. Joints are to be designed to prevent the backfilling from escaping.

9.2.4 Auxiliary anchors at the top of steel sheet piling structures

9.2.4.1 General

For structural and economic reasons, the anchors to a waterfront sheet pile wall are, in general, not connected at the top of the wall, but rather at some distance below the top. This reduces the span of the wall between anchor and fixed support and, thus, the bending moment. In addition, this contributes to redistributing the passive earth pressure.

In such cases, the section above the anchors is frequently provided with auxiliary anchors at the top to secure the position of the top of the sheet piling and reduce its deflection. Auxiliary anchoring is not considered in the design of the sheet piling.

9.2.4.2 Considerations for positioning auxiliary anchors

The height of the section above the main anchors determines whether auxiliary anchoring is necessary and depends on various factors, such as the flexural stiffness of the sheet piling, the magnitude of the horizontal and vertical imposed loads and the demands that port operations place on the alignment of the top of the sheet pile wall.

When a waterfront structure is directly subjected to crane loads, auxiliary anchors should be added as near as possible to the top (unless it would be better to position the main anchors near the top of the wall, so that they can carry the loads directly). As a rule, loads on

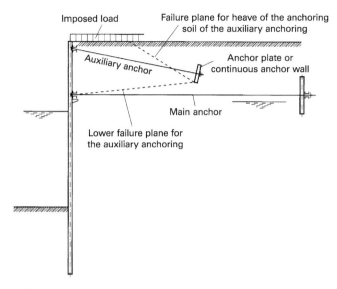

Figure 9.9 Sheet piling structure with single row of anchors plus auxiliary anchoring.

the section above the anchors caused by mooring hooks also call for auxiliary anchoring. Bollards are usually anchored separately.

9.2.4.3 Design of auxiliary anchors

Auxiliary anchors are calculated using an equivalent structural system in which the section above the anchor is considered to be fixed at the level of the main anchor. The loads as per Section 4.2.5 are applied to this system. Auxiliary anchors are connected via a waling.

With regard to the stresses due to aligning the top of the sheet piling and the need to withstand lighter vessel impacts, the auxiliary anchor waling should be stronger than theoretically required and is normally the same as the main anchor waling.

Auxiliary anchors are connected to anchor plates or continuous anchor walls, the stability of which must be checked for both heave of the anchorage soil and the lower failure plane. The lower failure plane begins at the base of the anchor plate/wall and ends at the main anchor connection (Figure 9.9). The analysis is carried out as per Section 9.3.

9.2.4.4 Work on site

It is advisable to dredge the bottom of the watercourse in front of the quay wall only after the auxiliary anchoring has been installed. If dredging is carried out beforehand, the top of the sheet pile wall may move uncontrollably, meaning that later adjustment with auxiliary anchoring alone may not always be successful.

9.2.5 Sheet piling anchors in unconsolidated, soft cohesive soils

9.2.5.1 General

The calculation for sheet pile quay walls in unconsolidated, soft cohesive soil is covered in Section 8.2.15. In principle, the settlement and wall deformations of sheet piling are greater in unconsolidated, soft cohesive soils than in soils with a higher strength.

The specific characteristics of soft cohesive soils must also be considered when anchoring such walls so that settlement and differential settlement do not result in unexpected loads.

As observations of completed structures have revealed, the shaft of a round steel tie rod is dragged downwards by the settlement of the backfill. The shaft itself, however, hardly cuts into soft cohesive soils.

Anchor connections at a quay wall whose base is situated in load-bearing soil (end bearing foundation), therefore, undergo significant curvature when the soil behind the wall settles. Inclinations of anchor rods of up to 1 : 3 – where the anchor rods were originally installed horizontally – have been noticed at quay walls of moderate height. The same applies to anchor connections to structures on deep foundations, anchor walls or pile bents/trestles.

Even when, in special cases, sheet pile quay walls are not embedded in load-bearing soil, it must nevertheless be assumed that the base area of such "floating" walls is located in soils that are stiffer than the soils above. It is to be expected that the soil settles around the anchorages of such walls and that the angles of anchor connections change.

In the case of anchor walls on a "floating" foundation, the differential settlement between a "floating" main wall and an anchor wall is generally small.

The settlement of unconsolidated cohesive soils can also be significantly different along the anchor itself. The anchor must, therefore, be able to bend without the normal stress due to the planned anchor load and the additional stresses from the bending exceeding the strength of the anchor.

Upset round steel tie rods with rolled threads have proved to be a viable solution for anchoring sheet pile walls in unconsolidated, soft cohesive soils. This is because such tie rods exhibit a greater elongation and are more flexible than non-upset round steel tie rods for the same load-bearing capacity.

9.2.5.2 Connecting round steel tie rods to quay walls

Hinged connections are used to attach round steel tie rods to both quay and anchor walls. In order to provide the end of the anchor with sufficient flexibility for the increased rotational movement about the hinge, which is to be expected in soft cohesive soils, there must be a large enough gap between the webs of the channel sections forming the waling. However, the spacing required to accommodate the settlement frequently exceeds the structurally acceptable dimension. In such cases, the anchor must be attached below the waling so that it can rotate freely regardless of the spacing of the channels. The unhindered flow of forces from the wall to the anchor must then be ensured by strengthening the sheet piling or additional measures on the waling.

9.2.5.3 Connecting round steel tie rods to "floating" anchor walls

In general, the usual gap between the two channel sections of a waling is sufficient to allow the free rotation of the anchor connection at a "floating" anchor wall when the anchor passes between the channels and is connected to a compression waling behind the wall with a hinged joint (Figure 9.10).

Hinged anchor connections are also required for structures and components on end bearing foundations.

9.2.5.4 Waling design

The properties of soft, unconsolidated cohesive soils can change dramatically over very short distances. Consequently, areas with a particularly low load-bearing capacity should not be excluded, even if they have not been investigated by means of boreholes and penetrometer tests at the usual spacing (Section 3.2.2).

In order to accommodate the additional loads from differential settlement, walings on sheet pile walls in these soil types must be designed to be stronger than in other cases.

In general, U 400 channels of steel grade S 235 JR should be used for the walings, and S 355 J2 for larger structures, even if smaller sections will suffice structurally. Reinforced concrete walings must have at least the same load-carrying capacity as steel walings made from U 400 channels. RC walings are divided into sections 6.00–8.00 m long. They must be provided with joggle joints to prevent horizontal movement (Section 9.2.3).

9.2.5.5 Anchors at pile trestles or "floating" anchor walls

If displacement of the quay wall head is not permissible, the anchors must be fixed to pile trestles or non-sway structures or structural members. If displacement of the quay wall can be accepted, the anchors can also be tied to a "floating" anchor wall. To this end, the horizontal pressures in front of the anchor wall must be limited in accordance with the permissible displacement of the wall. Where local experience relating to the displacement of the wall and the mobilised passive earth pressure is lacking, appropriate soil mechanics tests and, if necessary, loading tests should be carried out and the results assessed by a geotechnical expert.

9.2.5.6 Design of anchors for "floating" anchor walls

If the terrain is filled with sand up to the level of the port operations area, a "floating" anchorage design as shown in Figure 9.10 is recommended. The soil with a low load-bearing capacity in front of an anchor wall is excavated to just below the anchor connection and replaced by a compacted bed of sand of sufficient width. The anchors can be laid in trenches that are either filled with carefully compacted sand or suitably excavated soil.

The anchor wall is then connected off-centre, so that the permissible horizontal loads are not exceeded in either the sand fill or the area of the soft soil. A uniformly distributed bedding can be assumed both above and below the anchor connection, the magnitude of which is the result of the equilibrium conditions related to the anchor connection. The anchorage according to Figure 9.10 also compensates for irregularities in the bedding.

To prevent excess water pressure acting on the anchor wall, weepholes must be included in the anchor wall to equalise the hydrostatic pressure.

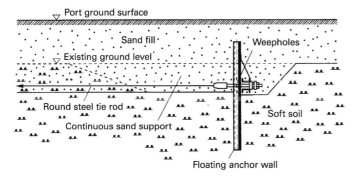

Figure 9.10 "Floating" anchor wall with eccentric anchor connection.

9.2.5.7 Verifying the stability of "floating" anchor walls

For "floating" anchor walls, the stability at the lower failure plane must be verified for both the initial state of the unconsolidated soil as well as for all intermediate states, and the final state of the consolidated soil as per Section 8.2.6. In the case of soft cohesive soil whose shear strain in a triaxial test to DIN 18137-2 is > 10%, the degree of utilisation is to be reduced according to DIN 4084.

9.2.6 Design of protruding quay wall corners with round steel tie rods

9.2.6.1 Guidance on anchorage design

A protruding corner in sheet piling held by anchors running diagonally from quay wall to quay wall leads to high tension loads being transferred to the walings. Waling splices are particularly problematic (Section 8.2.11). The anchors located furthest from the quay wall corner are subjected to the highest tension loads.

Diagonal tension loads require a corresponding design of the walings and their connection to the wall. Steel sheet piling can handle the components of the anchor load in the plane of the wall but only with a large embedment depth.

Therefore, anchorage systems for protruding quay wall corners with diagonal anchors across the corner are technically awkward and also costly. It is also not easy to predict the effects of unintended loads, e. g. due to varying subsoil properties.

Anchorages for protruding quay wall corners are, therefore, not recommended.

9.2.6.2 Recommended crossing anchors

Tensile forces acting diagonally at protruding corners in sheet pile walls can be avoided through a crosswise arrangement of anchors tied to rear anchor walls (Figure 9.11). The quay wall and the anchor walls, thus, form a robust corner section that is stable in itself. However, the crossing anchors require the levels of the walings and anchor positions to be offset, so that there is sufficient clearance between the anchors where they cross.

Edge bollards around protruding quay wall corners should have their own anchorages.

9.2.6.3 Walings

Walings to sheet pile walls are in the form of steel tension members and are shaped to match the alignment of the quay wall. The transition from the walings of a corner section to the walings of the quay wall should be arranged to allow independent movement. The walings are, therefore, connected by means of splices with slotted holes.

Freedom of movement is also necessary for anchor wall walings at the connections with the quay wall and the intersection with the other anchor wall.

Anchor wall walings are in the form of compression members made from steel or reinforced concrete.

9.2.6.4 Anchor walls

Anchor walls at the corner continue through to the quay wall (Figure 9.11). Staggered embedment lengths are used, gradually increasing to the depth of the watercourse bottom alongside the quay wall. This arrangement reduces the risk that, in the case of a vessel impact at a corner (corners are always particularly vulnerable), the backfilling in the corner block will only be washed out as far as the anchor walls.

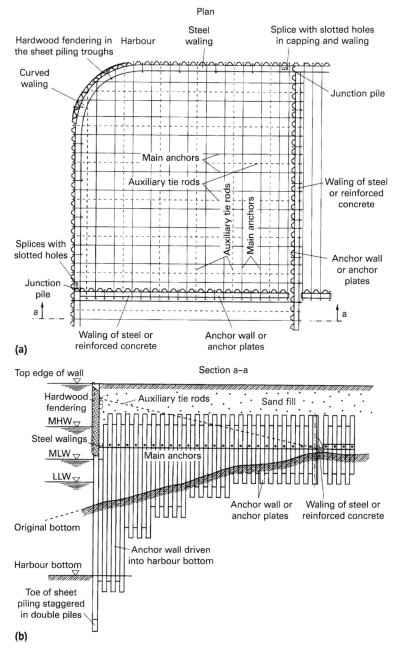

Figure 9.11 Anchoring a protruding corner in a sheet pile quay wall for a seaport: (a) plan view; (b) section a–a.

Embedment to the depth of the watercourse bottom is also recommended when the round steel tie rods are anchored to anchor plates, e. g. made from reinforced concrete, instead of continuous anchor walls.

9.2.6.5 Fendering in sheet pile troughs

Ships and quay wall corners can be protected by fitting, for example, marine timbers or plastic sections into the sheet pile troughs at protruding corners. This fendering should project about 5 cm beyond the outer face of the sheet piling (Figure 9.11).

9.2.6.6 Rounded and reinforced concrete wall corners

Since protruding quay wall corners are particularly vulnerable to damage by waterborne traffic, they should be rounded off if possible and also strengthened by a strong reinforced concrete wall, if necessary.

9.2.6.7 Protective dolphins at corners

If the waterborne traffic allows it, protruding quay wall corners can be protected against vessel impact by placing elastic dolphins or a guidance structure in front of the exposed corner.

9.2.6.8 Stability verification

Verification of stability in accordance with Section 9.3 is carried out individually for each quay wall segment. A special check for the corner section is not necessary if the anchors of the quay walls are carried through to the other wall, as shown in Figure 9.11.

9.2.7 Configuration and design of protruding quay wall corners with raking anchor piles

9.2.7.1 General

Protruding quay wall corners are particularly vulnerable to damage from vessel impacts. In many cases, quay wall corners are also equipped with heavy-duty bollards at the ends of the quay wall to secure large ships as per Section 7.2.5.2. In addition, they are also equipped with the necessary fenders having a higher energy absorption capacity than the adjacent quay wall sections. Overall, protruding quay wall corners must be robust and as rigid as possible.

The anchorage of protruding quay wall corners with raking anchor piles is covered below, as this type of anchor essentially complies with the requirement for a particularly rigid and robust construction.

9.2.7.2 Design of the corner structure

The design of the corner structure is primarily determined by the structural design of the adjoining quay walls, the difference in ground levels and the angle enclosed by the wall sections forming the corner. In addition, the design to be selected is very much influenced by the existing water depth and the in situ soil conditions.

The raking piles to be used to anchor the corner structure should always be connected perpendicularly to the quay wall. They must, therefore, overlap in the anchoring zone. In order to ensure that the anchor piles can be installed properly and without colliding, a minimum clearance must be maintained at the points where they cross. Whereas the clear spacing of crossing piles above the level of the watercourse bottom can be kept comparatively small (usually about 25–50 cm), a similar spacing is not sufficient for long piles and difficult driving conditions. The clearance at crossing points below the level of the watercourse bottom should, therefore, be at least 1.0 m, although 1.5 m is preferable. In soils in which a more significant deviation of longer piles is to be expected during driving, the clearance should be at least 2.5 m. Steel wings (if fitted) must always be taken into account when calculating the clearance between piles at crossing points.

In order to be able to comply with the minimum clearance requirements at crossing points, the spacing and inclination of the steel piles must also be varied accordingly. However, the positions of the piles within a group should be kept fairly uniform because of the different load-bearing behaviour of piles at different angles.

Should deep foundations prove necessary at a quay wall corner for heavily loaded bollards or other items, such as bracing structures for conveyor belts, etc., the construction of a special reinforced concrete corner section with a slab supported on a deep foundation is recommended in most cases. The slab requires a pinned support at the sheet pile wall.

This solution is also suitable for quay wall corners where the minimum spacing of piles required for driving in the crossover area cannot be achieved by adjusting the positions of the piles (Figure 9.12). In such corner designs, it is expedient to position the piles required in the corner area towards the rear of the slab. They then lie in a plane that is different from that of the piles of the neighbouring quay wall sections, thus making it easier to maintain adequate clearances at crossing points. As this solution requires additional compression piles at the rear edge of the slab to accommodate the vertical component of the anchor force, such designs are more costly. However, this is a reliable method of construction without any particular risks.

The designs described in Sections 9.2.6.5 and 9.2.6.7 for round steel tie rods also apply to protruding quay wall corners anchored with raking piles.

9.2.7.3 Use of 3D images to visualise pile positions

In order to anticipate driving difficulties, 3D images of difficult corner situations should be produced during the planning phase of a project (Figure 9.13). The colour coding of individual anchor positions means it is very easy to highlight the critical anchor crossing points and then find optimum solutions.

During construction on site, the 3D model should be updated with all the actual positions of anchors already installed to determine the necessary corrections due to driving deviations.

9.2.7.4 Verifying the stability of corner sections

The stability of corner sections must be verified for all structural parts of the corner anchorages. To this end, each wall at the corner must be considered separately. At corners with additional loads, e. g. due to corner structures, bollards, fenders and other equipment, it must be verified that the piles are also able to accommodate these additional forces safely.

If major changes to positions of piles are required during construction, their effects must be verified by means of supplementary calculations.

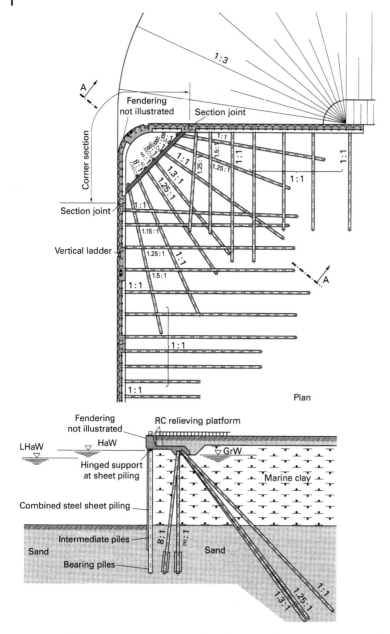

Figure 9.12 Example of the construction of a protruding quay wall corner with steel anchor piles.

9.2.8 Prestressing of high-strength steel anchors for waterfront structures

9.2.8.1 General

Anchorages for sheet piling waterfront structures, also for the subsequent securing of other structures such as walls on pile trestles, normally make use of non-prestressed anchors in steel grades S 235 JR, S 235 J2 or S 355 J2.

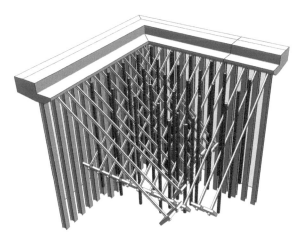

Figure 9.13 3D image of protruding quay wall corner with raking anchors rotated down into position.

However, in certain situations it can be beneficial to prestress anchors with a high proportion of their calculated anchor force. Such anchors must be made from high-strength steel.

The prestressing of anchors made from high-strength steels can be useful for limiting displacements, for example, particularly in the case of structures with long anchors, or when building near structures sensitive to settlement or when subsequently connecting sheet pile walls driven beforehand.

Furthermore, the prestressing can achieve a redistribution of the passive earth pressure when the sheet pile wall is backfilled with either medium density non-cohesive soil or stiff cohesive soil. Redistributing the passive earth pressure decreases the span moment and increases the anchor force.

Permanent anchors manufactured from high-strength steel must be protected against corrosion. Any existing special recommendations and standards, e. g. for grouted anchors to DIN EN 1537, must be complied with.

9.2.8.2 Effects of anchor prestressing on active earth pressure

Prestressing the anchors of sheet piling structures always reduces wall deformations towards the water side, especially in the upper part of the quay wall. A high prestress redistributes the active earth pressure upwards. Consequently, the resultant of the active earth pressure can shift from the bottom third point of the wall height above the watercourse bottom to about $0.55h$ – with a corresponding increase in the anchor force. This active earth pressure redistribution is especially pronounced at quay walls that extend above the level of the anchors.

When the intention of prestressing is to achieve an active earth pressure distribution deviating from the classic distribution according to Coulomb, the anchors must be prestressed to about 80% of the characteristic anchor load determined for design situation DS-P.

9.2.8.3 Time of prestressing

The prestressing of the anchors may not begin until the respective prestressing forces can be accommodated without appreciable, undesirable movements of the structure or its members. This presumes appropriate backfilling behind the wall. The prestressing forces must then be transferred from the structure into the backfilling as intended.

Owing to the fact that when tensioning the anchors, the prestress in adjacent – already prestressed – anchors is at least partly lost due to the redistribution of loads in the soil, the anchors must be prestressed beyond the design value so that, after the adjacent anchors have been tensioned, they still possess their planned prestress. Random checks should be made on prestressing forces during construction, so that any required corrections can be made.

Prestressing in several steps can avoid the redistribution of loads and, therefore, also the short-term overloading of the anchors. However, this complicates operations.

DIN EN 1537 is the main standard that applies to pressure-grouted anchors.

9.2.8.4 Further guidance

Where anchor prestressing is only carried out for certain parts of a waterfront structure, it should be borne in mind that the freedom of movement of the wall as a whole varies locally. The prestressed zones act as fixed points, which are subjected to correspondingly higher spatial active earth pressures. The increased active earth pressure must be taken into account in the structural analyses for the respective wall areas and their anchors.

Prestressed anchors must remain permanently accessible, so that, if required, the prestressing can be checked and corrected. Furthermore, anchor ends should have hinged connections.

Since bollards are only loaded from time to time, their anchors should not be made from prestressed high-strength steel but instead from firmly tightened (but not prestressed) round steel tie rods made from grade S 235 JR, S 235 J2 or S 355 J2. The latter grades exhibit only a minor elongation under load.

9.2.9 Hinged connections between driven steel piles and steel sheet piling structures

9.2.9.1 General

Sheet pile walls are subjected to bending loads due to active earth pressure, which results in wall rotations near the anchor connections. These rotations are transferred to the piles when they are connected in accordance with Section 9.2.3. The outcome is corresponding bending stresses in the pile in addition to the intended stresses due to the anchor force. Settlement and/or subsidence may create additional bending loads on the piles.

By contrast, a hinged connection for the driven steel piles allows essentially unrestrained mutual rotation of the sheet piling and the pile, so that the pile and the pile connection are only subjected to the intended pile loads and can, therefore, be designed economically.

A hinged pile connection detail must be designed in accordance with the principles of structural steelwork.

9.2.9.2 Guidance for designing a hinged connection

The pile connection's ability to rotate can be achieved through the use of single or double hinge pins or through the plastic deformation of a structural member designed for this purpose (plastic hinge). A combination of hinge pin and plastic hinge is also possible. When designing for plastic hinges, the following points should be considered:

1. Plastic hinges should be located at a sufficient distance from butt and fillet welds, so that the steel is not stressed to yield in the vicinity of weld seams. Lateral fillet welds should be located in the plane of the force or plane of the tension element, so that they do not peel off. Otherwise, peeling-off is to be prevented by other measures.
2. Weld seams transverse to the tensile force of the steel pile can act as metallurgical notches and should, therefore, be avoided.
3. Non-structural erection seams executed in difficult welding positions without observing proper welding techniques increase the risk of failure.
4. At difficult connections, also hinged connections, the recommendation is to calculate the likely plastic hinge cross-section for the intended normal forces in conjunction with possible additional stresses, (Section 9.2.3). When designing plastic hinges, DIN EN 1993-1 is to be taken into consideration.
5. Abrupt changes in stiffness, e.g. flame-cut notches in a pile and/or metallurgical notches due to cross-welds, as well as sudden increases in the steel cross-section, e.g. due to very thick, welded splice plates, are to be avoided, especially in the potential plastic hinge regions of steel piles in tension, because they can lead to sudden failure without prior deformation.

Examples of hinged connections for steel piles are shown in Figures 9.14–9.21.

9.2.9.3 Work on site

Depending on the local conditions and the design, steel piles may be installed either before or after the sheet piling. If the location of the connection depends on the geometry, as would be the case if the connection must be made in the trough of a sheet pile or on a bearing pile in combined sheet piling, it is important that the upper end of each anchor pile is as close as possible to its intended position. This is best accomplished if the steel piles are installed after the sheet piling. However, the design of the connection must always be such that certain deviations and rotations can be compensated for and accommodated.

If the steel pile is driven directly above the top of the sheet piling, or through an opening in the sheet piling, the sheet piling can then provide effective guidance for driving the anchor pile. A driving opening can also be provided by burning off the upper end of a double pile, hoisting it clear and later returning it to its original position and welding it in place.

A certain degree of adjustment is possible at the head of steel piles whose upper end is not embedded in the soil at the time of making the connection to the sheet pile wall.

Anchor piles that are longer than required should be provided to allow for cutting off the top end in the case of damage to the microstructure caused during driving.

Where possible, slots for connection plates should not be cut in sheet piles and/or steel piles until after they have been driven to their final depth.

9 Anchorages

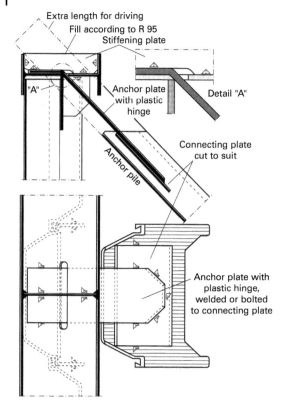

Figure 9.14 Hinged connection between lightweight steel pile and lightweight steel sheet piling by means of a splice plate plus a plastic hinge.

9.2.9.4 Connection details

A hinged connection to sheet piling is generally placed in the trough, especially with interlocks on the centroid axis, or, for combined sheet piling, on the web of the bearing piles.

In the case of smaller anchor forces, the steel pile may also be connected to the capping beam on top of the sheet piling (Figure 9.14) or to a waling behind the sheet piling by means of splice plate plus plastic hinge. Special attention must be paid to the danger of corrosion in such cases, especially for waterfront structures for cargo-handling and at berths.

Tension elements made from round steel rods (Figure 9.16), steel plates or wide steel flats (tension splice plates) are frequently fitted between the connection in the sheet pile trough or bearing pile web and the upper end of the pile (Figures 9.17 and 9.18). Connections consisting of a threaded steel rod, nut, plate washer and hinged plate have the advantage that they can be tensioned.

In addition to the hinged connection in the sheet pile trough, at the capping beam or on the web of the bearing pile, an additional hinge may be included near the top end of the anchor pile in special cases.

This solution is depicted in Figure 9.18 for the case of double bearing piles. This type of connection can also be used for single bearing piles. The (flame-cut) slots in the flanges of the bearing piles should be extended far enough below the connection plates in order to allow sufficient freedom of movement for pile rotations and to rule out any restraint forces that can arise as a result of unwanted fixity should there be contact between plates and pile. Care is also required to ensure that the intended hinge effect is not impaired by

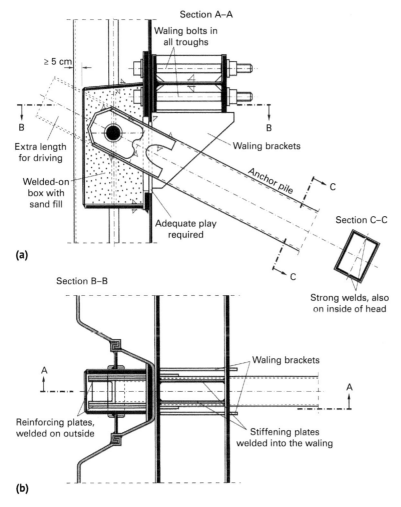

Figure 9.15 Hinged connection between steel pile and heavy steel sheet piling by means of a hinge pin.

incrustations, sintering or corrosion around the connection. This is to be checked for each individual case and considered in the structural design.

The steel pile can also be driven through an opening in a sheet pile trough and connected via a hinge detail and supporting plates welded to the sheet piling (Figure 9.15).

If the connection is made in the trough of a sheet pile wall on the water side, all parts must terminate well behind the line of the wall – at least 20 mm is recommended wherever this is possible. The aim of this is to ensure that if a vessel should come into contact with the sheet pile wall, neither vessel nor pile head are damaged. Furthermore, the point at which the pile penetrates the sheet piling should be carefully protected against soil escaping and/or being washed out (e. g. by means of a sand-filled box as shown in Figure 9.15).

Depending on the design selected, preference should be given to those connections that can be largely fabricated in the works and exhibit adequate tolerances (Figure 9.19). Extensive modifications on site are expensive and are, therefore, to be avoided if at all possible.

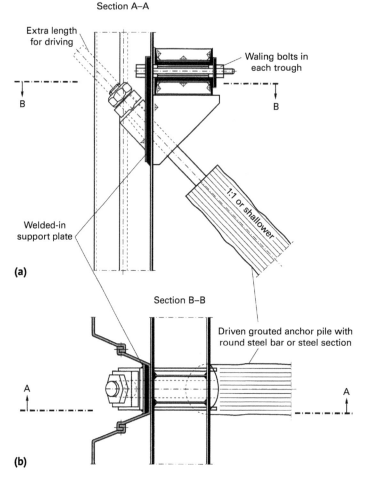

Figure 9.16 Hinged connection between driven grouted steel pile and heavy steel sheet piling: (a) section A–A; (b) section B–B.

With a connection such as the one shown in Figure 9.19, all load-bearing weld seams on the bearing pile can be welded on in the horizontal flat position in the fabrication works. However, this solution is only advisable when the jaw bearing plates are attached after driving the steel piles and when the lengths of the bearing piles can be specified exactly. This sort of connection prepared in the works is, therefore, impossible where bearing piles have to be extended or shortened.

In the solution shown in Figure 9.20, the connection between steel pile and sheet piling is created by loops that enclose a tubular waling welded into the sheet piling. Care is required here because the free rotation of the connection is prevented by friction between loops and tube. The loops must, therefore, be designed to allow for the adequate load-bearing reserves required for the resulting unequal load distribution. As a rule, deflections of the steel piles and displacement of the pile heads are to be expected, and so the connection has to be designed to transfer shear forces as well as pile tensile forces. This can be achieved with a bracket-type extension of the piles through to the tubular waling (Figure 9.20).

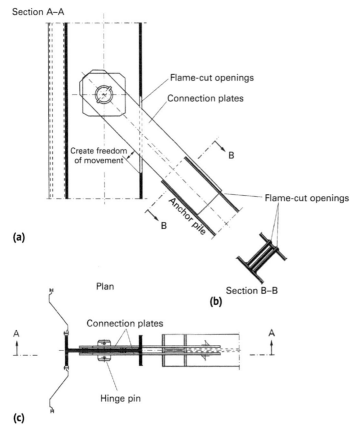

Figure 9.17 Hinged connection (hinge pin) between steel pile and combined sheet piling with single bearing piles: (a) section A–A; (b) section B–B; (c) plan view.

Figure 9.21 depicts a hinged connection between a small injection pile and a sheet pile wall. As such connections are not generally covered by standards, a national technical approval may be required.

Corrosion protection of the steel tendon in the ground takes the form of a ribbed plastic tube into which cement mortar is injected. The outboard pile connection is protected against corrosion by enclosing it in a fitting filled with a plastic compound. The connection to a reinforced concrete waling is regulated by the national technical approval for the injection pile.

9.2.9.5 Verifying the load-bearing capacity of the connection

All pile connections should be designed for the calculated pile forces plus 15%. Loads from the water side due to, for example, berthing pressure, ice pressure, mining subsidence, etc., can at times reduce any tension in the pile, even transform it into compression.

If necessary, the stresses in the pile connection and the pile itself should be checked to ensure they can be accommodated. A buckling analysis should be carried out on free-standing piles. Ice impact must also be taken into account in some cases.

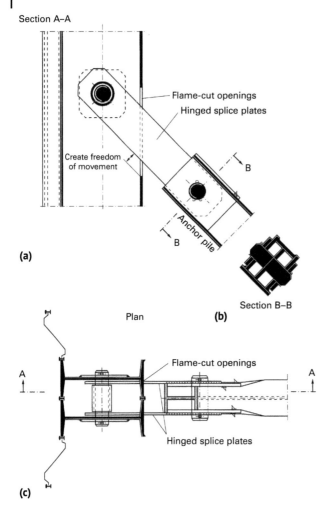

Figure 9.18 Hinged connection (hinged splice plates) between steel pile and combined sheet piling with double bearing piles: (a) section A–A; (b) section B–B; (c) plan view.

Wherever possible, the connection should lie at the intersection of the sheet piling and pile axes (Figures 9.14–9.21). If the pile connection deviates substantially from this, additional moments in the sheet piling must be assumed.

The vertical and horizontal components of the force must also be taken into account in the connection to the sheet piling and – where not every load-bearing wall element is anchored – in the waling and its connections. If a vertical load due to soil surcharge is to be expected, this must also be allowed for in the reaction forces and in the verification of the load-bearing capacity of connections. This is always the case when deflection of the piles is to be expected.

When the angle between pile and sheet piling changes as a result of pile deflection, then the ensuing changes to the tensile and shear forces at the anchor connection must also be considered in the analyses.

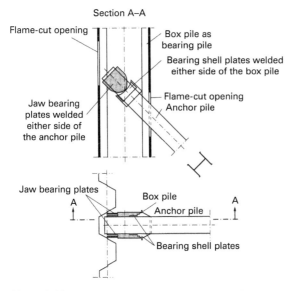

Figure 9.19 Hinged connection between steel pile and combined steel sheet piling using jaw bearings/bearing shells – section A–A.

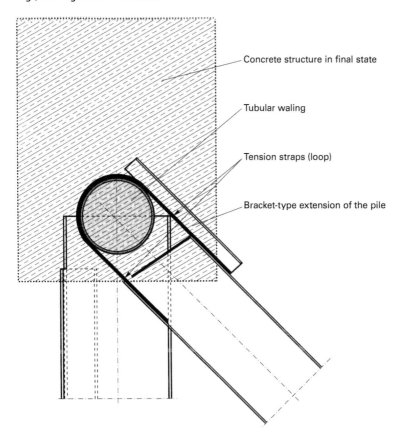

Figure 9.20 Connection between a steel pile and combined sheet piling using loops.

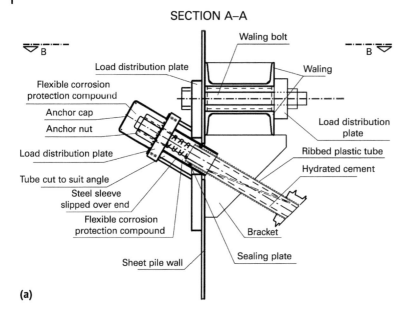

(a)

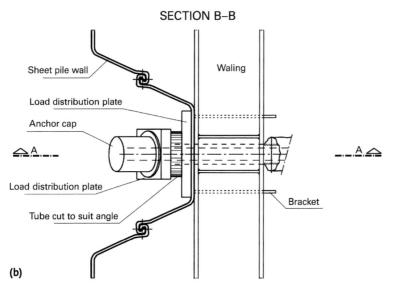

(b)

Figure 9.21 Hinged connection between small injection pile with double corrosion protection and steel sheet piling: (a) section A–A; (b) section B–B.

Where pile connections are located in the troughs of sheet pile walls, the horizontal force component is to be transferred into the sheet pile web via plate washers of sufficient width (Figure 9.16). The weakening of the sheet piling cross-section by the pile penetration should be considered. It may be necessary to reinforce the sheet piling in the area around a connection.

Connections for hinged anchorages must be positioned to facilitate the transfer of tensile and shear forces in a constant flow. If the flow of forces is not obvious in the case of difficult,

9.3 Verification of stability for anchoring at the lower failure plane

9.3.1 Stability at the lower failure plane for anchorages with anchor walls

Stability at the lower failure plane is verified using the method proposed by Kranz (1953). This method works on the basis of taking a section behind the retaining wall, along the lower failure plane and behind the anchor wall. The lower failure plane has a convex form and runs between the base of the anchor wall and the fulcrum of the sheet piling in the soil; it is approximated by a straight line in the analysis. In the case of quay walls simply supported in the soil, the fulcrum of the wall is the base of the wall. The analysis of stability at the lower failure plane determines the minimum anchor length required for carrying the anchor force.

Figure 9.22 shows the forces in the body of soil between anchor wall, quay wall, lower failure plane and ground surface (sliding body FDBA).

Notation for Figure 9.22:

ϑ inclination of lower failure plane
G_k total characteristic weight of sliding body FDBA plus imposed load if applicable
$E_{a,k}$ characteristic active earth pressure (increased active earth pressure if applicable)
$F_{U1,k}$ characteristic hydrostatic pressure at section AF between soil and retaining wall
$F_{U2,k}$ characteristic hydrostatic pressure at anchor wall DB
$F_{U3,k}$ characteristic hydrostatic pressure at lower failure plane FD
Q_k characteristic resultant force at lower failure plane due to normal force and maximum possible friction force (therefore inclined at angle φ_k to normal to failure plane)
C_k characteristic cohesion force at lower failure plane (with its magnitude dependent on the characteristic value of the cohesion and the length of the lower failure plane)
$E_{1,k}$ characteristic active earth pressure with imposed load on anchor wall DB
P_k characteristic anchor force

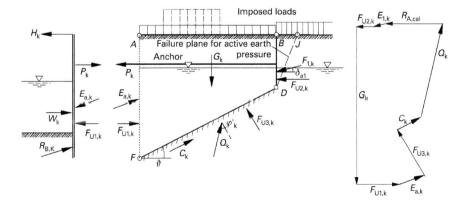

Figure 9.22 Analysis of stability at the lower failure plane.

In the case of the characteristic anchor force, we must distinguish between the component $P_{G,k}$ due to permanent actions and the component $P_{Q,k}$ due to variable actions.

The analysis must be carried out for exclusively permanent loads and also for permanent plus variable loads. In the latter case, the components due to variable loads are to be applied in an adverse position. The force $P_{Q,k}$ resulting from these components is to be identified separately.

Stability at the lower failure plane is given when

$$P_{G,k} \cdot \gamma_G \leq \frac{R_{A,cal}}{\gamma_{Ep}}$$

where $R_{A,cal}$ is determined from the polygon of forces as per Figure 9.22 for exclusively permanent loads, and

$$P_{G,k} \cdot \gamma_G + P_{Q,k} \cdot \gamma_Q \leq \frac{R_{A,cal}}{\gamma_{Ep}}$$

where $R_{A,cal}$ is determined from the polygon of forces as per Figure 9.22 for permanent plus variable loads.

The following DIN 1054 partial safety factors are applied when analysing stability at the lower failure plane:

γ_G partial safety factor for permanent actions.
γ_Q partial safety factor for variable actions.
γ_{Ep} partial safety factor for passive earth pressure.

The analysis of stability at the lower failure plane is based on the concept that transferring the anchor force into the soil causes a failure body to become established behind the retaining wall and that this body is bounded by the quay wall, the anchor wall and the lower failure plane. In this situation, the maximum possible shear resistance at the lower failure plane is exploited, whereas the limiting value for the reaction force at the base is not achieved. Force $R_{A,cal}$ is the characteristic anchor force that can be accommodated by the sliding body FDBA assuming the full shear strength of the soil. The definition of the exploitation of the anchor force is equivalent to the exploitation of the shear strength of the soil.

Equilibrium of the applied moments is not considered in the verification of stability at the lower failure plane either. This is because only the resultants of the actions transferred via the boundaries of the sliding body are included in the analysis. It is sufficiently accurate to replace the lower failure plane by the straight line DF as the critical failure plane.

If the flow force of any groundwater flowing in the sliding body (water table descending downwards towards sheet piling) is to be taken into account, then the hydrostatic pressures on quay wall, anchor wall and lower failure plane must be determined using a flow net according to Section 3.4. The pressures are added together to form resultants at the respective boundary surface of the sliding body.

9.3.2 Stability at the lower failure plane in unconsolidated, saturated cohesive soils

Verification of stability at the lower failure plane for quay walls and their anchors in unconsolidated cohesive soils is carried out as per Section 9.3.1. The active earth pressure is determined for the unconsolidated, saturated case as per Section 3.5.7. The characteristic cohesion force $C_{u,k}$ is effective at the lower failure plane. The angle of internal friction is to be taken as $\varphi_u = 0$ for unconsolidated, saturated, virgin cohesive soils.

9.3.3 Stability at the lower failure plane with varying soil strata

Verification of stability at the lower failure plane in varying soil strata is carried out as per Section 9.3.1. The sliding body as shown in Figure 9.23 is divided by imaginary vertical planes passing through the intersections of the lower failure plane with the boundaries of the strata. This splits the sliding body FDBA into as many parts as there are strata intersected by the lower failure plane. Equilibrium of forces is then carried out for all individual parts in turn. If parts of the lower failure plane pass through cohesive soil, a cohesion force is applied to these sections (cohesion is neglected in the polygon of forces in Figure 9.23).

The active earth pressures in the vertical sections between the individual parts are assumed to act parallel to the surface.

The following characteristic forces act on the sliding bodies in Figure 9.23:

$G_{1,k}$ total weight of sliding body F_1DBB_1, plus imposed load, if applicable
$G_{2,k}$ total weight of sliding body FF_1B_1A, plus imposed load, if applicable
$E_{a,k}$ active earth pressure (across all soil strata)
P_k anchor force
$F_{U1,k}$ hydrostatic pressure between soil and retaining wall AF
$F_{U2,k}$ hydrostatic pressure on anchor wall DB
$F_{U3,k}$ hydrostatic pressure on lower failure plane in section FF_1
$F_{U4,k}$ hydrostatic pressure on lower failure plane in section F_1D

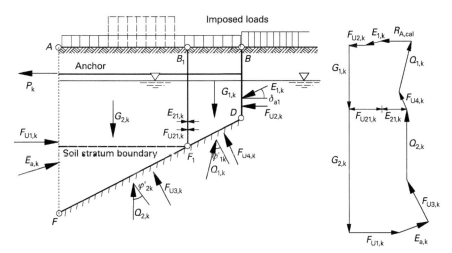

Figure 9.23 Verification of stability for the lower failure plane in stratified soil.

$F_{U21,k}$ hydrostatic pressure on vertical separation plane F_1B_1
$E_{1,k}$ active earth pressure with imposed load on anchor wall DB
$E_{21,k}$ earth pressure in vertical separation plane F_1B_1

The stability at the lower failure plane is determined using the inequalities in Section 9.3.1.

9.3.4 Verification of stability at the lower failure for a quay wall fixed in the soil

The aforementioned analysis of stability at the lower failure plane is also sufficiently accurate for sheet pile walls fixed in the soil. With this type of wall, the lower failure plane runs between the base of the anchor wall and the point of zero shear in the wall's fixity zone. This point coincides with the position of the greatest fixed-end moment. Its position can, therefore, be taken from the sheet piling calculation.

The active earth pressure in this case is only determined down to the theoretical base of the sheet piling and the anchor force is taken from the sheet piling analysis for the fixed wall.

9.3.5 Stability at the lower failure plane for an anchor wall fixed in the soil

If an anchor wall is fixed in the ground, then Section 9.3.4 suggests that the lower failure plane should continue to the theoretical base at the level of the point of zero shear in the fixed-end zone of the anchor wall.

9.3.6 Stability at the lower failure plane for anchors with anchor plates

If anchors are connected to separate anchor plates with a clear spacing a, an imaginary equivalent anchor wall placed at a distance $1/2 \cdot a$ in front of the anchor plates is assumed when analysing stability at the lower failure plane.

9.3.7 Verification of safety against failure of anchoring soil

Analysis of the safety against failure of the anchoring soil is intended to verify that the design values of the resisting horizontal forces in front of anchor plates/wall from the underside of the anchor plates/wall to the surface of the ground are at least equal to or greater than the sum of the horizontal components of the design value of the anchor force, the design value of the active earth pressure acting on the anchor wall and any excess water pressure present.

The active and passive earth pressures acting on the anchor wall or individual anchor plates is determined according to DIN 4085. A non-permanent action (imposed load on ground surface) may only be applied when it has an unfavourable effect. This is usually the case with imposed loads applied behind the anchor wall or anchor plates. The groundwater level is to be applied at its most unfavourable level.

When calculating the passive earth pressure on the anchor wall, the only angle of inclination to be considered should be the one for which the condition $\sum V = 0$ is satisfied at the anchor wall (sum of all applied vertical forces including dead load, with earth surcharge corresponding to vertical component of passive earth pressure).

In the case of simply supported anchor plates and walls, the anchor connection is generally located in the middle of the height of the plate/wall. Analyses of prefabricated inclined anchors should be carried out in a similar way.

9.3.8 Stability at the lower failure plane for quay walls anchored with piles or grouted anchors at one level

Quay walls are also anchored with piles or grouted anchors that transfer the anchor force into the soil via skin friction. Basically, we distinguish between three groups of anchor elements:

- Steel piles with and without grouted skin according to Section 9.2.
- Grouted micropiles to DIN EN 14199 and DIN EN 1536.
- Grouted anchors to DIN EN 1537.

The required lengths of these anchor elements are determined as part of the stability verification at the lower failure plane as per Figure 9.24.

Notation for Figure 9.24 (see Figure 9.22 for forces on failure body):

l_a length of anchor element.

l_r required minimum anchor length or nominal grouting length for a grouted anchor calculated from the design value of pile force P_d and the design value of skin friction T_d for the pile ($l_r = P_d/T_d$).

T_d design value for skin friction, calculated from the design value of the pull-out resistance $R_{t,d}$ and the force transfer length l_0 in a tensile test ($T_d = R_{t,d}/l_0$).

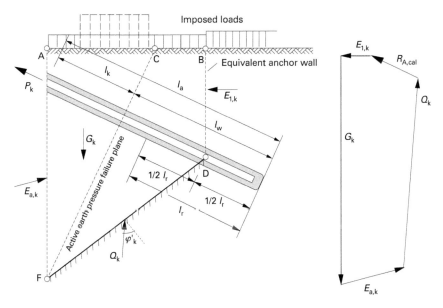

Figure 9.24 Verification of stability for the lower failure plane when using piles and grouted anchors.

l_k upper pile length not effective structurally (which begins at the head of the pile and ends at the active earth pressure failure plane or at the upper edge of the load-bearing soil if this is deeper).

l_w structurally effective anchor length (which extends from the active earth pressure failure plane or the upper edge of the load-bearing soil to the end of the grouted anchor or pile – excluding toe of pile); $l_w \geq l_r$ must always apply, and $l_w \geq 5.00$ m for grouted skin anchors.

In this case, the sliding body is limited by an equivalent anchor wall that is assumed to be vertical in the centre of the force transfer length l_r of the anchor element. In the case of grouted anchors, the force transfer length l_r is the nominal length of the grout.

The active earth pressure $E_{1,k}$ at the equivalent anchor wall is always assumed to act parallel with the surface.

If the anchor spacing a_a is greater than half the calculated force transfer length ($a_a > 1/2 l_r$), then the possible anchor force $R_{A,cal}$ must be reduced by the ratio of half the calculated force transfer length and the anchor spacing:

$$R^*_{A,cal} = R_{A,cal} \cdot \frac{1/2 \cdot l_r}{a_A}$$

The possible anchorage force reduced in this way is usually smaller than the anchorage force that must be actually accommodated when the pull-out resistance of the anchorage elements mobilised behind the base of the equivalent wall is not applied in the analysis.

This analysis represents a considerable simplification. A more accurate analysis is required when the anchorages are so short that the calculated force transfer length l_r is the same as the anchorage length l_w outside the active sliding wedge ($l_w = l_r$) (Heibaum 1991).

A more accurate – but also more complex – analysis is possible by varying the inclination of the failure plane to find the most unfavourable failure plane while taking into account the remaining potential pull-out resistance behind the base of the equivalent anchor wall (in line with DIN 4084) (Heibaum 1987). The characteristic value of the pull-out force transferred from the force transfer length behind the failure plane to the unaffected soil should be used in this approach.

The actual anchorage force depends on the structure and must be verified through tests. It is important to remember here that for piles, the skin friction acts over the entire length of the pile and can result in a higher permissible pile load being calculated than can be transferred via the actually effective force transfer length. In the case of small-diameter grouted piles, this can be dealt with by flushing clean the structurally ineffective area of the pile (see *Recommendations on Piling*, EA-Pfähle, 2012).

9.3.9 Stability at the lower failure plane for quay walls with anchors at more than one level

If several rows of piles or grouted piles are to be used, the sliding bodies to be analysed for stability at the lower failure plane are bounded by a failure plane through each of the centroids of the force transfer paths.

In a design with several anchor levels, if the lower failure plane intersects a pile or grouted pile before or in the force transfer path, the force that can be transferred behind the failure

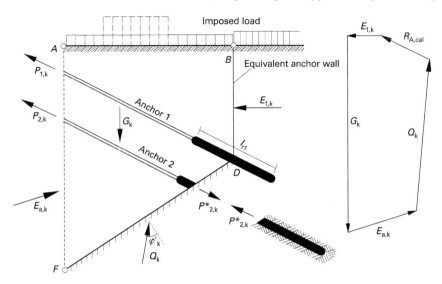

Figure 9.25 Verification of stability at the lower failure plane when using several anchors.

plane in undisturbed soil may be considered (force P^*_{2k} in Figure 9.25). The separated anchor force P^*_{2k} may be determined from a uniform distribution of the anchor force $P_{2,k}$ over transfer path l_r.

In the analysis, we must distinguish between the permanent components $P_{G,k}$ and the variable components $P_{Q,k}$ of the anchor force.

Stability at the lower failure plane is assured when

$$\sum \left(P_{G,k} - P^*_{G,k}\right) \cdot \gamma_G \leq \frac{R_{A,cal}}{\gamma_{Ep}}$$

where $R_{A,cal}$ is determined from the polygon of forces using exclusively permanent loads, and

$$\sum \left(P_{G,k} - P^*_{G,k}\right) \cdot \gamma_G + \sum \left(P_{Q,k} - P^*_{Q,k}\right) \cdot \gamma_Q \leq \frac{R_{A,cal}}{\gamma_{Ep}}$$

where $R_{A,cal}$ is determined from the polygon of forces using permanent and variable loads, where

$\sum(P_{G,k} - P^*_{G,k})$ sum of all permanent components of characteristic anchor forces minus the forces transferred to the undisturbed soil behind the failure plane;

$\sum(P_{Q,k} - P^*_{Q,k})$ sum of all variable components of characteristic anchor forces minus the forces transferred to the undisturbed soil behind the failure plane.

(Safety factors γ_i as given in Section 9.3.1.)

9.3.10 Safety against slope failure

The verification of stability at the lower failure plane in conjunction with the analyses required for anchor walls and plates to verify adequate safety against the failure of the anchoring soil replace the slope failure investigation to DIN 4084 normally required.

Irrespective of the stability at the lower failure plane, the overall stability according to DIN 4084 must still be verified if there are unfavourable soil strata (soft strata beneath the anchoring zone) or high loads behind the anchor wall or equivalent anchor wall or particularly long anchors are being used.

References

Achmus, M., Kaiser, J. and Wörden, F.T. (2005). Bauwerkserschütterungen durch Tiefbauarbeiten, Grundlagen, Messergebnisse, Prognosen. *Mitteilungen Institut für Grundbau, Bodenmechanik und Energiewasserbau (IGBE), Universität Hannover* 61.

Alberts, D. (2001). Korrosionsschäden und Nutzungsdauerabschätzung an Stahlspundwänden und -pfählen im Wasserbau. *1st conf. "Korrosionsschutz in der maritimen Technik"*, Germanischer Lloyd, Hamburg.

Alberts, D. and Heeling, A. (1996). Wanddickenmessungen an korrodierten Stahlspundwänden; statistische Datenauswertung. *Mitteilungsbl. BAW* 75, Karlsruhe.

Alberts, D. and Schuppener, B. (1991). Comparison of ultrasonic probes for the measurement of the thickness of sheet-pile walls, In: *Field Measurements in Geotechnics* (ed. by Sørum), Rotterdam: Balkema.

Baumaschinen-LärmVO: 15. Verordnung zur Durchführung des BImSchG vom 10.11.1986 (Baumaschinen-LärmVO).

Binder, G. (2001). Probleme der Bauwerkserhaltung – eine Wirtschaftlichkeitsberechnung. BAW-Brief Nr. 1, Karlsruhe.

Blum, H. (1931). *Einspannverhältnisse bei Bohlwerken*. Berlin: Ernst & Sohn.

Clasmeier, H.-D. (1996). Ein Beitrag zur erdstatischen Berechnung von Kreiszellenfangedämmen. *Mitteilung des Instituts für Grundbau und Bodenmechanik, Universität Hannover* 44.

Clausen, C.J.F., Aas, P.M. and Karlsrud, K. (2005). Bearing capacity of driven piles in sand, the NGI approach. *Int. Symp. Front. Offshore Geotechn. (ISFOG2005)*, Perth, pp. 677–681.

Dynamit Nobel (1993). *Sprengtechnisches Handbuch*. Troisdorf: Dynamit Nobel Aktiengesellschaft (ed.).

EA-Pfähle (2012). *Recommendations on Piling* (ed. by Deutsche Gesellschaft für Geotechnik e. V.). Berlin: Ernst & Sohn.

Fages, R. and Gallet, M. (1973). *Calculations for Sheet Piled or Cast in Situ Diaphragm Walls*. Civil Engineering and Public Works Review.

Fages, R. and Bouyat, C. (1971). Calcul de rideaux de paroismouileeset de palplanches. *Travaux* 439: 49–51;

Fages, R. and Bouyat, C. (1971). Calcul de rideaux de paroismouileeset de palplanches. *Travaux* 441: 38–46.

Graff, M., Klages, D. and Binder, G. (2000). Mikrobiell induzierte Korrosion (MIC) in marinem Milieu. *Mater. Corros.* 51: 247–254.

DIN Deutsches Institut für Normung e. V. (Hrsg.) (2011): *Handbuch Eurocode 7 – Geotechnische Bemessung, Band 1: Allgemeine Regeln*, 1st edn. Berlin: Beuth.

Heibaum, M. (1987). Zur Frage der Standsicherheit verankerter Stützwände auf der tiefen Gleitfuge. Diss. TH Darmstadt, Fachbereich Konstruktiver Ingenieurbau. Erschienen in: Franke, E. (ed.): Mitteilungen des Instituts für Grundbau, Boden- und Felsmechanik der TH Darmstadt, No. 27.

Heibaum, M. (1991). Kleinbohrpfähle als Zugverankerung – Überlegungen zur Systemstandsicherheit und zur Ermittlung der erforderlichen Länge. In: *Institut für Bodenmechanik, Felsmechanik und Grundbau der Technischen Universität Graz (organiser.): Bohrpfähle und Kleinpfähle – Neue Entwicklungen (6th Christian Veder Coll.)*. Graz TU: Institute of Soil Mechanics.

Hein, W. (1990). Zur Korrosion von Stahlspundwänden in Wasser. *Mitteilungsbl. BAW 67*, Karlsruhe.

Henke, S. (2011). Numerical and experimental investigations of soil plugging in open-ended piles. Tagungsband zum Workshop Ports for Container Ships of Future Generations. *Veröffentlichungen des Instituts für Geotechnik und Baubetrieb der TU Hamburg-Harburg* 22: 97–122.

Henke, S. (2012). Large deformation numerical simulations regarding the soil plugging behaviour inside open-ended piles. *31st Int. Conf. on Ocean Offshore Arct. Eng.*, Rio de Janeiro, Brazil, digitally published under OMAE2012-83039.

Henke, S. and Grabe, J. (2008). Numerische Untersuchungen zur Pfropfenbildung in offenen Profilen in Abhängigkeit des Einbringverfahrens. *Bautechnik* 85 (8): 521–529.

Jardine, R., Chow, F., Overy, R. and Standing, J. (2005). *ICP Design Methods for Driven Piles in Sand and Clays*. London: Thomas Telford.

Kempfert, H.-G., Lüking, J. and Mardfeldt, B. (2011). Zum Ansatz von Verbundspannungen bei Verpressmörtelpfählen. *Bauingenieur* 86 (11): 464–474.

Kranz, E. (1953). *Über die Verankerung von Spundwänden*. 2nd edn. Berlin: Ernst & Sohn.

Laumans, Q. (1977). Verhalten einer ebenen, in Sand eingespannten Wand bei nichtlinearen Stoffeigenschaften des Bodens. *Baugrundinstitut Stuttgart, Mitteilung* 7.

Lehane, B.M., Schneider, J.A. and Xu, X. (2005). The UWA-05 method for prediction of axial capacity of driven piles in sand. *Int. Symp. Front. Offshore Geotechn. (ISFOG2005)*, Perth, 683–689.

Mahutka, K.-P., König, F. and Grabe, J. (2006). Numerical modelling of pile jacking, driving and vibro driving. *Proc. Int. Conf. on Numer. Simul. Constr. Process. Geotech. Eng. Urban Environ. (NSC06)*, Bochum, (ed. by T. Triantafyllidis), Rotterdam: Balkema, pp. 235–246.

Meek, J.W. (1995). Der Spitzenwiderstand von Stahlrohrpfählen. *Bautechnik* 72 (5): 305–309.

Os, P.J. van (1976). Damwandberekening: Computermodel of Blum. *Polytech. Tijdschr. Ed. B* 6: 367–378.

Randolph, M.F. (2003). Science and empiricism pile foundation design. *Géotechnique* 53 (10): 847–875.

Rausche, F., Likins, G. and Klingmüller, O. (2011). Zur Auswertung dynamischer Messungen an großen offenen Stahlrohrpfählen. Pfahl-Symposium 2011, Braunschweig. *Mitteilungen des Instituts für Grundbau und Bodenmechanik, TU Braunschweig* 94: 491–507.

Counsel Directive 79/113/EEC of 19 December 1978 on the approximation of the laws of the Member States relating to the determination of the noise emission of construction plant and equipment (OJ L 33, 8 Feb 1979, pp. 15–30).

Schenk, W. (1968). Verfahren beim Rammen besonders langer, flachgeneigter Schrägpfähle. *Bauingenieur* 5.

Sherif, G. (1974). *Elastisch eingespannte Bauwerke. Tafeln zur Berechnung nach dem Bettungsmodulverfahren mit variablen Bettungsmoduli*. Berlin, München, Düsseldorf: Ernst & Sohn.

Weißenbach, A. (1985). *Baugruben, Teil II, Berechnungsgrundlagen*, 1. Nachdruck. Berlin: Ernst & Sohn.

Wirsbitzki, B. (1981). *Kathodischer Korrosionsschutz im Wasserbau*. Hamburg: Hafenbautechnische Gesellschaft e. V.

Standards and Regulations

DIN 13-1:1999-11 ISO general purpose metric screw threads – Part 1: Nominal sizes for coarse pitch thread

DIN 1054: Subsoil – Verification of the safety of earthworks and foundations – Supplementary rules to DIN EN 1997-1

DIN 4017: Soil – Calculation of design bearing capacity of soil beneath shallow foundations

DIN 4084: Soil – Calculation of embankment failure and overall stability of retaining structures

DIN 4085: Subsoil – Calculation of earth-pressure

DIN 4150: Vibrations in buildings

DIN 18137: Soil, investigation and testing – Determination of shear strength

DIN 18196: Earthworks and foundations – Soil classification for civil engineering purposes

DIN 18800: Steel structures

DIN 19703: Locks for waterways for inland navigation – Principles for dimensioning and equipment

DIN 19704-1: Hydraulic steel structures – Part 1: Criteria for design and calculation.

DIN 45669: Measurement of vibration immissions.

DIN EN 756: Welding consumables – Solid wires, solid wire-flux and tubular cored electrode-flux combinations for submerged arc welding of non alloy and fine grain steels – Classification.

DIN EN 996: Piling equipment – Safety requirements.

DIN EN 1536: Execution of special geotechnical work – Bored piles.

DIN EN 1537: Execution of special geotechnical works – Ground anchors.

DIN EN 1990: Eurocode: Basis of structural design.

DIN EN 1992 Eurocode 2: Design of concrete structures.

DIN EN 1993 Eurocode 3: Design of steel structures.

DIN EN 1995 Eurocode 5: Design of timber structures.

DIN EN 1997 Eurocode 7: Geotechnical design.

DIN EN 10025: Hot rolled products of structural steels.

DIN EN 10219: Cold formed welded structural hollow sections of non-alloy and fine grain steels.

DIN EN 10248: Hot-rolled steel sheet piling.

DIN EN 10249: Cold-formed steel sheet pile.

DIN EN 12063: Execution of special geotechnical work – Sheet-pile walls

DIN EN 12699: Execution of special geotechnical works – Displacement piles.

DIN EN 14199: Execution of special geotechnical work – Micropiles.

DIN EN 16228-1/7:2022-07 Drilling and foundation equipment – Safety

DIN EN ISO 2560: Welding consumables – Covered electrodes for manual metal arc welding of non-alloy and fine grain steels.

DIN EN ISO 12944: Paints and varnishes – Corrosion protection of steel structures by protective paint systems.

IN EN ISO 14171:2016-12 Schweißzusätze – Massivdrahtelektroden, Fülldrahtelektroden und Draht-Pulver-Kombinationen zum Unterpulverschweißen von unlegierten Stählen und Feinkornstählen – Einteilung

DIN EN ISO 14341: Welding consumables – Wire electrodes and weld deposits for gas shielded metal arc welding of non alloy and fine grain steels – Classification.

DIN EN ISO 14688: Geotechnical investigation and testing –Identification and classification of soil.

DIN EN ISO 15614: Specification and qualification of welding procedures for metallic materials – Welding procedure test.

Supplementary provisions to DIN EN 1537:2014-07, Execution of special geotechnical works – Ground anchors.

Supplementary provisions to DIN EN 12699:2001-05, Execution of special geotechnical work – Displacement piles.

Supplementary provisions to DIN EN 14199:2012-01, Execution of special geotechnical works – Micropiles.

10
Quay walls and superstructures in concrete

10.1 General

Concrete is the building material of choice for many waterfront structure applications. It is used for building retaining walls and gravity structures, as well as individual parts of waterfront structures, e. g. superstructure elements. Concrete can be cast in many different shapes, which means that a diverse range of structures can be built economically using tried-and-tested methods of construction.

In addition to using cast-in-place and/or ready-mixed concrete, it is also possible to employ (partially) precast concrete elements, which opens up a huge range of options for concrete as a construction material for waterfront structures. The proper design, execution of construction works and maintenance are crucial for functionality and durability.

Besides static loads, waterfront structures are also subjected to changing water levels, soils and water that are aggressive to concrete, ice, vessel impacts and chemical attacks. Thermal actions plus creep and shrinkage must also be taken into account. It is, therefore, not sufficient to design the concrete components of waterfront structures for static loads only.

When designing concrete waterfront structures, it is particularly the high mechanical loads, the effects from fluctuating temperatures and the effects of chemicals on the surface of the concrete that have to be considered, in addition to the loads due to earth pressure, water pressure and quayside operations.

Waterfront structures are almost always large-scale engineering work. In the case of large, bulky mass concrete elements, the heat of hydration can lead to significant temperature differences between the core and the surface of such elements, which brings with it a certain cracking risk due to restraint stresses. Therefore, to achieve a robust, durable structure, the mix design of the mass concrete is crucial.

The intended design life must be taken into account. The design life for waterfront structures can be shorter than that for standard engineering work or waterway engineering structures.

Additional relevant factors here are the need for straightforward building operations without complicated formwork and the simple integration of piles, sheet pile walls, etc. This can lead to structural and constructional measures that exceed the minimum requirements of DIN EN 1992-1-1.

10 Quay walls and superstructures in concrete

In the case of quay walls made from sheet piles with a reinforced concrete superstructure, the reinforced concrete cross-section shall overhang the front edge of the sheet piles on the water side by about 15 cm. For reasons of corrosion protection, the underside of the reinforced concrete superstructure should be at least 1 m above the high-tide level or the mean water level. The transition between the sheet pile wall and the superstructure should be sloped at about 2:1, so that vessels and load-handling equipment cannot be trapped underneath. In addition, it is worthwhile providing sheet metal protection aligned flush with both the sheet pile wall and the superstructure (see Figure 7.52, for example).

All measures and requirements must be summarised in a specification that forms the basis for the design and must be agreed to by client, designer, checking engineer and the authorities responsible.

10.2 Construction materials

10.2.1 Concrete

The strength and deformation characteristic values must be taken from DIN EN 1992-1-1, Section 3.1. The concrete properties must be specified in compliance with DIN EN 1992-1-1, Section 4, and according to DIN EN 206-1 in conjunction with DIN 1045-2.

When choosing concretes, the relevant exposure classes are critical and must take account of the ambient conditions. The exposure classes govern the minimum requirements for concrete quality, concrete cover and the criteria regarding maximum crack widths.

For typical elements in waterfront structures, allocations to the exposure classes are shown in Figures 10.1 and 10.2 for both seawater and freshwater.

The exposure class for chemical attack by natural soils and groundwater is determined according to DIN EN 206-1, Table 2, in accordance with DIN 4030.

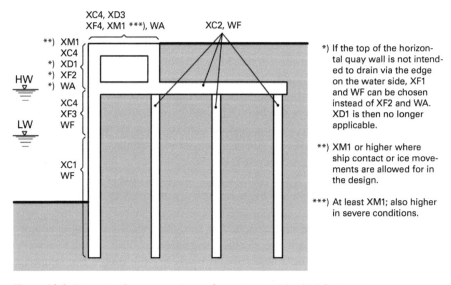

Figure 10.1 Example of exposure classes for a quay wall in tidal freshwater.

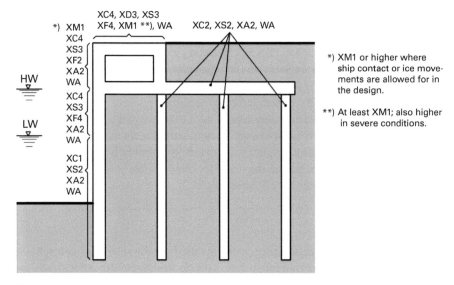

Figure 10.2 Example of exposure classes for a quay wall in seawater.

Complying with the requirements of exposure class XF4 can present problems in the case of large, bulky elements because of the measures required to reduce the heat development. These may affect the type and amount of cement and the use of concrete admixtures and can have adverse effects on the frost resistance, which must be verified using the CDF method. Here, the choice of suitable raw materials is very important. If the actions to be expected at the location and the intended design life mean that a lower frost resistance is acceptable for the concrete, then requirements that are different from those in the standard may be specified in individual cases to suit the specific location and type of structure. Building legislation normally requires an individual project approval in such cases.

To alleviate this conflict of aims, in more recent waterfront structure projects, a zoned casting method has been used, with different concrete mixes being employed for the core and surface zones of the concrete element (Morgen et al. 2005, Morgen 2012).

The strength classes chosen depending on the exposure class, the minimum cement content, the maximum water/cement ratio and other requirements are defined in DIN EN 206-1 in conjunction with DIN 1045-2. For large, bulky elements (smallest dimension ≥ 80 cm) requirements according to the DAfStb directive on bulky concrete components apply. ZTV-W LB 215 contains further regulations for waterfront structures made from plain and reinforced concrete.

Concrete for load-bearing elements in waterfront structures should normally be grade C20/25 or higher. The water/cement ratio of concrete must not exceed 0.65 for waterfront structures.

Where significant longitudinal expansion of concrete elements is expected (see Section 10.3.4) and large cross-sections are in use, attention must be paid to employing low-shrinkage concrete with low heat of hydration development.

Important factors for the durability of the structure are dense-structured concrete, intensive curing and sufficient concrete cover to embedded steel.

The concrete cover should be larger than that given in DIN EN 1992-1-1 and at least $c_{min} = 50$ mm, with a nominal cover $c_{nom} = 60$ mm. With respect to limiting crack widths under service load, see Section 10.3.5 of DIN EN 1992-1-1.

Allocation to exposure class XA depends on the aggressiveness of the soil and the water as given in the geotechnical report. The casting of a trial element might be advisable in the case of challenging structures with complicated construction conditions. Formwork liners that drain water away from the surface zones can be used to improve the surface quality of the concrete. The use of such liners must be agreed upon with the client.

Concrete for deep foundations and diaphragm walls should generally be cast using the tremie method. Additional information on the use of this method can be found in the *Guide to tremie concrete for deep foundations* (EFFC and DFI 2018).

10.2.2 Steel reinforcement

Steel reinforcement and reinforcing steel are the terms used for the steel used to reinforce concrete elements. Reinforcing steel is a construction product manufactured specifically for use in reinforced concrete, with properties that have been optimised for that application.

Steel reinforcement is available in the form of individual bars or welded meshes and must comply with the requirements of DIN EN 10080. Special steel reinforcement products are covered by national technical approvals.

The detailing of steel reinforcement must suit the structural and constructional requirements. It is always important to ensure that the reinforcement can be easily installed and that there is sufficient space between the bars to enable the concrete placement and compaction, and this is even more critical for concrete elements with irregular geometric forms. For the surfaces of reinforced concrete waterfront structures, it is generally necessary to provide reinforcement in both directions, so that it forms a grid.

10.3 Design and construction

10.3.1 Construction joints

Construction joints are to be avoided wherever possible. Unavoidable construction joints are to be planned prior to commencing concreting works and designed in such a way that shear forces can be properly transferred and the durability of the element is not adversely affected by the inclusion of the joint. This calls for all construction joints to be carefully prepared and finished off accordingly (e. g. the pre-installation of grouting hoses for subsequent sealing of the joint).

Corrugated expanded metal or similar materials along with cement slurry must be completely removed prior to concreting the adjoining element. It is good practice to expose the grains of aggregate in the concrete cast first by means of high-pressure water jets and to apply a preliminary mix with a high cement content prior to concreting the second component to serve as a cushion and to improve adhesion. The requirements regarding the roughness of joint faces are defined in DIN EN 1992-1-1. When it comes to waterfront structures, compliance with the requirements for the "indented" category is generally necessary for surfaces not cast against formwork and "rough" for surfaces cast against formwork. Vertical construction joints should be arranged in a staggered pattern to improve the robustness of concrete elements.

10.3.2 Expansion joints

Where long, linear structures are to be built, it is always necessary to weigh up whether the structure should be built with or without expansion joints. Expansion joints place high demands on construction details and work on site and are costly to maintain.

Expansion joints are generally designed to accommodate the changes in length of the parts of the structure. If the intention is that the individual parts of the structure support each other in the transverse direction, then the expansion joints must include shear keys (indentations) or dowels. Expansion joints must always be designed to prevent washout of any backfill. The effects of the prevailing environmental conditions and material fatigue must be taken into account when using permanently elastic materials for the joints.

Components that cross expansion joints, e. g. crane rails for container cranes, will require bridging beams or similar measures. The transition between superstructure or capping beam and uninterrupted sheet pile wall must be designed to allow movement at the joint (see Section 7.2.12.4).

10.3.2.1 Formwork
Elaborate formwork should be avoided if at all possible in areas affected by tides and waves. Concreting works in areas affected by tides and waves should be carried out in periods of calm weather where this is possible. Formwork for building quay walls should be secured against uplift and wave slamming.

Permanent formwork can be used for areas where access is difficult. Such formwork must be permanently connected to the concrete structure.

10.3.3 Jointless construction

More and more quay walls are being built without joints in order to avoid the disadvantages of expansion joints, which include difficult, fault-prone construction details and structural disadvantages, such as the need to resist horizontal forces (Morgen et al. 2005).

When designing waterfront structures without joints, it is particularly important to take into account the restraint stresses due to the dissipation of heat of hydration, differential settlement, shrinkage and climate-induced temperature fluctuations (Morgen 2012).

10.3.4 Crack width limitation

Crack widths must be limited if the proper use and durability of reinforced concrete waterfront structures are not to be adversely affected. By restricting crack width to $w_k \leq 0.25$ mm, it can generally be assumed that the reinforcement is adequately protected against corrosion, and that the concrete is sufficiently impermeable to water. Cracks that are wider than the permissible width must be injected with a suitable material to permanently seal them.

Stricter requirements for crack width limitation or special measures may be necessary where unfavourable local conditions prevail, e. g. attack by chlorides, or where water flows through separating cracks or prestressing steel is being used.

Crack widths in large cross-sections can be limited by using minimum reinforcement in accordance with the code of practice supplied by the Federal Waterways Engineering & Research Institute (BAW-Merkblatt 2011), which is currently being revised (E-BAW-Merkblatt, 2020). In addition, or as an alternative, the reinforcement can be protected by a cathodic corrosion protection system. Advice on such systems can be found in HTG (1994).

Restraint stresses can be reduced by planning the concreting sequence, so that parts of the structure already built do not prevent the deformations of subsequent concrete pours due to shrinkage and the dissipation of heat of hydration. This can be achieved for larger elements, e.g. a pier structure, by casting beams and slabs in a single concreting operation.

10.4 Forms of construction

10.4.1 Concrete walls

10.4.1.1 General

Provided that they are designed and built properly, concrete walls represent an alternative to sheet pile walls for waterfront structures. Besides their economic and technical advantages, concrete walls have the benefit that they can be built essentially without vibration and with low noise.

However, building concrete walls in open water is a difficult procedure. Therefore, they are mainly used where they can be constructed from the land side, and the soil can be excavated in front of the wall afterwards. Bored pile walls and diaphragm walls are types of cast-in-place concrete wall that can be considered here.

The high stiffness and – compared with sheet pile walls – ensuing small deformations of bored pile walls and diaphragm walls mean that they are normally to be designed for higher, active earth pressure. The earth-pressure design approach depends in each case on the deformation behaviour of the wall. See Section 8.2.5 for the angle of earth pressure. Redistribution of the earth pressure is possible, see Section 8.2.3.2. To consider the deformation realistically, it is necessary to take into account not only the stiffness of the wall but also the flexibility of the anchorages and the base support.

10.4.1.2 Bored pile walls

DIN EN 1536, DIN SPEC 18 140 and the *Recommendations on Piling* (EA-Pfähle, 2012) apply for the construction of cast-in-place concrete bored pile walls, which can be built as contiguous or secant pile walls.

The centre-to-centre spacing of the individual piles in a secant bored pile wall is smaller than the pile diameter. The unreinforced primary piles are constructed first, and then the reinforced secondary piles are built between these but cutting into them by an amount that is generally equal to 10–15% of the pile diameter, but at least 10 cm. The overlap at the top of the pile should be chosen such that an adequate overlap is guaranteed over the full length of the pile while allowing for construction tolerances. The concrete mix for the primary piles should be designed so that the strength of those piles at the time of drilling the holes for the secondary piles does not exceed 3–10 MPa, depending on the power of the piling rig. When constructed properly, secant bored pile walls are technically impermeable to water. Loads in the longitudinal direction of the wall can be resisted by a capping beam that spreads the loads or, to some extent, also by the wall acting as a plate, provided the intersecting surfaces are sufficiently rough.

In a contiguous bored pile wall, the centre-to-centre spacing of the individual piles is, for technical reasons, slightly larger than the pile diameter. As a rule, every pile is reinforced. The watertightness of a contiguous bored pile wall can only be achieved through additional measures, and the wall cannot be assumed to act as a shear wall.

The construction of a bored pile wall calls for highly accurate drilling operations with good guidance of the casing. Boreholes for bored pile walls are formed by rotary drilling or grab excavation methods. Obstacles encountered during drilling are overcome by chiselling them or by core drilling through them. Below groundwater level, drilling must be generally carried out with a water-filled casing to prevent a heave failure, with associated loss of soil, at the base of the borehole.

The arrangement of the reinforcement inside each pile should be radially symmetric because it is impossible to rule out unintentional rotation of the reinforcement cage while extracting the casing. Unintentional lifting of the reinforcement cage can be prevented by forming the base of the cage appropriately and by employing a limited maximum aggregate size taking into account the space between the cage and the casing. To rule out deformations of the reinforcement cage and to comply with the concrete cover requirements, it is necessary to brace each reinforcement cage at suitable intervals and provide spacers. Guidance for detailing the pile reinforcement can be found in ZTV-ING (2019). Connections to adjoining parts of the structure can be achieved by including appropriate starter bars at the top of each pile.

If the piles are not integrated into adequately stiff superstructure elements, then waler beams are generally required to resist the anchor force and spread the loads. Waler beams are unnecessary for anchored secant or contiguous bored pile walls, provided that there is an anchor at every second pile for secant pile walls and every second gusset between piles for contiguous pile walls. The latter is valid in ground with at least medium-dense granular soil and/or with at least semi-solid cohesive soil. Irrespective of this, tension-resistant walings of sufficient length must be provided at the ends of such walls.

10.4.1.3 Diaphragm walls

DIN EN 1538 applies for the construction of diaphragm walls, which are built by placing concrete and reinforcement in a trench excavated in the ground and held open by a supporting fluid. With good joints between the individual sections, a diaphragm wall is practically impermeable to water. That it acts as a shear wall cannot be assumed in general for such walls. Precast concrete elements can be suspended in the trench where high demands are placed on the quality of the wall surface, although their use is rather limited to depths of 12–15 m owing to the heavy loads.

DIN 4126 requires verification of the stability of the open trench for a diaphragm wall during the construction stage when it is filled with the supporting fluid. The critical groundwater level must be established when building diaphragm walls in tidal regions. Where it is expected that the critical groundwater level will be exceeded, open trenches must be filled in good time. As a residual film of supporting fluid on the steel reinforcement cannot be ruled out, moderate bond conditions may be assumed for horizontal reinforcement.

A clay or bentonite suspension is typically used as the supporting slurry; where applicable, polymer solutions or hybrid systems can also be used (EFFC and DFI, 2019). Groundwater with a certain level of salinity can alter the ion balance of the supporting slurry unfavourably, resulting in flocculation, which decreases the supporting effect of the fluid. Therefore, in certain situations, it will be necessary to carry out suitability tests to check the salt content of the water, the soil conditions and other relevant factors before commencing work on site. The same applies to soils with chemical contamination or when excavating through organic soil strata. The test methods for supporting fluids and their constituent materials are covered by DIN 4127.

Concentrations of reinforcement and openings that might hinder the flow of the concrete must be avoided when detailing the reinforcement cage. An adequate number of large spacers should be included to guarantee the necessary concrete cover. Reinforcement cages should include adequate bracing; see ZTV-ING (2019).

10.4.2 Retaining walls

Retaining walls are generally bulky reinforced concrete structures constructed in an open and dry excavation. Cantilever retaining walls with a platform ("heel") at the base on the earth side have proved worthwhile (Figure 10.3). With its large base and material-savings compared with a gravity retaining wall, the cantilever retaining wall often represents an economic alternative.

Owing to their large dimensions, retaining walls are bulky, heavyweight structures. The specific requirements regarding the mass concrete must be taken into account during design and construction.

Retaining walls require level ground with a suitable bearing capacity. To counteract the risk of soil erosion caused by water seeping below the wall, it is necessary to check for filter stability in the soil directly beneath the foundation. Drains can be installed to reduce high water pressure differences, see Section 7.3. Soil strata at the foundation level with an inadequate bearing capacity or unstable filter action must be removed beforehand and replaced by a suitable material (Section 5.9). To prevent this soil replacement layer from settling on fine-grained, non-cohesive soil, it might be necessary to compensate the pore volume of the soil replacement layer with a suitably graded gravel mix. Apart from that, a graded gravel filter could be laid between the foundation layer and the undisturbed soil, possibly together with a geotextile.

10.4.3 Block-type construction

Waterfront structures can be built from stacks of unconnected large-format concrete blocks; see Figure 10.4. The dimensions and weights of the individual blocks depend on the production and transport options available, the equipment available for installing them on site and the circumstances expected during construction and quayside operations. A small number of large blocks is preferred from the economic viewpoint because that minimises the times for setting up and striking formwork, transportation and installation. Blocks with an effective installation load of 600–800 kN are frequently chosen when the blocks have to be installed from a floating barge.

Various materials can be used for the backfill to a quay wall made up of concrete blocks. A filter layer that is permanently effective at preventing washout should be installed between the backfill of coarse material and the final fill above this.

10.4 Forms of construction | 489

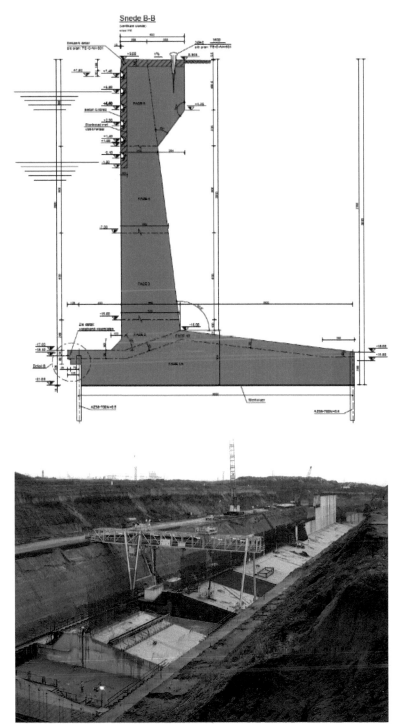

Figure 10.3 Example of a cantilever retaining wall.

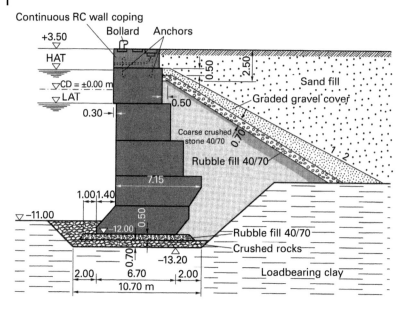

Figure 10.4 Section through a quay wall built from large blocks.

Block-type construction is mainly preferred in areas with considerable wave action. Besides the need for a heavy-duty equipment, expensive diver operations are also necessary to supervise the work and ensure that the blocks are laid properly.

Gravity walls made up of concrete blocks must be checked for overturning and sliding at all joints plus for ground failure and slope failure. Owing to the flexibility of this form of construction, the active earth pressure may be taken into account in all analyses. The remarks regarding the foundation level given in Section 10.4.2 apply here as well.

The joints between the individual blocks are usually left permeable because, in the past, permanent sealing has proved to be impossible. With appropriate joints and a backfill with a high permeability, the water level is quickly the same on both sides of the wall. In these cases, the water pressure acting on the quay wall may be considered with half the significant wave height at the most unfavourable level. Where uncertainties exist, drains can be installed, but they must also work reliably in the event of wave impact.

It is necessary to assess the effects of the water pressure in the joints between the blocks for incoming and outgoing waves. In particular, the pressure drop during outgoing waves is delayed in narrow joints, and this must be taken into account in the calculations.

Block-type construction is unsuitable for conditions in which breaking waves cannot be ruled out (approximately $d < 1.5H$, Sections 4.3.5 and 4.3.6.2).

The cross-sectional form of a wall built from blocks should be such that the bearing pressure beneath the base is distributed as uniformly as possible for permanent actions. This can be achieved with a base that has a projecting toe on the water side and a projecting heel on the lower blocks on the land side. To ensure a uniformly distributed bearing pressure at all stages of construction, the recommendation is to backfill behind the blocks as work progresses. A joint opening up due to the characteristic stresses caused by permanent and variable actions is permissible up to the centroid axis.

The capping beam can generally be designed as a beam on an elastic foundation. Where a range of settlement is to be expected, the internal forces at the top of the wall should be determined through a sensitivity analysis of the foundation stiffness. Quay walls built from blocks must be founded on a levelled bed of stones, gravel or sand with an adequate bearing capacity.

The individual blocks must be shaped and laid so that damage during installation is avoided. If the blocks are stacked vertically, one above the other, which is recommended for subsoil that is vulnerable to settlement, then excessively wide joints can only be avoided through elaborate, costly measures. However, such wide joints are acceptable with a suitable backfill material. Alternatively, the blocks can be interlocked by means of tongue and groove or I-shaped joints. If continuous vertical joints are to be avoided, then this can be achieved, for example, by laying the blocks at an angle of 10–20° to the vertical. The base can be formed by, for example, blocks laid horizontally, a sunken box caisson or similar elements, and wedge-shaped blocks can be used for the transition. The latter can also be used when it is necessary to correct the angle of the wall. Installing the blocks at an angle minimises the widths of the joints between the individual blocks but increases the number of block types. In this approach, all the blocks have tongue and groove-type interlocks on their side faces. The projecting tongue is positioned on the outside of the blocks that have already been laid, so that the grooves of the next blocks slide down into position around this tongue.

A reinforced concrete capping beam is usually provided along the top of a wall of concrete blocks. This beam compensates for differential settlement along the length of the wall and provides a base for quayside equipment. If there are crane rails on top of such a wall, then measures to prevent local differential settlement will be required at the joints.

Further information on construction can be found in Zdansky (2002).

10.4.4 Box caissons

Box caissons are hollow reinforced concrete elements to which additional ballast can be added, if necessary, to prevent them from moving in the water. After being floated into position and sunk, they are filled inside with suitable material and backfilled outside. Box caissons can be used to form wave-breaking structures but can also function as waterfront structures when backfilled. In the case of elongated structures made up of several separate caissons, it has proved worthwhile to add a reinforced concrete superstructure after installing the box caissons to give the structure additional stiffness and form the top section of the front wall. This helps to compensate for differential settlement and horizontal displacement.

Apart from verifying stability for the final condition, it is also necessary to check the stability while being floated into position and to check the ultimate limit state STR during launching, floating into position, sinking and backfilling. For the final condition, it is also necessary to examine the risk of soil being washed out from beneath the caisson by water seeping underneath, a situation that cannot be completely ruled out.

In a change to DIN 1054, an open joint due to the characteristic stresses caused by permanent and variable actions is not permitted between the underside of a caisson and the soil. The remarks regarding the foundation level given in Section 10.4.2 apply here as well.

The roughness of the underside must be taken into account when checking for sliding. When the degree of roughness is coordinated with the grain size of the foundation layer, the angle of friction of the base can be assumed to be equal to the internal angle of friction φ, of the material of the (soil replacement) foundation layer. This figure should be reduced to $2/3\varphi'$ for a smooth underside.

Reinforced concrete box caissons are cast in a dry or floating dock and subsequently floated into position.

The joint between two adjacent box caissons should be formed in such a way that the differential settlement expected can be accommodated without the risk of damaging the concrete element. A tongue and groove arrangement continuous over the full height of the caisson may only be used when the relative movements of adjacent caissons are small, even if the joint is properly sealed.

It is necessary to guarantee that the backfill to a backfilled structure cannot be washed out in the final condition. The solution shown in Figure 10.5 has proved worthwhile. Here, four pillars are formed on the side wall of each caisson, such that they form three chambers once the caissons have been sunk into position. The outer two chambers are filled with a mix of sand and gravel with stable filter properties, and the middle chamber is carefully filled with concrete after all settlement has subsided. The use of a textile formwork system can be considered in some circumstances.

Alternatively, it is possible to drive make-up sheet piles (which exhibit a certain "elasticity") that are threaded into cast-in sheet pile interlocks. The recommendation is to provide a pair of such piles and to fill the space in between; see Figure 10.6. Concrete can be used as the filling only where little differential settlement is expected, otherwise a mixture of sand and gravel is to be used.

10.4.5 Open caissons

Open caissons are hollow, reinforced concrete elements whose bottom edge is shaped to cut into the subsoil. As the soil within the caisson is excavated, so it sinks into the ground as a result of local ground failures on the inside caused by the angle of the bottom edge due to the self-weight of the caisson overcoming the skin friction exerted by the subsoil. Compressed-air caissons have an additional, enclosed working chamber at their base, in which an overpressure can be generated to suppress the ingress of water and enable excavation work to be carried out in the dry. It is, therefore, possible to enter this chamber to clear from obstacles, for example. Once the caisson has reached the desired depth, a concrete base slab is added to an open caisson, or the working chamber of a compressed-air caisson is filled with concrete (*Geotechnical Engineering Handbook* 2018, Section 3.8).

The design principles given in Section 10.4.4 apply accordingly to open caissons.

Open caissons can be installed from the land or from the water. Compressed-air caissons with their enclosed working chamber can float without assistance. However, open caissons require the help of jack-up platforms or floats to reach their final positions. Open caissons are simpler and less expensive to install than compressed-air caissons and can be positioned at much greater depths. However, they cannot be positioned as accurately.

Open caissons should be designed for the unavoidable dissimilar support conditions around the bottom edge during the sinking procedure.

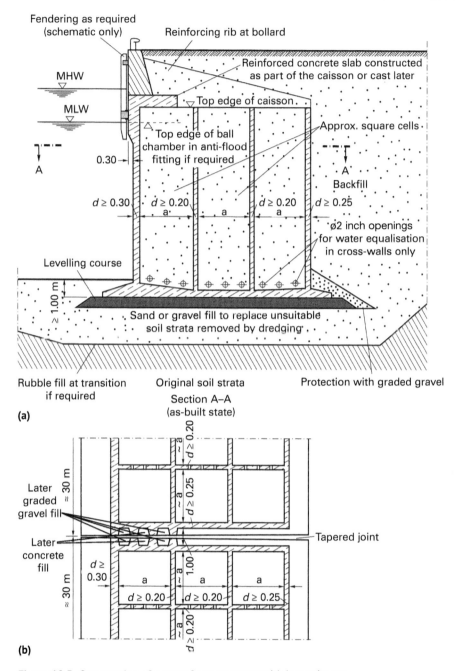

Figure 10.5 Construction of a waterfront structure with box caissons.

The bottom edge and concrete-filled working chamber of a compressed-air caisson achieve a good interlock with the subsoil. This means that an open joint due to the characteristic stresses caused by permanent and variable actions is permissible between the

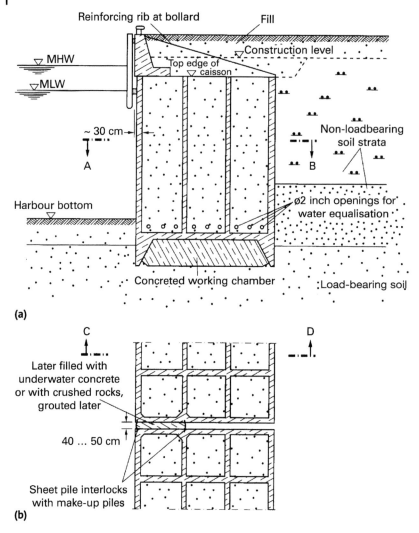

Figure 10.6 Construction of a quay wall with compressed-air caissons and subsequent basin dredging.

underside of the caisson and the soil. However, in contrast to DIN 1054, the eccentricity should be limited to one-quarter of the width of the base.

After being floated into position, an open caisson is sunk to the existing or deepened foundation level. Rough levelling is generally sufficient for this, because the shape of the narrow bottom edge means that it can easily sink into the soil, which makes any minor unevenness of the foundation surface irrelevant. The bottom edge should be formed by cast-in steel parts or made from high-strength concrete (at least C80/95) or steel fibre-reinforced concrete. Jetting lances along the bottom edge and facing inwards can help to loosen the soil to be excavated (Figure 10.7, section C-D, illustrated on water side). Compressed-air caissons should be installed with a 40–50 cm gap between the individual elements, open caissons with a 60–80 cm gap.

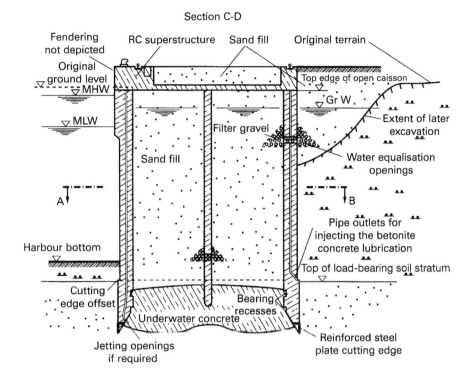
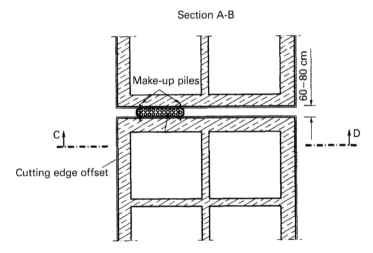

Figure 10.7 Design for a quay wall built from open caissons with subsequent basin dredging.

The bottom edges of the cross-walls of open caissons must terminate at least 0.5 m above the outer bottom edge in order to prevent unintentional support during the sinking procedure. To transfer the loads to the underwater concrete base, dependable bearing pockets (recesses), which can be easily cleaned after sinking, should be cast into outer walls and cross-walls.

Open caissons that are rectangular in plan do not stand as evenly on their bottom edges as those that are circular in plan because of the funnel-shaped excavation. This results in a higher risk of deviating from the intended position. If a rectangular form is necessary, a compensate size should be preferred.

The frictional resistance during the sinking procedure depends on the properties of the subsoil and the construction of the open caisson. It is generally sufficient when the vertical forces acting on the open caisson can overcome a skin friction of $20\,kN/m^2$ on the sides of the open caisson. Experience shows that application of lubricant to the sides can reduce this to below $10\,kN/m^2$. Thick walls should be chosen for open caissons because the options for adding ballast are limited.

References

BAW-Merkblatt (2011). *Rissbreitenbegrenzung für frühen Zwang in massiven Wasserbauwerken (MFZ)*. Bundesanstalt für Wasserbau.

E-BAW-Merkblatt (2020). *Rissbreitenbegrenzung für Zwang in massiven Wasserbauwerken (MRZ)*. Bundesanstalt für Wasserbau.

EA-Pfähle (2012). *Recommendations on Piling* (ed. by Deutsche Gesellschaft für Geotechnik e. V.). Berlin: Ernst & Sohn.

HTG (1994). *Kathodischer Korrosionsschutz für Stahlbeton*. Hamburg: Hafenbautechnische Gesellschaft e. V. (HTG).

Morgen, K. Thaden, H. von and Vollstedt, H.-W. (2005). *Fugenlose Überbauten für die Containerkajen CT3a und CT4 in Bremerhaven, Beton- und Stahlbetonbau 12/2005*, pp. 1003–1011. Berlin: Ernst & Sohn.

Morgen, K. (2012). *Fugenlose Kaimauern am Beispiel Bremerhaven und am neuen deutschen Tiefwasserhafen in Wilhelmshaven, Beton- und Stahlbetonbau Häfen und Kaianlagen, Spezial 2012*. Berlin: Ernst & Sohn.

Witt, K.J. (2018). *Geotechnical Engineering Handbook*, 8th edn., Pt. 3, Sect. 3.8. Berlin: Ernst & Sohn.

Zdansky, V. (2002). Kaimauern in Blockbauweise. *Bautechnik* 79 (12): 857–864.

ZTV-ING (2019). Zusätzliche Technische Vertragsbedingungen und Richtlinien für Ingenieurbauten (ZTV-ING), (ed. by Bundesanstalt für Straßenwesen, BASt). BASt.

ZTV-W LB 215 (2012). Zusätzliche Technische Vertragsbedingungen für Wasserbauwerke aus Beton und Stahlbeton. Bundesministerium für Verkehr, Bau & Stadtentwicklung.

Standards and regulations

DIN 1045: Concrete, reinforced and prestressed concrete structures.

DIN 1054: Subsoil – Verification of the safety of earthworks and foundations – Supplementary rules to DIN EN 1997-1.

DIN 4030: Assessment of water, soil and gases for their aggressiveness to concrete.

DIN 4126: Stability analysis of diaphragm walls.

DIN 4127: Earthworks and foundation engineering – Test methods for supporting fluids used in the construction of diaphragm walls and their constituent products.

DIN EN 206: Concrete – Specification, performance, production and conformity.

DIN EN 1536: Execution of special geotechnical work – Bored piles.

DIN EN 1538: Execution of special geotechnical work – Diaphragm wall.

DIN EN 1992: Eurocode 2: Design of concrete structures
DIN EN 10080: Steel for the reinforcement of concrete – Weldable reinforcing steel – General.
DIN SPEC 18140: Supplementary provisions to DIN EN 1536:2010-12, Execution of special geotechnical works – Bored piles.
EFFC and DFI (2018). Guide to tremie concrete for deep foundations, 2nd edn.
EFFC and DFI (2019). Guide to supporting fluids for deep foundations.

11 Pile bents and trestles

11.1 General

A pile bent (planar) or trestle (spatial) is made up of a group of piles supporting a slab. To bridge over abrupt changes in the ground level, sheet pile walls are specified for planar structures on the water side and, if necessary, on the land side. Groups of piles in three dimensions (pile trestles) allow the construction of working platforms that are either free-standing or cantilever out from the bank.

Piles can be exposed to reversed loads (tension/compression) caused by line pull, lateral crane impacts, tidal effects, etc. The suitability of the chosen pile system should, therefore, be verified for such loads.

The horizontal head deformations of the pile bents or trestles covered in the following sections should always be determined as well and checked for compatibility with serviceability requirements. It is important to ensure that the anchor forces correspond to the deformation potential of the structure.

In general, anchor forces should be transferred into the anchor elements via the shortest route. Provided that the subsoil conditions are suitable, this design principle is best achieved with horizontal anchors fixed to anchor plates (or fixed to pile bents in the case of deeper load-bearing strata) or by anchoring with raking piles.

11.2 Configuration and design of a pile bent

11.2.1 General

There are two main types of pile bent for dealing with changes in ground level:

1. Exposed pile bents above an underwater slope with land and water-side sheet pile walls bridging over an abrupt change in the ground level (Figure 11.1). To determine the largest bending moments and anchor forces in each case, gradients below the slab of $1:4$ as well as $1:10$ should be analysed.
   The corresponding partial safety factors should be applied to dead loads, variable loads and water pressure.

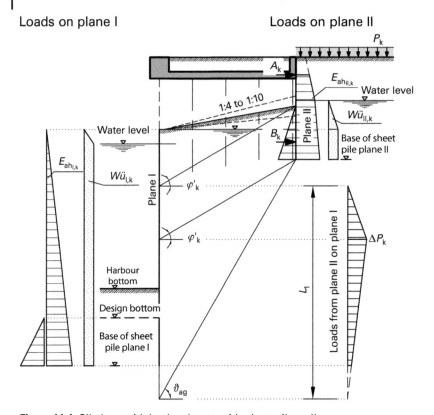

Figure 11.1 Pile bent with land and water-side sheet pile walls.

2. A pile bent with water-side sheet pile wall
   - Either as a strengthening structure in front of or above the existing pile bent to increase the design depth, with an existing embankment generally remaining intact. The sheet pile wall should continue to a certain depth, but at least down to the base of the embankment, and – to prevent excess water pressure from building up behind it – should not be impermeable (Figure 11.1).
   - As a structure for a new pile bent with a fully backfilled sheet pile wall. The slab supported on the piles shields the earth pressure on the sheet pile wall against surcharges (Figure 11.2).

Pile bents are one-dimensional, linear structures in the longitudinal direction and are, thus, planar structures for the purpose of structural design. Consequently, actions and resistances can be defined per linear metre or per system module length.

11.2.2 Earth pressure loads

The reader is referred to Section 3.5 for information on stresses in sheet pile walls due to earth pressure.

Plane II (Figure 11.1), i.e. the plane of a sheet pile wall, if present on the land side, or a vertical plane intersecting the rear edge of the superstructure, is usually taken as the reference plane for the active earth pressure due to the soil behind the structure (reference

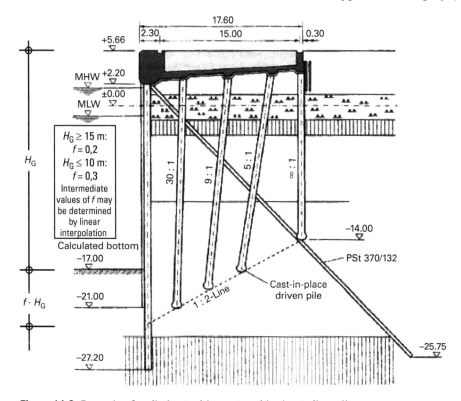

Figure 11.2 Example of a pile bent with a water-side sheet pile wall.

plane on the land side). The load of the land-side sheet pile wall (plane II) on the water-side sheet pile wall (plane I) can be derived from the following equation:

$$B_k = E_{ah_{II},k} + W"_{u_{II,k}} - A_k$$

$$\Delta P_k = \frac{2 \cdot B_k}{L_1}$$

11.2.3 Load due to excess water pressure

The excess water pressure acting on the water-side sheet pile wall depends on the soil conditions, the depth of soil behind the wall and the presence of a drainage system, as well as other factors. In the case of new structures with a water-side sheet pile wall only and soil extending up to the underside of the slab, the excess water pressure is assumed to act directly on the sheet pile wall according to Section 3.3.2. Where the ground surface is below the outer water level for land-side sheet pile walls, and the new wall has permanently effective openings for equalising the water level, the excess water pressure is again according to Section 3.3.2. The flow force to be estimated is the difference between the water level behind the wall and the outer water level. As a precaution, an excess water pressure from the outer water level plus half the height of the waves expected in the basin should be assumed to act directly on the water-side sheet pile wall. As a rule, a water level difference of 0.5 m is sufficient as a characteristic value.

11.2.4 Load path for piles

For economic reasons, the relatively high compression pile loads below a slab are often carried by cast-in-place driven piles. Where they carry the loads primarily by way of end bearing, the actions on the quay wall due to the pile loads can be neglected if the piles are installed at a rake of 1 : 2 or less, as shown in Figure 11.2. This applies to both single pile bents, as well as several rows of compression piles.

Should the pile loads be carried via surface friction and end bearing, they generate actions on the quay wall, which must then be multiplied by partial safety factors. These, too, can be neglected if the centroid of the point of application of the pile forces is below a straight line sloping at 1 : 2, as shown in Figure 11.2.

The magnitude of the actions acting on the quay wall due to the pile forces is calculated in accordance with DIN 4085 and depends, in particular, on the location of the piles in relation to the wall, the rake of the piles and the properties of the in situ subsoil.

When designing the compression piles, it should be remembered that they are subjected to lateral pressure due to deflection of the quay wall and are, thus, also subjected to bending. The magnitude of the load due to lateral pressure decreases with the distance from the quay wall and can be avoided if the main deformations of the sheet pile wall have already occurred when the piles are constructed.

The construction of the piles causes further compaction of the backfilled or in situ soil, provided that such soil can be compacted further. This increases the soil's bearing capacity and reduces the active earth pressure accordingly. However, during construction of the piles in the direct vicinity of the wall, an additional local compaction earth pressure acts, which can cause further deflection of the wall.

Concentrated actions from bollards and fenders can be distributed proportionately over the design cross-section within a section of the structure due to the plate effect of the slab. Where high loads acting at a point are carried by a group of piles for structural reasons, the load-bearing capacity of the group of piles should also be verified.

Determining the internal forces for piles and superstructure for the planar case is relatively straightforward:

- The structural system of the slab strip with the piles placed in its plane can be modelled by an elastically supported continuous beam. The piles are thus represented here by springs in the direction of the pile axis, and the associated spring stiffness is derived from the pile data. It must be guaranteed that the distance between the piles is large enough to avoid them influencing each other when carrying loads, and that the interaction between slab and piles is negligible.
- Where piles have a very high axial stiffness, e.g. cast-in-place driven piles, a rigid support for the slab can be assumed as well.
- As an equivalent to the elastically supported continuous beam, the system can be modelled as a plane frame consisting of several pile legs and a cross-member representing the superstructure.
- Fixed or hinged supports for the piles in the soil and at the slab, lateral bedding (and, hence, bending of the pile) and axial bedding can be handled with standard software.
- Calculations for waterfront structures with a "rigid" superstructure are, therefore, only appropriate in special cases, e.g. where old quays with solid pile head blocks must be re-analysed.

Special construction methods may be required in difficult driving conditions or where obstacles in the subsoil suggest that the planned driving of the quay wall and pile bents may not appear to be feasible.

When verifying the anchorage of a sheet pile wall driven afterwards in front of an existing structure, it should be taken into account that the anchoring of the existing structure is also loaded by additional earth, water and line pull loads. It is necessary to analyse how the loads are distributed between the old and new anchors. In order to keep the displacements small and to avoid overloading the existing anchors, prestressed anchors can be used for a sheet pile wall driven afterwards in front of an existing structure. The displacement of the anchor connection should be specified based on the local circumstances.

11.3 Design of pile trestles

A pile trestle can be modelled as a three-dimensional frame supporting a superstructure in the form of a slab. Pile trestles are primarily used when soil with at least a limited bearing capacity is only available at greater depths, when a lightweight structure is required in regions with severe seismic activity and when the waterfront structures and other port structures do not make use of walls but, instead, make use of embankments or, exclusively, piles, so that waves can pass through practically unobstructed in order to avoid long-period oscillations in the basin or to dissipate the wave energy. The loads acting on the slab are distributed to all the piles in the pile trestle through the load-bearing capacity of the superstructure acting as both plate (= in-plane forces) and slab (= out-of-plane forces) to achieve an effective load transfer.

11.3.1 Free-standing pile trestles

Slabs supported on free-standing pile trestles are generally made from reinforced concrete, whereas driven steel sections are normally chosen for the foundation piles. Foundation piles made from steel can also meet the requirements for icy conditions (Figure 11.3).

Some of the foundation piles should be arranged as pile trestles, so that the load on the slab can be accommodated horizontally. This simultaneously reduces the bending moments in all piles where otherwise no further external horizontal actions with a large influence occur, e. g. lateral pressure on piles due to flowing soil masses, strong currents, ice pressure, ice impact, etc.

Should the aforementioned assumptions apply, then assuming that the piles are hinged at the slab and at their base is sufficiently accurate for the calculations, even if the piles are not actually constructed as such. In order to deal with any unwanted fixity, special construction details will be needed at the head of the pile. This may be necessary where large changes in the length of the slab occur due to temperature changes.

Displacements of the head of a pile due to shrinkage during construction can, therefore, be kept under control by constructing sections of the slab with wide contraction joints in between. These joints are closed off with structural concrete after the shrinkage has abated. Restraint stresses due to shrinkage are generally kept within safe limits by employing such joints.

504 | 11 Pile bents and trestles

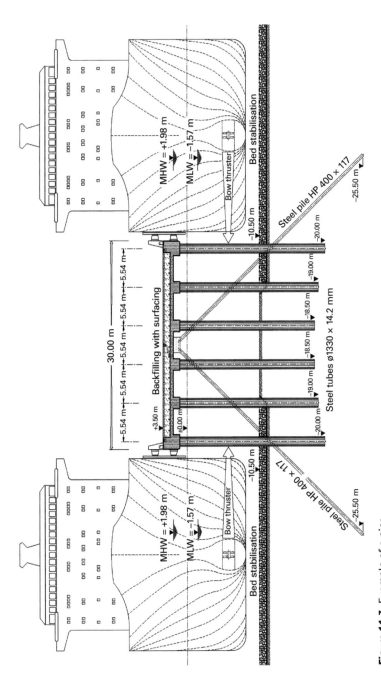

Figure 11.3 Example of a pier.

The depth of the slab is determined by calculating the effective bending moments and shear forces at the supports on the foundation piles. Such slabs are normally only 50–75 cm thick and, so, should be considered to be flexible with respect to the spring stiffness of the piles. The use of precast or partly precast concrete elements has proved to be beneficial here.

Transferring larger bending moments caused by loads on the slab can be avoided by positioning the piles to suit the load transfer. The actions applied at the respective loading position should largely be carried by the directly adjacent piles.

Slabs should not be directly traversed by vehicles. Covering the slab with a sand fill and paving the surface to suit the traffic loads has operational and constructional benefits because pipes, cables, ducts, etc., can be accommodated within the sand layer, and the loading on the slab and the piles due to localised actions are lower due to the load being spread within the sand compared with slabs traversed directly. At the same time, the substructure need not be designed for any dynamic actions caused by vehicles.

A sand fill with a thickness of about 1.0 m or more will, in most cases, be deep enough to accommodate all the services within the filling.

If the actions due to road and rail traffic are assumed to be uniformly distributed (basic situation 2 or 3 according to Section 4.2), the depth of the sand fill must be at least 1.0 or 1.5 m, respectively.

11.3.2 Special structures designed as pile trestles

Sharp bend and corner structures, as well as breakwaters and jetty heads in the form of pile trestles, are special structures for which the foundation piles have to be positioned in accordance with the actual site conditions. This is also the case where, for example, the layout of Ro/Ro ramps requires projections or different heights for the front edge of the quay.

In such cases, it is normally no longer possible to arrange the piles in the plane of the pile bent. Instead, the piles must be installed with a large number of intersections and rakes intersecting in three dimensions. In addition, the connections to the superstructure can be at different heights.

11.3.3 Structural system and calculations

A superstructure supported on a pile trestle can be properly modelled as an elastic slab on elastic supports. From a structural point of view, and with regard to the spread of the loads, the slab is an elastic plane frame that is supported on – likewise – elastic piles with or without a column head. According to Section 11.2, each pile can be represented by an elastic spring acting in the direction of the pile axis and the pile lengths set as elastic lengths between the hinge points.

The wave pressure on the piles can be calculated in accordance with Section 4.7. Loading on the slab from below due to "wave slamming" can also be assumed to be as set out in Section 4.7.9.4.

The most beneficial arrangement of the piles below the slab is when all the support moments and the pile loads on one pile axis are roughly equal. However, this cannot be achieved in all situations due to the actions from cranes, line pull and vessel impact.

11.3.4 Construction guidance

The following are just some points that should be complied with to ensure that the pile trestles are as economical as possible; see also Agatz and Lackner (1977):

- As far as possible, the berthing forces of large vessels should be absorbed completely by fenders plus heavy-duty fender panels positioned in front of the pile trestle. If necessary, the fenders can also be supported off the slab.
- In the vicinity of the fenders, a heavy-duty mooring bollard at the top of the wall can be combined with the fender structure.
- Local horizontal loads such as a line pull or vessel impact are distributed through the slab, which is very stiff in its plane, to all the piles in a group.
- Fender piles should be specified for small waterborne traffic in order to protect the structural piles and the hulls of the vessels.
- Craneway beams can be included as structural components in the reinforced concrete slab.
- Vertical loads from crane operations can be carried by additional piles along the line of the crane, if necessary.
- In order to minimise the effect on the bending moments, changes of direction of the slab should only be positioned above rows of piles.
- In tidal areas, it is expedient to place the slab at a sufficient height above mean high tide in order to be independent from the normal tide levels during the construction of the pile trestle.
- When designing the formwork for the piled slab, the effects of waves should also be taken into account.
- Rows of vertical and raking piles should be arranged offset with respect to each other.
- If the slab is divided into blocks, these are generally connected to one another through horizontal joggle joints.
- Horizontal actions in the longitudinal direction are carried by pile bents in the middle of the block with raking piles as shallow as possible.
- Horizontal actions in the transverse direction are carried by pile bents on the longitudinal axis of the structure with raking piles as shallow as possible.
- Resisting all the horizontal actions in the manner described above minimises the stresses in the slab and the piles.
- Slabs for large transhipment bridges can be concreted on sliding or moving formwork mainly supported on vertical piles.
- The fixity zones of steel foundation piles at the slab must be protected against corrosion, especially in areas with a high risk of corrosion (saltwater, brackish water).

Further measures are required for parts that deviate from the standard cross-section and, thus, disrupt progress on site, e. g. connections for the aforementioned raking pile bents.

- Where possible, raking piles should only be driven from the slab through dedicated driving openings after concreting and then connected to the pile bent with local reinforced concrete plugs in a second concreting operation.
- The contraction joints referred to in Section 11.3.1 can also be used as driving openings, which for this purpose might need to be widened locally.

- The stability of blocks divided by joints must be verified. To ensure stability, temporary bracing across the openings and contraction joints may be required under certain circumstances.

11.4 Design of pile bents and trestles in earthquake zones

11.4.1 General

When designing pile bents and trestles for earthquake zones, it should be remembered that the superstructure – including fill materials, imposed loads and supported structures – is accelerated by the effects of the earthquake in such a way that additional, horizontal inertial forces arise, which load the structure and its foundations. Therefore, in principle, the weight of pile bent/trestle should be kept as small as possible in seismic regions. In the case of pile bents/trestles with a relieving slab, it is necessary to check whether the benefit of including such a slab could possibly be negated during an earthquake by the ensuing horizontal inertial forces due to the large mass of concrete at a relatively high level, which is associated with this form of construction.

The reader is referred to Section 3.5.14 for more information on how earthquakes can affect pile bents/trestles. Structures that are particularly tall and narrow must also be checked for the resonance that amplifies seismic amplitudes.

11.4.2 Active and passive earth pressures, excess water pressure, variable loads

The information in Sections 3.5.14.3–3.5.14.5 applies accordingly. However, in the event of an earthquake, it should be borne in mind that variable loads and soil dead loads assumed to act behind the slab should be applied at a shallower angle due to the horizontal earthquake acceleration, and that the relieving slab is, therefore, less effective.

11.4.3 Resisting the horizontal inertial forces of the superstructure

The horizontal inertial forces due to an earthquake can act in any direction. At right-angles to the waterfront structure they can generally be readily accommodated by raking piles. Arranging additional pile bents in the longitudinal direction of the structure can, however, present problems in certain circumstances.

If the backfill soil to a water-side sheet pile wall extends up to the underside of the slab, it is advantageous to resist the horizontal loads acting in the longitudinal direction through the bending of the piles in the subsoil. However, this effect has to be mobilised by displacing the piles against the soil. To ensure their serviceability and for constructional reasons, the displacements should not be greater than approximately 3 cm, and the bedding should be limited correspondingly.

Where superstructures are built over embankments, the earthquake load due to active earth pressure is significantly lower than that for backfilled structures. Slabs built above such slopes should be designed to be as light as possible in order to minimise the horizontal inertial forces from earthquakes.

References

Agatz, A. and Lackner, E. (1977). *Erfahrungen mit Grundbauwerken*. Berlin: Springer.

Standards and Regulations
DIN 4085: Subsoil – Calculation of earth pressure.

12
Dolphins

12.1 Design and construction

12.1.1 Dolphins – purposes and types

Dolphins are designed to allow ships to berth and moor safely. They are also used to protect waterfront structures and guide waterborne traffic. Breasting dolphins must absorb the impacts of the berthing process. Mooring dolphins should be designed to take account of the effects of line pull forces, wind loads and currents. Breasting dolphins generally also serve as mooring dolphins. Dolphins as part of a crash barrier are sacrificial structures and are designed to meet specific requirements. They are not included in this recommendation.

Dolphins can be designed as single piles, a row of piles (also named fender rack) or a group of piles (cluster dolphins), see Figure 12.1. The type chosen depends on the loads to be accommodated, the sections and materials available, the subsoil, the function of the dolphin, etc. Dolphins can be fitted with fenders to reduce the contact forces between a vessel and the dolphin.

12.1.2 Layout of dolphins

At least two breasting dolphins are needed to moor a vessel. The distances between dolphins should be chosen so that the straight part of the hull (called the parallel body) of the smallest vessel using the berth is in contact with at least two dolphins. At the same time, the distance between dolphins should ensure that the ensuing lever arms for the mooring lines are not too short. In practice, a dolphin spacing of 25–40% of the vessel's length has proved suitable.

To cover all likely vessel lengths, more than two dolphins are normally required at a berth. The outer dolphins are normally more heavily loaded than the inner ones as the largest vessels always berth at the outer dolphins. The inner dolphins can therefore be designed for lower loads. However, they should still be protected against overloading. This can be done, for example, by setting the line of the inner dolphins back from that of the outer dolphins.

The spacing of dolphins designed to protect waterfront structures should be close enough to prevent any contact between vessel and structure. A dolphin spacing equal to 15% of the length of the shortest vessel is generally sufficient. The distance between a fully deflected dolphin and a quay wall should be at least 0.5 m. The same distance should be maintained

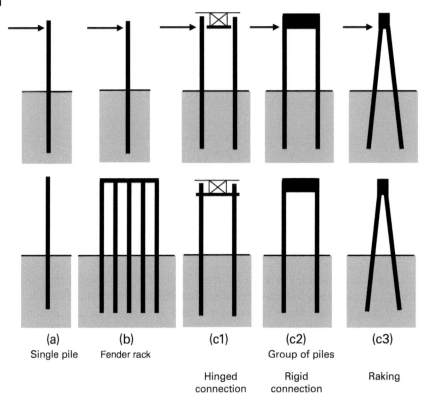

Figure 12.1 Types of dolphin, side elevations (top) and front elevations (bottom).

for each part of the vessel's hull except for the point of contact, with an unfavourable listing angle of a minimum of 3° being assumed. Providing a dolphin pile with a small outward rake can help to meet both these requirements.

If the dolphin is fitted with a fender, this should be placed at a height that allows all vessels to berth safely for all combinations of water level and loading conditions. With a view to the simple construction and maintenance of the key parts of the structure, such parts should be above the water level for as much of the time as is possible.

12.1.3 Equipment for dolphins

Breasting dolphins are fitted with bollards, mooring hooks and/or quick release hooks. Bollards are proven and reliable mooring devices that require almost no maintenance. Slip hooks have the advantage that mooring lines – even those under tension – can be released in emergencies if required, which is impossible with bollards. This can be especially necessary at berths where hazardous materials are being loaded/unloaded, which is why slip hooks are often specified in such cases. However, as the mechanism can also be released accidentally, which can result in damage, e.g. the rope becoming entangled in a vessel's propeller, slip hooks should really only be used where the safety requirements of particular transhipment operations dictate this.

Where bollards with break-off bolts are mounted on dolphins, high dynamic loads occur once the bollard is torn off. These loads must be taken into account when designing the dolphin.

The edges of dolphin topside deck should be fitted with a nosing (rope protection) to prevent chaffing hawsers. Such edge protection can, for example, consist of steel pipes placed wherever the ropes could make contact with the dolphin.

Dolphins are often connected by walkways (called catwalks) to ensure quick and safe operations. Walkways must be designed to safely absorb any potential deformation of the dolphins during berthing manoeuvres. Ladders that reach down to the lowest water level should also be fitted, so that anyone falling into the water can climb out again. The ladders can at the same time provide access to a pilot boat, for instance. It should be possible to access dolphins safely from the land via a walkway that can also serve as an escape route.

Breasting dolphins are often fitted with fenders. These are attached with bolts on a steel plate mounted on the dolphin cap. The connection should ensure that precipitation and spray water can run off easily and that mooring lines cannot get caught.

Where the lighting from the adjacent quayside is inadequate, each dolphin should have its own lighting on the top to ensure safe functioning at night as well. However, the lighting should not obscure or be confused with navigation lights.

Systems are available to control and monitor the berthing and mooring process on the dolphins. These include systems for measuring the berthing speed, which activate an alarm when a critical speed is exceeded, or slip hooks that continually measure the tensile forces and regulate the tensioning of the ropes, also dolphin position loggers that provide conclusive information in the event of a dolphin being damaged. Such systems have proved particularly beneficial for berths that handle hazardous goods, as well as in unprotected areas. The relevant figures are displayed on large, illuminated panels. The current trend is towards GPS-controlled portable systems that do not require expensive installations on the berths and can be read regardless of location.

12.1.4 Advice for selecting materials

The energy absorption capacity of tubular pile dolphins can be increased while saving materials by manufacturing the dolphin from tubular sections with different wall thicknesses and grades of material. The use of fine-grained structural steel also increases the energy absorption capacity. The high permissible contact loads, as well as the large deformations associated with such high loads, result in an increased absorption of energy. Data on materials are given in Section 8.1.2.1.

It is worthwhile building the topmost part of a dolphin from weldable low-strength, fine-grained structural steel or structural steel with a strength $\leq 355\,\text{N/mm}^2$, so that it is easy to weld on bracing members and other structural components.

Section 8.1.4 applies to all welding work accordingly. The information given in Section 8.1.10 applies to the corrosion protection of steel piles.

12.2 Detailed design

12.2.1 Stiffness of the system

For the purpose of design, a distinction should be made between rigid and flexible dolphins. Rigid dolphins accommodate actions without any significant deformation, whereas flexible dolphins exhibit greater deformations under the effect of loads.

The stiffness of a dolphin is, therefore, an important aspect in its design. It is derived from the interaction between the pile (or group of piles), the fender (where fitted) and the subsoil. The resulting overall stiffness can, therefore, be significantly non-linear. Figure 12.2 shows the design model for a dolphin as well as typical stress–strain curves for individual components and the overall system.

12.2.2 Structural behaviour

Mooring dolphins are loaded by line pull and contact forces, which, in turn, are significantly affected by wind and wave loads. These loads can be assumed to be static for the purpose of design.

Breasting dolphins, on the other hand, are loaded by forces arising from ships' berthing manoeuvres. Such dolphins are not designed by prescribing a load but by specifying the berthing energy of a vessel, which corresponds to the integral of the stress–strain diagram of the dolphin's deformation (Figure 12.2). The magnitude of the resulting horizontal reaction force F_R is essentially determined by the system stiffness. A greater stiffness results in smaller deformations and larger forces; less stiffness has the opposite effect. There is no explicit solution to this design problem. It is the responsibility of the engineer to reach an optimum compromise for each particular situation, taking into account the fact that the berthing force is limited by the permissible pressure on the hull of the vessel (Section 12.2.7.1) and that the dolphin deformation should not exceed approximately 1.5 m (Section 12.2.7.2).

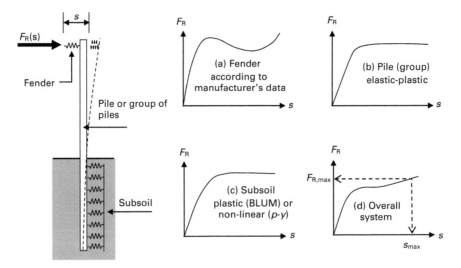

Figure 12.2 Overview of the structural system of a dolphin with fender and typical stress–strain relationships of individual components (a)–(c); the overall system (d).

Breasting dolphins in particular are often designed as flexible elements because of the need to limit the reaction force from berthing to avoid the vessel being damaged. Breasting dolphins must, nevertheless, absorb the kinetic energy of the berthing vessel either fully or partially. The energy absorbed by the deformed dolphin is called the energy absorption capacity and, according to Figure 12.2d, is as follows:

$$A = \int_0^{s_{max}} F_R(s) ds$$

where

R energy absorption capacity (internal work) of dolphin [kN m]
$F_R(s)$ horizontal reaction force (berthing force) between vessel and dolphin as a function of deflection s at the level of the force application point [kN]
s_{max} maximum deflection of dolphin at the level of the force application point [m]

For linear load-deformation behaviour, this equation can be simplified to

$$A = \frac{1}{2} \cdot F_{R,max} \cdot s_{max}$$

where

$F_{R,max}$ horizontal reaction force (characteristic berthing force) related to deflection s_{max} [kN]

The horizontal reaction force is transferred to the dolphin pile and, in turn, to the subsoil via the embedment of the pile. For raking piles or groups of piles with a rigid connection (c2 and c3 in Figure 12.1), normal forces can also arise, which can be very large if the spacing of the piles is small compared with the vertical distance from the bottom to the force application point.

The magnitude of the horizontal soil bearing pressure due to the loads to be transferred to the soil depends on the deformations and can increase up to the point where the soil fails. The horizontal soil bearing pressure can be calculated using different models (Section 12.2.5).

12.2.3 Actions

12.2.3.1 Loads due to berthing manoeuvres

The energy of berthing manoeuvres that must be resisted by the dolphin can be determined as shown in Section 7.4.1.4. The berth configuration factor of the waterfront structure should be taken as $C_c = 1.0$ (open structure).

The level of the vessel impact force transferred to the dolphin depends on its form of construction, the vessel dimensions and the water levels, and may differ depending on the design parameters (force, deflection and stresses).

The berthing force arising from berthing manoeuvres is determined by the stiffness of the dolphin (Section 12.2.5). For precise calculations, the flexibility of the vessel's hull must be

Table 12.1 Partial safety factors for verifying the ultimate limit state of a dolphin.

	Actions	Resistances	
		Soil	Steel
	γ_Q	$\gamma_{R,e}$	γ_M
Loads from berthing manoeuvres	1.00	1.00	1.00
Mooring forces (line pull) and contact forces	1.20	1.15	1.10
Wave, wind and current loads	1.20	1.15	1.10
Ice loads (see also Section 4.12)	1.00	1.10	1.10

taken into account. This can be allowed for by reducing the kinetic energy of the vessel (C_s factor, Section 7.4.1.4.2).

As the stiffness of the system can only be determined appropriately for characteristic loads and resistances, breasting dolphins are designed using characteristic values. Table 12.1 lists the partial safety factors to be used.

12.2.3.2 Mooring and contact forces

Vessels moored to dolphins are subjected to winds, currents and wave loads, which have to be resisted by the dolphin. This gives rise to tension loads (mooring forces) or compression loads (lean-on forces), depending on the location of the dolphin in relation to the vessel. The tensile forces can be directed upwards at an angle of up to 45°.

Wind loads acting on vessels can be calculated as shown in Section 4.8. Recommendations for estimating current loads can be found in DNV (2010).

Mooring and contact forces in sheltered harbour/port areas, i. e. locations with no significant wave effects (sea state and swell), can be determined using the information set out in Section 4.9.

The wave loads on vessels moored at unsheltered berths can become critical. In these cases, the design loads for the dolphin can be determined by simulating the vessel movements induced by the sea state over time. The effects on the fatigue strength must be taken into account (see Section 12.2.7.3).

Unsheltered jetties alongside navigation channels can be subjected to additional loads caused by passing vessels. Seelig and Flory have set out design approaches for determining these loads (Naval Facilities Engineering Service Center 2005).

The resulting mooring and lean-on forces on individual dolphins are derived from the applicable equilibria of forces and moments. Mooring forces are limited by the load-bearing capacity of the vessel's own on-board mooring equipment (ropes and winches). The winches normally fail at 60% of the rope load-bearing capacity. This load must be accommodated by mooring dolphins:

$$F_T = 0{,}6 \cdot n \cdot F_{rope}$$

where

F_T critical tensile force on bollard [kN]
n number of ropes pulling on dolphin simultaneously in same direction [–]
F_{rope} tensile strength of ropes of critical vessel, also called minimum breaking load (MBL) [kN]

12.2.3.3 Other actions

If there is no vessel berthed at the dolphin, current and wave forces act directly on the dolphin pile or dolphin topside/deck. These forces can be determined as shown in Section 4.7. Even if these actions are generally smaller than the actions due to berthing manoeuvres or the mooring and contact forces of moored vessels, in exceptional cases, they can affect the fatigue strength of the dolphin due to their cyclic nature.

Ice loading on dolphins can be estimated as described in Section 4.12, with any vertical loads due to any adhering ice being included in the calculations as well. Ice loading should be given particular attention if pontoons are permanently attached to a dolphin. In this case, the ice loading acts on the dolphin via the pontoon and can, therefore, be considerably greater than ice loading acting on the dolphin only.

Mooring dolphins also accommodate vertical loads that are transmitted through friction between vessel and dolphin due to the effects of waves, tides, loading/unloading, etc. These loads can become critical for the design of the dolphins and the stresses in the vessel's hull.

12.2.4 Safety concept

In principle, elastic design is used for dolphins; see Section 12.2.7.3 for more details. When verifying the ultimate limit state of the dolphin, the partial safety factors given in Table 12.1 are recommended. For the serviceability limit state, the characteristic actions and resistances should be used.

12.2.5 Soil–structure interaction and the resulting design variables

12.2.5.1 Overview

The design variables for dolphins are derived from the deformation-dependent interaction between the soil and the dolphin. Earth pressure at rest acting on a non-loaded dolphin is rotationally symmetric and is, thus, cancelled out for the dolphin itself. The dolphin is pushed against the soil by the loads of berthing or moored vessels. This increases the lateral soil stresses in the loading direction beyond the steady-state earth pressure until a critical tensile stress is reached. At the same time, the lateral soil stresses on the opposite side of the dolphin are reduced. The dolphin is thus subjected to bending.

Two fundamentally different approaches have proved effective for modelling the deformation-dependent interaction between dolphin and soil.

The traditional method according to Blum (1932) estimates the spatial passive earth pressure E_{ph} in front of the dolphin as a lateral stress (Figure 12.3a). For a homogeneous non-cohesive subsoil in the embedment area, this stress can be calculated using traditional earth pressure theory. Cohesive and/or stratified non-cohesive subsoils require further factors to be taken into account when determining the spatial passive earth pressure. The Blum method, therefore, assumes the lateral soil stress to be the critical stress. This method, thus, normally provides an upper bound for the bending and shear loads on a dolphin. However, the method is of only limited use when verifying whether a dolphin complies with serviceability requirements (deformations). Section 12.2.5.2 contains further details of Blum's method.

The $p-y$ method is a modulus of subgrade reaction based on non-linear load-deflection curves (Figure 12.3b). Here, the lateral soil stresses are assumed to correspond to the de-

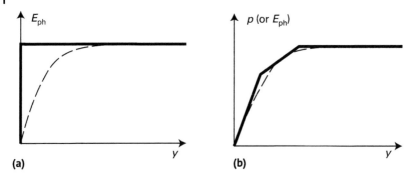

Figure 12.3 Idealised load–displacement diagrams for the dolphin foundation (soil–structure interaction): (a) Blum model; (b) non-linear modulus of the subgrade reaction method.

formations of the dolphin over the embedment depth, which vary with the load. The p–y method, thus, supplies more realistic values for the load on the soil and the dolphin, and also allows the deflections of the dolphin to be calculated. Section 12.2.5.3 contains further guidance on the p–y method.

Comparative calculations (Rudolph et al. 2011) have shown that both methods yield fundamentally comparable results for breasting and mooring dolphins in cohesive and non-cohesive subsoils with a strength $c_u < 96\,\text{kN/m}^2$. In general, when modelling the soil with a lower stiffness, the p–y method supplies larger deformations than the Blum method does. Consequently, for breasting and lead-in dolphins, lower forces – and, thus, generally more economic structures – are the outcome.

However, for stiff cohesive subsoils with $c_u > 96\,\text{kN/m}^2$, the two methods can yield significantly different results. In this case, smaller structural member dimensions are calculated with the Blum method compared with the p–y method. Therefore, for dolphins in stiff cohesive subsoils, the recommendation is to use the p–y method for design.

The embedment depth determined for a dolphin consisting of a group of piles is greater than that for a comparable number of individual piles. Section 12.2.5.4 contains further guidance on this.

12.2.5.2 The Blum method

According to Blum, the embedment of the dolphin in the subsoil should be modelled as a fixed support with a parabolic distribution of the passive earth pressure and an equivalent force C applied at the theoretical base (Figure 12.4). Contrary to the original approach according to Blum, the recommendation is to apply the characteristic passive earth pressure as a spatial passive earth pressure E^r_{ph} in accordance with DIN 4085:

$$E^r_{ph,k} = E^r_{pgh,k} + E^r_{pch,k} + E^r_{pph,k}$$

Here, $E^r_{ph,k}$ is the total characteristic passive earth pressure due to dead load, cohesion and surcharges acting from the bottom of the watercourse down to the calculated embedment depth t. The higher spatial passive earth pressure, compared with the approach set out above, is taken into account through equivalent dolphin widths or shape factors. The equivalent dolphin width is calculated as depending on the angle of internal friction of the soil and the embedment depth of the dolphin, as shown in Eqs. (74)–(77) of DIN 4085.

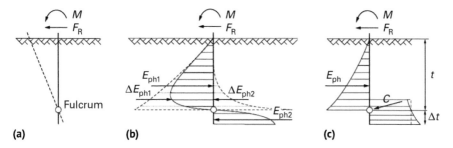

Figure 12.4 System idealisation for spatial passive earth pressure according to Blum:
(a) rotation of the wall; (b) anticipated stress distribution; (c) equivalent load model after Blum.

The curved line of the spatial passive earth pressure is calculated from the differentiation of the spatial active earth pressure in accordance with Eq. (78) (DIN 4085) for depth z, with Eqs. (74)–(77) (DIN 4085) being applied and derived. In contrast to the calculation for the resultant active earth pressure, a distinction must be made between a position near the surface ($d/z < 0.3$) and a deeper position ($d/z \geq 0.3$), where

d   diameter of dolphin or width of dolphin perpendicular to direction in which the loads act (for cluster dolphins the distance between outside edges of outer piles).
z   depth in the soil.

The approach and the result of this differentiation can be found, for example, in Rudolph et al. (2011).

In cohesive subsoils the passive earth pressure should be calculated using the undrained shear parameters φ_u and c_u if the load is applied quickly compared with the consolidation of the soil. The submerged unit weight of the relevant soil strata is assumed to be $\gamma'_{k,l}$.

The process is more complex in stratified subsoils, as the different shear parameters and unit weights of the upper strata must be taken into account. Further guidance on determining the spatial earth pressure coordinates for homogeneous and stratified subsoils can be found in, for example, Rudolph et al. (2011). The first step in the design is to calculate the required embedment depth t from the moment equilibrium at the base ($\sum M_{\text{base}} = 0$). Equivalent force C is derived from the equilibrium of the horizontal forces as the difference between the mobilised spatial passive earth pressure and the forces acting while neglecting the active earth pressure:

$$C_{h,k,\text{Blum}} = E^r_{ph,\text{mob}} - \sum F_{h,k,i}$$

where

$C_{h,k,\text{Blum}}$ horizontal component of Blum equivalent force
$\sum F_{h,k,i}$ total of characteristic horizontal actions
$E^r_{ph,\text{mob}}$ $= E^r_{ph,k}/(\gamma_Q \cdot \gamma_{Ep})$; horizontal component of mobilised spatial passive earth pressure
γ_Q partial safety factor for actions
$\gamma_{R,e}$ partial safety factor for passive earth pressure

The equivalent force can be inclined at an angle of up to $\delta_{c;k} = +1/3\varphi$ with respect to a normal to the dolphin as long as equilibrium of vertical forces is assured Rudolph et al.

(2011). Verification that the vertical forces can be accommodated is carried out with the reduced C force value by including an allowance for the embedment depth in accordance with Section 8.2.5.4:

$$\Delta t = \frac{1}{2} \cdot C_{h,k,Blum} \cdot \gamma_Q \cdot \frac{\gamma_{R,e}}{e^r_{ph,k}}$$

where

$e^r_{ph,k}$ ordinate of characteristic spatial earth pressure at level of equivalent force C

Similarly as with sheet pile walls, the passive earth pressure can be assumed to have the maximum possible angle of inclination. This means that in calculations using curved failure planes, an angle of up to $\delta_p = -\varphi'_k$ is possible. When calculating with straight failure planes (which is permitted for an angle of internal friction $\varphi' \leq 35°$), the angle can be as much as $\delta_p = 2/3\varphi'_k$ when the sum of the characteristic vertical forces is equal to zero (Figure 12.5). Otherwise, a shallower angle must be assumed. For mooring dolphins, a vertical component of the line pull acting unfavourably upwards should be considered. The follow-

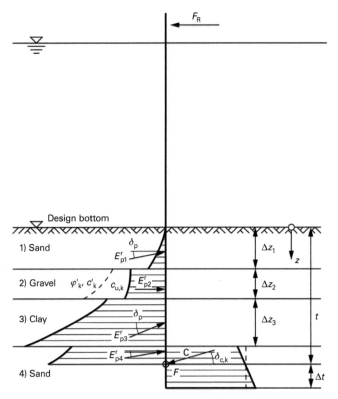

Figure 12.5 Estimating the spatial passive earth pressure and equivalent force C in stratified subsoils (example).

ing characteristic actions acting vertically downwards can be assumed:

- The self-weight of the dolphin.
- The submerged weight of the body of soil surrounded by the dolphin.
- The perpendicular skin friction on the lateral surfaces below the embedment depth down to the theoretical base parallel with the direction of movement of the dolphin.
- For $\delta_{c,k} > 0$, the vertical component $C_{v,k}$ of the Blum equivalent force.

12.2.5.3 The p–y method

The $p-y$ method is a modulus of subgrade reaction method using non-linear load-deformation curves. It is based on experiments conducted in the 1970s and 1980s on horizontally loaded piles of up to 60 cm in diameter (Matlock 1970, Reese et al. 1975, O'Neill and Murchinson 1983). The $p-y$ method has since come into use for the design of horizontally loaded piles with significantly greater diameters for offshore projects. In contrast to Blum's ultimate limit state method, the soil deformations required to mobilise the spatial passive earth pressure are taken into account, so that it is possible to account for the characteristic soil–structure interaction.

The $p-y$ curves describe the relationship between the passive earth pressure p and the displacement needed to mobilise this. The formulas differ for static and cyclic loads and also differ in dependence on the type of soil and the stress conditions (depth).

Further information on determining the $p-y$ curves for sandy soils and soft cohesive soils can be found in API (2000), DNV (2010) and EN ISO 19 902. Information on stiff cohesive subsoils can be found in Reese et al. (1975). Figure 12.6 shows typical examples of $p-y$ curves.

The passive earth pressure p from the $p-y$ curves is a characteristic value. The partial safety factors given in Table 12.1 should be used for ultimate limit state analyses. Checks can be carried out on the basis of EC 7 and the supplementary national regulations (National Annex to German Eurocode 7 (NAD); DIN 1054, Section 8.5.2) for piles loaded transversely to the pile axis. No local analysis of the tension in the soil ($\sigma_{h,k} \leq e_{ph,k}$) is required here because it is automatically satisfied by the $p-y$ curves. In a global analysis of the tension in the soil, the design value of the mobilised lateral soil force $F_{h,d}$ up to the point of zero displacement may not be greater than the design value of the maximum passive earth pressure $E_{ph,d}$ to be mobilised:

$$F_{h,d} = \int p \cdot dz \cdot \gamma_Q \leq E_{ph,d} = \int p_u^* \cdot dz / \gamma_{R,e}$$

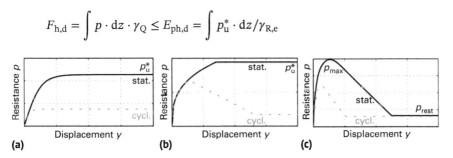

Figure 12.6 Qualitative $p-y$ curves for (a) sand near the surface; (b) soft cohesive soil; (c) stiff cohesive soil for static (continuous line) and cyclic (dotted line) loads.

Here, p_u^* is the maximum value of the $p-y$ curve. Where the shape of the $p-y$ curve follows Figure 12.6b, $p_u^* = p_{max}$ must be applied if the displacement y is smaller than that required to achieve p_{max}, otherwise $p_u^* = p_{rest}$.

Unlike the Blum method, the $p-y$ method does not supply the embedment depth of the dolphin directly. This must normally be determined iteratively. To do this, a secondary condition must be introduced. This could be, for example, the embedment depth that satisfies the ultimate limit state analysis according to EC 7 (Section 7.7, load case ULS).

Alternatively, the embedment depth can be determined by limiting the horizontal deformation of the dolphin. A common and proven approach in practice is to determine the deformations of the dolphin for different embedment depths. The deflection reduces degressively as the depth of embedment increases. The embedment depth is then set at the depth at which the deflection no longer decreases significantly. Occasionally, the embedment depth is set at the depth at which the deflection curve becomes vertical. This leads to very large embedment depths, which are unnecessary for stability.

The $p-y$ curves can generally be used for designing dolphins for static loads, but a check should be carried out to establish whether the lateral support must be reduced for repetitive cyclic loads (see Section 12.2.7.3). Further guidance on designing dolphins using the $p-y$ method plus examples of application can be found in, for example, Rudolph et al. (2011).

12.2.5.4 Soil–structure interaction for cluster dolphins

Cluster dolphins are groups of piles with a pile spacing that is less than six times the pile diameter. When designing such dolphins, the overlapping of the areas from which the spatial passive earth pressure is generated means that the passive earth pressure of an individual pile cannot be used.

Reduction of the active earth pressure compared with that of the individual pile depends on the pile spacing, the pile diameter and the arrangement of the piles. It can be estimated in accordance with, for example, *Recommendations on Piling* (EA-Pfähle 2012) or DIN 4085. Further methods are described in the literature, e. g. Mokwa (1999).

12.2.6 The required energy absorption capacity of breasting dolphins

12.2.6.1 General remarks on the energy absorption capacity of dolphins

The energy absorption capacity of a dolphin, or rather the deformation energy of the dolphin (plus fender if fitted), must be as large as the kinetic energy of the berthing vessel (Section 7.4.1.4). At the same time, neither the permissible stresses in the components of the dolphin, nor the deformation limits may be exceeded. The reader is referred to Section 12.2.7 for more information.

Vessels cannot always be manoeuvred to the middle of a berth. When designing breasting dolphins, the energy absorption capacity should, therefore, always be calculated for a distance $e = 0.1 \cdot l \le 15$ m (parallel to the row of fenders) between the vessel's centre of gravity and the centre of the berth. For tankers, an eccentricity of the transfer connections (manifold) with respect to a vessel's centre of gravity may need to be allowed for as well.

12.2.6.2 Special guidance for seaports

Where a ship is manoeuvred to a dolphin berth with the help of tugboats, it can be assumed that it will hardly move in the direction of its longitudinal axis. Therefore, the longitudinal component of the berthing velocity can generally be ignored when calculating the berthing energy. As an approximation, the velocity vector v for large vessels can be assumed to be perpendicular to distance r ($\alpha = 90°$) (Figure 7.65).

In practice, all dolphins at a berth are generally of the same type. In principle, the inner dolphins could also be designed for lower loads than the outer ones. In this case, it should then be taken into account that the berthing angle for large vessels docking with tug assistance is generally smaller than that for smaller vessels that dock without tugs.

12.2.6.3 Special guidance for inland ports

For dolphins at inland ports, using the approach of Section 7.4.1.4 often leads to the kinetic energy of the vessel being overestimated. This can result in dimensions of dolphins that are uneconomic.

The required energy absorption capacity of dolphins for inland ports should, therefore, be governed by economic considerations. A low energy absorption capacity leads to lightweight dolphins with lower investment costs but also to potentially higher maintenance costs due to damage to the dolphins. In the end, this can be more economical than specifying a heavy structure with a large energy absorption capacity. The choice between the two options must take account of local conditions (the likelihood of damage to the dolphins).

For the standard vessel sizes in use on German inland waterways, according to ETAB (1996), an energy absorption capacity of 70–100 kNm, occasionally also 120 kNm, has proved to be expedient.

12.2.7 Other calculations

12.2.7.1 Berthing/hull pressure

The maximum permissible berthing force $F_{R,max}$ between a vessel and a breasting dolphin is determined by the permissible berthing pressure of the vessel. Where no more detailed information is available, the maximum berthing pressures (under unfavourable berthing conditions) for dolphins with fender panels can be taken from Table 12.2 in dependence on the loading capacity of the ship.

Higher berthing pressures may be permitted, if it can be proved that the load can be accommodated by the plates and frames of the vessels using the berth.

For gas tankers, the permissible berthing pressure may be lower than the values given in Table 12.2. Vessels with rubbing strips and steel belts all round, e.g. ferries, require special constructional measures for fenders.

12.2.7.2 Deformations

Unless particular circumstances dictate otherwise, e.g. a greater energy absorption capacity than actually required at berths for large vessels, the dolphin's maximum deflection s_{max} should not exceed approximately 1.5 m. This is because otherwise the impact between ves-

Table 12.2 Recommended maximum berthing pressures at berths with fender panels.

Capacity (dwt)	≤ 20 000	40 000	60 000	80 000	100 000	≥ 120 000
Pressure (kN/m²)	400	350	300	250	200	150

sel and dolphin becomes so soft that the skipper can no longer recognise clearly enough when the vessel touches the dolphin.

Furthermore, the advice given in Section 12.1.2 should be followed.

12.2.7.3 Utilisation of material

Breasting dolphins are generally designed elastically, exploiting the yield strength. The unused plastic reserve acts as a safety margin for unforeseen berthing procedures.

Owing to the wind and wave loads frequently acting on a vessel, mooring and breasting dolphins may need to be designed for fatigue as well (see Section 12.2.3.2). At least twice the number of load cycles for the anticipated design life should be assumed here (EN ISO 19902).

12.2.7.4 Design depth

Local scouring (e. g. due to currents flowing around the dolphin or propeller wash) should be taken into account in the design. The reader is referred to Hoffmans and Verheij (1997) for more information on scouring depth. The scouring depth should be taken as the design depth over the whole area.

References

Blum, H. (1932). Wirtschaftliche Dalbenformen und deren Bemessung. *Bautechnik* 10 (5): 50–55.

EA-Pfähle (2012). *Empfehlungen des Arbeitskreises „Pfähle"*, (Hrsg. Deutsche Gesellschaft für Geotechnik e. V.). Berlin: Ernst & Sohn.

ETAB (1996). *Empfehlungen des Technischen Ausschusses Binnenhäfen: Dalben in Binnenhäfen – Erfahrungen mit Bau, Instandhaltung und Benutzung (B29)*, April 1996. Bundesverband Öffentlicher Binnenhäfen und Hafenbautechnische Gesellschaft e. V.

Hoffmans, G.J.C.M. and Verheij, H.J. (1997). *Scour Manual 1997*. Rotterdam/Brookfield: Balkema.

Matlock, H. (1970). Correlations for Design of Laterally Loaded Piles in Soft Clay, OTC 1204.

Mokwa, R.L. (1999). Investigation of the resistance of pile caps to lateral loading. PhD thesis, Faculty of the Virginia Polytechnical Institute and State University.

Naval Facilities Engineering Service Center (2005). TR-6056-OCN Mooring Loads due to parallel passing ships.

O'Neill, M.W. and Murchinson, J.M. (1983). Fan Evaluation of $p-y$ Relationships in Sands. A report to the American Petroleum Institute.

Reese, L.C., Cox, W.R. and Koch, F.C. (1975). Field Testing and Analysis of Laterally Loaded Piles in Stiff Clay, OTC 2312.

Rudolph, C., Mardfeldt, B. and Dührkop, J. (2011). Vergleichsberechnungen zur Dalbenbemessung nach Blum und mit der $p-y$-Methode. *Geotechnik* 34 (4): 237–251.

API (2000). Recommended Practice for Planning, Designing and Constructing Fixed Offshore Platforms – Working Stress Design, American Petroleum Institute.

DNV (2010). Det Norske Veritas: Offshore Standard DNV-OS-J101. Design of Offshore Wind Turbine Structures.

Standards and Regulations

DIN 1054: Subsoil – Verification of the safety of earthworks and foundations – Supplementary rules to DIN EN 1997-1

DIN 4085: Subsoil – Calculation of earth-pressure

DIN EN 1997 Eurocode 7: Geotechnical design

EN ISO 19 902 (2008). Petroleum and natural gas industries – Fixed steel offshore structures

13
Operation, maintenance and repair of waterfront structures

13.1 Operation of waterfront structures

13.1.1 General

Key elements in the life of a waterfront structure are regular inspections, evaluations and records of the condition of the structure. The inspections described in this chapter – and the assessments of the load-bearing capacities of existing waterfront structures based on those inspections – are intended to reveal when and to what extent a waterfront structure should be repaired, upgraded or deconstructed and replaced. Owners and users will, therefore, be able to arrange the necessary measures in good time.

13.1.2 Building information modelling (BIM)

In the future, building information modelling (BIM) will play an important role in collating and processing the information about a structure that has been systematically recorded (which is described in this chapter).

BIM is a cooperative way of working that uses digital models of a structure to collate and manage the data relevant to that structure over its life cycle in a consistent manner. That data can then be readily exchanged between all persons involved and made available for further processing. Digital models link the geometric and semantic data that was previously stored in a decentralised fashion. BIM allows all the information relevant to the whole life cycle of a structure to be brought together in one model.

If a waterfront structure is to be planned using BIM, then the recommendation is to take into account all the aspects described in this chapter right from the very start. Such an approach creates the conditions for adding the data necessary for the use of the structure to the existing model. Besides the model of the structure, it should also be possible to refer to a model of the subsoil – consisting of digital models of terrain, soil stratification and hydrology – throughout the lifecycle of the structure.

13.2 Inspecting waterfront structures

As with other hydraulic engineering structures, regular inspections of waterfront structures are necessary (while taking account of the various potential risks and the robustness of the structural design), so that any damage that could influence the load-bearing capacity and serviceability of the structure is identified in good time. On the basis of such inspections, the persons responsible for maintenance can, thus, assume responsibility for guaranteeing the safety and soundness of the structure and its fitness for purpose.

Careful, regular and consistent inspections of the structure also help in the planning and management of maintenance work. Higher refurbishment costs or the premature replacement of a structure can, therefore, be avoided.

Structures can be categorised according to their structural design, or rather their structural properties (robustness), and the consequences of any damage. This is set out in more detail in the guidance contained in VDI Guideline 6200 (2010) and Argebau (2006). Using this classification, it is then possible to plan and specify the nature and regularity of inspections and examinations of waterfront structures or port/harbour structures that do not need to comply with DIN 1076 or other provisions (see also Table 13.1).

Irrespective of the potential hazards or a structure's robustness or structural design, taking into account the individual case based on DIN 1076 for civil engineering structures, *a structural inspection of a waterway structure* consists of:

- *Structural check* – a close inspection of all the parts of the structure, including those difficult to access, by experienced engineers who assess the structural, constructional and hydromechanical conditions in relation to the requirements imposed by the structure's use and can instruct divers where necessary.
- *Intermediate structural checks* – intensive, comprehensive visual inspections of the structure carried out by experienced engineers.
- *Structural monitoring* – intensive, comprehensive visual inspections of the structures carried out by experienced staff familiar with the structural behaviour and functions of the structure.
- *Structural survey* – examination of the structure for any obvious damage by experienced personnel familiar with the structural behaviour and functions of the structure.

The following provisions and regulations applicable to certain areas should be referred to when planning and carrying out structural inspections (listed in alphabetical order):

- Argebau (2006), Conference of German Building Ministers: "Hinweise für die Überprüfung der Standsicherheit von baulichen Anlagen durch den Eigentümer/Verfügungsberechtigten".
- BAW Recommendations (2017) "thickness measurements of sheet piles in harbours and waterways (ESM)", Federal Waterways Engineering and Research Institute.
- BAW Code of Practice (2010): Asset Inspection (MBI), Federal Waterway Engineering and Research Institute.
- BAW Code of Practice (2018b): Classifying Waterway Construction Damages (MSV), Federal Waterway Engineering & Research Institute.
- DIN 1076:1999-11 "Engineering structures in connection with roads – Inspection and test".

- DIN 19702:2013-02 "Solid structures in hydraulic engineering Bearing capacity, serviceability and durability".
- PIANC Report (2004): "Inspection, maintenance and repair of maritime structures exposed to material degradation caused by a saltwater environment", MarCom Report of WG 17.
- PIANC Report (1998): "Life cycle management of port structures – General principles", Report of WG 31.
- PIANC-Report (2006): "Maintenance and renovation of navigation infrastructure", InCom Report of WG 25.
- RÜV-Richtlinie (2008). für die Überwachung der Verkehrssicherheit von baulichen Anlagen des Bundes.
- VDI Guideline 6200 (2010): "Structural safety of buildings – Regular inspections".
- VV-WSV 2101 (2009/2010): Structural inspection.
- VV-WSV 2301 (2019): Dam inspection.

13.2.1 Documentation

Structural inspection files and logbooks form the basis of the structural inspections. They contain all the information required for an inspection, including the structure's key design data together with sketches, as-built drawings, as-built calculations, measurement software, measurement results (e.g. soundings, measurements taken during construction and reference measurements after completion), information on corrosion protection, reports, information on repairs and the results of previous inspections (test, monitoring and survey reports).

Checklists or task sheets are suitable for the systematic recording of the condition of a structure, especially if the as-built structure is very heterogeneous in terms of its construction, design and the materials used. Task sheets should contain guidance on preparing and carrying out structural inspections and should also describe typical signs of damage for each component and material used. The structural, constructional and hydromechanical conditions should be taken into account. As preparation for inspections, the task sheets can also include important information about the inspection intervals and activities specific to that structure, such as the use of divers, lifting gear, vehicles, etc.

The recommendation is to set standard criteria and assessment principles to ensure maximum consistency when assessing damage. To assess the condition of waterway structures, these principles are summarised in, for example, the BAW Code of Practice "Classifying Waterway Construction Damages" (BAW 2018b). VDI Guideline 6200 (2010) "Structural safety of buildings – Regular inspections" also provides guidance on changes to and ascertaining the properties of the building materials used

The use of especially developed software to record the structural inspections also supports the standard presentation of reports and gives those responsible for maintenance the opportunity to investigate further any damage found. Examples include the WSV-Pruf software developed by the Wasserstraßen- und Schifffahrtsverwaltung des Bundes (Federal Waterways and Shipping Administration) and the SIB-Bauwerke software developed by the Federal Highway Research Institute, although this is very much geared towards highway structures (e.g. bridges).

13.2.2 Structural inspections

13.2.2.1 Structural checks

During a structural check, all parts of the structure, including those difficult to access, should be examined closely with the help of any equipment required to do so. Where necessary, individual parts of the structure should be carefully cleaned before the structural check, so that any hidden damage can be identified. The nature and scope of the structural check should be specified for each structure in accordance with the local conditions, e. g. with the help of task sheets.

Depending on their relevance and the boundary conditions plus the robustness of the structure and its design (load-bearing systems, damage processes and advance warning of failure), inspections of waterfront structures particularly at risk should include checking and measuring the following:

- Location and extent of damage to a quay wall, using divers where necessary.
- Condition and functionality of drainage systems.
- Condition of previous repairs.
- Condition of corrosion protection coatings.
- Condition of cathodic protection system.
- Settlement and subsidence behind the quay wall.
- Soundings of watercourse bed in front of the quay wall.
- Seals at joints and connections.
- Movements at joints and supports.
- Damage to concrete (including steel reinforcing bars).
- Measurements of horizontal movements (including head deformations) and vertical movements (settlement and uplift).
- Measurements of residual wall thickness or average and maximum rusting (for implementation, evaluation and damage, see the BAW Recommendations (2017), Chapter 7).

Further measurements may be necessary for particular structures, e. g. measurement of anchor forces, inclinometer readings, potential field measurements, etc.

The inspection report should describe any damage found and assess the causes, as well as the load-bearing capacity and serviceability of the structure. The structural check can include structural analyses, depending on the residual wall thicknesses of sheet piling. The inspection report should also set out any further action required.

13.2.2.2 Intermediate structural checks and structural monitoring

Intermediate structural checks and structural monitoring involve carrying out intensive, comprehensive visual inspections of all accessible parts of the structure where this is feasible without draining or using inspection equipment. However, use can be made of inspection and access facilities that are already available on the structure. The nature and scope of the structural monitoring should be specified for each particular structure in dependence on the local boundary conditions, e. g. with the help of task sheets. As a minimum, the water side of the structure should be subjected to an intensive visual inspection from a boat during low water. For waterfront structures particularly at risk, the following abnormalities are generally relevant for structural monitoring, depending on the structural and constructional design of the structure (Table 13.1):

- Damage or changes to surfaces.
- Subsidence, settlement, displacements.
- Changes to joints and connections.
- Damage to or missing equipment.
- Improper usage.
- Functionality of drainage system.
- Damage to sheet pile walls.
- Scouring or accretions in front of a sheet pile wall.

The monitoring report should describe any damage found and assess the causes and the load-bearing capacity and serviceability of the structure. The monitoring report should also set out any further action that is required.

13.2.2.3 Structural survey

During a structural survey, the structure should be closely inspected without any significant aids such as survey vehicles, scaffolding, etc. However, any available survey equipment can be used and any accessible voids in the structure subjected to a visual inspection. The fitness for purpose, overall condition of the structure and any abnormalities should be investigated.

As a minimum, the water side of the structure should be subjected to an intensive visual inspection during low water.

The following items in particular should be verified during a structural survey:

- Unusual changes to the structure, significant changes to damage.
- Significant damage to or missing equipment.
- Significant concrete spalling, noticeable cracks.
- Deformations to or displacements of the structure visible to the naked eye.
- Unusual water discharges.
- Damage to embankments or slopes, scouring, accretions.

Any significant damage or unusual changes identified and any further actions necessary should be described in the structural survey report.

13.2.3 Inspection intervals

Inspections carried out no less frequently than every 6 years has proved effective for waterway structures (see VV-WSV 2101) and engineered highway structures (DIN 1076). For waterfront structures particularly at risk, the recommendation is, therefore, to carry out inspections at least every 6 years, depending on the structural design and construction details (see Table 13.1). The structure should be checked for the first time during the acceptance procedure for the structure. A further check should then be carried out before the end of the warranty period for any claims relating to defects in accordance with VOB/B, the German construction contract procedures. Levelling and alignment should be included, and reference levels and positions taken. Further follow-up measurements should be specified as part of the structural checks.

At the very latest 3 years after a structural check, all waterfront structures that require such checks (see Table 13.1) should undergo an intermediate structural check.

Table 13.1 Inspection intervals for waterfront structures.

Robustness to VDI 6200	Risk potential to VDI 6200	Structural survey[a]	Structural monitoring[b]	Intermediate structural check	Structural check[c]
RC 1–RC 4	Low (CC1)	Annually	—	—	—
RC 3–RC 4	Moderate (CC2)	Annually	Every 3 years	—	—
RC1 –RC 2		Annually	—	3 years after the structural check	Every 6 years

a) Corresponds to structural monitoring according to DIN 1076 and "inspection by the owner" according to VDI 6200.
b) Corresponds to the simple check according to DIN 1076 and the "inspection by an experienced person" according to VDI 6200.
c) Corresponds to the main check according to DIN 1076 and the "detailed inspection by a very experienced person" according to VDI 6200.

The recommendation for waterfront structures with a particular risk potential but without complexity is to carry out structural monitoring at least every 3 years (Table 13.1). A structure without complexity is characterised by the fact that the inspection tasks for such a structure can be described in just a few points, its features are clearly identifiable and a general briefing on its load-bearing behaviour is possible.

The structural survey set out in Section 13.2.2.3 should normally be carried out annually for all waterfront structures. A structural survey is normally sufficient for most waterfront structures due to the robustness of such structures and the consequences of any damage caused (see VDI Guideline 6200). This recommendation is based on experience of ports, harbours and quay facilities, which has been acquired over many years. Reference values relating to the inspection intervals can be found in Table 13.1.

Inspections might need to be carried out at more frequent intervals than those given in the table depending on the relevance, condition and economic significance of the structure.

13.2.4 Structural monitoring supported by measurements

Structural monitoring based on measurements can be used to supplement or replace the inspection methods described above for recording and assessing damage. For example, soil deformations below load-distribution plates, anchor loads and water levels (on water and land sides) related to the structural cross-section can be recorded for quayside facilities. Appropriate sensors can also be installed to measure line forces or monitor concrete degradation. Additional sampling is recommended for individual parts of the structure, e. g. extraction of cores from concrete structures.

The assessment of results obtained from structural inspections, measurements to support structural monitoring plus further samples taken from the structure can be used to build models for forecasting any expected changes to the condition of the structure. Such data also form the basis for long-term, preventive maintenance.

13.3 Assessing the load-bearing capacity of an existing waterfront structure

In principle, the as-built calculations for a structure are valid and should be referred to when assessing the stability of the structure, provided that:

- Those calculations are correct.
- There have been no structurally relevant changes to loads or usage.
- Additional, structurally relevant knowledge has not become available (e.g. numerical models, load paths).
- The structure does not exhibit any damage or abnormalities.

Should one or more of the above points no longer apply, then an assessment of the load-bearing capacity must be carried out. The recalculation guidelines for quay walls (Hamburg Port Authority 2019) contains a step-by-step procedure based on the recalculation guidelines for bridges (NRR 2011). For massive hydraulic structures and gates in hydraulic steel structures, the BAW codes of practice "Assessment of the load-bearing capacity of existing, massive hydraulic structures (TbW)" (2016) and "Assessment of the load-bearing capacity of existing gates in hydraulic steel structures (TbVS)" (2018a) are to be considered.

Examining and evaluating as-built documents forms the basis for a verifiable assessment of load-bearing capacity. In addition, an inspection of the structure should be carried out according to Section 13.2 to establish its actual condition. This work should include an estimate of to what extent the assumptions in the as-built calculations apply and whether the existing structure corresponds to the as-built documents. The on-site inspection, including estimates of residual wall thicknesses, is especially important for the assessment of the load-bearing capacities of waterfront structures.

For the verifications, it is important to compare the original load assumptions with those of today's standards and to adjust them if necessary. Material parameters should be allocated to the year in which the structure was built. The aforementioned publications contain appropriate reference values for dead loads (TbW and Hamburg Port Authority 2019) and tensile and compressive strengths for concrete (TbW), steel reinforcement (TbW), structural steel (NRR 2011) and steel for sheet piles (Hamburg Port Authority 2019).

A geotechnical specialist should assess the subsoil using the soil parameters and soil profiles. If the extent and/or quality of the soil investigations is inadequate, then new investigations must be carried out and evaluated.

The recalculation guidelines for quay walls devised for the Port of Hamburg (Hamburg Port Authority 2019) can be used to assess load-bearing capacities. Table 13.2 presents the step-by-step procedure described in those guidelines.

In step 0, the safety concept employed in the as-built calculations is applied. This step serves to record the momentary situation and check the as-built calculations. As part of this, the original calculation model is recalculated in step 0.1 in order to check the as-built calculations for potential discrepancies. Step 0.1 also serves to calibrate the system with respect to modern computerised methods. No analyses are carried out; instead, internal forces are merely compared.

The current situation, including any damage that has occurred and any changes to the loads (e.g. loss of wall thickness and higher water pressures), is applied to the as-built

Table 13.2 Step-by-step recalculation procedure.

Step		Description		
0				
	0.1	Assessment of as-built calculations	Original safety concept and load assumptions	As-built data
	0.2		Current situation taking into account structural damage and current water pressure	
	0.3		As for 0.2, plus taking into account Increased water depth, elastic-plastic stress calculations and reduced imposed loads	
1		Fictitious new construction	Current technical codes of practice	Investigations necessary, e. g. level measurements, cone penetration tests
2				
	2.1		Taking into account structural damage, design situation according to current safety concept	
	2.2		As for 2.1, plus design for all load case combinations for the accidental design situation	
	2.3		Forecast of residual design life	
3		Compensatory measures	Load reduction, changes to the structure	Measures required
4		Scientific methods	Measurements, elaborate methods of calculation	Methods from practice and research

calculations in step 0.2. The residual safety factors are calculated with respect to yield or ultimate strength.

Any increases in water depth that have occurred over the years and characteristic soil parameters are additionally applied in step 0.3. Furthermore, the existing stresses are determined by way of elastic-plastic analyses. In this step, the imposed load is decreased successively until a residual factor safety of 1.0 is reached.

Step 0 does not involve any new measurements or soil investigations, apart from the foregoing inspection(s), and, therefore, it is possible to reach conclusions quickly regarding necessary actions and/or priorities for further measures. It cannot be used for classification into verification classes. However, the results form the basis for specifying in which step the actual new calculations are to be carried out based on current codes of practice. The semi-probabilistic safety concept (partial safety factors method) forms the basis for all further steps.

Steps 1 and 2 require further investigations to be carried out, especially with respect to water levels and the subsoil. Current standards are applied here, especially calculating an addition to the nominal depth of water. A successful analysis allows the residual design life to be estimated. Otherwise, the end of the design life might have been reached, and steps 3 and/or 4 have to be applied.

Step 3 calls for further relieving measures, possibly also structural changes. Step 4 requires measurements and complex calculation methods based on latest practice and research. The complexity of the methods and the depth of investigation increase with every step.

In line with the recalculation guidelines for road bridges, after completing appropriate investigations involving at least step 1, and generally step 2 as well, the structure can be classified into a verification class. This classification specifies how further analyses are to be carried out and whether compensatory measures must be taken. Crucial here is whether and at what step the analysis was completed. All results must be presented in an explanatory report.

This approach is described in detail in the recalculation guidelines for quay walls (Hamburg Port Authority 2019). Recording the actual condition, carrying out analyses and compensatory measures are all fully covered.

13.4 Repairing concrete waterfront structures

The contents of the previous recommendations R 194 and R 195 with respect to surveys of and repairs to hydraulic engineering structures have been omitted from this current EAU edition. Instead, the reader should consult the valid standards and regulations, primarily ZTV-W LB 219 (2017 and ZTV-ING Teil 3 Massivbau (2019)), when carrying out repairs to concrete components in hydraulic engineering structures.

The essential and indispensable prerequisite for determining the need for repairs and planning repair measures is a comprehensive survey to establish the actual condition of the concrete components in question. Based on that, a forecast of future developments that covers the intended design life must be prepared. Comprehensive basic information on the planning and execution of surveys of large-scale hydraulic engineering structures can be found in Westendarp et al. (2015 and BAW Empfehlung (2019)).

13.5 Upgrading and deconstructing existing waterfront structures

This section describes potential upgrading measures that can be used to improve the load-bearing capacities of existing waterfront structures. Such upgrading measures should normally subsequently be monitored using measurement technology.

13.5.1 Upgrading measures

Embankment
The addition of a frontal embankment increases the passive earth pressure and reduces the span width of the structural system in the case of a sheet pile wall. This upgrading measure requires a sufficient depth of water.

Drainage systems
One of the principal loads on a waterfront structure is water pressure. Current water levels might be totally different to those that prevailed when the structure was built, which means that different loads act on the structure (see Section 13.3). Therefore, drainage systems, with redundancy if possible, should be included and maintained at regular intervals. A further option for sheet pile walls is to cut slots, holes or openings in the wall. Measures must be taken to prevent the subsidence that can occur in this case.

Relieving measures
The vertical loads can be reduced by replacing materials with lighter alternatives, e. g. expanded clay aggregate. This reduces the bearing pressure on the subsoil and also the horizontal earth pressure in the case of backfilling.

Scour protection
Scour protection frequently consists of a layer of soil above the design bottom that is not taken into account in the calculations. Its effectiveness depends on its thickness. The stability of the waterfront structure is, therefore, not affected by scour caused by vessels and currents. Alternative forms of scour protection are stone fill, interlocking systems and concrete, for example. These can reduce the depth of protection required, or rather, increase the depth of water available.

Earth pressure shielding
Shielding slabs, vertical pile elements and similar measures on the land side of a waterfront structure can reduce the active earth pressure and, thus, relieve the loads on the structure. The design and construction of such measures must take into account the total system, i. e. the existing waterfront structure plus the intended shielding arrangements.

Ground improvement
Ground improvement measures can be carried out on the land and/or water side. The aim here is to improve the soil parameters and, thus, increase or decrease the earth pressure, as required. Ground improvement measures can take the form of sand piles, grout-injected stone columns and mixed-in-place methods, for instance. Soil replacement is another form of ground improvement.

Anchorages
The installation of additional anchors can relieve existing anchorages. In addition, installation of these at a different level can reduce the bending stresses in the waterfront structure. The actual conditions must be taken into account when designing additional anchors.

Upgrading the cross-section
Steel components (sheet piles, anchors, pile head details) can be upgraded by, for example, welding on additional plates.

Reinforced concrete components are generally upgraded by subsequently installing additional steel reinforcement.

It becomes necessary to replace an existing waterfront structure by a new structure when it reaches the end of its design life or in the case of a major change of use.

13.5.2 Deconstruction in conjunction with replacement measures

As vessels become larger and larger and require ever larger facilities to handle them, it is increasingly necessary to build replacement structures practically in the same place as existing waterfront structures. Wherever possible, the aim is to minimise the interaction between the existing structure and the new load-bearing structure. As a rule, this is achieved by building on top of and/or positioning the new quay wall in front of the old wall, provided that there is enough space on the water side. In harbour areas that have historically grown, this procedure leads to various quay structures built on top of and/or in front of each other.

The consequence of this is that extensive deconstruction of the old structures becomes necessary, or they have to be penetrated numerous times by the new supporting elements of the new quay wall. This work calls for accurate information about the nature of the existing quay wall, how it was designed and constructed and the condition of the materials used, e. g. types, strengths and degree of damage. Planning the deconstruction of the existing structure, like the design of the new structure, must take account of states of partial deconstruction. Where it is necessary to remove old, deep piles or sheet piles completely, the success of this work depends very much on the methods and equipment used, as well as the experience of the contractor appointed to carry out the work. Extensive geotechnical and structural surveys to establish the condition of the existing structure must be carried out beforehand.

Drilling techniques using special drill bits have proved worthwhile when drilling through new and old structural elements. The new parts of the quay wall should be able to adapt flexibly to suit the existing structure during the construction work.

References

Argebau (2006). Hinweise für die Überprüfung der Standsicherheit von baulichen Anlagen durch den Eigentümer/Verfügungsberechtigten. *Building Ministers Conference*, 2006.

BAW-Empfehlung (2017). *Spundwanddickenmessungen in Häfen und an Wasserstraßen: Grundlagen, Planung, Durchführung, Auswertung und Interpretation (ESM)*. Federal Waterways Engineering & Research Institute.

BAW-Empfehlung (2019). *Instandsetzungsprodukte – Hinweise für den Sachkundigen Planer zu bauwerksbezogenen Produktmerkmalen und Prüfverfahren*. Federal Waterways Engineering & Research Institute.

BAW-Merkblatt (2010). *Bauwerksinspektion (MBI)*. Federal Waterways Engineering & Research Institute.

BAW-Merkblatt (2016). *Bewertung der Tragfähigkeit bestehender, massiver Wasserbauwerke (TbW), Ausgabe Juli 2016*. Federal Waterways Engineering & Research Institute.

BAW-Merkblatt (2018a). *Bewertung der Tragfähigkeit bestehender Verschlüsse im Stahlwasserbau (TbVS) – Ausgabe 2018*. Federal Waterways Engineering & Research Institute.

BAW-Merkblatt (2018b). *Schadensklassifizierung an Verkehrswasserbauwerken (MSV)*. Federal Waterways Engineering & Research Institute.

Federal Ministry of Transport, Building & Town Development (2011). Richtlinie zur Nachrechnung von Straßenbrücken im Bestand (Nachrechnungsrichtlinie –NRR).

Hamburg Port Authority (AöR) (ed.) (2019). Nachrechnungsrichtlinie für Kaimauern.

PIANC Report (1998). Life cycle management of port structures – General principles. Pianc Bulletin No. 99, Brussels.

PIANC Report (2004). Inspection, maintenance and repair of maritime structures exposed to material degradation caused by a salt water environment. MarCom Report of WG 17.

PIANC Report (2006). Maintenance and renovation of navigation infrastructure. InCom Report of WG 25.

Westendarp, A. et al. (2015). Erhaltung und Instandsetzung von massiven Verkehrswasserbauwerken. *Betonkalender* 104: 187–246.

ZTV-W LB 219 (2017). Zusätzliche Technische Vertragsbedingungen – Wasserbau (ZTV-W) für die Instandsetzung der Betonbauteile von Wasserbauwerken (Leistungsbereich 219).

ZTV-ING Teil 3 Massivbau (2019). Zusätzliche Technische Vertragsbedingungen und Richtlinien für Ingenieurbauten (ZTV-ING) Massivbau, Sections 4 & 5.

Standards and Regulations

DIN 1076: Engineering structures in connection with roads – Inspection and test.

DIN 19702: Solid structures in hydraulic engineering – Bearing capacity, serviceability and durability.

RÜV-Richtlinie für die Überwachung der Verkehrssicherheit von baulichen Anlagen des Bundes (2008).

VDI-Richtlinie 6200 (2010). Structural safety of buildings – Regular inspections.

VV-WSV 2101: Bauwerksinspektion. Herausgegeben vom Bundesminister für Verkehr, Bonn, erhältlich bei der Drucksachenstelle der Wasser- und Schifffahrtsdirektion Mitte, Hannover, 2009/2010.

VV-WSV 2301: Damminspektion. Herausgegeben vom Bundesminister für Verkehr, Bonn, erhältlich bei der Drucksachenstelle der Wasser- und Schifffahrtsdirektion Mitte, Hannover, 2019.

Appendix A
Notation

The most important symbols and abbreviations used in the text, figures and equations are listed below. As far as possible, they match those used in the Eurocode.

All symbols are also explained in the respective passages of text.

- A.1 Symbols for variables, sorted according to
- A.1.1 Latin lower-case letters
- A.1.2 Latin upper-case letters
- A.1.3 Greek letters
- A.2 Subscripts and indices
- A.3 Abbreviations
- A.4 Designations for water levels and wave heights

Appendix A Notation

A.1 Symbols for variables

A.1.1 Latin lower-case letters

Symbol	Definition	Unit
a	Geometric information, length, tidal range, etc.	m
	Acceleration	m/s²
b	Geometric information, width	m
	Pore water pressure parameter	1/m
c	Wave propagation velocity	m/s
	Spring constant	kN/m, MN/m
	Concrete cover	mm
	Cohesion, e. g.:	kN/m², MN/m²
c_c	Apparent cohesion, capillary cohesion	
c_u	Undrained cohesion	
	Geometric factor, e. g.:	1
c_B	Shape factor	
D	Geometric information, e. g.:	m, mm
	Thickness or diameter	
	Particle size	
	Embedment depth	
e	Void ratio	1
	Active earth pressure	kN/m², MN/m²
	Eccentricity	m
f	Deflection	M
	Frequency	1/s
	Material strength, e. g.:	kN/m², MN/m²
f_u	Tensile strength	
f_y	Yield strength/point/stress	
g	Gravitational acceleration	m/s²
h	Geometric information, height	m
i	Hydraulic gradient	1
k	Permeability coefficient	m/s
	Wave number	1/m
	Radius of gyration of a ship	m
	Creep rate	mm
k_s	Modulus of subgrade reaction	kN/m³, MN/m³
l	Geometric information, length	m
m	Mass	t
n	Porosity	1
	Number	1
p	Permanent load (per unit area or per length)	kN/m², kN/m
q	Variable load (per unit area or per length)	kN/m², kN/m
	Flow rate	m³/(s m)
	Compressive strength, e. g.:	kN/m², MN/m²

Symbol	Definition	Unit
q_b	End bearing pressure	
q_s	Skin friction	
q_u	Uniaxial compressive strength	
r	Radius	m
s	Geometric information, e.g. displacement, settlement	m, cm
t	Depth	m
	Time	s, h, d, a
u	Horizontal component of velocity of water particles	m/s
	Pore water pressure	kN/m², MN/m²
v	Velocity	m/s
w	Water/hydrostatic pressure	kN/m², MN/m²

A.1.2 Latin upper-case letters

Symbol	Definition	Unit
A	Energy capacity	kNm, MNm
	Area	m²
	Accidental action	kN
B	Reaction force in soil	kN, MN, kN/m, MN/m
C	Equivalent force for soil reaction (after Blum)	kN/m
	Coefficient	1
	Factor, e.g.:	
C_D	Resistance factor for flow pressure	
C_e	Eccentricity factor	
C_m	Mass factor	
C_M	Resistance factor for flow acceleration	
C_S	Stiffness factor	
D	In situ density (soil), e.g.:	1
D_{pr}	Proctor density	1
E	Modulus of elasticity/Young's modulus	kN/m², MN/m²
	Energy	kJ
	Earth pressure force	kN, MN, kN/m, MN/m
	Action effect, load	kN, MN, kN/m, MN/m
E_s	Stiffness modulus	kN/m², MN/m²
F	Force, action, e.g.:	kN, MN, kN/m, MN/m
F_c	Cohesive force	
F_s	Flow force	
F_u	Pore water pressure force	
F_w	Wind load	
F_z	Tensile force	
G	Dead load, self-weight, permanent vertical action	kN, MN, kN/m, MN/m
H	Horizontal load	kN, MN, kN/m, MN/m
I	Wave height	m
	Second moment of area	m⁴
	Condition index, e.g.:	1
I_c	Consistency index	
I_D	Relative in situ density	
I_P	Plasticity index	
K	Earth pressure coefficient	
M	Moment	kNm, MNm, kNm/m, MNm/m
N	Normal force	kN, MN, kN/m, MN/m
N_{10}	Number of blows per 10 cm penetration (penetrometer test)	1
P	Anchor force, load	kN, MN, kN/m, MN/m
	Probability	1
Q	Variable action	kN, MN, kN/m, MN/m m³/s
	Flow rate	

Symbol	Definition	Unit
R	Reaction force, resistance force	kN, MN, kN/m, MN/m
	Resistance	kN, MN, kN/m, MN/m
R_B	Reaction force in soil	kN, MN, kN/m, MN/m
R_C	Equivalent force for soil reaction (after Blum)	kN/m
$\ldots R_e$	Reynolds number	1
S	Parameter, e.g.: standard surface finish	1
	First moment of area	m^3
T	Wave period	s
	Temperature	°C, K
	Force, e.g. shear force	kN/m^2, MN/m^2
T_c	Cohesive force	
U	Uniformity coefficient	1
V	Vertical load	kN, MN, kN/m, MN/m
	Volume	m^3
W	Hydrostatic force	kN, MN, kN/m, MN/m
	Section modulus	m^3

A.1.3 Greek letters

Symbol	Definition	Unit
α	Angle of inclination (watercourse bottom, wall)	°
	Moment reduction factor	1
α_T	Coefficient of thermal expansion	1/°C
β	Angle of inclination (slope, terrain)	°
γ	Partial safety factor	1
	Unit weight, e. g.:	kN/m³
γ'	Buoyant/submerged unit weight	
Δ	Increase, decrease, change	1
δ	Angle of interface (e. g. wall) friction, e. g.:	°
δ_a	Angle of inclination of active earth pressure	
δ_p	Angle of inclination of passive earth pressure	
ε	Strain	1
η	Modification factor	1
Θ, ϑ	Angle of inclination of failure plane	°
κ	Reflection coefficient	1
λ	Hoop tension factor, coefficient	1
μ	Coefficient of friction	1
	Correction factor	1
ν	Stiffness factor	1
ζ	Reduction factor	1
	Breaker index (for waves)	1
ρ	Density	t/m³
σ	Stress, normal stress, e. g.:	kN/m², MN/m²
σ'	Effective (normal) stress	
σ_v	Equivalent stress	
τ	Shear stress	kN/m², MN/m²
Φ	Impact factor, dynamic factor	1
φ	Angle of internal friction	°
Ψ	Combination factor	1
ω	Angular frequency	1/s

A.2 Subscripts and indices

Symbol	Definition
a	Active
abs	Absolute
at	Atmospheric
b	Base/bottom
c	Cohesion
	Compression
cal	Calculatory/arithmetically
crit	Critical
d	Design value
	Dry
dyn	Dynamic
dst	Destabilising
e	Eccentric
eff	Effective
E	Fixity
f	Fracture
g	Due to permanent actions
h	hoRizontal
k	Characteristic value
kin	Kinetic
m	Mean, average
max	Maximum
min	Minimum
mob	Mobilised
p	Passive
	Persistent
pl	Plastic
pr	Proctor
q	Due to variable actions
r	Resultant
	Resistance
red	Reduced
rep	Representative
s	Layer, stratum
	Shaft
	Flow, current
stat	Static

Symbol	Definition
stb	Stabilising
t	Tension
	Transient
tot	Total
tr	Transverse
u	Undrained
	Bottom
	Uniaxial
ü	Excess pore water pressure
upl	Uplift
v	Vertical
	Comparative
w	Water, water level
y	Yield point

A.3 Abbreviations

Symbol	Definition
dwt	Deadweight tonnage
GT	Gross tonnage
DS	Design situation, e. g.:
DS-P	Persistent design situation
DS-T	Transient design situation
DS-A	Accidental design situation
DS-E	Earthquake design situation
EQU	Limit state for a loss of equilibrium of the structure or ground regarded as a rigid body, where the strength of the material and the subsoil are not critical for the resistance
GC	Geotechnical category
GEO	Limit state of the ground:
GEO-2	Limit state of the ground for which design approach 2 is used
GEO-3	Limit state of the ground for which design approach 3 is used
GRT	Gross register tonnage (now replaced by gross tonnage, GT)
HPI	High-pressure injection (jet grouting)
HYD	Limit state of failure caused by hydraulic gradients in the ground, e. g. hydraulic heave, internal erosion or piping
NN	*Normalnull* (obsolete official vertical datum for elevations, levels)
LWZ	Low water zone
SLS	Serviceability limit state
SPZ	Splash zone
STR	Limit state of failure or very large deformations of the structure or its individual parts, including the foundations, piles, basement walls, etc., where the strength of the material is critical for the resistance
TEU	Twenty feet equivalent unit
ULS	Ultimate limit state
UPL	Limit state for a loss of equilibrium of the structure or the ground as a result of uplift by water pressure (buoyancy) or other vertical actions
PIZ	Permanent immersion zone
IZ	Intertidal zone

A.4 Water levels and wave heights

Non-tidal water levels

GW	Groundwater level, water table
HaW	Normal harbour water level
LHaW	Lowest harbour water level
HHW	Highest high water level
HW	High water level
MHW	Mean high water level
MW	Mean water level
MLW	Mean low water level
LW	Low water level
LLW	Lowest low water level
HNW	Highest navigable water level

Tidal water levels

HAT	Highest astronomical tide level
MHWS	Mean high water spring tide level
MHW	Mean high tide level
MW	Mean tide level
$T\frac{1}{2}W$	Half tide level
$MT\frac{1}{2}W$	Mean half tide level
MLW	Mean low tide level
MLWS	Mean low water spring tide level
LAT	Lowest astronomical tide level
CD	Chart datum (corresponds roughly to MLWS)

Wave heights

H_b	Height of breaking wave
H_d	Design wave height
H_m	Mean wave height
H_{max}	Maximum wave height
H_{rms}	Root-mean-square wave height
$H_{1/3}$	Mean of 33 % of highest waves
$H_{1/10}$	Mean of 10 % of highest waves
$H_{1/100}$	Mean of 1 % of highest waves

List of Advertisers

		Page
ArcelorMittal Commercial RPS S.a.r.l	4221 Esch Sur Alzette, Luxembourg	Bound insert
Friedr. Ischebeck GmbH	58256 Ennepetal, Germany	436a
Heinrich Hirdes GmbH	21079 Hamburg, Germany	132a
iwb Ingenieure Infrastruktur GmbH & Co. KG	20459 Hamburg, Germany	Bookmark
Ramboll Deutschland GmbH	22763 Hamburg, Germany	0a
Sellhorn Ingenieurgesellschaft mbH	20459 Hamburg, Germany	204b
SteelWall ISH GmbH	82166 Gräfelfing, Germany	Bound insert
terra infrastructure GmbH	45127 Essen, Germany	326a
WTM Engineers GmbH	20459 Hamburg, Germany	2nd cover page